Artificial Intelligence

An authoritative and accessible one-stop resource, the first edition of *An Introduction to Artificial Intelligence* presented one of the first comprehensive examinations of AI. Designed to provide an understanding of the foundations of artificial intelligence, it examined the central computational techniques employed by AI, including knowledge representation, search, reasoning and learning, as well as the principal application domains of expert systems, natural language, vision, robotics, software agents and cognitive modelling. Many of the major philosophical and ethical issues of AI were also introduced. This new edition expands and revises the book throughout, with new material to augment existing chapters, including short case studies, as well as adding new chapters on explainable AI, big data and deep learning, temporal and web-scale data, statistical methods and data wrangling. It expands the book's focus on human-centred AI, covering gender, ethnic and social bias, the need for transparency, intelligent user interfaces, and designing interactions to aid machine learning. With detailed, well-illustrated examples and exercises throughout, this book provides a substantial and robust introduction to artificial intelligence in a clear and concise coursebook form. It stands as a core text for all students and computer scientists approaching AI.

Artificial Intelligence

Humans at the Heart of Algorithms

Second Edition

Alan Dix with Janet Finlay

CRC Press
Taylor & Francis Group
Boca Raton London New York

CRC Press is an imprint of the
Taylor & Francis Group, an **informa** business

A CHAPMAN & HALL BOOK

Designed cover image: Alan Dix

First edition published 2025
by CRC Press
2385 NW Executive Center Drive, Suite 320, Boca Raton FL 33431

and by CRC Press
4 Park Square, Milton Park, Abingdon, Oxon, OX14 4RN

CRC Press is an imprint of Taylor & Francis Group, LLC

© 2025 Alan Dix

First edition published by CRC Press 1996

ISBN: 9780367536879 (hbk)
ISBN: 9780367515980 (pbk)
ISBN: 9781003082880 (ebk)

DOI: 10.1201/ 9781003082880

Typeset in Minion
by Deanta Global Publishing Services, Chennai, India

Contents

SECTION II   **Data and Learning**

CHAPTER 5 ■ Machine Learning   59

CHAPTER 6 ■ Neural Networks   74

Preface

The first edition of this textbook was written 25 years ago. A lot has changed since then, both in artificial intelligence and for the authors. Janet Finlay, who was the first author then, has now shifted her full-time focus to another non-human intelligence ... dogs, and her influence is still felt in the pages that follow, not least in the various canine examples.

A major driver for the first edition was the lack of an appropriate textbook for an introductory course on artificial intelligence (AI) at the University of York where Janet and I were at the time. In the first edition, we said that our aim was to create a book that gave a sound introduction to technical aspects of AI without assuming too much background knowledge beforehand, especially for those coming to AI from areas other than computer science. Then and still today, most books on AI either assume too much technical knowledge or provided a very limited coverage of the subject.

In the introduction to the first edition we wrote, "*It is clear then that although the goals and emphases of AI may have changed over time, the subject is far from dead or historical.*" This was written as AI was entering its long 'winter' and hence had almost an apologetic tone. How things have changed!

Over the years we did think (indeed the publisher suggested we think) about a second edition, but for many years, while there was much new work in the frontiers of AI, the fundamentals were relatively stable. This started to change with developments around big data and the web, but it has really been in the last few years that we have seen more substantial changes, and so it felt an appropriate time for a new edition including new topics such as deep learning, big data, the Semantic Web, large-language models and explainable AI.

This edition, while substantially expanding the material covered, still seeks to follow the same principles as the first edition, providing accessible coverage of the key areas of AI in such a way that it will be understandable to those with only a basic knowledge of mathematics and computer science. The huge growth of data science and the ubiquity of AI mean that today this approach is more important than ever, and I hope that the new material in this new edition has followed this principle as well as the first.

The book takes a pragmatic approach to AI, looking at how AI techniques are applied to various application areas, and includes both more traditional symbolic AI and sub-symbolic AI including neural networks and deep learning. It covers both general principles such as reasoning and machine learning and also more specific techniques for areas such as computer vision, language understanding and the web.

Ultimately all AI impacts humans directly or indirectly. This was always true, and in the first edition we wrote, "*It may not be long before AI is an integral part of all our lives.*" This is now far more clear as aspects of AI permeate nearly every area of life from online shopping to smart cities. So, the book attempts to highlight both the design issues involved in having AI work alongside humans and also the social, ethical and philosophical challenges raised by AI. The book concludes with a brief peek into possible futures for AI, although, given the pace of change, these futures may be upon us before you read this.

This book does not claim to be comprehensive: there are many books on the market which give more detailed coverage in specific areas. However, it does attempt to give a broad view of AI that is accessible to a wide audience and yet opens up more technical aspects. Throughout it attempts to give the reader a sense of the overall feel of the area not just how to do it but also when, where and why to use particular techniques.

Educators can use the book to support a one-semester introductory module spending approximately one week each on Chapters 2 to 8 and selected further chapters. Alternatively it can be used as a longer course covering most of the chapters, again at around one week per chapter. You are encouraged to include some material on social, human or philosophical aspects, both to bring the topic to life and most critically because the questions of how AI fit into wider society are some of the most pressing for everyone.

If you are an AI professional, this book will primarily be useful to give you a grandstand view of the area, helping you to understand the field as a whole, and identify the particular topics you need to know about in more detail. Having identified these areas use the recommended reading at the end of each chapter or web resources to dig deeper.

For the more casual reader, after the introductory chapter you might like to skip to Chapters 19, 20 or 23 to address some of the 'why' questions for AI and then step back to delve into topics in the earlier chapters given that context. Indeed educators might also want to take this approach with their classes.

Look at the book's website https://alandix.com/aibook / for lots of support material including code examples, videos and teaching resources.

ACKNOWLEDGEMENTS

I would like to thank family, friends and colleagues who have helped in many ways and especially Alex Blandin for extensive insightful comments on the text, Fiona for her personal support and help ensuring the text is readable and coherent, and, of course, Janet without whom the first edition and in turn this fresh edition would never have happened. Any remaining faults or omissions are my own.

Alan Dix
Cardiff Metropolitan University, Cardiff, Wales, UK

About the Author

Alan Dix is Professorial Fellow at Cardiff Metropolitan University and Professor Emeritus at Swansea University. He started his academic career as a mathematician and was part of the British team to the International Mathematical Olympiad in 1978. However, he is best known for his work in human–computer interaction (HCI), including writing one of the key textbooks in the area. He was elected to the ACM SIGCHI Academy in 2013 and is a Fellow of the Learned Society of Wales. Outside academia, Alan has been co-founder of two dot-com era tech companies, developed intelligent lighting and worked in local government and even submarine design.

In every role, Alan seeks to understand and innovate in all aspects where people and technology meet. He has often been prescient in recognising the implications of digital technology, in 1990 writing the first paper on privacy within the HCI literature and in 1992 predicting the potential danger of social, ethnic and gender bias in black-box machine learning algorithms.

Alan writes and talks extensively on the connections between artificial intelligence and human issues, both in terms of individual user interfaces and also social implications of technology. This has included leading the algorithmic social justice theme within the UK Not-Equal programme and participating in the European TANGO project on synergistic human–AI decision making. He has also worked on fundamental models of emotion in AI and practical applications of genetic algorithms in engineering.

His other books include *Human–Computer Interaction* (with Janet Finlay, Gregory Abowd and Russell Beale), one of the key international textbooks in the area; *TouchIT: Understanding Design in a Physical-Digital World* (with Steve Gill, Devina Ramduny-Ellis and Jo Hare) on the design of physical-digital products; and *Statistics for HCI: Making Sense of Quantitative Data.* He is also completing two books in the CRC/Taylor & Francis "AI for Everything" series: *AI for Social Justice* (with Clara Crivellaro) and *AI for HCI.*

Introduction

1.1 WHAT IS ARTIFICIAL INTELLIGENCE?

Artificial intelligence (AI) is many different things to different people. It is likely that everyone who picks up this book has their own, albeit perhaps vague, notion of what it is. As a concept, AI has long captured the attention and imagination of journalists and novelists alike, leading both to popular renditions of current AI developments and futuristic representations of what might be just around the corner. Television and film producers have followed suit, so that AI is rarely far from the public eye. Robots, computers that talk to us in our own language and AI entrepreneurs are continually in the media, though there is some division as to whether these developments will provide us with benign servants or sinister and deadly opponents.

But outside the media furore, what is AI all about? Unfortunately there is no single answer: just like in the media representation, it very much depends upon who you talk to.

1.1.1 How Much Like a Human: Strong vs. Weak AI

One of the major divides in AI is between *strong* and *weak* AI:

strong AI – There are those who view AI in high-level terms as the study of the nature of intelligence and, from there, how to reproduce it. Computers are therefore used to model intelligence in order to understand it. Within this group there are those who believe that human intelligence is essentially computational and, therefore, that cognitive states can be reproduced in a machine. Others use computers to test their theories of intelligence: they are interested less in replicating than in understanding human intelligence. For either of these groups, it is vital that the techniques proposed actually reflect human cognitive processes.

weak AI – On the other hand, there are those who view AI as a discipline that provides engineering techniques to solve difficult problems. Whether these techniques reflect human cognition or indicate actual intelligence is not important. To this group the success of an AI system is judged on its behaviour in the domain of interest. It is not necessary for the machine to exhibit general intelligence.

A third set of people, who fall somewhere between the previous two, want to develop machines that not only exhibit intelligent behaviour but are able to learn and adapt to their environment in a way similar to humans. In striving towards this, it is inevitable that insights will be gained into the nature of human intelligence and learning, although it is not essential that these are accurately reproduced.

This book takes this third, more pragmatic line, leaving it to you to decide which side of the strong/weak argument you want to adopt.

1.1.2 Top-down or Bottom-up: Symbolic vs. Sub-symbolic

Another major divide is between those who address the problem top-down or bottom-up:

DOI: 10.1201/9781003082880-1

symbolic AI – Most of early AI addressed intelligence top-down, starting with high-level human abilities such as logical reasoning and then building systems that emulated these. Crucially they can be seen as symbol manipulation systems, where symbols are tokens that represent quite complex concepts such as 'human', 'block' or 'move'.

sub-symbolic AI – On the other hand, there are methods that start off with very simple models inspired by human neurons, and by combining many millions or billions of these aim to achieve intelligent behaviour bottom-up. For many today, these artificial neural networks are what first come to mind when they hear "AI".

In addition there are methods that are clearly sub-symbolic in nature, but not based on neural networks. Some take their inspiration from other natural life forms or processes that seem to exhibit 'intelligent' overall behaviour, even when the individual components are not deemed (very) intelligent. This has included emulating ant colonies, the immune system, evolutionary development and crystal formation. These in-between systems may include representations that have a symbolic nature (e.g. rules) but operate in a bottom-up fashion. Others apply purely mathematical or statistical approaches to very large datasets, for example, the algorithms underlying web search.

In the brief AI history later in this chapter, we will see (spoiler alert) a general arc of dominance in the field from symbolic to sub-symbolic methods. However, it is likely that the next major AI steps will combine the two.

1.1.3 A Working Definition

So, can we derive a definition of AI that encompasses some of these ideas? A working definition may go something like this:

> AI is concerned with building machines that can act and react appropriately, adapting their response to the demands of the situation. Such machines should display behaviour comparable with that considered to require intelligence in humans.

Such a definition incorporates learning and adaptability as general characteristics of intelligence but stops short of insisting on the replication of *human* intelligence. Indeed

it can sometimes be more useful to think of AI as Alien Intelligence – something that behaves intelligently, but not necessarily as we know it.

1.1.4 Human Intelligence

What types of behaviour would meet this definition and therefore fall under the umbrella of AI? Or, perhaps more importantly, what types of behaviour would not? It may be useful to think about some of the things we consider to require intelligence or thought in human beings. A list would usually include conscious cognitive activities: problem solving, decision making, reading and mathematics. Further consideration might add more creative activities: writing and art. We are less likely to think of our more fundamental skills – language, vision, motor skills and navigation – simply because, to us, these are automatic and do not require conscious attention. But consider for a moment what is involved in these "everyday" activities. For example, language understanding requires recognition and interpretation of words, spoken in many different accents and intonations, and knowledge of how words can be strung together. It involves resolution of ambiguity and understanding of context. Language production is even more complex. One only needs to take up a foreign language to appreciate the difficulties involved – even for humans.

On the other hand, some areas that may seem to us very difficult, such as arithmetic calculation, are in fact much more formulaic and therefore require only the ability to follow steps accurately. Such behaviour is not inherently intelligent, and computers are traditionally excellent as calculators. However, this activity would not be classed as artificial intelligence. Of course, we would not want to suggest that mathematics itself does not require intelligence! For example, problem solving and interpretation are also important in mathematics, and these aspects have been studied as domains for AI research.

There are also some "grey" areas, activities that require skill and strategy when performed by humans but that can, ultimately, be condensed to a search of possible options (albeit a huge number of them). Game playing is a prime example of such an activity. In the early days, chess and other complex games were very much within the domain of humans and not computers and were considered a valid target for AI research. But today computers can play chess at grandmaster level, largely due to their huge memory capacity. Some would say that such brute force techniques are not true AI; however, when AlphaGo beat the Go Grandmaster Lee Sedol, commentators described AlphaGo's game as 'beautiful' [189]. It is hard not to think of this as intelligent, even if it is not clear how it achieves what it does.

1.1.5 Bottom-up and Top-down

However, it should be noted that many of the things that make us intelligent, especially in more creative ways, are not easily explainable by us. A mathematician may be able to justify each step in a proof, but not the 'gut feeling' that led to formulating a particular lemma that was crucial to the proof. In day-to-day life these two aspects of human problem solving and activity work together. On the one hand is the fast emotional or subconscious thinking that leads to instant reactions and rapid decisions when time is limited. On the other hand is our more leisurely logical and also imaginative thinking. These are often called 'System 1' (fast, subconscious, emotional) and 'System 2' (slow, conscious, rational) [151]. Early symbolic AI focused almost exclusively on System 2 processes, whereas systems such as AlphaGo and neural networks, which rely on more sub-symbolic methods, in general are more like highly competent System 1 thinking. This book will consider both of these aspects of AI, and in later chapters we see how they can be in various ways integrated.

However, before we move on to look in more detail at the techniques and applications of AI, we will pause to consider how it has developed up to now.

1.2 HUMANS AT THE HEART

As a theoretical discipline there will be some who study AI purely for its own sake, just because it is fun to create algorithms that do cool things. There will also be a few on the extreme end of strong AI who study it for the sake of the artificial entities that are being created, maybe to stand up for the rights of coming sentient artificial life forms.

However, for the majority, AI is being developed and used because it does something in the human world, to solve problems in engineering, medicine, law or day-to-day life. Whether or not AI is like a human in terms of the intelligence it portrays, it is *for humans*.

Indeed, every AI-based system will, in the end, need to work with people. Sometimes this will be very explicit, such as a chatbot, in others virtually invisible, for example in an engine management system.

In the latter case it may be possible for the ultimate human user to ignore or be ignorant of the AI aspect. It is a black box that does a job; so long as it is reliable and performs well, how it does that job doesn't matter. Similarly the programmer or engineer creating the system in the black-box is only concerned with meeting a specification. Even such systems may need to be comprehensible by an engineer or lawyer if there is a malfunction or accident, that is the behaviour may need to be explainable (see Chapter 21). Humans will be involved over a long timescale in the creation and oversight of the system, but, for day-to-day use, the AI and the human can operate separately.

However, more often the boundaries are less clear, with levels of active interaction or mutual influence. For example, when a media website suggests films you might like to watch, you may often take it for granted, but sometimes the intelligence of the underlying algorithm, or lack of it, is obvious.

We will deal with the more direct human contact in Part IV; however, you will find examples throughout the book. If you get into the technical aspects of AI, there may be times when you just want to get your head down, buried in the algorithmic details. However, when you do, from time to time take an opportunity to step back and think about the wider picture, how the systems you are creating fit into a wider human and technical and human environment. If you don't, you might find yourself producing something that is wonderful in itself but useless or even dangerous for the purpose it is intended.

1.3 A SHORT HISTORY OF ARTIFICIAL INTELLIGENCE

AI is not a new concept. The idea of creating an intelligent being was proposed and discussed in various ways by writers and philosophers centuries before the computer was even invented. The earliest writers imagined their "artificial" beings created from stone: the Roman poet Ovid wrote a story of Pygmalion, the sculptor, whose statue of a beautiful woman was brought to life (the musical *My fair lady* is the more recent rendition of this fable). Much later, in the age of industrial machines and the discovery of almost magical qualities of electrical phenomena, Mary Shelley had Victor Frankenstein manufacture a man from separate biological components and bring him to life through electricity. By the 1960s, fiction was beginning to mirror the goals of the most ambitious AI researcher. In Arthur C. Clarke's *2001*, we find the computer HAL displaying all the attributes of human intelligence, including self-preservation. Other films, such as *Terminator* and *Ex Machina*, present a vision of cyborg machines almost indistinguishable from humans.

Early philosophers also considered the question of whether human intelligence can be reproduced in a machine. In 1642, Descartes argued that, although machines (in the right guise) could pass as animals, they could never pass as humans. He went on to identify his reasons for this assertion, namely that machines lack the ability to use language and the ability to reason. Interestingly, although he was writing at a time when clocks and windmills were among the most sophisticated pieces of machinery, he had identified two areas that still occupy the attention of AI researchers today and that are central to one of the first tests of machine intelligence proposed for computers, the Turing test.

Precursors of AI can be seen in the development of first mechanical and later electronic devices for various forms of specialised calculations, including landmark systems such as Babbage's Difference Engine for calculating polynomials and the Bombe for decrypting Enigma Machine messages. Even these would, at the time, seem to mimic certain human traits that would have been regarded as requiring human thought. As these machines became programmable, the step from calculation to computation brought yet more thought-like potential. This was foreseen by Ada Lovelace when she described the fundamental difference between the Difference Engine and Babbage's later Analytical Engine:

> *The Difference Engine can merely tabulate, and is incapable of developing, the Analytical Engine can either tabulate or develope.* [176]

She noted the potential for "*symbolic results*", but she also cautioned about the "*possibility of exaggerated ideas*" about the capabilities of the Analytical Engine, a sentiment that might have been better heeded by AI commentators over the years.

Turing and the Turing Test

To find the start of modern AI many look more than one hundred years after Babbage and Lovelace, to 1950, when computers were still basically large numeric calculators. In that year, a British mathematician, Alan Turing,

wrote a now famous paper entitled *Computing machinery and intelligence*, in which he posed the question "*can machines think?*" [276]. His answer to the question was to propose a game, the 'imitation game', as the basis for a test for machine intelligence. His test is now known as the Turing test.

His proposal was as follows. Imagine you have a person able to communicate with two others, one male and one female, only through a teletype or computer link. This person must try to distinguish the male from the female simply by asking questions, while those being questioned try to fool the interrogator about their gender. In the Turing test, one of the human participants is replaced by a computer. If the computer is able to convince the interrogator it is the human, Turing argued, it can be said to be intelligent.

The Turing test has been widely criticised since it was proposed, and in Chapter 23 of this book we will look in detail at some of the objections to it. However, the main criticism is that it takes a behavioural rather than a psychological view of intelligence. Regardless of the criticisms the Turing test is significant as it provided the first concrete goal for AI research.

1.3.1 The Development of AI

In 1956 about a dozen researchers met at Dartmouth College. They represented a number of academic disciplines including mathematics, electrical engineering and psychology, but they had one interest in common: using the powerful tool of the computer to simulate human intelligence. It was here, at the Dartmouth Workshop, that the name "Artificial Intelligence" was coined.

In the early days of AI, research concentrated on applications that were considered to involve intelligence and reasoning, mainly game playing and theorem proving. During the decade following the Dartmouth Workshop there were significant developments in search techniques, pattern matching, heuristics and constraint satisfaction, which are still important today. However, these early applications proved to be solvable using rapid searches through all possible options, rather than the selection strategies and reasoning applied by humans, and attention turned to the problem of modelling and reproducing true understanding.

From the mid-1960s researchers concentrated on applications such as perception, language understanding and general problem solving. One such application, ELIZA [299], was hailed by some as a breakthrough in language understanding. ELIZA appeared, on the surface, to understand and respond intelligently to human conversation, albeit in the slightly artificial context of Rogerian psychoanalysis. "Patients" would communicate with ELIZA via a keyboard, and the program would respond in such a way that many patients believed that they were talking to a human. A conversation with ELIZA would run something like this (machine responses in upper case):

> TELL ME ABOUT YOUR PROBLEM
> men are all alike
> HOW DO YOU FEEL ABOUT MEN?
> none of them can be trusted
> NOT ANY?
> well very few
> COULD YOU BE MORE PRECISE PLEASE?
> my grandfather was trustworthy
> TELL ME ABOUT YOUR GRANDFATHER
> he is dead
> HOW LONG HAS HE BEEN DEAD?
> a couple of years – he was my mother's father
> TELL ME ABOUT YOUR MOTHER
> she is very nice
> CAN YOU THINK OF A SPECIFIC EXAMPLE?
> she is always glad to see me
> IS YOUR FATHER ALSO ALWAYS GLAD TO SEE YOU?
> most of the time
> HOW DID YOUR PARENTS TREAT YOU AS A CHILD?

Such a conversation may appear convincing, but does ELIZA therefore pass the Turing test? In 1966, the program's developer, Joseph Weizenbaum, wrote a paper to dispel this notion (at the same time casting some doubt on the validity of the Turing test) [299]. ELIZA actually works on the basis of some clever tricks of substitution, pattern matching and random selection. The user's sentence is searched for key words that are then either substituted into a new sentence by ELIZA or used to fire an associated response. If more than one possible response is available, ELIZA chooses between them randomly. In spite of appearances, therefore, ELIZA embodies no *understanding* of the English language. As Weizenbaum says, it shows "*how easy it is to create and maintain the illusion of understanding*" [299]. Once these tricks are understood it is very easy to fool ELIZA:

> HOW DID YOUR PARENTS TREAT YOU AS A CHILD?
> they gave me my grandfather clock
> TELL ME ABOUT YOUR GRANDFATHER
> …

However, research continued into language understanding in the early 1970s and another significant system emerged: SHRDLU. SHRDLU (the name comes from six of the 12 most commonly used letters in English) was developed by Terry Winograd [301]. The system can understand and comply with quite complex sentences, but there is a catch: it is restricted to the domain of simulated blocks. Although this may seem limited, SHRDLU still far surpassed any other system of the time.

Consider an instruction such as *"Find a block that is taller than the one you are holding and put it on top of the red box."* What knowledge is required to interpret such a sentence? First you need to understand the concepts of relative sizes. Then you need to interpret the reference in the second clause: to what does "it" refer? Then you need to understand relative position and differentiate by colour. SHRDLU was able to interpret such instructions through the use of stored knowledge and was one of the applications of this period that led to the development of a number of methodologies for knowledge representation (discussed in Chap. 2).

1.3.2 The Physical Symbol System Hypothesis

In 1976 Newell and Simon proposed a hypothesis that has become the basis of research and experimentation in AI: the physical symbol system hypothesis [210]. The hypothesis states that

> A physical symbol system has the necessary and sufficient means for general intelligent action.

So what does this mean? A *symbol* is a token that represents something else. For example, a word is a symbol representing an object or concept. The symbol is physical, although the thing represented by it may be conceptual. Symbols are physically related to each other in *symbol structures* (e.g. they may be adjacent). In addition to symbol structures, the system contains operators or processes that transform structures into other structures, for example copying, adding and removing them. A *physical symbol system* comprises an evolving set of symbol structures and the operators required to transform them. The hypothesis suggests that such a system is able to model intelligent behaviour.

Without a clear definition of human intelligence, the only way to test this hypothesis is by experimentation: choose an activity that requires intelligence and devise a physical symbol system to solve it. Computers are a good means of simulating the physical symbol system and are therefore used in testing the hypothesis. It is not yet clear whether the physical symbol system hypothesis will hold in all areas of intelligence. It is certainly supported by work in areas such as game playing and decision making, but in lower-level activities, such as vision, sub-symbolic approaches (such as neural networks) often prove to be more useful. However, this in itself does not disprove the physical symbol system hypothesis, since it is clearly possible to solve problems in alternative ways.

The physical symbol system hypothesis is important as the foundation for the belief that it is possible to create artificial intelligence. It also provides a useful model of human intelligence that can be simulated and therefore tested.

1.3.3 Sub-symbolic Spring

By the late 1970s, while the physical symbol system hypothesis provided fresh impetus to those examining the nature of intelligent behaviour, some research moved away from the "grand aim" of producing general machine understanding and concentrated instead upon developing effective techniques in restricted domains. Arguably this approach has had the most commercial success, producing, among other things, the expert system (see Chap. 18).

The 1980s saw a period of great optimism within AI. In 1982 Japan launched the Fifth Generation Computer Project, aiming to become a world leader in supercomputing and artificial intelligence [206, 253]. This led to rival initiatives in the UK, US and pan-EU. The amounts were staggering for the time, the UK's Alvey Programme alone was 500 million pounds sterling [214]. The central focus of the Fifth Generation Computer Project was on logic programming, that is traditional knowledge-rich AI. However, the funding also allowed other areas to flourish.

The development of artificial neural networks, in the late 1980s and early 1990s, modelled on the human brain, was hailed by some as the basis for genuine machine intelligence and learning. Neural networks, or "connectionist" systems, initially proved effective in small applications, but many have huge resource requirements. Traditional AI researchers were slow to welcome the connectionists, being sceptical of their claims and the premises underlying neural networks.

In one example, a recognition system used neural networks to learn the properties of a number of photographs taken in woodland. Its aim was to differentiate between those containing tanks and those without. After a number of test runs in which the system accurately picked out all the photographs of tanks, the developers were feeling suitably pleased with themselves. However, to confirm their findings they took another set of photographs. To their dismay the system proved completely unable to pick out the tanks. After further investigation it turned out that the first set of photographs of tanks had been taken on a sunny day while those without were cloudy. The network was not classifying the photographs according to the presence of tanks at all but according to prevailing weather conditions! Since the "reasoning" underlying the network is difficult to examine such mistakes can go unnoticed.

The Great AI Winter

To some extent expectations of AI were over-hyped. Progress in some areas was rapid but often hit limitations. This led to a period of around 15 years, often called the AI Winter, when funding and enthusiasm died, and it seemed as if progress had slowed. If you talked to many in AI during this period, they would feel they were having little impact outside the research lab.

In fact, many of the areas that would have once been seen as part of core AI were progressing steadily during this time, notably vision, speech and robotics, but this did not counter the overall sense of a subject in the doldrums.

Historians of the Dark Ages in Europe (500–1000 AD) tell us that it was far from a period of anarchy and ignorance, but one where knowledge flourished in pockets, often at the outer fringes. Similarly historians of AI can pick up the threads of work that flourished during this AI winter. In particular, while it was not originally seen as an AI revolution, the web changed everything.

The seeds of the impact of the web on AI were (in retrospect) evident from the late 1990s and early 2000s.

As domestic use of the internet grew, recommendation systems for shopping services grew. These were initially seen as purely statistical algorithms, rather than AI, but the often uncanny ability of "*other people bought ...*" to recommend pertinent books and music had the feel of intelligence – recall our definition "*considered to require intelligence in humans*".

Similarly the success of search engines, notably Google's PageRank algorithm [32], harnessed the growing volume of human-produced material on the web to produce results that were not just sensible but often prescient. Around the same time Berners-Lee and others proposed the semantic web with an aim to make the material on the web machine-readable and hence available for large-scale intelligent reasoning [25].

1.3.4 AI Renaissance

It is no hyperbole to describe as seismic the impact of AlphaGo's defeat of Go grandmaster Lee Sedol in 2016 [17]. Go had been considered an almost impossible challenge for AI requiring true human insight and intelligence.

Although the win had been unexpected, the resurgence had begun some years earlier, not least with the success of IBM's Watson at the quiz game Jeopardy! [101], even leading some to warn against overblown optimism and the possibility of a return to another AI winter [128]. It would be good to be able to say that this resurgence was purely due to the ingenuity of AI researchers creating new and more powerful algorithms. There is truth in this, but the reality is more prosaic – more about speed and scale than science.

The technology giants did have access to human intellectual capital and funding for internal AI initiatives. However, they also had access to vast amounts of data and computational power, the twin enablers of deep learning. Well before AlphaGo, Google researchers had written about "*The unreasonable effectiveness of data*", describing the way big data analysis based on simple word concurrence was able to tackle issues that had been thought to need knowledge-rich natural language understanding [122].

However, the availability of big data and massive computation has in turn led to new approaches and algorithms. Some are brute force, but others are highly creative ways of harnessing that power, for example general adversarial learning approaches inspired by game playing.

Furthermore the application scale has continued to yield new surprises. In particular, GPT-4 and other large-language models are based on deep neural networks trained on billions of documents. They behave as if they have grammatical and even semantic abilities, despite being based solely on low-level weights. This has led to a scale-based version of the physical symbol system hypothesis, where it is believed that simply making bigger and bigger deep neural networks will lead to AI systems that exhibit true general intelligence.

1.3.5 Moving Onwards

At the end of the introduction to the first edition of this book we said that "*it may not be long before AI is an integral part of all our lives*". Of course, this is now the case: personalised product recommendations and news on the web, voice-operated home automation systems, face recognition on our phones and autonomous vehicles on the road.

Of course this ubiquity has led to its own challenges, not least ethical issues including gender, ethnic and social bias in the outputs of many machine learning algorithms and the need for explainable AI.

Time will tell how far the trend towards bigger and bigger machine learning will take us. Even if it does continue to be successful, there will be increasing needs for 'small AI' for environmental reasons (large AI models consume lots of power) and also social (only the rich can afford big AI).

It does seem likely that the next step however will require a re-integration of the more knowledge-rich techniques of 'traditional' symbolic AI and the hugely successful but hard to interpret sub-symbolic systems.

1.4 STRUCTURE OF THIS BOOK – A LANDSCAPE OF AI

The chapters that follow will take a relatively pragmatic approach to AI and are divided into five parts:

Part I. Knowledge-Based AI – This is rooted in classic areas of AI covering knowledge representation, reasoning and search. While these are mostly concerned with symbolic AI, they are also important basics for those interested in neural networks and machine learning as they give a conceptual vocabulary with which to interpret more emergent features.

Part II. Data and Learning – This covers various forms of machine learning, that is techniques where the AI algorithm works from data and creates its own rules. This includes some more traditional algorithms, various forms of neural networks, statistical techniques and architectural decisions for deep learning. It will consider more conceptual and theoretical issues for any form of machine learning as well as techniques for dealing with big data and practical 'data wrangling'.

Part III. Specialised Areas – This part considers a number of specific areas: games, natural language processing (NLP), vision, time-varying signals and media, robotics, agents and the web. Each of these has some specialised methods and algorithms, but also there is much overlap and many cross-cutting lessons. It is well worth dipping your toes into all the areas even if you have one specific focus.

Part IV. Humans at the Heart – Every computer system and every system involving AI will have some impact on people. However, some applications and some issues have more direct impact than others. In this part we consider such areas including expert and decision support systems, methods for designing human interactions with intelligent systems, issues of bias, privacy and explainable AI, more cognitively inspired AI and critical philosophical, social and ethical questions. We see that there are still many opportunities for using understanding of human cognition to inspire AI and for AI to help us understand some of the profound aspects of being human.

Part V. Looking Forward – The book ends with a short glimpse into the possible future directions of AI: what is current, upcoming and maybe coming next. In such a fast-moving area, some of this will undoubtedly look dated by the time the book is even printed. However, looking back to the epilogue of the first edition, 25 years ago, there are issues that are still valid today. So, clearly, some challenges are likely to take longer to come to fruition ... maybe ones you would like to address yourself.

I

Knowledge-Rich AI

CHAPTER 2

Knowledge in AI

2.1 OVERVIEW

Knowledge is vital to all intelligence. In this chapter we examine four key knowledge representation schemes looking at examples of each and their strengths and weaknesses. We consider how to assess a knowledge representation scheme in order to choose one that is appropriate to our particular problem. We discuss the problems of representing general knowledge and changing knowledge.

2.2 INTRODUCTION

Knowledge is central to intelligence. We need it to use or understand language, to make decisions, to recognise objects, to interpret situations and to plan strategies. We store in our memories millions of pieces of knowledge that we use daily to make sense of the world and our interactions with it.

Some of the knowledge we possess is factual. We know what things are and what they do. This type of knowledge is known as declarative knowledge. We also know how to do things: procedural knowledge. For example, if we consider what we know about the English language, we may have some declarative knowledge that the word *tree* is a noun and that *tall* is an adjective. These are among the thousands of facts we know about the English language,

However, we also have procedural knowledge about English. For example, we may know that in order to provide more information about something we place an adjective before the noun.

Similarly, imagine you are giving directions to your home. You may have declarative knowledge about the location of your house and its transport links (e.g. "my house is in Golcar", "the number 301 bus runs through Golcar", "Golcar is off the Manchester Road"). In addition you may have procedural knowledge about how to get to your house ("Get on the 301 bus").

Another distinction that can be drawn is between the specific knowledge we have on a particular subject (domain-specific knowledge) and the general or "common-sense" knowledge that applies throughout our experience (domain-independent knowledge). The fact "the number 301 bus goes to Golcar" is an example of the former: it is knowledge that is relevant only in a restricted domain – in this case Huddersfield's transport system. New knowledge would be required to deal with transport in any other city. However, the knowledge that a bus is a motorised means of transport is a piece of general knowledge which is applicable to buses throughout our experience.

General or common-sense knowledge also enables us to interpret situations accurately. For example, imagine someone asks you "Can you tell me the way to the station?". Your common-sense knowledge tells you that the person expects a set of directions; only a deliberately obtuse person would answer literally "yes"! Similarly there are thousands if not millions of "facts" that are obvious to us from our experience of the world, many acquired in early childhood. They are so obvious to us that we wouldn't normally dream of expressing them explicitly. Facts about age: a person's age increments by one each year, children are always younger than their parents, people don't live much longer than 100 years; facts about the way that substances such as water behave; facts about the physical properties of everyday objects and indeed ourselves – this is the general or "common" knowledge

DOI: 10.1201/9781003082880-3

that humans share through shared experience and that we rely on every day.

Just as we need knowledge to function effectively, it is also vital in artificial intelligence. As we saw earlier, one of the problems with ELIZA was lack of knowledge: the program had no knowledge of the meanings or contexts of the words it was using and so failed to convince for long. So the first thing we need to provide for our intelligent machine is knowledge. As we shall see, this will include procedural and declarative knowledge and domain-specific and general knowledge. The specific knowledge required will depend upon the application. For language understanding we need to provide knowledge of syntax rules, words and their meanings, and context; for expert decision making, we need knowledge of the domain of interest as well as decision-making strategies. For visual recognition, knowledge of possible objects and how they occur in the world is needed. Even simple game playing requires knowledge of possible moves and winning strategies.

2.3 REPRESENTING KNOWLEDGE

We have seen the types of knowledge that we use in everyday life and that we would like to provide to our intelligent machine. We have also seen something of the enormity of the task of providing that knowledge. However, the knowledge that we have been considering is largely experiential or internal to the human holder. In order to make use of it in AI we need to get it from the source (usually human but can be other information sources) and represent it in a form usable by the machine. Human knowledge is usually expressed through language, which, of course, cannot be accurately understood by the machine. The representation we choose must therefore be both appropriate for the computer to use and allow easy and accurate encoding from the source.

We need to be able to represent facts about the world. However, this is not all. Facts do not exist in isolation; they are related to each other in a number of ways. First, a fact may be a specific instance of another, more general fact. For example, "Spotty Dog barks" is a specific instance of the fact "all dogs bark" (not strictly true but a common belief). In a case like this, we may wish to allow property inheritance, in which properties or attributes of the main class are inherited by instances of that class. So we might represent the knowledge that dogs bark and that Spotty Dog is a dog, allowing us then to deduce

FIGURE 2.1   Four knights: how many moves?

by inheritance the fact that Spotty Dog barks. Secondly, facts may be related by virtue of the object or concept to which they refer. For example, we may know the time, place, subject and speaker for a lecture and these pieces of information make sense only in the context of the occasion by which they are related. And of course we need to represent procedural knowledge as well as declarative knowledge.

It should be noted that the representation chosen can be an important factor in determining the ease with which a problem can be solved. For example, imagine you have a 3×3 chess board with a knight in each corner (as in Figure 2.1). How many moves (i.e. chess knight moves) will it take to move each knight round to the next corner?

Looking at the diagrammatic representation in Figure 2.1, the solution is not obvious, but if we label each square and represent valid moves as adjacent points on a circle (see Figure 2.2), the solution becomes more obvious: each knight takes two moves to reach its new position, so the minimum number of moves is eight.

In addition, the granularity of the representation can affect its usefulness. In other words, we have to determine how detailed the knowledge we represent needs to be. This will depend largely on the application and the use to which the knowledge will be put. For example, if

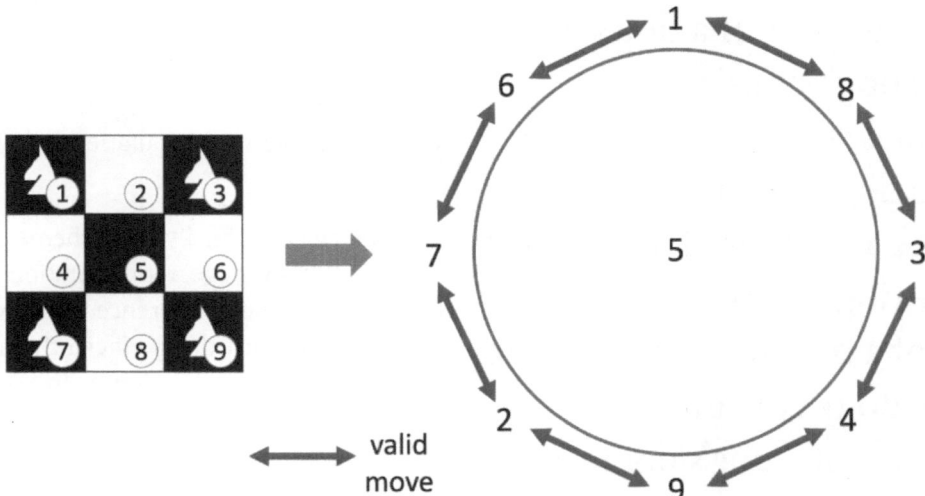

FIGURE 2.2   A different representation makes the solution clearer.

we are building a knowledge base about dog family relationships for a pedigree management system, we may include a representation of the definition of the relation "cousin" (given here in English but easily translatable into logic, for example):

> an offspring of a sibling of a dog's parent
> is the dog's cousin

However, this may not be enough information; we may also wish to know the gender of the cousin. If this is the case, a more detailed representation is required. For a female cousin:

> a daughter of a sibling of a dog's parent
> is the dog's cousin,

or a male cousin:

> a son of a sibling of a dog's parent
> is the dog's cousin.

Similarly, if you wanted to know which side of the family a dog's cousin belongs, you would need different information; from the dog's sire's (father's) side:

> an offspring of a sibling of a dog's sire (father)
> is the dog's cousin,

or its dam's (mother's) side:

> an offspring of a sibling of a dog's dam (mother)
> is the dog's cousin.

A full description of all the possible variations is given in Figure 2.3. Such detail may not always be required and therefore seem unnecessarily complex, but some simple reasoning systems may need this form of expansion. Note too as we are more precise, or at least exhaustive, we may miss or, if we are observant, notice cases that might slip between or challenge our definitions, for example dog surrogacy,

There are a number of knowledge representation methods that can be used. Later in this chapter we will examine some of them briefly and identify the areas for which each is best suited. In later chapters of the book we will see how these methods can be used in specific application areas. But what makes a good knowledge representation scheme, and how can different schemes be evaluated against one another? Before going on to consider specific approaches to knowledge representation, we will look in more detail at what features a knowledge representation scheme should possess.

2.4 METRICS FOR ASSESSING KNOWLEDGE REPRESENTATION SCHEMES

We have already looked at some of the factors we are looking for in a knowledge representation scheme. However, we can expand upon these and generate some metrics by which to measure the representations available to us. The main requirements of a knowledge representation scheme can be summarised under four

a daughter of a sister of a dog's dam
  is the dog's cousin
a daughter of a sister of a dog's sire
  is the dog's cousin
a daughter of a brother of a dog's dam
  is the dog's cousin
a daughter of a brother of a dog's sire
  is the dog's cousin
a son of a sister of a dog's dam
  is the dog's cousin
a son of a sister of a dog's sire
  is the dog's cousin
a son of a brother of a dog's dam
  is the dog's cousin
a son of a brother of a dog's sire
  is the dog's cousin

FIGURE 2.3   Full list of variants of the relationship "cousin".

headings: *expressiveness*, *effectiveness*, *efficiency* and *explanation*.

- expressiveness. We have already considered some of the types of knowledge that we might wish to represent. An expressive representation scheme will be able to handle different types and levels of granularity of knowledge. It will be able to represent complex knowledge and knowledge structures and the relationships between them. It will have means of representing specific facts and generic information (e.g. by using variables). Expressiveness also relates to the *clarity* of the representation scheme. Ideally, the scheme should use a notation that is natural and usable both by the knowledge engineer and the domain expert. Schemes that are too complex for the latter to understand can result in incorrect knowledge being held, since the expert may not be able to critique the knowledge adequately. In summary, our representation scheme should be characterised by completeness and *clarity* of expression.

- effectiveness. The second measure of a good representation scheme is its effectiveness. In order to be effective, the scheme must provide a means of inferring new knowledge from old. It should also be amenable to computation, allowing adequate tool support.

- efficiency. Thirdly, the scheme should be efficient. The knowledge representation scheme must not only support inference of new knowledge from old but must do so efficiently in order for the new knowledge to be of use. In addition, the scheme should facilitate efficient knowledge gathering and representation.

- explicitness. Finally, a good knowledge representation scheme must be able to provide an explanation of its inferences and allow justifications of its reasoning. The chain of reasoning should be explicit.

In the rest of this chapter we will use these four metrics to compare the effectiveness of the techniques we will consider.

2.5 LOGIC REPRESENTATIONS

Logic representations use expressions in formal logic to represent the knowledge required. Inference rules and proof procedures can apply this knowledge to specific problems. First-order predicate calculus is the most common form of logic representation, with Prolog being the most common language used to implement it.

Logic is appealing as a means of knowledge representation, as it is a powerful formalism with known inference procedures. We can derive a new piece of knowledge by proving that it is a consequence of knowledge that is already known. The significant features of the domain can be represented as logical assertions, and general attributes can be expressed using variables in logical statements. It has the advantage of being computable, albeit in a restricted form.

So how can we use logic to represent knowledge? Facts can be expressed as simple propositions. A proposition is a statement that can have one of two values: true or false. These are known as truth values. So the statements *It is raining* and *I am hungry* are propositions whose values depend on the situation at the time. If I have just eaten dinner in a thunderstorm, then the first is likely to be true and the second false. Propositions can be combined using operators such as *and* ($\land$) and *or* ($\lor$). Returning

P	Q	P ∧ Q	P ∨ Q
T	T	T	T
T	F	F	T
F	T	F	T
F	F	F	F

FIGURE 2.4   Truth values for simple logic operators.

to our dining example, we could combine the two statements: *It is raining* and *I am hungry* (which for convenience we will express as P ∧ Q). The truth value of the combined propositions will depend upon the truth values of the individual propositions and the operator connecting them. If the situation is still as it was, then this combined propositional statement will be false, since one of the propositions (Q) is false.

Figure 2.4 shows a truth table that defines the truth values of *and* and *or*.

Do note however that the use of 'and' and 'or' in day-to-day speech is not always the same as in formal logic. For example, if asked about food preferences, you might reply "I like sausage and ice cream" or "I'd like ice cream or I'd like sausage", but neither would be taken to mean you'd like both on the same plate. Note too that the word 'like' in the two sentences means something different, in the first 'enjoy' in the second 'want to eat now', but this requires an understanding of context. So, when expressing natural language in formal logic take care that the translation may not be direct.

Propositional logic is limited in that it does not allow us to generalise sufficiently. Common elements within propositions cannot be used to make inferences. We need to be able to extract such common elements as parameters to the propositions, in order to allow inferences with them. Parametrised propositions give us predicate logic. For example, if we wish to represent our knowledge of the members of Thunderbirds' *International Rescue* organisation, we might include such facts as

> *father(Jeff, Virgil)*
> *father(Jeff, Alan)*

to mean *Jeff is the father of Virgil* and *Jeff is the father of Alan,* respectively. *Father* is the predicate here and *Jeff, Virgil* and *Alan* are parameters. In predicate logic, parameters can also include variables. For example,

> *father(Jeff, x)*

where x is a variable that can be instantiated later with a value – the name of someone of whom Jeff is the father.

Quantifiers (universal and existential) allow the scope of the variable to be determined unambiguously. For example, in the statement above, we do not know for certain that there is value for x; that is, that Jeff is indeed someone's father (ignoring the two earlier facts for a moment). In the following statement we use the existential quantifier, ∃ (read as 'there exists'), to express the fact that Jeff is the father of at least one person:

> $\exists x$: *father(Jeff, x)*

Similarly we can express rules that apply universally using the universal quantifier, ∀ (read as 'for all'):

> $\forall x \forall y$: *father(x, y)* ∨ *mother(x, y)*
> → *parent(x, y)*
> $\forall x \forall y \forall z$: *parent(x, y)* ∧ *parent(x, z)*
> → *sibling(y, z)*

The first of these states that for all values of x and y if x is the father of y or (∨) the mother of y, then x is the parent of y. The second uses this knowledge to say something about siblings: for all values of x, y and z, if x is the parent of y (i.e. the father or the mother), and (∧) x is the parent of z, then y and z are siblings.

Inference methods allow us to derive new facts from existing facts. There are a number of inference procedures for logic, but we can illustrate the principle using the simple rule that we can substitute a universally quantified variable with any value in its domain. So, given the rule about parenthood and the facts we already know about the family from *International rescue*, we can derive new facts as shown below.

Given

> $\forall x \forall y$: *father(x, y)* ∨ *mother(x, y)*
> → *parent(x, y)*
> *father(Jeff, Virgil)*
> *father(Jeff, Alan)*

we can derive the facts (by substitution)

> *parent(Jeff, Virgil)*
> *parent(Jeff, Alan)*

Similarly, given

$$\forall x \, \forall y \, \forall z \, : \, parent(x, y) \wedge parent(x, z)$$
$$\rightarrow sibling(y, z)$$
$$parent(Jeff, Virgil)$$
$$parent(Jeff, Alan)$$

we can derive the fact

$$sibling(Virgil, Alan)$$

Facts and rules such as these can be represented easily in Prolog. However, predicate logic and Prolog have a limitation, which is that they operate under what is known as the closed world assumption. This means that we assume that all knowledge in the world is represented: the knowledge base is complete. Therefore any fact that is missing is assumed to be false. Prolog uses a problem-solving strategy called negation as failure, which means that it returns a result of *false* if it is unable to prove a goal to be true. This relies on the closed world assumption [233]. Such an assumption is useful when all relevant facts are represented but can cause problems when the knowledge base is incomplete.

In summary, logic is

- expressive: it allows representation of facts, relationships between facts and assertions about facts. It is relatively understandable. Prolog is less expressive since it is not possible to represent logical negation explicitly. This in turn leads to less clarity.

- effective: new facts can be inferred from old. It is also amenable to computation through Prolog.

- efficient: the use of predicates and variables makes the representation scheme relatively efficient, although computational efficiency depends to a degree on the interpreter being used and the programmer.

- explicit: explanations and justifications can be provided by backtracking.

2.6 PROCEDURAL REPRESENTATION

Logic representations, such as we have been looking at, are declarative: we specify what we know about a problem or domain. We do not specify how to solve the problem or what to do with the knowledge. Procedural approaches, on the other hand, represent knowledge as a set of instructions for solving a problem. If a given condition is met, then an associated action or series of actions is performed. The production system is an example of this [209].

A production system has three components:

- a database of facts (often called working memory)

- a set of production rules that alter the facts in the database. These rules or *productions* are of the form

    IF <condition> THEN <action>

- an interpreter that decides which rule to apply and handles any conflicts.

2.6.1 The Database

The database or working memory represents all the knowledge of the system at any given moment. It can be thought of as a simple database of facts that are true of the domain at that time. The number of items in the database is small: the analogy is to human working memory, which can hold only a small number of items at a time. The contents of the database change as facts are added or removed according to the application of the rules.

2.6.2 The Production Rules

Production rules are operators that are applied to the knowledge in the database and change the state of the production system in some way, usually by changing the content of the database. Production rules are sometimes called *condition–action* rules, and this describes their behaviour well. If the condition of a rule is true (according to the database at that moment), the action associated with the rule is performed. This may be, for example, to alter the contents of the database by removing a fact or to interact with the outside world in some way.

Production rules are usually unordered, in the sense that the sequence in which the rules will be applied depends on the current state of the database: the rule whose condition matches the state of the database will be selected. If more than one rule matches, then conflict resolution strategies are applied. However, some production systems are programmed to apply rules in order, so avoiding conflict (this is itself a conflict resolution strategy).

1. IF <client working? is unknown>
   THEN ask "Are you working?"
         read WORKING
         remove <client working? is unknown>
         add <client working? is WORKING>
2. IF <client working? is YES> and <salary is unknown>
   THEN ask "What is your salary?"
         read SALARY
         remove <salary is unknown>
         add <salary is SALARY>
3. IF <client working? is YES> and <salary is SALARY> and SALARY > (5 * AMOUNT REQUESTED)
   THEN grant loan of AMOUNT REQUESTED
         clear database
         finish
4. IF <client working? is YES> and <salary is SALARY> and SALARY ≤ (5 * AMOUNT REQUESTED)
   THEN grant loan of (SALARY/5)
         clear database
         finish
5. IF <client working? is NO> and <client student? is unknown>
   THEN ask "Are you a student?"
         read STUDENT
         remove <client student? is unknown>
         add <client student? is STUDENT>
6. IF <client working? is NO> and <client student? is YES>
   THEN discuss student loan
         clear database
         finish
7. IF <client working? is NO> and <client student? is NO>
   THEN refuse loan
         clear database
         finish

FIGURE 2.5   Production system rules for assessing a loan application.

2.6.3 The Interpreter

The interpreter is responsible for finding the rules whose conditions are matched by the current state of the database. It must then decide which rule to apply. If there is more than one rule whose condition matches, then one of the contenders must be selected using strategies such as those proposed below. If no rule matches, the system cannot proceed. Once a single rule has been selected the interpreter must perform the actions in the body of the rule. This process continues until there are no matching rules or until a rule is triggered which includes the instruction to stop.

The interpreter must have strategies to select a single rule where several match the state of the database. There are a number of possible ways to handle this situation. The most simple strategy is to choose the first rule that matches. This effectively places an ordering on the production rules, which must be carefully considered when writing the rules. An alternative strategy is to favour the most specific rule. This may involve choosing a rule that matches all the conditions of its contenders but that also contains further conditions that match, or it may mean choosing the rule that instantiates variables or qualifies a fact.

For example,

> IF <salary is high> and < age > 40>

is more specific than

> IF <salary is high>.

Similarly

> IF <salary > £40,000>

is more specific than

> IF <salary > £20,000>

since fewer instances will match it.

2.6.4 An Example Production System: Making a Loan

This production system gives advice on whether to make a loan to a client (its rules are obviously very simplistic, but it is useful to illustrate the technique). Initially the database contains the following default facts:

> <client working? is unknown>
> <client student? is unknown>
> <salary is unknown>

and a single fact relating to our client:

> <AMOUNT REQUESTED is £2000>

which represents the amount of money our client wishes to borrow (we will assume that this has been added using other rules). We can use the production system to find out more information about the client and decide whether to give this loan. Figure 2.5 shows a set of rules that could be used to determine this.

Imagine our client is working and earns £7500. Given the contents of the database, the following sequence occurs:

1. Rule 1 fires since the condition matches a fact in the database. The user answers YES to the question, instantiating the variable WORKING to YES. This adds the fact <client working? is YES> to the database, replacing the fact <client working? is unknown>

   - Database contents after rule 1 fires:
     <client working? is YES>
     <client student? is unknown>
     <salary is unknown>
     <AMOUNT REQUESTED is £2000>

2. Rule 2 fires instantiating the variable SALARY to the value given by the user. This adds this fact to the database, as above.

   - Database contents after rule 2 fires:
     <client working? is YES>
     <client student? is unknown>
     <salary is £7500>
     <AMOUNT REQUESTED is £2000>

3. Rule 4 fires since the value of SALARY is less than five times the value of AMOUNT REQUESTED. This results in an instruction to grant a loan of SALARY/5, that is £1500. The system then clears the database to the default values and finishes.

This particular system is very simple and no conflicts can occur. It is assumed that the interpreter examines the rule base from the beginning each time.

To summarise, we can consider production systems against our metrics:

- expressiveness: production systems are particularly good at representing procedural knowledge. They are ideal in situations where knowledge changes over time and where the final and initial states differ from user to user (or subject to subject). The approach relies on an understanding of the concept of a working memory, which sometimes causes confusion. The modularity of the representation aids clarity in use: each rule is an independent chunk of knowledge, and modification of one rule does not interfere with others.

- effectiveness: new information is generated using operators to change the contents of working memory. The approach is very amenable to computation.

- efficiency: the scheme is relatively efficient for procedural problems, and their flexibility makes it transferable between domains. The use of features from human problem solving (such as short-term memory) means that the scheme may not be the most efficient. However, to counter this, these features make it a candidate for modelling human problem solving.

- explicitness: production systems can be programmed to provide explanations for their decisions by tracing back through the rules that are applied to reach the solution.

2.7 NETWORK REPRESENTATIONS

Network representations capture knowledge as a graph, in which nodes represent objects or concepts and arcs represent relationships or associations. Relationships can be domain specific or generic (see below for examples).

Networks support property inheritance. An object or concept may be a member of a class and is assumed to have the same attribute values as the parent class (unless alternative values override). Classes can also have subclasses that inherit properties in a similar way. For example, the parent class may be *Dog*, which has attributes such as *has tail*, *barks* and *has four legs*. A subclass of that parent class may be a particular breed, say *Great Dane*, which consequently inherits all the attributes above, as well as having its own attributes (such as *tall*). A particular member (or instance) of this subclass, that is a particular Great Dane, may have additional attributes such as colour. Property inheritance is overridden where a class member or subclass has an explicit alternative value for an attribute. For example, *Rottweiler* may be a subclass of the parent class *Dog* but may have the attribute *has no tail*.

Alternatives may also be given at the instance level: *Rottweiler* as a class may inherit the property *has tail*, but a particular dog, whose tail has been docked, may have the value *has no tail* overriding the inherited property.

Semantic networks are an example of a network representation. A semantic network illustrating property inheritance is given below. It includes two generic relationships that support property inheritance: *is-a* indicating class inclusion (subclass) and *instance* indicating class membership. The network of classes, subclasses and their properties and relations is sometimes called an ontology.

Property inheritance supports inference, in that we can derive facts about an object by considering the parent classes. For example, in the *Dog* network in Figure 2.6, we can derive the facts that a Great Dane has a tail and is carnivorous from the facts that a dog has a tail and a canine is carnivorous, respectively. Note, however, that we cannot derive the fact that a Basenji can bark since we have an alternative value associated with Basenji. Note also how the network links together information from different domains (dogs and cartoons) by association.

Network representations are useful where an object or concept is associated with many attributes and where relationships between objects are important. Considering them against our metrics for knowledge representation schemes:

- expressiveness: they allow representation of facts and relationships between facts. The levels of the hierarchy provide a mechanism for representing general and specific knowledge. The representation is a model of human memory, and it is therefore relatively understandable.

- effectiveness: they support inference through property inheritance. They can also be easily represented using Prolog and other AI languages making them amenable to computation.

- efficiency: they reduce the size of the knowledge base, since knowledge is stored only at its highest level of abstraction rather than for every instance or example of a class. They help maintain consistency in the knowledge base, because high-level properties are inherited by subclasses and not added for each subclass.

- explicitness: reasoning equates to following paths through the network, so the relationships and inference are explicit in the network links.

Some kinds of knowledge or data you encounter will be very clearly networks, for example links in web pages or friend connections in a social network. Others you may build up based on analysing the meaning of data (such as the semantic network above) or derived from other forms of information, for example connecting words that frequently occur close to one another in text. These are all symbolic networks, where the individual nodes have a well-defined meaning (a web page, person or word). However, you will also encounter sub-symbolic networks, including different forms of neural network, that is where the nodes in the network have no predefined meaning. We will discuss these in Chapter 6.

2.8 STRUCTURED REPRESENTATIONS

In structured representations information is organised into more complex knowledge structures. *Slots* in the structure represent attributes into which values can be placed. These values are either specific to a particular instance or default values, which represent stereotypical information. Structured representations can capture complex situations or objects, for example eating a meal

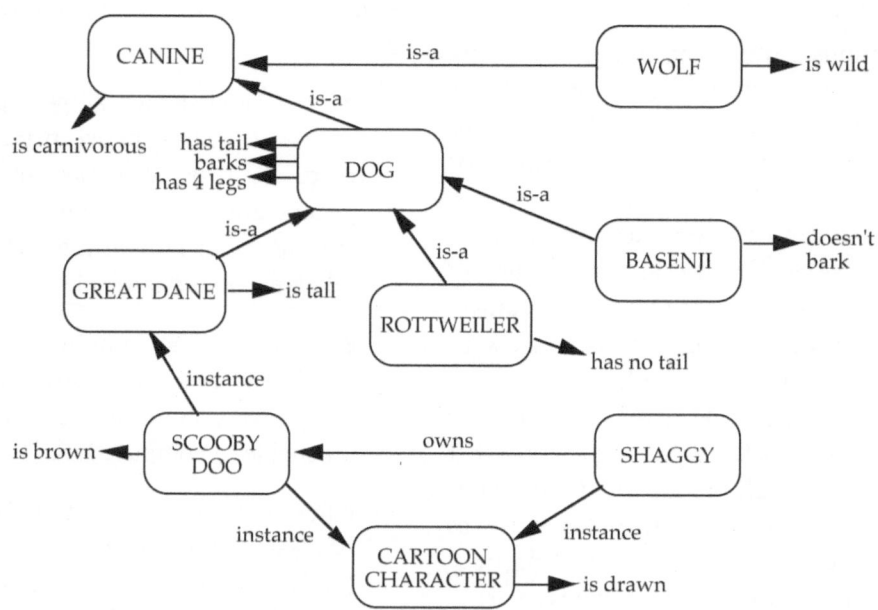

FIGURE 2.6   A fragment of a semantic network.

in a restaurant or the content of a hotel room. Such structures can be linked together as networks, giving property inheritance. Frames and scripts are the most common types of structured representation.

2.8.1 Frames

Frames are knowledge structures that represent expected or stereotypical information about an object [196]. For example, imagine a supermarket. If you have visited one or two, you will have certain expectations as to what you will find there. These may include aisles of shelves, freezer banks and check-out tills. Some information will vary from supermarket to supermarket, for example the number of tills. This type of information can be stored in a network of frames where each frame comprises a number of slots with appropriate values. A section of a frame network on supermarkets is shown in Figure 2.7.

In summary, frames extend semantic networks to include structured, hierarchical knowledge. Since they can be used with semantic networks, they share the benefits of these, as well as

- expressiveness: they allow representation of structured knowledge and procedural knowledge. The additional structure increases clarity.

- effectiveness: actions or operations can be associated with a slot and performed, for example, when-

ever the value for that slot is changed. Such procedures are called *demons*.

- efficiency: they allow more complex knowledge to be captured efficiently.

- explicitness: the additional structure makes the relative importance of particular objects and concepts explicit.

2.8.2 Scripts

A script, like a frame, is a structure used to represent a stereotypical situation [243]. It also contains slots that can be filled with appropriate values. However, where a frame typically represents knowledge of objects and concepts, scripts represent knowledge of events. They were originally proposed as a means of providing contextual information to support natural language understanding (see Chap. 13).

Consider the following description:

> Alison and Brian went to the supermarket. When they had got everything on their list they went home.

Although it is not explicitly stated in this description, we are likely to infer that Alison and Brian paid for their selections before leaving. We might also be able

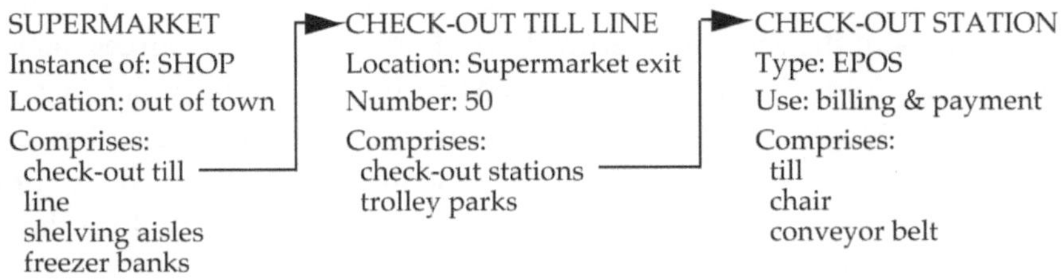

SUPERMARKET | CHECK-OUT TILL LINE | CHECK-OUT STATION
Instance of: SHOP | Location: Supermarket exit | Type: EPOS
Location: out of town | Number: 50 | Use: billing & payment
Comprises: | Comprises: | Comprises:
    check-out till line | check-out stations | till
    shelving aisles | trolley parks | chair
    freezer banks | | conveyor belt

FIGURE 2.7 Frame representation of supermarket.

to fill in more details about their shopping trip: that they had a trolley and walked around the supermarket, that they selected their own purchases, that their list contained the items that they wished to buy. All of this can be inferred from our general knowledge concerning supermarkets, our expectations as to what is likely to happen at one. Our assumptions about Alison and Brian's experience of shopping would be very different if the word *supermarket* was replaced by *corner shop*.

It is this type of stereotypical knowledge that scripts attempt to capture, with the aim of allowing a computer to make similar inferences about incomplete stories to those we were able to make above. Schank and colleagues developed a number of programs during the 1970s and 1980s that used scripts to answer questions about stories [243]. The script would describe likely action sequences and provide the contextual information to understand the stories.

A script comprises a number of elements:

- *entry conditions*: these are the conditions that must be true for the script to be activated.

- *results*: these are the facts that are true when the script ends.

- *props*: these are the objects that are involved in the events described in the script.

- *roles*: these are the expected actions of the major participants in the events described in the script.

- *scenes*: these are the sequences of events that take place.

- *tracks*: these represent variations on the general theme or pattern of the script.

For example, a script for going to a supermarket might store the following information:

Entry conditions:	supermarket open, shopper needs goods, shopper has money
Result:	shopper has goods, supermarket has less stock, supermarket has more money
Props:	trolleys, goods, check-out tills
Roles:	shopper collects food, assistant checks out food and takes money, store manager orders new stock
Scenes:	selecting goods, checking out goods, paying for goods, packing goods
Tracks:	customer packs bag, assistant packs bag

Scripts have been useful in natural language understanding in restricted domains. Problems arise when the knowledge required to interpret a story is not domain specific but general, "common-sense" knowledge. Charniak [46] used children's stories to illustrate just how much knowledge is required to interpret even simple descriptions. For example, consider the following excerpt about exchanging unwanted gifts:

> Alison and Brian received two toasters at their engagement party, so they took one back to the shop.

To interpret this we need to know about toasters and why, under normal circumstances, one wouldn't want two; we also need to know about shops and their normal exchange policies. In addition, we need to know about engagements and the tradition of giving gifts on such occasions. But the situation is more complicated than it appears. If instead of toasters Alison and Brian had received two gift vouchers, two books or two £20 notes, they would not have needed to exchange them. So the rule that one doesn't want two of something only applies to certain items. Such information is not specific to engagements: the same would be true of birthday presents, wedding presents or Christmas presents. So in which script do we store such information?

This is indicative of a basic problem of AI, which is how to provide the computer with the general, interpretative knowledge that we glean from our experience, as well as the specific factual knowledge about a particular domain. We will consider this problem in the next section.

Scripts are designed for representing knowledge in a particular context. Within this context the method is expressive and effective, except as we have seen in representing general knowledge, but it is limited in wider application. Similarly, it provides an efficient and explicit mechanism for capturing complex structured information within its limited domain.

2.9 GENERAL KNOWLEDGE

Most knowledge-based systems are effective in a restricted domain only because they do not have access to the deep, common knowledge that we use daily to interpret our world. Few AI projects have attempted to provide such general knowledge, the CYC project begun at MCC, Texas, by Doug Lenat [164], being a notable exception.

The CYC project aimed to build a knowledge base containing the millions of pieces of common knowledge that humans possess. It was originally envisaged as a ten-year project involving many people, meticulously encoding the type of facts that are "obvious" to us, facts at the level of "all men are people" and "children are always younger than their parents". To us, expressing such facts seems ludicrous; for the computer they need to be represented explicitly.

The project set out to investigate whether it is possible to represent such common-sense knowledge effectively in a knowledge base and also considers the problems of building and maintaining large-scale knowledge bases. Its critics claimed that it would be a waste of time and money, since such knowledge can only be gained by experience, for example the experiences children have through play. However, CYC can derive new knowledge from the facts provided, effectively learning and generalising from its, albeit artificial, experience.

Although the original 10-year timescale was massively optimistic, CYC has confounded its critics, and the open and research versions (OpenCYC and ResearchCYC) have been widely used [68]. Furthermore other projects have picked up aspects of this including

the YAGO/YAGO2 project that draws knowledge from a variety of sources including DBpedia [136]. Another ambitious project was FreeBase, which was crowdsourced collection of interlinked data based on standard ontologies and relationships and provided a publicly available API. The original startup was acquired by Google and used to pump-prime its own KnowledgeGraph used for the information boxes attached to many web searches, but sadly closed the original service. Happily, the FreeBase data was made available in various forms including to the WikiData project [219].

We will see in Chapters 8 and 17 how the availability of large data sources, especially on the web, has allowed alternative ways to acquire (apparent) knowledge based on text mining. It is likely that ongoing and future initiatives will leverage the large curated resources such as OpenCyc alongside data mining techniques.

2.10 THE FRAME PROBLEM

Throughout this chapter we have been looking at knowledge representation schemes that allow us to represent a problem at a particular point in time: a particular *state*. However, as we will see in subsequent chapters, representation schemes have to be able to represent sequences of problem states for use in search and planning. Imagine the problem of moving an automatic fork lift truck around a factory floor. In order to do this we need to represent knowledge about the layout of the factory and the position of the truck, together with information dictating how the truck can move (perhaps it can only move if its forks are raised above the ground). However, as soon as the truck makes one movement, the knowledge has changed and a new state has to be represented. Of course, not all the knowledge has changed; some facts, such as the position of the factory walls, are likely to remain the same. The problem of representing the facts that alter from state to state as well as those that remain the same is the essence of the frame problem [185].

In some situations, where keeping track of the sequence of states is important, it is infeasible to simply store the whole state each time – doing so will soon use up memory. So it is necessary to store information about what does and does not change from state to state. In some situations even deciding what changes is not an easy problem. In our factory we may describe bricks as being on a pallet which in turn is by the door:

$on(pallet, bricks)$

$by(door, pallet)$

If we move the pallet, then we infer that the bricks also move but that the door does not. So in this case at least the relationship *on* implies no change, but *by* does imply a change (the pallet is no longer by the door).

A number of solutions have been proposed to the frame problem. One approach is to include specific frame axioms which describe the parts that do not change when an operator is applied to move from state to state. So, for example, the system above would include the axiom

$$on(x, y, s_1) \land move(x, s_1, s_2) \rightarrow on(x, y, s_2)$$

to specify that when an object, y, is on object x in state s_1, then if the operation move is applied to move x to state s_2, then object y is still on object x in the new state. Frame axioms are a useful way of making change explicit but become extremely unwieldy in complex domains.

An alternative solution is to describe the initial state and then change the state description as rules are applied. This means that the representation is always up-to-date. Such a solution is fine until the system needs to backtrack in order to explore another solution. Then there is nothing to indicate what should be done to undo the changes. Instead we could maintain the initial description but store changes each time an operator is applied. This makes backtracking easy since information as to what has been changed is immediately available, but it is again a complex solution. A compromise solution is to change the initial state description but also store information as to how to undo the change.

There is no ideal solution to the frame problem, but these issues should be considered both in selecting a knowledge representation scheme and in choosing appropriate search strategies. We will look at search in more detail in Chapter 4.

2.11 KNOWLEDGE ELICITATION

All knowledge representation depends upon knowledge elicitation to get the appropriate information from the source (often human) to the knowledge base. Knowledge elicitation is the bottleneck of knowledge-based technology. It is difficult, time consuming and imprecise. This is because it depends upon the expert providing the right information, without missing anything out. This in turn often depends upon the

person trying to elicit the knowledge (the knowledge engineer) asking the expert the right questions in an area that he or she may know little about.

To illustrate the magnitude of the knowledge elicitation problem, think of a subject that you know something about (perhaps a hobby, a sport, a form of art or literature, a skill). Try to write down everything you know about the subject. Even more enlightening, get a friend who is not expert in the subject to question you about it, and provide answers to the questions. You will soon find that it is difficult to be precise and exhaustive in this type of activity.

A number of techniques have been proposed to help alleviate the problem of knowledge elicitation. These include structured interview techniques, knowledge elicitation tools and the use of machine-learning techniques that learn concepts from examples. The latter can be used to identify key features in examples which characterise a concept. We will look in more detail at knowledge elicitation when we consider expert systems in Chapter 18.

2.12 SUMMARY

In this chapter we have seen the importance of an appropriate knowledge representation scheme and how we can assess potential schemes according to their expressiveness, effectiveness, efficiency and explicitness. We have considered four key representation schemes – logic, production rules, network representations and structured representations – looking at examples of each and their strengths and weaknesses. We have looked at the problems of representing general knowledge and changing knowledge. Finally, we have touched on the problem of knowledge elicitation, which we will return to in Chapter 18.

2.1 UK law forbids marriage between certain relatives (e.g. parents and children, brothers and sisters) but allows it between others (e.g. first cousins). Use a logic formalism to represent your knowledge about UK (or your own country's) marriage laws.

2.2 A pet shop would like to implement an expert system to advise customers on suitable pets for their circumstances. Write a production system to incorporate the following information (your system should elicit the information it needs from the customer).

A budgie is suitable for small homes (including city flats) where all the members of the family are out during the day. It is not appropriate for those with a fear of birds or who have a cat.

A guinea pig is suitable for homes with a small garden where the occupants are out all day. It is particularly appropriate for children. However, it will require regular cleaning of the cage.

A cat is suitable for most homes except high-rise flats, although the house should not be on a main road. It does not require exercise. Some people are allergic to cats.

A dog is suitable for homes with a garden or a park nearby. It is not suitable if all occupants are out all day. It will require regular exercise and grooming.

2.3 Construct a script for a train journey. (You can use a natural language representation but you should clearly indicate the script elements.)

2.4 Working in pairs, one of you should take the role of expert, the other of knowledge engineer. The expert should suggest a topic in which he or she is expert and the knowledge engineer should ask questions of the expert to elicit information on this topic. The expert should answer as precisely as possible. The knowledge engineer should record all the answers given. When enough information has been gathered, choose an appropriate representation scheme and formalise this knowledge.

FURTHER READING

F. Van Harmelen, V. Lifschitz, and B. Porter, editor. *Handbook of knowledge representation.* Elsevier, San Diego, CA, 2008.

Edited collection that covers a wide variety of knowledge representation techniques, including spatial reasoning and temporal logics.

G. A. Ringland and D. A. Duce. *Approaches to knowledge representation: An introduction.* John Wiley, Chichester, 1988.

This classic book explains the issues of knowledge representation in more detail than is possible here: a good next step. It is now out of print but available at: https://www.chilton-computing.org.uk/inf/pdfs/knowledge_representation.pdf

R. J. Brachman, H. J. Levesque, and R. Retier, editors. *Knowledge representation.* MIT Press Cambridge, MA, 1992.

A collection of papers that makes a good follow-on from the above covering fundamental research into representation for symbolic reasoning.

D. G. Bobrow and A. Collins, editors. *Representation and understanding: Studies in cognitive science.* Academic Press New York, 1975.

A collection of early papers on knowledge representation.

Reasoning

3.1 OVERVIEW

Reasoning is the ability to use knowledge to draw new conclusions about the world. Without it we are simply recalling stored information. There are a number of different types of reasoning, including induction, abduction and deduction. In this chapter we consider methods for reasoning when our knowledge is unreliable or incomplete. We also look at how we can use previous experience to reason about current problems.

3.2 WHAT IS REASONING?

Mention of reasoning probably brings to mind logic puzzles or "whodunit" thrillers, but it is something that we do every day of our lives. Reasoning is the process by which we use the knowledge we have to draw conclusions or infer something new about a domain of interest. It is a necessary part of what we call "intelligence": without the ability to reason we are doing little more than a lookup when we use information. In fact this is the difference between a standard database system and a knowledge-based or expert system. Both have information that can be accessed in various ways, but the database, unlike the expert system, has no reasoning facilities and can therefore answer only limited, specific questions.

Think for a moment about the types of reasoning you use. How do you know what to expect when you go on a train journey? What do you think when your friend is annoyed with you? How do you know what will happen if your car has a flat battery? Whether you are aware of

it or not, you will use a number of different methods of reasoning depending on the problem you are considering and the information that you have before you.

The three everyday situations mentioned above illustrate three key types of reasoning that we use. In the first case you know what to expect on a train journey because of your experience of numerous other train journeys: you infer that the new journey will share common features with the examples you are aware of. This is induction, which can be summarised as *generalisation from cases seen to infer information about cases unseen*. We use it frequently in learning about the world around us. For example, every crow we see is black; therefore we infer that all crows are black. If you think about it, such reasoning is unreliable: we can never prove our inferences to be true, we can only prove them to be false. Take the crows again. To prove that all crows are black we would have to confirm that all crows that exist, have existed or will exist are black. This is obviously not possible. However, to disprove the statement, all we need is to produce a single crow that is white or pink. So at best we can amass evidence to support our belief that all crows are black. In spite of its unreliability, inductive reasoning is very useful and is the basis of much of our learning. It is used particularly in machine learning, which we will meet in Chapter 5.

The second example we suggested was working out why a friend is annoyed with you, in other words trying to find an explanation for your friend's behaviour. It may be that this particular friend is a stickler for punctuality and you are a few minutes late to your rendezvous. You may therefore infer that your friend's anger is caused by your lateness. This uses abduction, the process of reasoning back from something to the state or event that caused it. Of course this too is unreliable; it may be that your friend is angry for another reason (perhaps you had

DOI: 10.1201/9781003082880-4

promised to telephone but had forgotten). Abduction can be used in cases where the knowledge is incomplete, for example where it is not possible to use deductive reasoning (see below). Abduction can provide a "best guess" given the evidence available.

The third problem is usually solved by deduction: you have knowledge about cars such as "if the battery is flat the headlights won't work"; you know the battery is flat so you can infer that the lights won't work. This is the reasoning of standard logic. Indeed, we could express our car problem in terms of logic: given that

$a = $ *the battery is flat*
$b = $ *the lights won't work*

and the axioms

$\forall x \; : \; a(x) \; \rightarrow \; b(x)$
$a(my\ car)$

we can deduce $b(my\ car)$. Note, however, that we cannot deduce the inverse: that is, if we know $b(my\ car)$ we cannot deduce $a(my\ car)$. This is not permitted in standard logic but is of course another example of abduction. If our lights don't work, we may use abduction to derive this explanation. However, it could be wrong; there may be another explanation for the light failure (e.g. a bulb may have blown).

Deduction is probably the most familiar form of explicit reasoning. Most of us at some point have been tried with syllogisms about Aristotle's mortality and the like. It can be defined as the process of deriving the *logically* necessary conclusion for the initial premises. So, for example,

Elephants are bigger than dogs

Dogs are bigger than mice

Therefore

Elephants are bigger than mice.

However, it should be noted that deduction is concerned with logical validity, not actual truth. Consider the following example; given the facts, can we reach the conclusion by deduction?

Some dogs are greyhounds

Some greyhounds run fast

Therefore

Some dogs run fast.

The answer is no. We cannot make this deduction because we do not know that all greyhounds are dogs. The fast greyhounds may therefore be the greyhounds that are not dogs. This of course is nonsensical in terms of what we know (or more accurately have induced) about the real world, but it is perfectly valid based on the premises given. You should therefore be aware that deduction does not always correspond to natural human reasoning.

3.3 FORWARD AND BACKWARD REASONING

As well as coming in different "flavours", reasoning can progress in one of two directions: *forwards* to the goal or *backwards* from the goal. Both are used in AI in different circumstances. Forward reasoning (also referred to as forward chaining, data-driven reasoning, bottom-up or antecedent-driven) begins with known facts and attempts to move towards the desired goal. Backward reasoning (backward chaining, goal-driven reasoning, top-down, consequent-driven or hypothesis-driven) begins with the goal and sets up subgoals which must be solved in order to solve the main goal.

Imagine you hear that a man bearing your family name died intestate a hundred years ago and that solicitors are looking for descendants. There are two ways in which you could determine if you are related to the dead man. First, follow through your family tree from yourself to see if he appears. Secondly, trace his family tree to see if it includes you. The first is an example of forward reasoning, the second backward reasoning. In order to decide which method to use, we need to consider the number of start and goal states (move from the smaller to the larger – the more states there are, the easier it is to find one) and the number of possibilities that need to be considered at each stage (the fewer the better). In the above example there is one start state and one goal state (unless you are related to the dead man more than once), so this does not help us. However, if you use forward reasoning, there will be two possibilities to consider from each node (each person will have two parents), whereas with backward reasoning there may be many more (even today the average number of children is 2.4; at the beginning of the century it was far more).

In general, backward reasoning is most applicable in situations where a goal or hypothesis can be easily generated (e.g. in mathematics or medicine), and where problem data must be acquired by the solver (e.g. a doctor asking for vital signs information in order to prove or disprove a hypothesis). Forward reasoning, on the other hand, is useful where most of the data is given in the problem statement but where the goal is unknown or where there are a large number of possible goals. For example, a system which analyses geological data in order to determine which minerals are present falls into this category.

3.4 REASONING WITH UNCERTAINTY

In Chapter 2 we looked at knowledge and considered how different knowledge representation schemes allow us to reason. Recall, for example, that standard logics allow us to infer new information from the facts and rules that we have.

Such reasoning is useful in that it allows us to store and utilise information efficiently (we do not have to store everything). However, such reasoning assumes that the knowledge available is complete (or can be inferred) and correct and that it is consistent. Knowledge added to such systems never makes previous knowledge invalid. Each new piece of information simply adds to the knowledge. This is called monotonic reasoning. Monotonic reasoning can be useful in complex knowledge bases since it is not necessary to check consistency when adding knowledge or to store information relating to the truth of knowledge. It therefore saves time and storage.

However, if knowledge is incomplete or changing, an alternative reasoning system is required. There are a number of ways of dealing with uncertainty. We will consider four of them briefly:

- non-monotonic reasoning
- probabilistic reasoning
- reasoning with certainty factors
- fuzzy reasoning

3.4.1 Non-monotonic Reasoning

In a non-monotonic reasoning system new information can be added that will cause the deletion or alteration of existing knowledge. For example, imagine you have invited someone round for dinner. In the absence of any other information you may make an assumption that your guest eats meat and will like chicken. Later you discover that the guest is in fact a vegetarian and the inference that your guest likes chicken becomes invalid.

We have already met two non-monotonic reasoning systems: abduction and property inheritance (see Chap. 2). Recall that abduction involves inferring some information on the basis of current evidence. This may be changed if new evidence comes to light, which is a characteristic of non-monotonic reasoning. So, for example, we might infer that a child who has spots has measles. However, if evidence comes to light to refute this assumption (e.g. that the spots are yellow and not red), then we replace the inference with another.

Property inheritance is also non-monotonic. An instance or subclass will inherit the characteristics of the parent class, unless it has alternative or conflicting values for that characteristic. So, as we saw in Chapter 2, we know that dogs bark and that Rottweilers and Basenjis are dogs. However, we also know that Basenjis don't bark. We can therefore infer that Rottweilers bark (since they are dogs and we have no evidence to think otherwise), but we cannot infer that Basenjis do, since the evidence refutes it.

A third non-monotonic reasoning system is the truth maintenance system or TMS [92]. In a TMS the truth or falsity of all facts is maintained. Each piece of knowledge is given a support list (SL) of other items that support (or refute) belief in it. Each piece of knowledge is labelled for reference, and an item can be supported either by another item being true (+) or being false (-). Take, for example, a simple system to determine the weather conditions:

(1) It is winter (SL ()())

(2) It is cold (SL (1+)(3-))

(3) It is warm

Statement (1) does not depend on anything else: it is a fact. Statement (2) depends on statement (1) being true and statement (3) being false. It is not known at this point what statement (3) depends on. It has no support list. Therefore we could assume that "it is cold" since we know that "it is winter" is true (it is a fact) and we have no information to suggest that it is warm (we can therefore assume that this is false). However, if "it is warm" becomes true, then "it is cold" will become false. In this way the TMS maintains the validity and currency of the information held.

3.4.2 Probabilistic Reasoning

Probabilistic reasoning is required to deal with incomplete data. In many situations we need to make decisions based on the likelihood of particular events, given the knowledge we have. We can use probability to determine the most likely cause.

Simple probability deals with independent events. If we know the probability of event A occurring (call it $p(A)$) and the probability of event B occurring ($p(B)$), the probability that both will occur ($p(AB)$) is calculated as $p(A) * p(B)$. For example, consider an ordinary pack of 52 playing cards, shuffled well. If I select a card at random, what is the likelihood of it being the king of diamonds? If we take event A to be the card being a diamond and event B to be the card being a king, we can calculate the probability as follows:

$$p(A) = \frac{13}{52} = 0.25$$
(there are 13 diamonds)

$$p(B) = \frac{4}{52} = 0.077$$
(there are four kings)

$$p(AB) = \frac{52}{2704} = \frac{1}{52} = 0.0192$$
(there is one king of diamonds)

However, if two events are interdependent and the outcome of one affects the outcome of the other, then we need to consider conditional probability. Given the probability of event A ($p(A)$) and that of a second event B which depends on it, $p(B|A)$ (B given A), the probability of both occurring is $p(A) * p(B|A)$. So, returning to our pack of cards, imagine I take two cards. What is the probability that they are both diamonds? Again, event A is the first card being a diamond, but this time event B is the second card also being a diamond:

$$p(A) = \frac{13}{52} = 0.25$$
(there are 13 diamonds)

$$p(B|A) = \frac{12}{51} = 0.235$$
(there are 12 diamonds left and 51 cards)

$$p(AB) = \frac{156}{2652} = 0.058$$

This is the basis of Bayes theorem and several probabilistic reasoning systems. Bayes theorem calculates the probabilities of particular "causes" given observed "effects".

The theorem is as follows:

$$p(h_i|e) = \frac{p(e|h_i)p(h_i)}{\sum_{j=1}^{n} p(e|h_j)p(h_j)}$$

where

$p(h_i|e)$ is the probability that the hypothesis h_i is true given the evidence e

$p(h_i)$ is the probability that h_i in the absence of specific evidence

$p(e|h_i)$ is the probability that evidence e will be observed if hypothesis h_i is true

n is the number of hypotheses being considered.

For example, a doctor wants to determine the likelihood of particular causes, based on the evidence that a patient has a headache. The doctor has two hypotheses, a common cold (h_1) and meningitis (h_2), and one piece of evidence, the headache (e), and wants to know the probability of the patient having a cold.

Suppose the probability of the doctor seeing a patient with a cold, $p(h_1)$, is 0.2 and the probability of seeing someone with meningitis, $p(h_2)$, is 0.000 001. Suppose also that the probability of a patient having a headache with a cold, $p(e|h_2)$, is 0.8 and the probability of a patient having a headache with meningitis, $p(e|h_2)$, is 0.9.

Using Bayes theorem we can see that the probability that the patient has a cold is very high:

$$
\begin{aligned}
p(h_1) &= \frac{0.8 \times 0.2}{(0.8 \times 0.2) + (0.9 \times 0.000\,001)} \\
&= \frac{0.16}{0.16 + 0.000\,000\,9} = 0.99
\end{aligned}
$$

In reality, of course, the cost of misdiagnosis of meningitis is also very high, and therefore many more factors would have to be taken into account.

Bayes theorem was used in the early expert system, PROSPECTOR [95], to find mineral deposits. The aim was to determine the likelihood of finding a specific mineral by observing the geological features of an area. PROSPECTOR was used to find several commercially significant mineral deposits. Bayesian reasoning especially in the form of Bayesian networks (Section 3.5), is heavily used in many AI applications, for example in medical diagnosis [177, 254].

In spite of such successful uses, Bayes theorem makes certain assumptions that make it intractable in many domains. First, it assumes that statistical data on the relationships between evidence and hypotheses are known, which is often not the case. Secondly, it assumes that the relationships between evidence and hypotheses are all independent. In spite of these limitations Bayes theorem has been used as the base for a number of probabilistic reasoning systems, including certainty factors, which we will consider next.

3.4.3 Certainty Factors

As we have seen, Bayesian reasoning assumes information is available regarding the statistical probabilities of certain events occurring. This makes it difficult to operate in many domains. Certainty factors are a compromise on pure Bayesian reasoning. The approach has been used successfully, most notably in the early expert system MYCIN [258]. MYCIN was a medical diagnosis system that diagnosed bacterial infections of the blood and prescribes drugs for treatment. Its knowledge was represented in rule form and each rule has an associated certainty factor.

For example, a MYCIN rule looks something like this:

If (a) the gram stain of the organism is gram negative and

(b) the morphology of the organism is rod and

(c) the aerobicity of the organism is anaerobic

then there is suggestive evidence (0.5) that identity of the organism is Bacteroides

In this system, each hypothesis is given a certainty factor (CF) by the expert providing the rules, based on his or her assessment of the evidence. A CF takes a value between 1 and -1, where values approaching -1 indicate that the evidence against the hypothesis is strong, and those approaching 1 show that the evidence for the hypothesis is strong. A value of 0 indicates that no evidence for or against the hypothesis is available.

A CF is calculated as the amount of belief in a hypothesis given the evidence ($MB(h|e)$) minus the amount of disbelief ($MD(h|e)$). The measures are assigned to each

rule by the experts providing the knowledge for the system as an indication of the reliability of the rule. Measures of belief and disbelief take values between 0 and 1. Certainty factors can be combined in various ways if there are several pieces of evidence. For example, evidence from two sources can be combined to produce a CF as follows:

$$CF(h|e_1, e_2) = MB(h|e_1, e_2) - MD(h|e_1, e_2)$$

where

$$MB(h|e_1, e_2) = MB(h|e_1) + \{MB(h|e_2)[1 - MB(h|e_1)]\}$$
(or 0 if $MD(h|e_1, e_2) = 1$)

and

$$MD(h|e_1, e_2) = MD(h|e_1) + \{MD(h|e_2)[1 - MD(h|e_1)]\}$$
(or 0 if $MB(h|e_1, e_2) = 1$)

The easiest way to understand how this works is to consider a simple example. Imagine that we observe the fact that the air feels moist (e_1). There may be a number of reasons for this (rain, snow, fog). We may hypothesise that it is foggy, with a measure of belief ($MB(h|e_1)$) in this being the correct hypothesis of 0.4. Our disbelief in the hypothesis given the evidence ($MD(h|e_1)$) will be low, say 0.1 (it may be dry and foggy, but it is unlikely). The certainty factor for this hypothesis is then calculated as

$$CF(h|e_1) = MB(h|e_1) - MD(h|e_1)$$
$$= 0.5 - 0.1 = 0.4$$

We then make a second observation, e_2, that visibility is poor, which confirms our hypothesis that it is foggy, with $MB(h|e_2)$ of 0.7. Our disbelief in the hypothesis given this new evidence is 0.0 (poor visibility is a characteristic of fog). The certainty factor for it being foggy given this evidence is

$$CF(h|e_2) = MB(h|e_2) - MD(h|e_2)$$
$$= 0.7 - 0.0 = 0.7$$

However, if we combine these two pieces of evidence, we get an increase in the overall certainty factor:

$$MB(h|e_1, e_2) = 0.5 + (0.7 * 0.5) = 0.85$$
$$MD(h|e_1, e_2) = 0.1 + (0.0 * 0.9) = 0.1$$
$$CF(h|e_1, e_2) = 0.85 - 0.1 = 0.75$$

Certainty factors provide a mechanism for reasoning with uncertainty that does not require probabilities. Measures of belief and disbelief reflect the expert's assessment of the evidence rather than statistical values. This makes the certainty factors method more tractable as a method of reasoning. Its use in MYCIN shows that it can be successful, at least within a clearly defined domain. However, in practice, despite some advantages, more direct Bayesian methods are more heavily used today.

3.4.4 Fuzzy Reasoning

Probabilistic reasoning and reasoning with certainty factors deal with uncertainty using principles from probability to extend the scope of standard logics. An alternative approach is to change the properties of logic itself. Fuzzy sets and fuzzy logic do just that.

In classical set theory an item, say a, is either a member of set A or it is not. So a meal at a restaurant is either expensive or not expensive and a value must be provided to delimit set membership. Clearly, however, this is not the way we think in real life. While some sets are clearly defined (a piece of fruit is either an orange or not an orange), many sets are not. Qualities such as size, speed and price are relative. We talk of things being very expensive or quite small.

Fuzzy set theory extends classical set theory to include the notion of *degree* of set membership. Each item is associated with a value between 0 and 1, where 0 indicates that it is not a member of the set and 1 that it is definitely a member. Values in between indicate a certain degree of set membership.

For example, although you may agree with the inclusion of Porsche and BMW in the set *FastCar*, you may wish to indicate that one is faster than the other. This is possible in fuzzy set theory:

$$FastCar = \left\{ \begin{array}{l} (Porsche\ 944, 0.9), \\ (BMW\ 316, 0.5), \\ (Vauxhall\ Nova\ 1.2, 0.1) \end{array} \right\}$$

Here the second value in each pair is the degree of set membership.

Fuzzy logic is similar in that it attaches a measure of truth to facts. A predicate, P, is given a value between 0 and 1 (as in fuzzy sets). So, taking an element from our fuzzy set, we may have a predicate

$$fastcar(Porsche\ 944) = 0.9$$

Standard logic operators, such as *and, or* and *not*, can be applied in fuzzy logic and are interpreted as follows:

$$P \wedge Q = min(P, Q)$$
$$P \vee Q = max(P, Q)$$
$$not\ P = 1 - P$$

So, for example, we can combine predicates and get new measures:

$$fastcar(Porsche\ 944) = 0.9$$
$$pretentiouscar(Porsche\ 944) = 0.6$$
$$fastcar(Porsche\ 944) \wedge$$
$$pretentiouscar(Porsche\ 944)$$
$$= 0.6$$

3.4.5 Reasoning by Analogy

Analogy is a common tool in human reasoning [123]. Given a novel problem, we might compare it with a familiar problem and note the similarities. We might then apply our knowledge of the old problem to solving the new. This approach is effective if the problems are comparable and the solutions transferable.

Analogy has been applied in AI in two ways: *transformational analogy* and *derivational analogy*. Transformational analogy involves using the solution to an old problem to find a solution to a new. Reasoning can be viewed as a state space search where the old solution is the start state and operators are used (employing means–ends analysis, for example) to transform this solution into a new solution.

An alternative to this is derivational analogy, where not only the old solution but the process of reaching it is considered in solving the new problem. A history of the problem-solving process is used. Where a step in the procedure is valid for the new problem, it is retained; otherwise it is discarded. The solution is therefore not a copy of the previous solution but a variation of it.

3.4.6 Case-based Reasoning

A method of reasoning which exploits the principle of analogy is case-based reasoning (CBR). All the examples (called cases in CBR) are remembered in a case base. When a new situation is encountered, it is compared with all the known cases and the best match is found. If the match is exact, then the system can perform exactly the response suggested by the example. If the match is not exact, the differences between the actual situation and

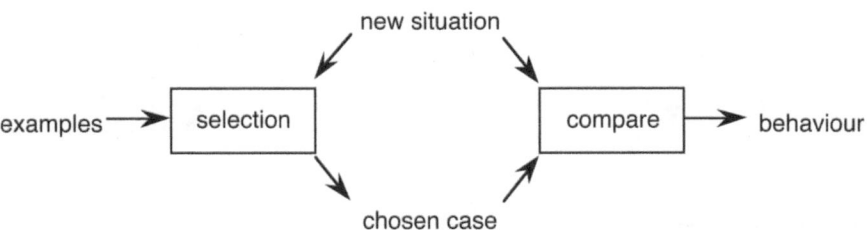

FIGURE 3.1   Case-based reasoning.

the case are used to derive a suitable response (see Figure 3.1).

Where there is an exact match, the CBR acts as a rote learning system, but where there is no exact match, the combination of case selection and comparison is a form of generalisation. The simplest form of CBR system may just classify the new situation, a form of concept learning. In this case, the performance of the system is determined solely by the case selection algorithm. In a more complicated system, the response may be some form of desired action depending on the encountered situation. The case base consists of examples of stimulus–action pairs, and the comparison stage then has to decide how to modify the action stored with the selected case. This step may involve various forms of reasoning.

Imagine we have the following situation:

situation: buy(fishmonger,cod),
              owner(fishmonger,Fred),
              cost(cod,£3)

The case base selects the following best match case:

stimulus: buy(postoffice,stamp),
             owner(postoffice,Dilys),
             cost(stamp,25p)
action: pay(Dilys,25p)

The comparison yields the following differences:

fishmonger → postoffice, cod → stamp,
Fred → Dilys, £3 → 25p

The action is then modified correspondingly to give "pay(Fred,£3)".

In this example, the comparison and associated modification are based on simple substitution of corresponding values. However, the appropriate action may not be so simple. For example, consider a blocks-world CBR (Figure 3.2). The situation is:

situation: blue(A), pyramid(A), on(A,table),
              green(B), cube(B), on(B,table),
              blue(C), ball(C), on(C,B),

The CBR has retrieved the following case:

stimulus: blue(X), pyramid(X), on(X,table),
             green(Y), cube(Y), on(Y,table),
             blue(Z), cube(Z), on(Z,table),
action: move(X,Y)

A simple pattern match would see that the action only involves the first two objects, X and Y, and the situation concerning these two objects is virtually identical. So, the obvious response is "move(A,B)". However, a more detailed analysis would show that moving the blue pyramid onto the green cube is not possible because, in the current situation, the blue ball is on it. A more sophisticated difference procedure could infer that a more appropriate response would be: move(C,table), move(A,B).

Note how the comparison must be able to distinguish irrelevant differences such as ball(C) vs. cube(Z) from significant ones such as on(C,B) vs. on(Z,table). This is also a problem for the selection algorithm. In practice there may be many attributes describing a situation, only a few of which are really important. If selection is based on a simple measure such as "least number of different attributes", then the system may choose "best match" cases where all the irrelevant attributes match, but none of the relevant ones! At the very least some sort of weighting is needed, similar to salience in human attention. For example, if one were developing a fault diagnosis system for a photocopier, the attributes would include the error code displayed, the number of copies required, the paper type, whether the automatic feeder was being used and so on. However, one would probably give the error code a higher weighting than the rest of the attributes. Where the comparison yields differences which invalidate the

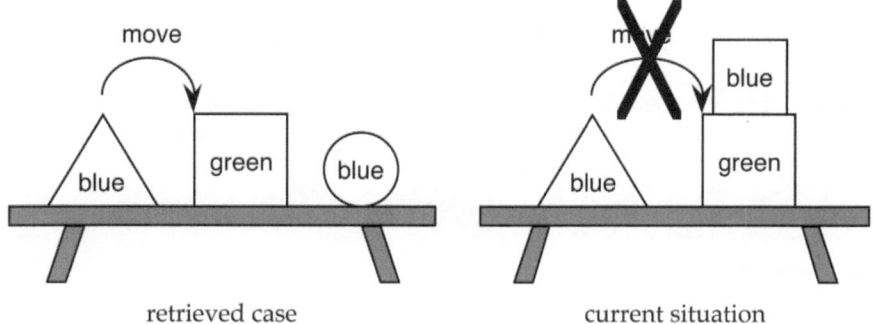

FIGURE 3.2   Modifying cases.

response given in the case and no repair is possible, the CBR can try another close match case. So, a good selection mechanism is important, but some poor matches can be corrected.

Case-based reasoning has some important advantages. The most important is that it has an obvious and clear explanation for the user: "the current situation is similar to this given case and so I did a similar response". Indeed, one option is to do no comparison at all, simply to present the user with similar cases and allow the user to do the comparison between the current situation and the selected cases. Arguably, because the human does the "intelligent" part, this is not really CBR but simply a case memory, a sort of database.

Another advantage of CBR is that it is not difficult to incorporate partial descriptions, in both the cases and the presented situations. This is because it is fairly easy to generalise measures of similarity to cases where some of the attributes are missing or unknown. For example, we could score +1 for each matching attribute, −1 for each non-match and 0 for any attributes that are missing from either the case or situation (weighted of course!). This is an important feature of CBR, as it is often the case that records are incomplete. For example, if we start to build a CBR based on past medical records, we will find that many symptoms are unrecorded – the doctor would not have taken the heart rate of someone with a skin complaint. Other reasoning methods can deal with such problems, but not so simply as CBR.

3.5 REASONING OVER NETWORKS

In Chapter 2, we saw that various forms of networks either arise naturally or can be ways of representing knowledge.

Some forms of reasoning over networks are about 'link chasing', following connections, perhaps of a particular kind, or through nodes that satisfy certain characteristics, until all nodes of a particular kind are found. For example, you might look in a social network for all friends, friends of friends, friends of friends of friends, etc., who all have snakes as pets. This would mean that when you organise a trip to the reptile house, you wouldn't leave anyone out. This is a form of search, and we'll look at this more in Chapter 4.

Networks can also be considered in terms of more complex patterns of links. For example, Figure 3.3 shows a simple pattern of four nodes representing four people where persons C and D are step-siblings. It consists of an undirected link 'married' connecting nodes A and B, and two directed links 'parent' connecting nodes C to node A and node D to node B, respectively. In addition the pattern describes links that shouldn't exist 'NOT parent'. This might be used to simply locate all step-siblings in a larger genealogical graph, or as part of a larger query or reasoning process. For example, one might create additional inferred 'step-sibling' links between all nodes that match in positions C and D in the graph.

Often graph algorithms involve more numerical weights or activations on links or nodes. You are likely to have used at least one such system. The PageRank algorithm [32], which is the basis of Google search, starts out by assigning a measure of importance to each page and then effectively 'shares' the importance of a web page to all the pages it is connected to links. The idea is that pages that are linked to from lots of important pages, or those that link to lots of quality pages, must themselves be important.

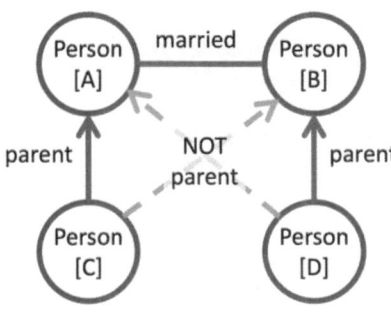

FIGURE 3.3 Graph pattern.

PageRank can be regarded as a form of spreading activation algorithm. These are based on parallels with the brain where if a particular concept is activated (say by showing a picture or speaking a word), other related ones clearly get some of this attention too, as you will react more quickly to them or spontaneously bring them to mind. Figure 3.4 shows an example of spreading activation between a collection of people and places. The node Vivi is initially activated (maybe the name appears in an email) and then some of this activation spreads to immediately connected nodes Costas and UoA, activation from UoA to Athens and then from Athens to Greece, but gets weaker with each additional step. There is also some spread back from Greece to Tripolis but spread through 1-to-m links (Greece has many cities) is weaker than along m-to-1 links (a person is usually in one university).

Bayesian methods and fuzzy reasoning have also been used as the basis of reasoning and learning networks. The former are particularly popular in the form of Bayesian networks with applications including medical diagnosis [267] and analysing gene expression [107, 295]

In a Bayesian network each node represents some feature or observable phenomenon and directed arrows represent causal links. Figure 3.5 shows a small example where the causal links are denoted as arrows between the nodes: being sunny influences whether you are likely to wear a coat; whether you feel hot will depend on both the weather and if you have a coat on; and finally whether you choose to eat ice cream will be based on if you feel hot and if it is sunny. Each node also has a probability table giving the conditional probability of the node being true depending on the causal nodes. Only the table for the 'feel hot' node is shown, showing that, for example, the conditional probability that you

feel hot if you are not wearing a coat and it is sunny is 0.7.

The network can be run 'forward' so that, for example, if you know it is sunny, you can work out the probability of eating ice cream. It can also be worked backwards using Bayes theorem, for example if we spot someone with a coat to work out the probability that is sunny. It can even be used in a mixed mode by fixing known nodes and then using the conditional probabilities forward where we know all the causal factors and backwards with Bayes theorem to fill gaps; for example if you know someone is eating ice cream (and some prior probabilities for everything else), you can work out the probability that they are wearing a coat.

The networks illustrated here and in Chapter 2 have all been small, but networks can also be very large – the web, social networks and neural networks may all contain billions of nodes. Chapter 8 will discuss some of the additional issues that arise when dealing with these very large networks.

3.6 CHANGING REPRESENTATIONS

As we saw in the four knights puzzle in Section 2.3, the choice of knowledge representation can make a huge difference to the ease of solving a particular problem. Sometimes there will be a single best solution, but at other times it may be necessary to use multiple representations as part of the same system. For example, the PageRank algorithm effectively uses a graph representation of the web in order to calculate the importance metric, but then this value is stored as a value for the page in a more database-record-like representation which is used when performing searches for content.

Changing representation can be expensive, especially for large datasets as it may involve complex manipulation of every data element. One solution is to maintain several representations of the same underlying data with some way to connect the two, for example, using shared identifiers. This means that reasoning algorithms can work on the representation that is most appropriate. However, if the underlying data changes, care is needed to keep the different representations consistent with one another.

Alternatively, there may be a single underlying representation, with ways in which it can be viewed as though it were different. Indeed, this is just what the PageRank algorithm does in practice, conceptually it is performing a spreading of page importance through the network, but

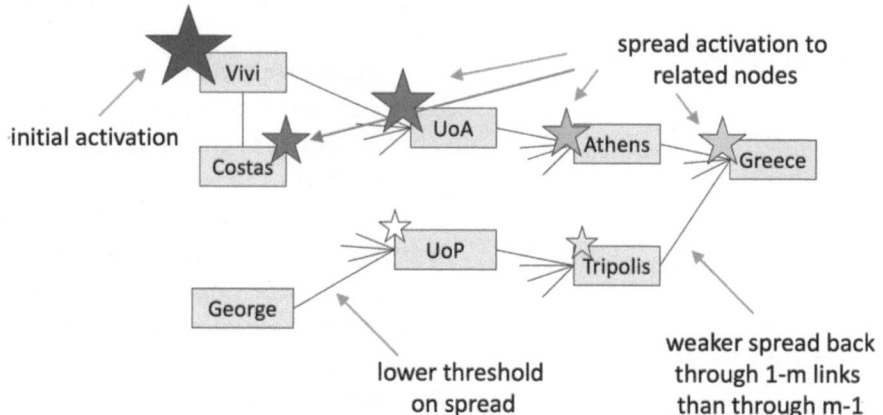

FIGURE 3.4   Spreading activation (from [89]).

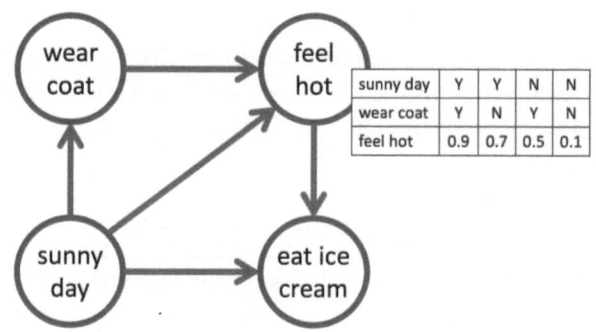

FIGURE 3.5   Bayesian network.

it achieves this through a series of passes through a more linear record structure in computer memory.

3.7 SUMMARY

In this chapter we have considered a number of different types of reasoning, including induction, abduction and deduction. We have seen that the knowledge that we are reasoning about is often incomplete and therefore demands reasoning methods that can deal with uncertainty. We have considered four approaches to reasoning with uncertainty: non-monotonic reasoning, probabilistic reasoning, reasoning with certainty factors and fuzzy reasoning. We have also considered analogical reasoning, case-based reasoning and reasoning over networks.

3.1 Distinguish between *deductive, inductive* and *abductive* reasoning, giving an example of the appropriate use of each.

3.2 Alison is trying to determine the cause of overheating in Brian's car. She has two theories: a leak in the radiator or a broken thermostat. She knows that leaky radiators are more common than broken thermostats: she estimates that 10% of cars have a leaky radiator while 2% have a faulty thermostat. However, 90% of cars with a broken thermostat overheat whereas only 30% overheat with a leaky radiator. Use Bayes theorem to advise Alison of the most likely cause of the problem.

3.3 Alison then checks the water level in Brian's car and notices it is normal. She knows that a car with a leaky radiator is very unlikely not to lose water (perhaps 1% chance), whereas water loss is not seen in 95% of cases of faulty thermostats. How would this new evidence affect your advice to Alison? (Use Bayes theorem again and assume for simplicity that all evidence is independent.)

FURTHER READING

J. Y. Halpern. *Reasoning about uncertainty.* MIT Press, Cambridge, MA, 2017.

Covers a wide variety of probabilistic methods in detail including Bayesian approaches.

H. T. Nguyen, C. Walker, and E. A. Walker. *A first course in fuzzy logic.* Chapman and Hall/CRC, Boca Raton, FL, 2018.

While still aimed at an introductory level, this book is entirely dedicated to fuzzy logics and fuzzy reasoning.

J. Kolodner. *Case-based reasoning.* Morgan Kaufmann, Burlington, MA, 2014.

Book-length treatment of case-based reasoning including methods to represent, index, rank and adapt cases.

G. Shafer and J. Pearl, editors. *Readings in uncertain reasoning.* Morgan Kaufmann, Los Altos, 1990.

A collection of classic articles which provide a useful introduction to reasoning with uncertainty. Now out of print but available at: http://www.glennshafer.com/books/rur.html

C. K. Riesbeck and R. Schank. *Inside case based reasoning.* Lawrence Erlbaum, Hillsdale, NJ, 1989.

A useful introduction to case-based reasoning.

Search

4.1 INTRODUCTION

When we want to solve a problem, we consider various alternatives, some of which fail to solve the problem. Of those that succeed, we may want to find the best solution or the easiest to perform. The act of enumerating possibilities and deciding between them is *search*. AI systems must search through sets of possible actions or solutions, and this chapter discusses some of the algorithms that are used. Before we go on to consider specific algorithms, we need to look at the sorts of problems that we are likely to face, as the appropriate algorithm depends on the form of the problem. The set of possible solutions is not just an amorphous bag but typically has some structure. This structure also influences the choice of search algorithm.

4.1.1 Types of Problem

State and Path

In some problems we are only interested in the state representing the solution, whereas in other cases we also want to know how we got to the solution – the path. A crossword puzzle is an example of the former: the important thing is that the crossword is eventually completed; the order in which the clues were solved is only of interest to the real crossword fanatic. The eight queens problem and solving magic squares are similar problems (see Figure 4.1). Typically with pure state-finding problems the goal state is described by some properties. In the case of the magic square, the states are the set of all 3×3 squares

filled in with numbers between 1 and 9, and the property is that each row, column and diagonal adds up to 15.

Mathematical theorem proving has been a major driving force in AI. If we consider this, we see that it is not only important that we solve the required theorem but that the steps we take are recorded – that is, the proof. Other path problems include various finding route problems, puzzles such as the Towers of Hanoi (Figure 4.2) and algorithms for planning actions such as means–ends analysis (Chap. 15). In all these problems we know precisely what the goal state is to be; it is only the means of getting there that is required. The solution to such problems must include not just a single goal state, but instead a sequence of states visited and the moves made between them. In some problems the moves are implicit from the sequence of states visited and can hence be omitted.

In fact, some route problems do not specify their goal state in advance. For example, we may want to find the fastest route from Zuata, Venezuela, to any international airport with direct flights to Sydney, Australia. In this case we want to find a route (sequence of places) where the goal state is a city that satisfies the property.

$$P(s) = \text{``}s \text{ has an international airport with direct flights to Sydney''}$$

The travelling salesman problem is more complex again. Imagine a salesman has to visit a number of towns. They must plan a route that visits each town exactly once and that begins and ends at their home town. They want the route to be as short as possible. Although the final state is given (the same as the start state), the important property is one of the whole path, namely that each place is visited exactly once. It would be no good to find a route which reached the goal state by going nowhere at all! The last chapter was all about the importance of the choice of representation. In this

DOI: 10.1201/9781003082880-5

A magic square is a square of numbers where each row, column and diagonal adds up to the same number. Usually the numbers have to be consecutive. So, for example, the 3×3 square would contain the numbers 1, 2, 3, 4, 5, 6, 7, 8 and 9. Here are some examples, one 3×3 square and one 4×4 square:

8	1	6
3	5	7
4	9	2

7	13	12	2
10	4	5	15
1	11	14	8
16	6	3	9

The 3×3 square is the simplest. (There are no 2×2 squares and the 1×1 square $\boxed{1}$ is rather boring!) So, when we talk about the "magic squares" problem in this chapter, we will always mean finding 3×3 squares.

The eight queens problem is another classic placing problem. In this case we must position eight queens on a chess board so that no queen is attacking another. That is, so that no queen is on the same row, column or diagonal as any other. There are similar problems with smaller numbers of queens on smaller chess boards: for example, a solution of the four queens problem is:

FIGURE 4.1   Magic squares and the eight queens problem.

example, it may well be best to regard the travelling salesman problem as a state problem where the state is a path!

Any Solution or Best Solution

When finding a proof to a theorem (path problem), or solving the magic square or eight queens problem (state problems), all we are interested in is finding some solution – any one will do so long as it satisfies the required conditions (although some proofs may be more elegant than others).

However, if we consider the travelling salesman problem, we now want to find the *shortest* route. Similarly, we may want to choose a colouring for a map that uses the fewest colours (to reduce the costs of printing) or sim-

ply be looking for the shortest path between two places. In each of these examples, we are not only interested in finding a solution that satisfies some property, we are after the *best* solution – that is, search is an optimisation problem. The definition of best depends on the problem. It may mean making some measure as big as possible (e.g. profit) or making something as small as possible (e.g. costs). As profits can be seen as negative costs (or vice versa), we can choose whichever direction is easiest or whichever is normal for a particular problem type.

For a state problem such as map colouring, the costs are associated with the solution obtained, whereas in a path problem it is a combination of the "goodness" of the final solution and the cost of the path:

$$\text{total cost} = cost(\text{route}) - benefit(\text{goal state})$$

In a monastery in deepest Tibet there are three crystal columns and 64 golden rings. The rings are different sizes and rest over the columns. At the beginning of time all the rings rested on the leftmost column, and since then the monks have toiled ceaselessly moving the rings one by one between the columns. It is said that when all the rings lie on the centre column the world will end in a clap of thunder and the monks can at last rest.

The monks must obey two rules:

1. They may move only one ring at a time between columns.
2. No ring may rest on top of a smaller ring.

With 64 rings to move, the world will not end for some time yet. However, we can do the same puzzle with fewer rings. So, with three rings the following would be a legal move:

But this would not be:

In the examples in this chapter, we will consider the even simpler case of two rings!

FIGURE 4.2   Towers of Hanoi.

However, one finds that for many path problems there is no second term; that is, all goal states are considered equally good.

In general, the specification of a problem includes both a property (or constraints), which must be satisfied by the goal state (and path), and some cost measure. A state (and path) that satisfies the constraints is said to be feasible, and a feasible state that has the least cost is optimal. That is, real problems are a mixture of finding any solution (feasibility) and finding the best (optimality). However, for simplicity, the examples within this chapter fall into one camp or the other. Where constraints exist in optimisation problems, they are often satisfied "by construction". For example, a constraint on map-colouring problems is that adjacent countries have different colours. Rather than constructing a colouring and then checking this condition, one can simply ensure as one adds each colour that the constraint is met.

Deterministic vs. Adversarial

All the problems considered so far have been deterministic, that is totally under the control of the problem solver. However, some of the driving problems of AI have been to do with game playing: chess, backgammon and even simple noughts and crosses (tic-tac-toe). The presence of an adversary radically changes the search problem: as the solver tries to get to the best solution (i.e. win), the adversary is trying to stop it! Most games are state based: although it is interesting to look back over the history of a game, it is the state of the chess board *now* that matters. However, there are some path-oriented games as well, for example bridge or poker, where the player needs to remember all past moves, both of other players and their own, in order to choose the next move.

Interaction with the physical environment can be seen as a form of game playing also. As the solver attempts to perform actions in the real world, new knowledge is found and circumstances may occur to help or hinder. If one takes a pessimistic viewpoint, one can think of the world as an adversary which, in the worst case, plays to one's downfall. (Readers of Thomas Hardy will be familiar with this world view!)

A further feature in both game playing and real-world interaction is chance. Whereas chess depends solely on the abilities of the two players, a game like backgammon also depends on the chance outcome of the throwing of dice. Similarly, we may know that certain real-world phenomena are very unlikely and should not be given too great a prominence in our decision making.

This chapter will only deal with deterministic search. Chapter 11 will deal with game playing and adversarial search.

Perfect vs. Good Enough

Finally, we must consider whether our problem demands the absolutely best solution or whether we can make do with a "good enough" solution. If we are looking for the best route from Cape Town to Addis Ababa, we are unlikely to quibble about the odd few miles. This behaviour is typical of human problem solving and is called satisficing. Satisficing can significantly reduce the resources needed to solve a problem, and when the problem size grows may be the only way of getting a solution at all.

There is a parallel to satisficing when we are simply seeking any solution. In such cases, we may be satisfied with a system that replies

YES   – here is your solution
NO   – there is no solution
SORRY – I'm not sure

In practice theorem provers are like this. In most domains, not only is it very expensive to find proofs for all theorems, it may be fundamentally impossible. (Basically, Gödel showed that in sufficiently powerful systems (like the numbers) there are always things that are true yet which can never be proved to be true [153].)

4.1.2 Structuring the Search Space

Generate and Test – Combinatorial Explosion

The simplest form of search is generate and test. You list each possible candidate solution in turn and check to see if it satisfies the constraints. You can either stop when you

reach an acceptable goal state or, if you are after the best solution, keep track of the best so far until you get to the end.

Figure 4.3 shows this algorithm applied to the 3 × 3 magic square. However, this is an extremely inefficient way to look for a solution. If one examines the solutions in the *lexicographic* order (as in the figure), the first solution is found only after rejecting 75 231 candidates. In fact, the whole search space consists of 9! = 362 880 possible squares of which only eight satisfy the goal conditions – and that is after we have been careful not to generate squares with repeated digits! This problem is called combinatorial explosion and occurs whenever there are a large number of nearly independent parameters.

In practice, only the most ill-structured problems require this sledge-hammer treatment. One can structure most problems to make the search space far more tractable.

Trees

The first square in Figure 4.3 fails because $1+2+3 \neq 15$. So does the second square, the third … in fact the first 720 squares all fail for exactly the same reason. Then the next 720 fail because $1+2+4 \neq 15$, etc. In each case, you do not need to look at the full square: the partial state is sufficient to fail it.

The space of potential magic squares can be organised into a tree, where the leaf nodes are completed squares (all 362 880 of them), and the internal nodes are partial solutions starting off at the top left-hand corner. Figure 4.4 shows part of this search tree. The advantage of such a representation is that one can instantly ignore all nodes under the one starting 123, as all of these will fail. There are 504 possible first lines, of which only 52 add up to 15 (the first being 1 5 9). That is, of 504 partial solutions we only need to consider 52 of them further – an instant reduction by a factor of 10. Of course, each of the subtrees under those 52 will be able to be similarly pruned – the gains compound.

There are many ways to organise the tree. Instead of doing it in reading order, we could have filled out the first column first, or the bottom right, and so on. However, some organisations are better than others. Imagine we had built the tree so that the third level of partial solution got us to partial solutions like the following:

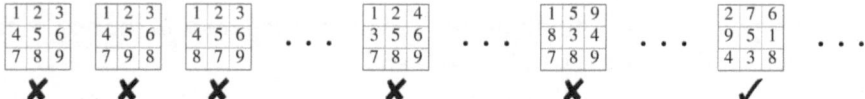

FIGURE 4.3   Generate and test – finding solutions to the magic square.

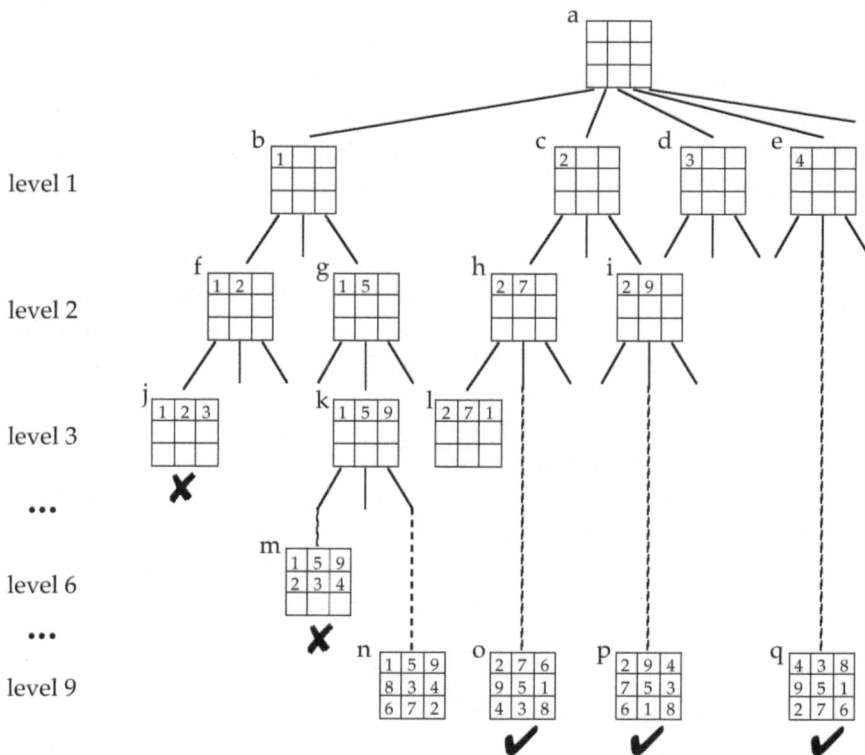

FIGURE 4.4   Magic square – search tree of potential solutions.

?	2	?
1	?	?
?	?	3

Clearly, we would not be able to prune the tree so rapidly. Choosing the best organisation for a particular problem is somewhat of an art, but there are general guidelines. In particular, you want to be able to test constraints as soon as possible.

Branching Factor and Depth

We can roughly characterise a tree by the number of children each node has – the branching factor – and the distance from the root of the tree to the leaves (bottom nodes) – the depth. The tree for magic squares has a branching factor of 9 at the root (corresponding to the nine possible entries at the top left), and a depth of 9 (the number of entries in the square). However, the branching factor reduces as one goes down the tree: at the second level it is 8, at the third level 7 and so on. For a game of chess, the branching factor is 20 for the first move (two possibilities for each pawn and four knight moves). For Go, played on a 19 × 19 board, the branching factor is 361! For a uniform tree, if the branching factor is b, there are b^n nodes at level n. That is, over 10 billion possibilities for the first four moves in Go – you can see why it took so long for AI Go-playing to be cracked!

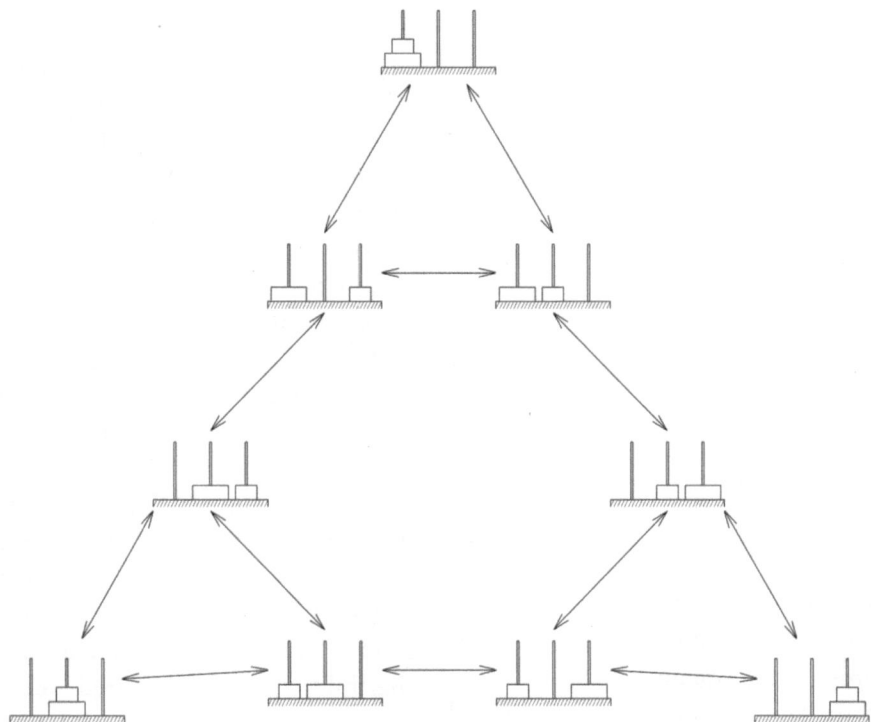

FIGURE 4.5   Towers of Hanoi: graph of possible states and moves.

Graphs

When one considers a problem consisting of states with moves between them, it is often the case that several move sequences get one from a particular start state to the same final state. That is, the collection of moves and states can be best thought of as a directed graph, where the nodes of the graph are the states and the arcs between nodes are the moves. Figure 4.5 shows the complete graph of states of the Towers of Hanoi (with only two rings!). Notice how even such a simple puzzle has a reasonably complex graph.

With the Towers of Hanoi, each arc is bidirectional, because each move between two states can be undone by a move in the reverse direction. This is not always so, for example when a piece is taken in chess; if the nodes represented states while making a cake, there would be no move backwards once the cake was cooked. When the arcs are directional, we can distinguish between the forward branching factor – the number of arcs coming from a node – and the backward branching factor – the number of arcs going to a node. If the backward factor is smaller than the forward factor, it suggests that searching

backwards from the goal state towards the start state may be more efficient than searching forwards.

Some algorithms search this graph directly. However, they will usually keep track of the path travelled through the graph as this will be part of the solution. For example, in the Towers of Hanoi puzzle, the path represents the moves to solve the puzzle. Similarly, Figure 4.6 shows a graph of states in a proof system. The states are addition formulae involving three variables, and the arcs are rewrites of the formulae using the associative (**A**) and commutative (**C**) laws of addition:

A: $L + (M + N) = (L + M) + N$
C: $M + N = N + M$

(Note that the commutative law is only applied to the outermost (unbracketed) addition in order to simplify the graph.)

If we wanted to prove, for instance, that

$$x + (y + z) = y + (z + x)$$

we could trace a path through the graph going clockwise. We begin at the start state $x + (y + z)$, apply first the commutative law getting us to state $(y + z) + x$ and then

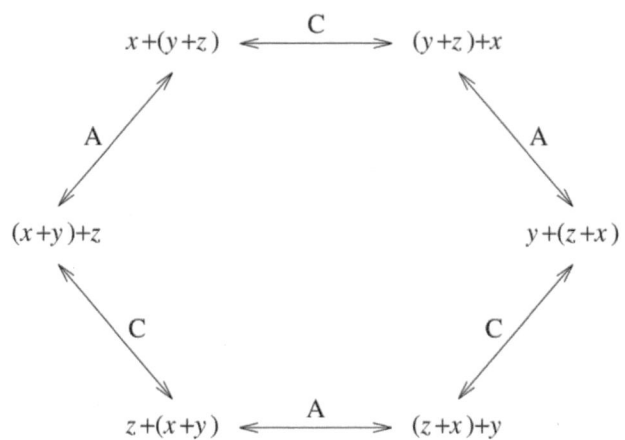

FIGURE 4.6   Addition proof graph.

the associative law getting us to the goal state $y + (z + x)$. The two steps (commutative law followed by associative law) constitute the proof of the equality.

Notice that there are several paths to the same goal state – we could have followed the graph anti-clockwise (ACAC). If we want to distinguish the paths more clearly, we can represent the graph by a tree. Figure 4.7 shows a portion of the proof tree for the same expression. The root is the start state and the children are those expressions that can be reached by applying one or other of the laws. The ellipses in the tree represent nodes where the tree "stutters": that is, the child node is the same as its parent. The figure stops expanding the tree when the goal state is reached, but in a sense the full tree reaches out below, as one could continue to apply arithmetic laws.

Note that in the tree representation, the goal state appears twice – once for each path to it. In fact, if one continued to expand the "stuttering" nodes one would uncover more goal states corresponding to "wasteful" proofs such as CCCA.

Adding Information

We have already seen that the way we organised the magic square search affected our ability to rule out nodes. This ability to detect that searching down certain paths of the tree is fruitless is a particularly simple (but very useful) form of heuristics. (Although arguably it is not a heuristic, as the information it provides is exact.) Heuristics are information that tells us something about

the future of our search, before we have investigated a path fully. Heuristics may tell us about the likelihood of finding a solution down a path, about how far we may have to search or how good the solution is likely to be. Heuristics are usually approximations – they give some indication but are not guaranteed to be right. Obviously the more accurate the heuristic and the more we *know* about its accuracy, the better it can inform our search. Section 4.3 is all about searching using heuristic information.

There are two major types of heuristic: those that tell us about a node – whether it is worth investigating further – and those which, when we are considering a node, suggest an order in which to search its children. Obviously information of the former category can be used to order the children, but only when the heuristic information for each child has been calculated. As this is sometimes expensive to do in its own right, or there may be an infinite number of children, a separate way of ordering the children may be required. The majority of the search algorithms in this chapter concentrate on the first type of heuristic. Furthermore these algorithms will simply use a heuristic evaluation function, a single number calculated for each node, which says how good or bad it is likely to be. Such heuristics are rather simple but can be surprisingly powerful. In Section 4.4 we will discuss more complex heuristics.

Virtual Trees and Real Trees

It is important to note that the trees and graphs that we have been discussing are not necessarily real. That is, they will not in general be constructed in the computer's memory. Indeed, given the size of the spaces (often infinite) they would be impossible to construct. Instead, they represent the space of possible solutions of which a system may only investigate a part. For example, we can imagine the graph of all chess games linked by possible moves. However, if we play chess, we do not by any means "construct" this graph in our heads and play using it. Neither will the algorithms we consider here!

4.2 EXHAUSTIVE SEARCH AND SIMPLE PRUNING

In this section we consider simple search algorithms that do not use heuristic information.

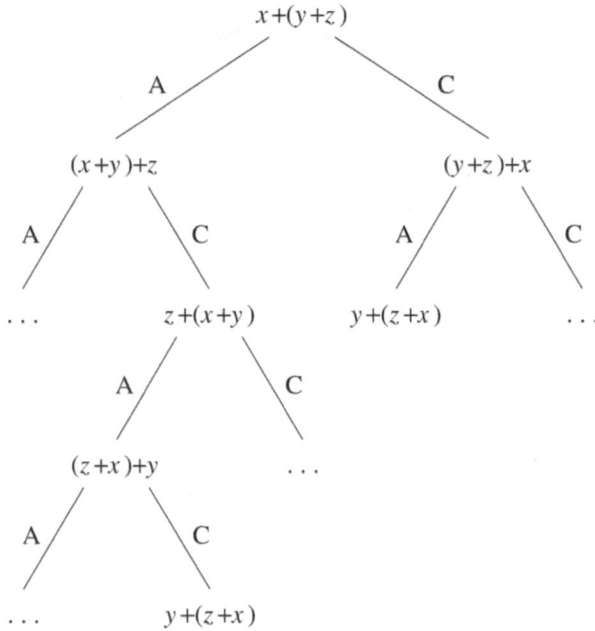

FIGURE 4.7 Addition proof tree.

4.2.1 Depth and Breadth First Search

Consider the following simple logic problem:

find x, y, z such that
$$(\neg x \wedge y) \vee (x \wedge \neg y \wedge \neg z) \text{ is true}$$
that is: (not x and y)
or (x and not y and not z)

Figure 4.8 shows a complete search tree for the problem based on choosing the variables in order. The tree is slightly ragged as the formula is true when x is *false* and y is *true* irrespective of the value of z. Note also that there are two solutions, marked with ticks.

We now consider two algorithms for searching this space of potential solutions: depth first and breadth first. Depth first search starts off at the root of the tree (the empty solution) and then works down the left-hand branch considering the partial solutions until it gets to a leaf. If this is not a goal state, it backs up and tries the next path down. That is, the algorithm tries to get as deep as possible as fast as possible, hence its name. Figure 4.9 shows the order in which this algorithm visits the nodes of the graph. In terms of the logic variables, one is considering them in the following order:

a:	$x = true$			– ?
c:	$x = true$	$y = true$		– ?
g:	$x = true$	$y = true$	$z = true$	– NO
h:	$x = true$	$y = true$	$z = false$	– NO
d:	$x = true$	$y = false$		– ?
i:	$x = true$	$y = false$	$z = true$	– NO
j:	$x = true$	$y = false$	$z = false$	– YES

In contrast, breadth first search moves back and forth through the search tree, only looking at the children of a node when all other nodes at a level have been examined. Figure 4.10 shows the order in which this algorithm visits the nodes, and the search progresses as follows:

a:	$x = true$		– ?
b:	$x = false$		– ?
c:	$x = true$	$y = true$	– ?
d:	$x = true$	$y = false$	– ?
e:	$x = false$	$y = true$	– YES

4.2.2 Comparing Depth and Breadth First Searches

Note that the two searches encounter a different goal state first. Often, one stops at the first goal state found – in this case, depth first and breadth first searches would return different solutions to the problem. Depth first al-

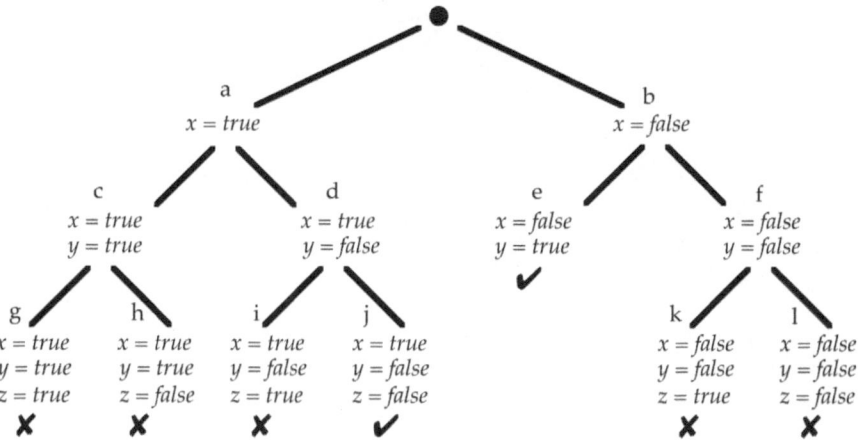

FIGURE 4.8   Tree of potential solutions for logic problem.

ways finds the first solution, reading the tree left to right, whereas breadth first finds the shallowest solution.

If we consider human problem solving, it is usually a mixture of depth first (looking at individual detailed options) and breadth first (considering the complete range of options at an abstract level). If anything, the tendency is towards depth first examination of a small part of the possible space, but this is combined with an almost uncanny ability to spot the right portion of the state space to explore. To some extent this ability is guided by heuristics, which enable us to make suitable choices. Algorithms that mimic this will be dealt with in Section 4.3.

To some extent depth first is the computationally easier method. However, breadth first searching has several advantages. First of all, it finds the shallowest solution. Often the depth in the tree is related to the complexity of the solution and hence shallowest is, in a sense, best. This is true, for example, of mathematical proofs, where a short proof is usually considered superior to a long one.

Even ignoring the issue of which solution is best, there are disadvantages to using depth first search. Consider the proof tree in Figure 4.7: the nodes represented by ellipses were those that stuttered; that is, the move reversed the effect of the previous move. If this were not detected, it would lead to an infinite search, for example continuously applying the commutativity axiom, and so moving back and forth for ever between the expressions $(x + y) + z$ and $x + (y + z)$. Figure 4.11 shows a similar tree for the Towers of Hanoi problem. In this case there

are nodes that stutter (marked with asterisks ***) but, in addition, paths that go on forever without repeating. For example, if we performed the moves 1–2, 1–3, 2–3, we would get to the state where all the rings are on the third column; another sequence of moves would return us to the initial state. One can avoid this terminal problem by keeping track of the states visited along the current path and backtracking whenever the current state is found on the path. In a more complex domain, infinite paths may exist that never repeat and, even where there are no infinite paths, the leftmost branch of the search tree could be immoderately large, making depth first search impractical.

4.2.3 Programming and Space Costs

From the previous discussion it would seem that breadth first search was a hands-down winner, except in the case where we were definitely seeking the leftmost solution in the search space. However, when we consider ease of programming and space costs, the situation is reversed.

To see this we will look at a simple implementation of the depth and breadth first algorithms. Both search algorithms must keep track of which nodes need to be examined next. This collection of nodes is known as the open list. The open list starts off containing only the root node. To search the tree the algorithm selects a node from the open list. The node is checked to see if it is a goal state: if it is, we have succeeded; if not, we add the children of the node to the open list and start again. The algorithm stops when the open list is exhausted. At that point the

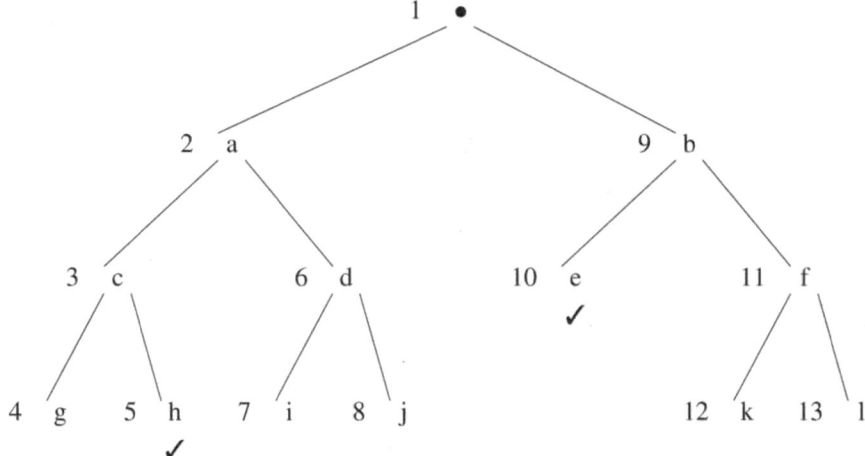

FIGURE 4.9   Depth first search – order of visiting nodes.

entire tree has been searched and therefore the algorithm reports failure.

The pseudocode for the algorithm is in Figure 4.12. (Note that this version of the algorithm does not check for repeated states.) The difference between depth and breadth first search is in the line marked **. If we add the children to the front of the open list (a stack), we get depth first search; if we add them to the end (a queue), we get breadth first search. It is truly amazing that such a small difference to the algorithm makes such a big difference to the order of the search.

Consider now a tree of depth d and branching factor b, and the largest open lists that can accumulate in the two algorithms. For depth first, the worst case is when it reaches the leftmost leaf node. At this point the open list contains the $b - 1$ siblings at each of the d levels. That is, the open list can contain up to $(b - 1) \times d$ nodes. The worst case for breadth first is when the algorithm is about to start looking at the leaf nodes. At this point all b^d leaf nodes will be in the open list – the space is exponential in the depth of the tree. So, space usage would discourage one from using breadth first search.

We turn now to ease of programming. In depth first search, the open list is a stack. By using recursion, either in procedural languages or in Prolog, we can effectively use the language's own run-time stack to give us depth first search almost for free. Indeed, Prolog's execution can be seen as a search process that is itself depth first (with consequent problems of infinite regress!).

4.2.4 Iterative Deepening and Broadening

We have seen that breadth first search may give an answer far faster if the search tree has some solutions closer to the root. However, breadth first search uses far more space, so much that searching large spaces will become prohibitive. One way to avoid this is an algorithm called iterative deepening. This is basically depth first search except with a maximum depth cut-off. The search is repeated with the depth increasing at each pass until a solution is found. If the depth is increased by one on each pass, the solution found will be precisely the same as that found by breadth first search. Like breadth first search it is immune to infinitely deep branches and hence is guaranteed to find a solution if one exists.

It seems as if iterative deepening would do a great deal of work, as it keeps searching the tree again and again. However, because of the exponential growth in the number of nodes at each level in the search tree, most of the work is done at the deepest level. Repeating work at higher levels has very little effect on the cost.

The worst case is when the tree is of constant depth. In this case, for a tree with branching factor b the extra work is only $1/(b - 1)$ of the normal breadth first time. For example, if $b = 6$, iterative deepening only takes 20 per cent longer. In contrast, if the tree has any infinite or very deep branches, the saving can be enormous.

Iterative deepening avoids the problems associated with very deep, or infinitely deep, branches. However, sometimes there is an infinite branching factor. For

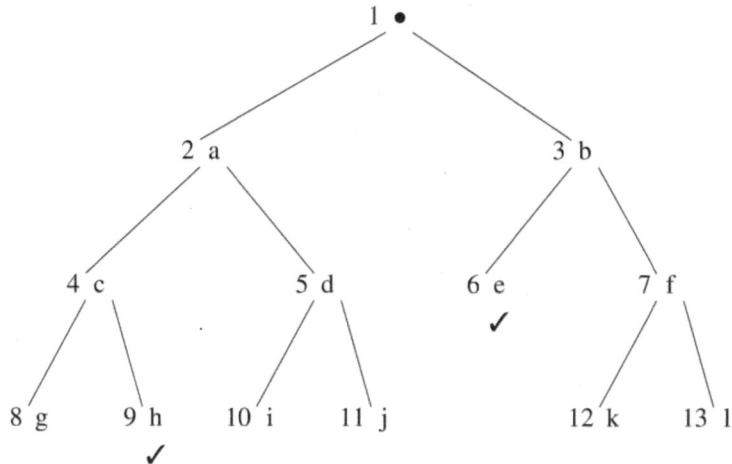

FIGURE 4.10   Breadth first search – order of visiting nodes.

example, we may want to find a positive integer solution to the equation $x^2 + y^2 = z^2$. As with the logic example at the start of this section, we could look at all the possibilities for x: 1, 2, 3, …– it might take some time! A second variation, iterative broadening, can deal with this problem by putting a bound on the number of children that are examined on each pass. Iterative broadening can be used on its own in conjunction with depth first or breadth first search, or combined with iterative deepening. In the latter case, one has to decide at each pass how much to increase both depth and branching cut-offs.

4.2.5 Finding the Best Solution – Branch and Bound

So far, we have only been concerned with finding the first solution. Now consider the case when we have cost associated with solutions. We cannot stop when we have found the first solution; instead we must keep track of the solution and its cost and then continue the search in case there is a better solution further on. We have to continue until the whole space is exhausted, all the time keeping track of the best solution encountered so far. This process could be combined with any of the search strategies we have encountered. However, every node must eventually be examined, so there is no advantage to using it with anything but depth first search.

If the cost function is associated solely with the final state, we can make no improvement to the algorithm without further heuristic guidance. However, if the path also has a cost, we can do somewhat better. We assume that the cost always increases with path length, as we shall do with all path costs. Examples of such costs include the distance travelled along a route, the time taken to perform actions between states, or a simple count of the moves taken. In fact all these costs are also additive and are the sum of the costs of each move; however, this is not necessary for the algorithm to work.

Imagine we have found a solution g with cost $c(g)$. We go on to look for a further solution and are about to examine a node at the end of a path p from the root of the tree. Now n and any state below n will have cost at least $c(p)$, so if $c(p) > c(g)$, it is not worth pursuing this path further – all nodes below it will exceed the current best cost. The algorithm resulting from this insight is called branch and bound. Figure 4.13 shows a search tree with costs associated with each path. Node d is optimal with path cost $3 + 1 = 4$. Assuming it is visited first, nodes below e and c need not be examined, as their partial path costs are $3 + 3 = 6$ and 5, respectively. Thus only the circled nodes are examined.

There are again variants of branch and bound associated with depth first, breadth first and iterative deepening. For the latter, the cut-offs can be based on the cost of the path rather than the depth. So long as the costs of the path increase suitably with path length, this will still be safe from infinitely deep branches.

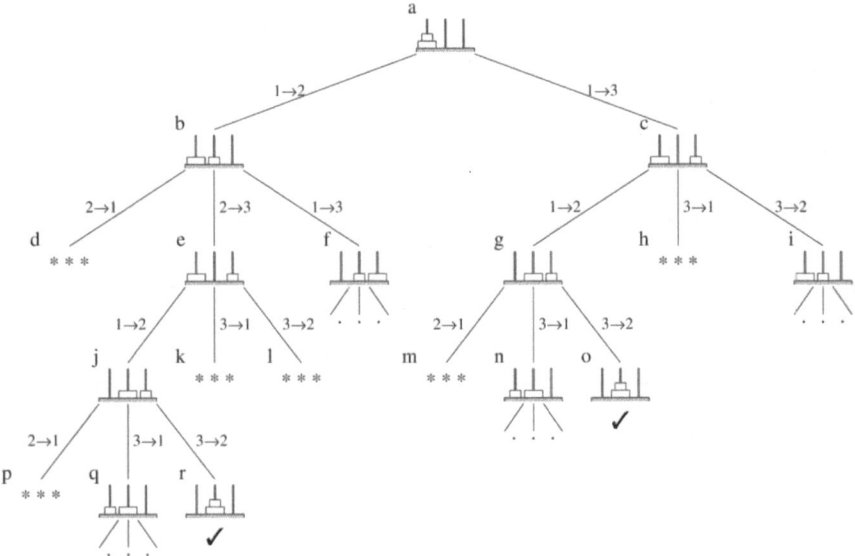

FIGURE 4.11 Towers of Hanoi – search tree.

4.2.6 Graph Search

Several of the example trees we have been searching have been trees of graph nodes, for example the arithmetic proof tree. Indeed, the problem of repeated states arose directly because of their graph-related nature. A tree generated from partial solutions, such as the magic square tree in Figure 4.4 or the logic problem in Figure 4.8, cannot have repeated states down a branch.

Notice also that, when a graph is considered as a tree, nodes can be repeated along different branches. This is seen in the Towers of Hanoi tree in Figure 4.11. It is not as serious a problem as repeats down a single branch, since it does not lead to infinite work. However, if the same node is repeated on different branches, time is wasted examining nodes that have already been searched.

To avoid infinite branches we checked that the new node was not already in the current path. A similar technique can be used with graph searching. In addition to the open list, we also keep a closed list, keeping track of all those nodes that have already been examined. If a node is in this list, it will not be examined again.

One can use branch and bound on graphs as well, simply by adding the cost check. However, a second visit to a node might be via a cheaper path. One therefore has to compare the new cost with the old one and, if cheaper,

remove the node from the closed list and add it again to the open.

Depending on where we add the children, we can search the graph in a depth or breadth first fashion. However, we cannot now avoid space costs as the closed list will expand until it includes the whole space. In addition to the space cost, this means that the lookup to see whether a state has been visited previously gets progressively more expensive. One option is to limit the size of the closed list, discarding some entries when it gets full. This will certainly reduce the space and time costs but leaves the possibility of repeated work and infinite loops.

4.3 HEURISTIC SEARCH

Recall from Section 4.1.2 that heuristics give us some information from a node part-way through a search about the nodes that lie beyond. Strictly, a heuristic could be any information but is most usually a simple number representing how good or bad that path is likely to be. In a state problem, the evaluation will usually only depend on the node itself $ev(n)$, but in a path problem it must also depend on the path to the node $ev(n, p)$.

Figure 4.14 shows the search tree for finding the way through a maze. This is a state problem, as we are not interested in the shortest way through, just any way. The start is marked with a bullet and the exit

```
Initialize Open=[Root]
while Open is not empty
  take first node N from Open
  if N is a goal state
        return N and SUCCESS!
  otherwise generate C: the set of children of N
                              (if it has any)
** add C to Open
if Open becomes empty with no success return FAILURE
```

FIGURE 4.12   Depth/breadth first search – generic algorithm.

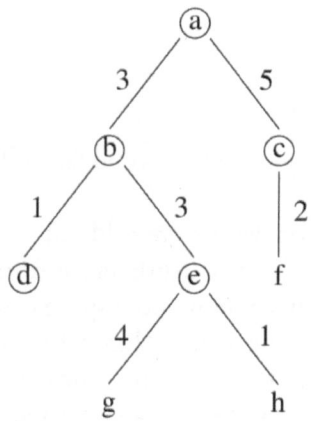

FIGURE 4.13   Search tree with path costs.

(goal state) is marked g. The rest of the letters mark the choice points in the maze. The figures in square brackets show the heuristic evaluation for each node. The evaluation function chosen is the distance from the node to the goal measured using the Manhattan block distance. (That is, the sum of the distances in the x and y directions.)

Notice the following about the maze search tree:

- *misleading directions* – At node a, it at first appears that b is the most promising direction. Unfortunately, it leads to a dead end.

- *local minima* – From node b nowhere looks any better; whatever path you take you appear (in terms of the heuristic) to get further from the goal. Hence

b is called a local minimum. A simple search might stop at b and never reach g, which is the global minimum.

- plateaux – The heuristic evaluation does not change between c and d; there is no sense of progress. In more complex problems there may be whole areas of the search space with no change of heuristic. When this happens, the heuristic ceases to give any guidance about possible direction.

- *getting worse to get better* – In order to progress towards the goal one has to get temporarily further away from it.

To be fair, a maze is designed to be hard to get through, and hence it is no wonder that a simple distance measure (or indeed any measure) is unhelpful. However, all these problems do occur in real scenarios and must be faced by any efficient search algorithm.

4.3.1 Hill Climbing and Best First – Goal-directed Search

We can use heuristic evaluation functions to improve basic depth first and breadth first searches. Both algorithms search the children of nodes in a left-to-right fashion. If instead we search for the child with the largest heuristic value first, we get hill climbing with backtracking and best first search. The difference between the two is that once hill climbing has chosen to follow a node it continues to do so in a depth first fashion, even if the heuristic value of the node's children is higher than the value for previously visited nodes. In contrast, best first

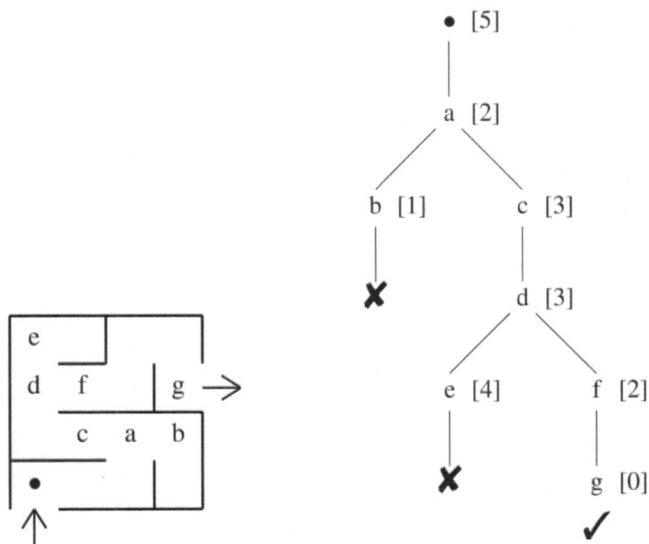

FIGURE 4.14 Maze search tree.

will consider all previously visited nodes at each stage, choosing the best so far.

Hill climbing is named after those situations where the biggest number is best (benefit rather than cost), and so the algorithm is constantly following the direction that gives the fastest rate of increase – a rather keen climber who always chooses the steepest path! (Perhaps in situations where lowest is best, we ought to use a downhill skier analogy?) However, the algorithm is prepared to back up and try a different path if the steepest ascent leads to a dead end, hence the addition "with backtracking". Because best first search considers the whole open list at each stage, it doesn't need to backtrack but consequently has to remember more nodes and to consider more possibilities at each step.

In the case of the maze tree, both algorithms would search the space in the same order, but this will not always be the case. Consider the tree in Figure 4.15. Like the maze tree, it represents some sort of goal-seeking search, where the heuristic is an estimated distance to a goal. The non-goal leaves are given an evaluation of 99 to represent the fact that they are no good at all! The heuristic evaluation function is rather better this time, especially as one gets closer to the goal nodes. This is typical of real-world examples. However, just as in the real world, it is not perfect – indeed, node f has quite a good evaluation, but no goal state is found beneath it.

Notice that the two algorithms reach different goal states. In fact, both algorithms are guaranteed to reach a goal state if one is there, as in the end they will both search the entire space. However, one hopes that the heuristic will so guide the search that a goal state is found when only a small portion of the space has been examined.

4.3.2 Finding the Best Solution – The A* Algorithm

Given good enough heuristics, hill climbing and best first searches can find a solution faster than exhaustive searches. However, when a solution is found, they cannot tell whether it is the best one. Consider the tree in Figure 4.15. The heuristic on the goal nodes represents how good they are. We see that the hill climbing algorithm gets to a suboptimal solution, l. In this case, best first does manage to find the optimal solution h, but this will not always be the case. If the heuristic had been a little less helpful and node d had had value 4, then best first would have found node e – again suboptimal.

The problem is that given a goal state and an open list, we cannot determine whether there are as yet unseen nodes with lower cost. The heuristic guides us to the good nodes but does not give enough information to guarantee we have found the optimum. Recall from Section 4.2.5 that branch and bound search did far better. It was able to prune whole areas of the search tree as unfruitful. This is because the cost of the path to a node serves as a lower bound on the cost of the nodes below it. If we have a heuristic function $ev(n, p)$ that has this property, we can have algorithms that guarantee an optimal solution.

A method of search that does this is the A* algorithm, which is effectively a modified form of best first. It is used especially on path problems where the cost of a path is the sum of the costs of the moves. However, it is not limited to such situations.

Rather than looking at the algorithm in detail, we shall simply consider the way it works in the case of real route finding on roads or around obstacles. We know that the shortest distance between two places must be at least as long as the straight-line distance. It may be longer if there is an obstacle in the way or if the roads are not straight, but, excluding cosmic worm-holes, it cannot be shorter.

Imagine we are looking for routes between Applethwaite and Gilby (see Figure 4.16). We have already found

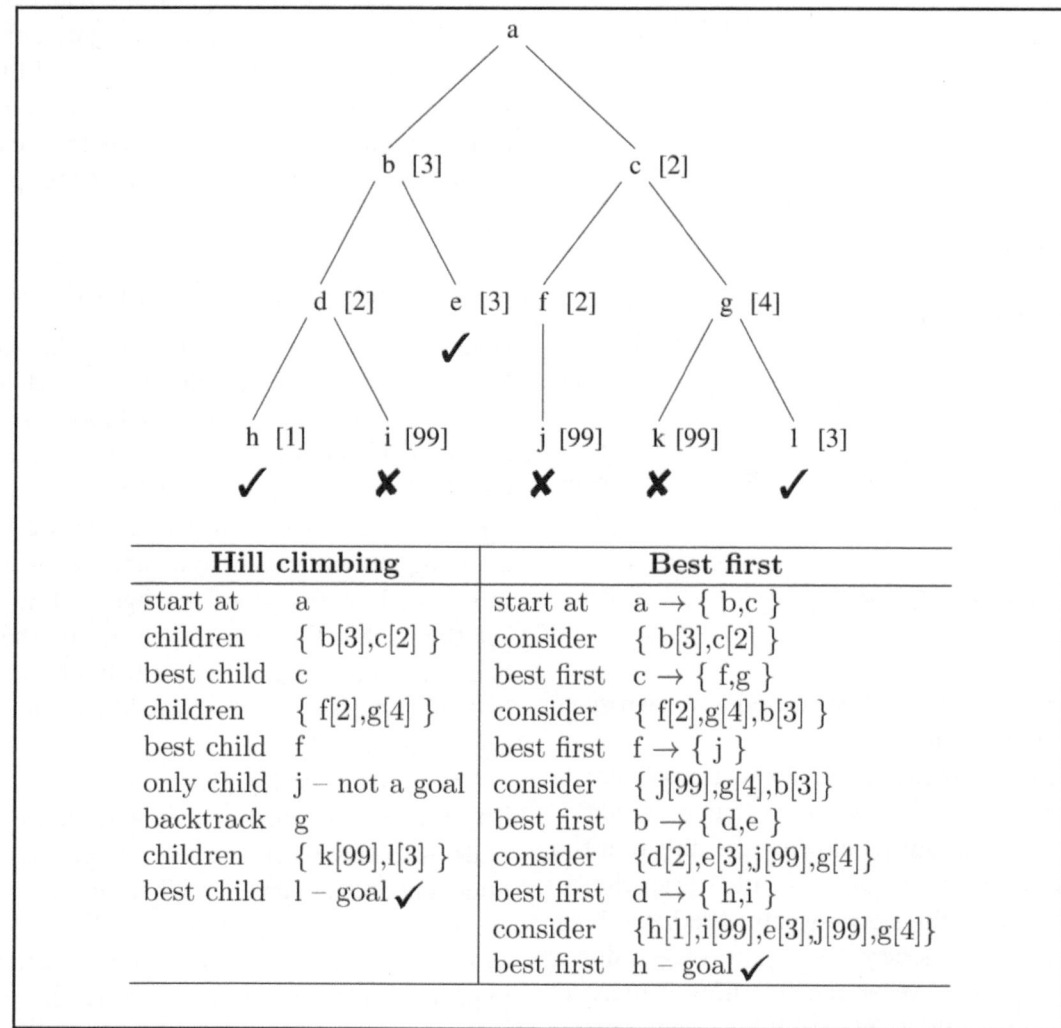

Hill climbing		Best first	
start at	a	start at	a → { b,c }
children	{ b[3],c[2] }	consider	{ b[3],c[2] }
best child	c	best first	c → { f,g }
children	{ f[2],g[4] }	consider	{ f[2],g[4],b[3] }
best child	f	best first	f → { j }
only child	j – not a goal	consider	{ j[99],g[4],b[3]}
backtrack	g	best first	b → { d,e }
children	{ k[99],l[3] }	consider	{d[2],e[3],j[99],g[4]}
best child	l – goal ✓	best first	d → { h,i }
		consider	{h[1],i[99],e[3],j[99],g[4]}
		best first	h – goal ✓

FIGURE 4.15   Trace of hill climbing and best first searches.

a route via Barton that is 62 miles long. We then go on to look for further routes. We find the shortest path from Applethwaite to Cardale is 17 miles, but we see that the straight-line distance from Cardale to Gilby is 50 miles. So, any route from Applethwaite to Gilby via Cardale must be at least 67 miles. As this is longer than the route we have already found, we can stop looking for routes via Cardale. That is, we have pruned the search tree at Cardale.

Using this sort of reasoning the A* algorithm can prune many fruitless paths but still guarantee to find the best solution.

Unfortunately, being a variant of breadth first, the A* algorithm inherits its storage problems for the open list.

However, there are depth first and iterative deepening versions of the algorithm that can be used to overcome the problem.

4.3.3 Inexact Search

Hill Climbing Revisited

As we noted in Section 4.1.1, we are often content with a good enough solution rather than the best. We saw that this would be the case if we used the first solution from best first or hill climbing. Furthermore, if all leaf nodes are feasible (although some are better than others), hill climbing will not need to do any backtracking. Thus we can use forgetful hill climbing; that is, we need only keep

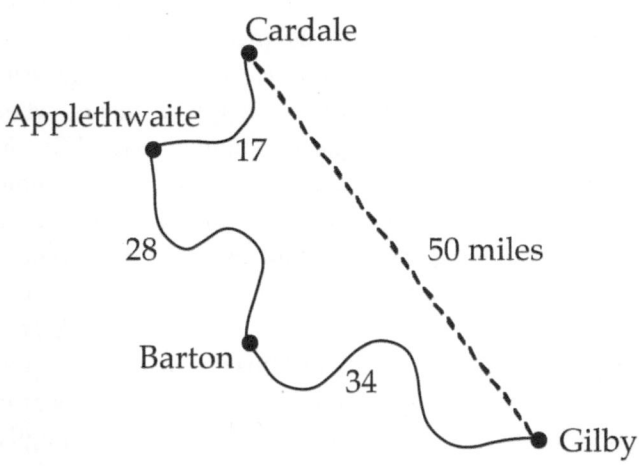

FIGURE 4.16 Using the A* algorithm.

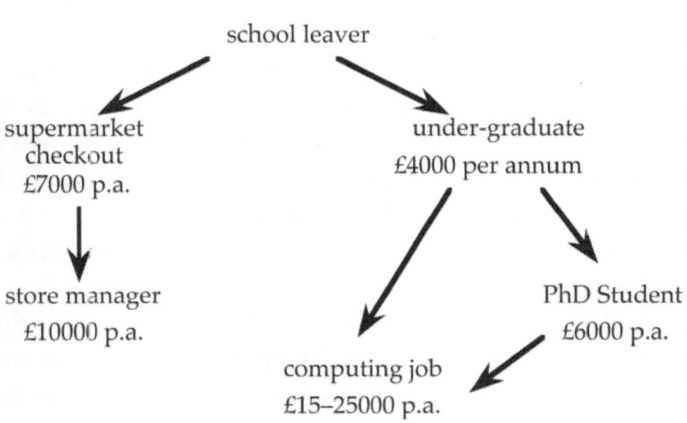

FIGURE 4.17 Graph of possible career moves.

downhill movement. One way to proceed is periodically to make random moves, or to start at several random positions and compare the outcome of hill climbing from each start position.

An advantage of hill climbing over the exact techniques discussed earlier is that it can be used in continuous as well as discrete domains. An example of this would be driving a car. The parameters to control include both discrete values (gear selected, choke on) and continuous ones (depression of accelerator, steering wheel position). Choosing continuous parameters, is beyond the scope of tree-searching techniques. On the other hand, traditional mathematical optimisation techniques deal *only* with continuous variables. However, hill climbing can be used with these rather difficult hybrid problems.

To apply hill climbing to the continuous part of the problem, one must choose some small step and look at the change in cost, or alternatively use the derivative of the cost function in that parameter. When one considers continuous parameters, a new phenomenon is encountered: the ridge. This is, like a real rocky ridge, a direction where the system is slowly moving uphill, but where it drops sharply downhill on either side. The problem is that if you are slightly off the ridge, the uphill direction is not to move along the ridge, but to ascend nearly directly up it. Unfortunately, the need to take discrete steps means that one frequently overshoots, leading to a wasteful zig-zag up the ridge, potentially missing it entirely. The equivalent problem where it occurs in discrete systems leads to a sequence of local minima occurring along the line of the ridge.

Simulated Annealing

A slightly more systematic approach is simulated annealing. At each step one considers a random move. If the move is uphill, one always follows that direction. However, even if the move is downhill, it is sometimes followed, with a probability that diminishes as the distance downhill increases. Slight downhill movements are likely, large ones less likely. As the process proceeds, the system is "cooled" – that is, it is made gradually less likely that a downhill step is taken. Basically when the system is hot the behaviour is almost a random walk; when it is cold, it is simple hill climbing. Note that although the discussion here compares it with hill

track of the current node and forget where we have been – no open list, no stack.

As the search space becomes very large, exhaustive search, even when guided by a heuristic, becomes impractical, and so this sort of inexact method is acceptable. Recall, however, the problems of hill climbing, plateaux and local minima (or maxima depending on the definition of best), making it hard to determine which direction to follow. Indeed, at a local minimum, one cannot even know if it is the best solution or not. Consider the career graph in Figure 4.17; the hill climbing school leaver would not get far!

In order to make progress in such domains (including running mazes), one needs to be prepared to accept some climbing, the algorithm is usually described with the best state having the lowest value, or "energy".

Simulated annealing is not guaranteed to find the global minimum but usually gets somewhere close. It can be made somewhat more systematic by keeping track of a fixed number of good past states. If the search seems to get stuck in the "lowlands" it can be restarted from one of these positions. You'll read more about this fitness 'landscape' in Chapter 9.

Simulated annealing has proved to be a very robust and flexible technique for solving problems that seem particularly intractable (e.g. timetabling), yet is very simple to program. However, one may feel uncomfortable using such a "random" algorithm for any critical application, such as medical diagnosis. In these cases it would need to be surrounded by firewalls to prevent mistakes.

Like hill climbing, simulated annealing is particularly powerful in hybrid problems where continuous and discrete parameters are mixed. The change in a continuous parameter can be chosen randomly, and the size of change can be arranged to reduce as the system is cooled, leading to smaller jumps and fine adjustment.

Genetic Algorithms

As the name suggests, genetic algorithms are based on an analogy with biotic genetics and natural selection. The problem state is coded into separate parameters (the genes). A random set of states (individuals) is then allocated. The system then goes through a series of generations. In each generation, some of the individuals die, some breed and occasionally some mutate:

- *death* – Some of the individuals are randomly killed (removed from the set). In order to simulate natural selection, those individuals that are considered good, as measured by the relevant cost function on the state, have a greater probability of survival.

- *breeding* – Pairs of individuals are chosen and a new individual formed by mixing the genes of its parents. The exact nature of this mixing will depend on the particular parametrisation chosen for the problem.

- *mutation* – As the original "gene pool" is not necessarily complete (or genes may be lost through deaths), occasionally parts of an individual's state may randomly change.

The idea is that the individuals that survive will gradually become better, as measured by the same costs that drive the natural selection. This algorithm has again proved fruitful in many domains where traditional techniques have found great difficulty. However, its success depends crucially on the choice of representation. To work the mixing of the genes from two good parents should lead (at least some of the time) to a good child. This is true where a problem consists of several almost independent parts. A good example of this is a crossword puzzle. Given a complete (but not necessarily correct) puzzle, we can measure goodness by the number of words that are in the English language. If one puzzle has good words in its top left and another good words in its bottom right, then combining the two is likely to lead to a reasonably promising solution. The parts of the puzzle are almost independent but not entirely so (as some words will cross between the two).

The particular advantage of genetic algorithms is that work spent on making one part right is not thrown away because another part is wrong. In a simple tree-based search, only the decisions near the root are re-used. Those near the leaves are constantly being discarded, even where different parts of the space have similar structure.

The language of genetics is helpful in discussing the general problem of search. The genotype is the internal description – in the case of search, the parameters in the state description. The phenotype is the external attribute – the goodness measure or cost of the state. Systematic progress can only be made in a search if the phenotype and genotype have a reasonably simple mapping. For example, simulated annealing and hill climbing rely on the fact that small changes in the genotype (state) will normally result in small changes in the phenotype (goodness measure). In addition, genetic algorithms rely on that mapping having reasonably good structural properties.

4.4 KNOWLEDGE-RICH SEARCH

The different algorithms that have been presented so far have depended upon the general type of problem (optimisation, game playing, etc.) but have been domain independent. They could be applied to solving crossword puzzles or controlling a chemical factory. Of course, the algorithms would not work equally well in all cases, but they are not intrinsically designed for a particular situation.

However, for many problems more domain knowledge is required, and the general algorithms must be tuned to the specific problem. There are obviously as many fine tunings as there are problems, so this section will briefly discuss two of the more general classes of problem: constraint satisfaction and means–ends analysis.

We have already explicitly assumed the use of domain knowledge in the heuristic evaluation function. Although this does in a sense embody knowledge about the domain, it is quite crude – a single number. Furthermore, it may be difficult to code knowledge about the goodness of a state into a number. We may be able to look at two states and say that one is better or more interesting than the other, but not be able to put a number to it. Note too that the word *heuristic* means more than just this evaluation function, but indeed any knowledge used to guide search.

In addition to the evaluation function, there have been several places where we have implicitly assumed that domain knowledge would or could be used:

- *ordering children* – In depth or breadth first search, we assumed that there was some ordering of the children of a node. A good choice of this ordering makes an enormous difference to the search efficiency. In minimax search, examining the best child first can double the depth to which it is possible to search in a given time. Even where an evaluation is being used, we may need extra guidance where there are many plateaux.

- *ordering the tree* – That is, choosing which decisions to make first in producing the tree. This is applicable to trees where the states represent partial solutions, and we have already seen in Section 4.1.2 how the order in which we expand the magic square search tree makes a big difference to our ability to prune impossible solutions.

In addition, in Chapter 11, we will consider the minimax algorithm for searching game tree searches. An important parameter in the minimax algorithm is the search horizon, which determines how deep the search looks down the game tree. The choice of search horizon will vary for different parts of the tree and embody a great deal of the knowledge of the particular game.

In this section, we will look at constraint satisfaction and see how a good order of search can be determined dynamically. In Chapter 15, we will see examples of knowledge-rich search algorithms for planning and route finding.

4.4.1 Constraint Satisfaction

Consider again the magic square tree shown in Figure 4.4. In fact, the magic square is an example of a constraint satisfaction problem. The goal is to have a state described by parameters $m_{11} \ldots m_{33}$, corresponding to the positions in the square:

m_{11}	m_{12}	m_{13}
m_{21}	m_{22}	m_{23}
m_{31}	m_{32}	m_{33}

These parameters are to be different integers in the range 1 to 9 and must satisfy the following constraints:

1. $m_{11} + m_{12} + m_{13} = 15$
2. $m_{21} + m_{22} + m_{23} = 15$
3. $m_{31} + m_{32} + m_{33} = 15$
4. $m_{11} + m_{21} + m_{31} = 15$
5. $m_{12} + m_{22} + m_{32} = 15$
6. $m_{13} + m_{23} + m_{33} = 15$
7. $m_{11} + m_{22} + m_{33} = 15$
8. $m_{13} + m_{22} + m_{31} = 15$

Constraints 1–3 say that the rows add up to 15; 4–6 say the same for the columns and 7 and 8 for the diagonals.

These constraints are arithmetic equalities, but constraints can also be inequalities or logical formulae. The logic problem in Section 4.2.1 is an example of the latter.

We can use the constraints to:

- *check* the correctness of a partial solution and hence prune fruitless branches.

- *calculate* some parameters from others. For example, once we know m_{11} and m_{12} we can calculate that

$$m_{13} = 15 - m_{11} + m_{12}$$

- *choose* which parameter to fix next.

In keeping with the idiom, the fixing of a parameter value can be thought of as adding a new (albeit simple) constraint. However, we must remember to distinguish those constraints that are given as part of the problem and those that are guesses and may thus be changed later (backtracking).

The first of these, checking, reduces the effective breadth of the tree, as it means some branches need not be examined. The second, calculation, reduces the effective depth as some choices are made "for free". These two can be accomplished using a general software method called constraint propagation. However, it is the third choice that is ultimately most powerful.

Recall how the particular order in which the magic square was searched led to rapid pruning of unfruitful paths, whereas an expansion that led to partial solutions of the form

?	2	?
1	?	?
?	?	3

was clearly unsuitable. The reason for this is that the chosen order filled in the parameters of constraints that could then be checked (in fact the third choice could have been made by calculation).

A general heuristics is to choose to fix parameters that will complete constraints. So, for example, once we have chosen to fix $m_{11} = 1$, there are only two more parameters required on constraints 1, 4 and 7. This suggests that we next choose to fix one of the other parameters in these constraints, say m_{12} (as in the tree) or m_{22}. A general heuristic is thus to choose a parameter that is in the constraint with the fewest free slots. Where this heuristic yields several possible parameters, we can choose one that reduces most other constraints. For example, this would suggest that for the first parameter we ought to choose m_{22}, as this is in four constraints as opposed to only three for m_{11}.

As one focuses on more specific problems, these general heuristics are also honed. In particular, we find that the choice of order can no longer be made *statically*. All the arguments we have used for the magic tree could be made without looking at a single node. We can look at the constraints and choose a search order (not square!) such as the following:

2	6	4
8	1	9
5	7	3

In other problems it may not be clear what order to choose until one has explicit information. This is particularly true where the constraints are complex logical formulae such as

$$(a \wedge b \wedge c \wedge d \wedge e \wedge f) \vee (g \wedge h)$$

Initially this has a lot of parameters in it and would be far down our list of interesting constraints. However, as soon as we begin to examine the branch with $a = false$, the whole left-hand side of the constraint becomes false and it reduces to $(g \wedge h)$, suggesting that we next fix g or h.

4.5 SUMMARY

Search problems can be classified in a number of ways:

- state or path based

- any or best solution

- deterministic or adversarial

- perfect solution or just good enough

Search spaces can be structured as trees or graphs. In some problems the interior nodes of a search tree may represent partial solutions. Trees can be characterised by their branching factor and their depth, either of which may be infinite.

Search can be guided by heuristic evaluation functions or by domain knowledge or can be virtually unguided.

Blind search algorithms include depth first search, breadth first search, iterative broadening, iterative deepening and branch and bound. Depth first search is simple to program and uses relatively little space compared with breadth first search but has problems with very deep branches. Iterative deepening and iterative broadening algorithms deal with problems of very (or infinitely) deep and very broad trees respectively. Where costs are associated with moves, branch and bound can reduce the number of nodes searched by pruning nodes of the search tree. But it is still guaranteed to find the best solution.

Heuristic evaluation functions can guide search. Hill climbing with backtracking and best first search use the heuristic value to choose the order to investigate nodes. But both must search the entire tree to be sure of finding the best solution. If the heuristic evaluation function gives a lower bound on the final term, the A* algorithm can prune nodes and so avoid searching all the tree, but still will get the best solution.

Exact methods are often impractical. Forgetful hill climbing can often find good solutions but suffers from problems caused by local mimima and plateaux in the search space. Genetic algorithms and simulated annealing use randomness in different ways to search complex spaces including problems with some discrete and some continuous parameters. They often find near-optimal solutions.

However, more knowledge is needed to tune algorithms for specific problem domains. Algorithms that include such knowledge include constraint satisfaction, as well as specialist algorithms discussed in later chapters.

4.1 In Section 4.1.1, it was said that for many path problems the cost was a function of the route only and not the goal state reached. Think of an example of a problem where both the goodness of the goal state and the cost of the path are important.

4.2 In Section 4.3.3 it was suggested that a genetic algorithm was a possible way to solve crossword puzzles. Find an online dictionary and extract all words of four letters. You are trying to produce 4×4 acrostics. That is, four lines of four characters so that each row and each column forms a word. The states will be lists of 16 characters, and goodness can be measured by the number of four-letter words (8 is perfect). For example, take the (incorrect) acrostic

```
P I N S
A M E O
I Q A N
L O T S
```

It has five correct words (pins, lots, pail, neat and sons) and three incorrect (ameo, iqan and imqo). Its goodness is therefore 5. Choose a method to combine two acrostics and use a genetic algorithm on it.

4.3 Mini-Sudoku only uses the numbers 1 to 4 on a 4 × 4 square where each row, each column, and each of the four corner squares has exactly one each of the digits 1 to 4. As with Sudoku, you are given a partially filled square and have to complete it. Here is your start square:

a. Draw a search tree that starts off by choosing values for the free squares beginning at the top left (so row 1 column 2 is the first blank to fill).

b. Now do a depth first search, but starting with the blank squares that are most constrained (have the most filled-in squares in the same column/row or small square).

FURTHER READING

J. Pearl. *Heuristics: Intelligent search strategies for computer problem solving.* Addison Wesley, Reading, MA, 1984.

A detailed, mathematically precise exposition of the classic search strategies and their application in AI. No longer in print, but the PDF is available at the Internet Archive: https://archive.org/details/heuristicsintell00pear/

S. Edelkamp and S. Schroedl. *Heuristic search: Theory and applications,* Morgan Kaufmann, Waltham, MA, 2011.

A more recent book on heuristic search, covering similar ground to Pearl and also other areas such as pragmatic issues and distributed algorithms.

II

Data and Learning

Machine Learning

5.1 OVERVIEW

In this chapter, we will see that machine learning is an important and necessary part of artificial intelligence. We will also discuss the general pattern of machine learning and some of the issues that arise. Several specific machine learning methods will be described: deductive learning, inductive learning and explanation-based learning. Most of the chapter will concentrate on two specific inductive learning algorithms: the version-space method and ID3. We will conclude with a description of an experimental system that uses machine learning in an intelligent database interface.

The techniques, algorithms and examples in this chapter focus on more traditional symbolic machine learning, and in this respect it is a bridge from Part I. However, when many people hear 'machine learning', they now immediately think of neural networks. We will deal with neural networks and deep learning separately in Chapters 6 and 8. However, they share many aspects with more traditional machine learning techniques, and in Chapter 9 we will return to some of the broader issues of machine learning.

5.2 WHY DO WE WANT MACHINE LEARNING?

One response to the idea of artificial intelligence is to say that computers can never think because they only do what their programmers tell them to do. Of course, it is not always easy to tell what a particular program will do (!), but at least given the same inputs and conditions it will do the same things – dependable if not predictable. If the program gets something right once, it will always get it right. If it makes a mistake once, it always makes the same mistake. In contrast, people tend to learn from their mistakes; attempt to work out why things went wrong; try alternatives. Also, we are able to notice similarities between things and so generate new ideas about the world we live in. An intelligence, however artificial or alien, that did not learn would not be much of an intelligence. So, machine learning is a prerequisite for any mature programme of artificial intelligence.

Of course, many practical applications of AI do not make use of machine learning. The relevant knowledge is built in at the start. Although perhaps fundamentally limited, such systems are useful and do their job. However, even where we do not require a system to learn "on the job", machine learning has a part to play. One of the most difficult problems in the building of expert system is capturing the knowledge from the experts. There are many knowledge elicitation techniques to aid this process (see Chap. 18), but the fundamental problem remains: things that are normally implicit, inside the expert's head, must be externalised and made explicit (Figure 5.1).

Using machine learning this problem can be eased. Experts may find it hard to say what rules they use to assess a situation, but they can usually tell you what factors they take into account. A machine learning program can take descriptions of situations couched in terms of these factors and then infer rules that match the expert's behaviour. The expert can then critique these rules and verify that they seem reasonable (it is easier to recognise correct rules than to generate them). If the rules are wrong,

DOI: 10.1201/9781003082880-7

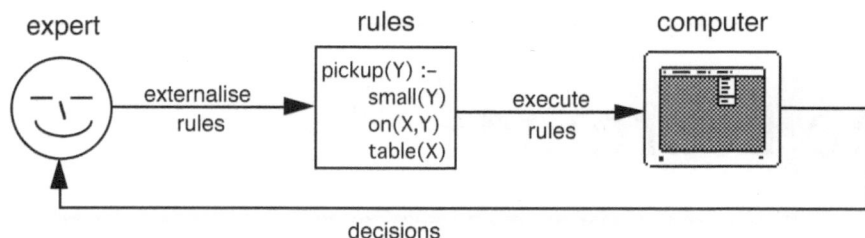

FIGURE 5.1 The knowledge elicitation bottleneck.

the expert may be able to suggest counter-examples that can guide further learning (Figure 5.2).

In addition there are many situations for which there is no expert knowledge. For example, Deep Mind used machine learning to control plasma within the Tokamak fusion reactor [262]. Although the scientists and engineers understand the basic physics, they could not themselves control the flow of plasma. Here the goal of machine learning is not to capture human knowledge but to generate new knowledge (or at least rules) based on examples or experience.

5.3 HOW MACHINES LEARN

In previous chapters we have discussed reasoning, knowledge representation and search. All are important for machine learning. In addition, there are various other factors that influence the choice and efficacy of a learning system, for example the amount of domain knowledge used by the system.

In this section we will look at several of these issues, which will be important when we look at particular learning algorithms later in this chapter. It will give a context to these algorithms, and we shall mention them where appropriate. We suggest that you revisit this section after reading the rest of the chapter. We'll start by looking at the phases in a typical machine learning system (Figure 5.3). The different issues will then be discussed in relation to the data and processes involved.

5.3.1 Phases of Machine Learning

Machine learning typically follows three phases:

training – A training set of examples of correct behaviour is analysed and some representation of the newly learnt knowledge is stored. This is often some form of rules.

validation – The rules are checked and, if necessary, additional training is given. Sometimes additional test data are used, but instead a human expert may validate the rules (as in Figure 5.2), or some other automatic knowledge-based component may be used. The role of the tester is often called the critic.

application – The rules are used in responding to some new situation.

These phases may not be distinct. Often there is no explicit validation phase; instead the learning algorithm guarantees some form of correctness. Also, in some circumstances, systems learn "on the job" – that is, the training and application phases overlap.

Obviously the training stage is the most important. It falls into two main types:

supervised learning – Here the training data comes pre-labelled with some form of classification or expected response. For example, this might be a set of images with a tag that says if it is urban or rural, or a collection of board position with the best next move to play. The aim of the machine learning algorithm in such cases is to emulate this behaviour to assign the correct tags or decisions to unseen inputs based on the examples.

unsupervised learning – Here the data is unlabelled and the machine learning algorithm creates its own labels or structure, for example clustering the data into groups.

In this chapter most of the examples will be supervised, but in the coming chapters we will see examples of unsupervised techniques including self-organising maps and clustering.

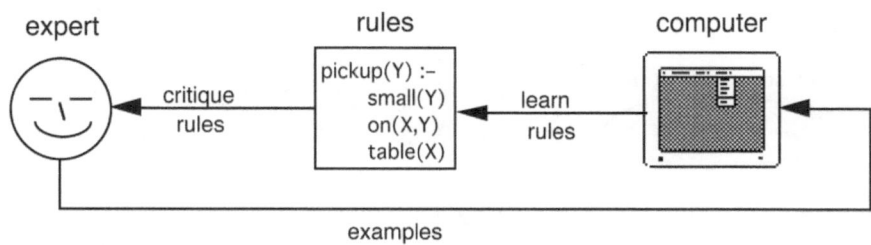

FIGURE 5.2   Machine learning avoids the bottleneck.

Some people add a third category **reinforcement learning**, which occurs when the learning has to happen while interacting with some kind of environment: real or virtual, such as playing a game or controlling a robot. This is often classed as a special case of unsupervised learning, but we shall see when we discuss reinforcement learning in more detail in Chapter 16, it also has elements of supervised learning.

In the rest of this section, we will look in turn at the inputs to training (the training set and existing background knowledge) and the outputs (the new knowledge learnt). First, however, we'll look at how rote learning fits into this picture.

5.3.2 Rote Learning and the Importance of Generalisation

The simplest kind of learning is rote learning. In this case examples of correct behaviour are stored, and when a new situation is encountered, it is matched with the learnt examples. If one of the examples matches, the relevant response is given. In this kind of learning there is no prior knowledge. Training consists simply of memorisation, and the output of training is just the stored training set. For example, the system may be given the following set of stimulus–response pairs:

24°C	–	75°F
−3°C	–	26°F
176°F	–	80°C
17°C	–	62°F
41°F	–	5°C
89°F	–	32°C
0°C	–	32°F

From these it might be able to respond to a stimulus "41°F" and give the response "5°C". However, it would not be able to respond to an unseen stimulus such as "15°C". Rote learning is clearly a very limited form of learning and is arguably not "real" learning at all.

Real learning involves some form of generalisation. We would like a system to infer that when a stimulus of the form "<a number>°C" is received, it should multiply the number by 9/5 and add 32. Note how this is not a simple arithmetic rule. The system would have to learn that different formulae should be used depending on whether the stimulus included "°C" or "°F". In fact, in most of the learning algorithms we will discuss, the rules learnt will be symbolic rather than numeric.

However, one should not underestimate the importance of rote learning. After all, the ability to remember vast amounts of information is one of the advantages of using a computer, and it is especially powerful when combined with other techniques. For example, heuristic evaluation functions are often expensive to compute; during a search the same node in the search tree may be visited several times and the heuristic evaluation wastefully recomputed. Where sufficient memory is available a rote learning technique called memorising can help. The first time a node is visited the computed value can be remembered. When the node is revisited, this value is used instead of recomputing the function. Thus the search proceeds faster, and therefore more complex (and costly) evaluation functions can be used.

5.3.3 Inputs to Training

In Figure 5.3, we identified two inputs to the training process: the training set and existing knowledge. Most of the learning algorithms we will describe are heavily example based; however, pure deductive learning (Section 5.4) uses no examples and only makes use of existing knowledge. There is a continuum (Figure 5.4) between knowledge-rich methods that use extensive domain knowledge and those that use only simple domain-independent knowledge. The latter is often implicit in the

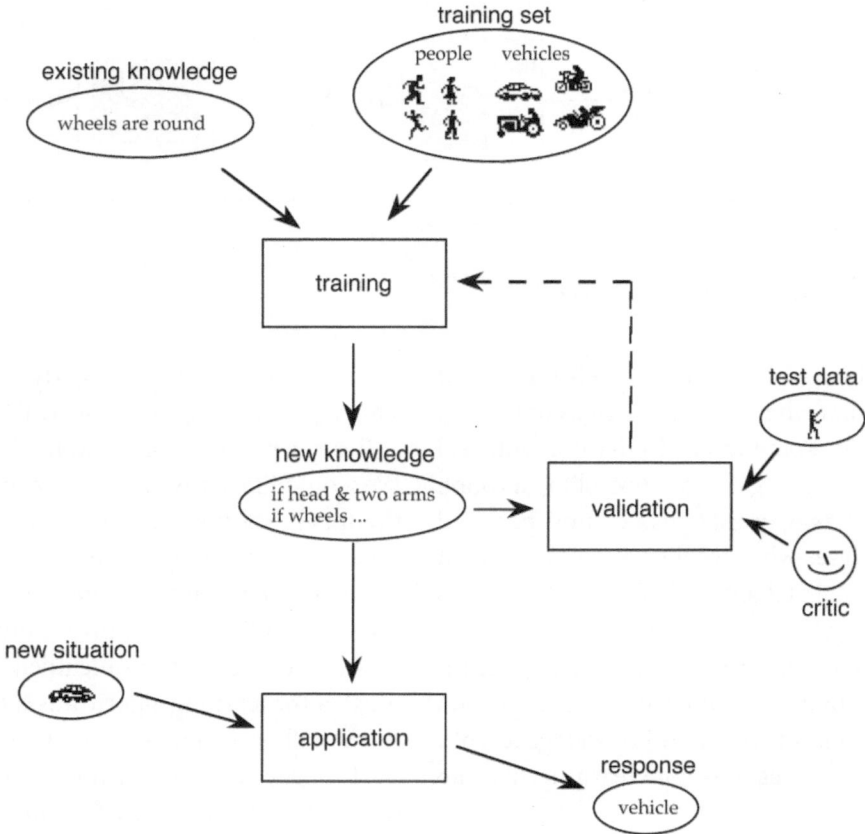

FIGURE 5.3   Phases of machine learning.

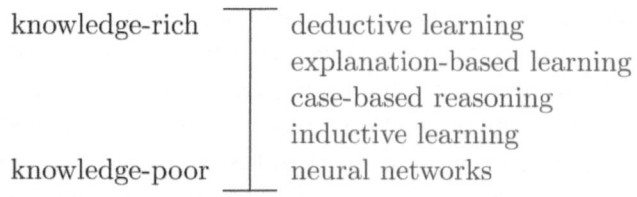

FIGURE 5.4   Knowledge continuum.

algorithms; for example, inductive learning is based on the knowledge that if something happens a lot, it is likely to be generally true.

Where examples are being used it is important to know what the source is. The examples may be simply measurements from the world, for example transcripts of grandmaster chess tournaments. If so, do they represent "typical" sets of behaviour or have they been filtered to be "representative"? If the former is true, then we can infer information about the relative probability from the frequency in the training set. However, unfiltered data may also be noisy, have errors, etc., and examples from the world may not be complete, since infrequent situations may simply not be in the training set.

Alternatively, the examples may have been generated by a teacher. In this case we can assume that they are a helpful set, covering all the important cases and including near miss examples. Also, one can assume that the teacher will not be deliberately ambiguous or misleading. For example, a helpful teacher trying to teach a relationship between numbers would not give the example (2, 2, 4), as this might be multiplication or addition.

Finally, the system itself may be able to generate examples by performing experiments on the world (for robots), asking an expert, or using an internal model of the world.

We also have to decide on a representation for the examples. This may be partly determined by the context, but often we will have some choice. Often the choice of representation embodies quite a lot of the domain knowledge.

A common representation is as a set of attribute values. For example, in Section 5.5.1, we will describe children's play tiles using four attributes: shape, colour, size and material. A particular example could be: triangle, blue, large, wood. In vision applications (see Chap. 12), the representation is often even cruder – simply a bitmap. On the other hand, more knowledge-rich learning often uses more expressive descriptions of the structure of the examples, using predicate logic or semantic networks.

5.3.4 Outputs of Training

To a large extent the outputs of learning are determined by the application. What is it we want to do with our new knowledge? Many machine learning systems are classifiers. The examples they are given are from two or more classes, and the purpose of learning is to determine the common features in each class. When a new unseen example is presented, the system then uses the common features to determine in which class the new example belongs. The new knowledge is thus effectively in the form of rules such as

>**if**   example satisfies condition
>**then** assign it to class X

In machine learning, this job of concept classification is often called concept learning (see Section 5.5.1). The simplest case is when there are only two classes, of which one is seen as the desired "concept" to be learnt and the other is everything else. In this case we talk about positive and negative examples of the concept. The "then" part of the rules is then always the same and so the learnt rule is simply a predicate describing the concept.

The form of this predicate, or of the condition part of a more complex rule, varies between machine learning algorithms. In some it is an arbitrary logical predicate, but more commonly its form is much simpler. In Section 5.5.1 we will consider predicates that are of the form

>attribute1 = value1 **and** attribute2 = value2
>**and ...**

That is, conjunctions of simple tests on attributes. In Section 5.5.2 more complex predicates in the form of decision trees will be considered. We will see that there is a trade-off between the allowable set of rules and the complexity of the learning process. The desire for simple rules is determined partly by computational tractability but also by the application of Occam's razor – always prefer simpler explanations: they are more likely to be right and more likely to generalise.

Not all learning is simple classification. In applications such as robotics one wants to learn appropriate actions. In this case, the knowledge may be in terms of production rules or some similar representation. More complex rules also arise in theorem provers and planning systems.

An important consideration for both the content and representation of learnt knowledge is the extent to which explanation may be required for future actions. In some cases the application is a black-box. For example, in speech recognition, one would not ask for an explanation of why the system recognises a particular word or not, one just wants it to work! However, as we shall see in Chapters 18 and 21, many applications require that the system can give a justification for decisions. Imagine you asked an expert system "is my aircraft design safe" and it said "yes". Would you be happy? Probably not. Even worse, imagine you asked it to generate a design – it might do a very good job, but unless it could justify its decisions would you be happy? Because of this, the learnt rules must often be restricted to a form that is comprehensible to humans. This is another reason for having a bias towards simple rules.

5.3.5 The Training Process

As we noted, real learning involves some generalisation from past experience and usually some coding of memories into a more compact form. Achieving this generalisation requires some form of reasoning. In Chapter 3, we discussed the difference between deductive reasoning and inductive reasoning. This is often used as the primary distinction between machine learning algorithms. Deductive learning works on existing facts and knowledge and deduces new knowledge from the old. In contrast, inductive learning uses examples and generates hypotheses based on similarities between them. In addition, abductive reasoning may be used and also reasoning by analogy (see Chap. 3).

Imagine we are analysing road accidents. One report states that conditions were foggy, another that visibility was poor. With no deductive reasoning it would be impossible to see the similarity between these cases. However, a bit of deduction based on weather knowledge would enable us to reason that in both cases visibility was poor. Indeed, abductive reasoning would suggest that visibility being poor probably means that it was foggy anyway, so the two descriptions are in fact identical. However, using this sort of reasoning is expensive both during learning and because it is dependent on having coded much of the background knowledge. If learning is being used to reduce the costs of knowledge elicitation, this is not acceptable. For this reason many machine learning systems depend largely on inductive reasoning based on simple attribute–value examples.

One way of looking at the learning process is as search. One has a set of examples and a set of possible rules. The job of the learning algorithm is to find suitable rules that are correct with respect to the examples and existing knowledge. If the set of possible rules is finite, one could in principle exhaustively search to find the best rule. We will see later in this chapter that the sizes of the search spaces make this infeasible. We could use some of the generic search methods from Chapter 4. For example, genetic algorithms have been used for rule learning. In practice, the structure of rules suggests particular forms of the algorithms. For example, the version-space method (Section 5.5.1) can be seen as a special case of a branch and bound search. This exhaustive search works because the rules used by version spaces are very limited. Where the rule set is larger exhaustive search is not possible and the search must be extensively heuristic driven with little backtracking. For example, the inductive learning algorithm ID3 discussed in Section 5.5.2 will use an entropy-based heuristic.

5.4 DEDUCTIVE LEARNING

Deductive Learning works on existing facts and knowledge and deduces new knowledge from the old. For example, assume you know that Alison is taller than Clarise and that Brian is taller than Alison. If asked whether Brian is taller than Clarise, you can use your knowledge to reason that he is. Now, if you remember this new fact and are asked again, you will not have to reason it out a second time, you will know it – you have learnt.

Arguably, deductive learning does not generate "new" knowledge at all, it simply memorises the logical consequences of what you know already. However, by this argument virtually all of mathematical research would not be classed as learning "new" things. Note that, whether or not you regard this as new knowledge, it certainly can make a reasoning system more efficient. If there are many rules and facts, the search process to find out whether a given consequence is true or not can be

very extensive. Memorisng previous results can save this time.

Of course, simple memorisation of past results would be a very crude form of learning, and real learning also includes generalisation. A proof system has been asked to prove that $3 + 3 = 2 \times 3$. It reasons as follows:

$$
\begin{aligned}
3 + 3 &= 1 \times 3 + 1 \times 3 \\
&\quad \text{(because for any number } n, 1 \times n = n) \\
&= (1 + 1) \times 3 \\
&\quad \text{(distributivity of} \times) \\
&= 2 \times 3
\end{aligned}
$$

Although this looks trivial, a real proof system might find it quite difficult. The step that uses the fact that 3 can be replaced by 1×3 is hardly an obvious one to use! Rather than simply remembering this result, the proof system can review the proof and try to generalise. One way to do this is simply to attempt to replace constants in the proof by variables. Replacing all the occurrences of "3" by a variable a gives the following proof:

$$
\begin{aligned}
a + a &= 1 \times a + 1 \times a \\
&\quad \text{(because for any number } a, 1 \times a = a) \\
&= (1 + 1) \times a \\
&\quad \text{(distributivity of} \times) \\
&= 2 \times a
\end{aligned}
$$

The proof did not depend on the particular value of 3; hence the system has learnt that in general $a + a = 2 \times a$. The system might try other variables. For example, it might try replacing 2 with a variable to get $3 + 3 = b \times 3$ but would discover that for this generalisation the proof fails. Hence, by studying the way it has used particular parts of a situation, the system can learn general rules. We will see further examples of deductive learning in Chapter 15, when we consider planning, and in Chapter 22, in the SOAR architecture. In this chapter, we will not look further at pure deductive learning, although explanation-based learning (Section 5.6) and case-based reasoning (Chap. 3) both involve elements of deductive learning.

5.5 INDUCTIVE LEARNING

Rather than starting with existing knowledge, inductive learning takes examples and generalises. For example, having seen many cats, all of which have tails, one might conclude that all cats have tails. This is of course a potentially unsound step of reasoning, and indeed Manx cats have no tails. However, it would be impossible to function without using induction to some extent. Indeed, in many areas it is an explicit assumption. Geologists talk about the "principle of uniformity" (things in the past work the same as they do now), and cosmologists assume that the same laws of physics apply throughout the universe. Without such assumptions it is never possible to move beyond one's initial knowledge – deductive learning can go a long way (as in mathematics) but is fundamentally limited. So, despite its potential for error, inductive reasoning is a useful technique and has been used as the basis of several successful systems.

One major subclass of inductive learning is concept learning. This takes examples of a concept, say examples of fish, and tries to build a general description of the concept. Often the examples are described using simple attribute–value pairs. For example, consider the fish and non-fish in Table 5.1.

TABLE 5.1 Fish and Non-fish.

	swims	has fins	flies	has lungs	is fish
herring	yes	yes	no	no	✓
cat	no	no	no	yes	✗
pigeon	no	no	yes	yes	✗
flying fish	yes	yes	yes	no	✓
otter	yes	no	no	yes	✗
cod	yes	yes	no	no	✓
whale	yes	yes	no	yes	✗

There are various ways we can generalise from these examples of fish and non-fish. The simplest description (from the examples) is that a fish is something that does not have lungs. No other single attribute would serve to differentiate the fish. However, it is dangerous to opt for too simple a classification. From the first four examples we might have been tempted to say that a fish was something that swims, but the otter shows that this is too general a description. Alternatively, we might use a more specific description. A fish is something that swims, has fins and has no lungs. However, being too specific also has its dangers. If we had not seen the example of the flying fish, we might have been tempted to say that a fish also did not fly. This trade-off between learning an over-general or overspecific concept is inherent in the problem.

Notice also the importance of the choice of attributes. If the "has lungs" attribute were missing, it would be impossible to tell that a whale was not a fish.

The two inductive learning algorithms described in detail in this section – version spaces and ID3 – are examples of concept learning. However, inductive learning can also be used to learn plans and heuristics. The final part of this section will look at some of the problems of rule induction.

5.5.1 Version Spaces

When considering the fish, we used our common sense to find the rule from the examples. In an AI setting we need an algorithm. This should take a set of examples such as those above and generate a rule to classify new unseen examples. We will look first at concept learning using version spaces, which uses examples to home in on a particular rule [199].

Consider again Table 5.1. Imagine we have only seen the first four examples so far. There are many different rules that could be used to classify the fish. A simple class of rules are those that consist of conjunctions of tests of attributes:

> **if**   *attribute1* = *value1*
>   **and** *attribute2* = *value2* …
> **then**   is a fish

Even if we restrict ourselves to these, there are seven different rules that correctly classify the fish in the first four examples:

R1.   **if**   swims = yes
      then   is a fish

R2.   **if**   has fins = yes
      then   is a fish

R3.   **if**   has lungs = no
      then   is a fish

R4.   **if**   swims = yes **and** has fins = yes
      then   is a fish

R5.   **if**   swims = yes **and** has lungs = no
      then   is a fish

R6.   **if**   has fins = yes **and** has lungs = no
      then   is a fish

R7.   **if**   swims = yes **and** has fins = yes
      and has lungs = no
      then   is a fish

If we only had the first four examples, what rule should we use? Notice how rules R1 and R2 are more general than rule R4, which is in turn more general than R7. (By more general, one means that the rule is true more often.) One option is to choose the most specific rule that covers

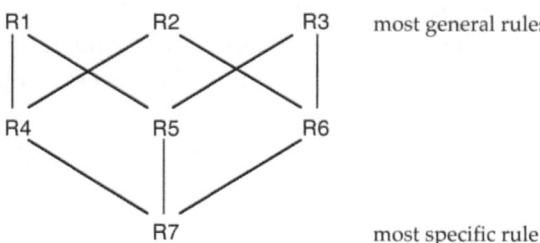

FIGURE 5.5   Rule lattice.

all the positive examples, in this case R7. Alternatively, we could look for the most general rule. Unfortunately, there is no single most general rule. The three rules R1, R2 and R3 are all "most" general in that there is no correct rule more general than them, but they are all "most" general in different ways. Figure 5.5 shows these rules as a lattice with the most general rules at the top and the most specific at the bottom.

Further examples may restrict this set of possible rules further. If one takes the next example, the otter, it swims, but is not a fish. Therefore rule R1 can be removed from the set of candidate rules. This gives rise to an algorithm:

1. start off with the set of all rules
2. for each positive example p
   2.1. remove any rules which p doesn't satisfy
3. for each negative example n
   3.1. remove any rules which n does satisfy
4. if there are no rules left FAIL
5. when there is one rule left it is the result

The only problem with this algorithm is that you have to keep track of all rules. If there are n attributes with m values each, then there are $(m + 1)^n$ rules! Clearly this is infeasible for any realistic problem.

Version spaces reduce this number by only keeping track of the most specific and most general rules: all the other possible rules lie somewhere between these. Positive examples change the set of most specific rules, forcing them to become more general in order to include the new examples. Negative examples change the set of the most general rules, forcing them to become more specific in order to exclude the new examples.

In addition, because we are looking for a single final rule we can further prune the two sets. After a positive example we examine the set of most general rules (G) and remove any that are not above (more general than) any of those in the set of most specific examples (S). Similarly, after a negative example we can prune S to remove any which are not below some rule in G.

An Example

Let's see how this would work when given the examples of tiles in Table 5.2. As a shorthand, rules will be represented by a tuple of the attributes they select. For example, the rule "**if** colour = blue **and** material = wood" is represented by the tuple (?,blue,?,wood). The question marks denote attributes which the rule doesn't test. The most general rule is (?,?,?,?), which doesn't care about any of the attributes.

TABLE 5.2   Example Tiles.

	shape	colour	size	material	
ex1	triangle	blue	large	wood	✓
ex2	square	blue	small	wood	✗
ex3	triangle	blue	small	plastic	✓
ex4	triangle	green	large	plastic	✗

After seeing the first example, the most specific rule is (triangle,blue,large,wood), which only matches ex1. The most general rule is (?,?,?,?), which matches anything. This is because we have not seen any negative examples yet and so cannot rule out anything. The state of the algorithm can thus be summarised:

set of most specific rules (S)
= { (triangle,blue,large,wood) }
set of most general rules (G)
= { (?,?,?,?) }

The second example is negative and so the set of most general rules must be modified to exclude it. However, the new most general rules should not contradict the previous examples, and so only those that are more general than all those in S are allowed. This gives rise to a new state:

set of most specific rules (S)
= { (triangle,blue,large,wood) }
set of most general rules (G)
= { (triangle,?,?,?), (?,?,large,?),
    (?,?,?,wood) }

The third example is positive. It does not satisfy (triangle,blue,large,wood), so S is generalised (again consistent with G):

set of most specific rules (*S*)
= { (triangle,blue,?,?) }

However, at this stage we can also use the pruning rules to remove the second two rules from (*G*), as neither is more general than (triangle,blue,?,?):

set of most general rules (*G*)
= { (triangle,?,?,?) }

Finally, we look at the fourth example, which is negative. It satisfies (triangle,?,?,?), so we must make *G* more specific. The only rule that is more specific than (triangle,?,?,?), but that is also more general than those in *S*, is (triangle,blue,?,?). Thus this becomes the new *G*. The set *S* is not changed by this new example.

set of most specific rules (*S*)
= { (triangle,blue,?,?) }
set of most general rules (*G*)
= { (triangle,blue,?,?) }

At this point *S* = *G*, and so we can finish successfully – which is just as well as we have reached the end of our examples!

Different Kinds of Rules – Bias

The version-space algorithm depends on being able to generate rules that are just a little more or less specific than a given rule. In fact, any class of rules which have a method of making them slightly more or less specific can be used, not just the simple conjunctions that we have dealt with so far. So, if an attribute has values that themselves have some form of generalisation hierarchy, then this can be used in the algorithm. For example, assume the shape attribute has a hierarchy as in Figure 5.6. We can then generalise from two rules (circle,?,small,?) (ellipse,?,small,?) to get (rounded?,small,?).

The rules can get even more complicated. With full boolean predicates generalisation can be achieved by adding disjunctions or turning constants into variables; specialisation by adding conjunctions or turning variables into constants. This sounds like a very general learning mechanism – but wait. If we allow more complicated rules, then the number of examples needed to learn those rules increases. If we are not careful, we end up with rules like

if *new example* = *ex*1 **or** *new example* = *ex*2
or ...

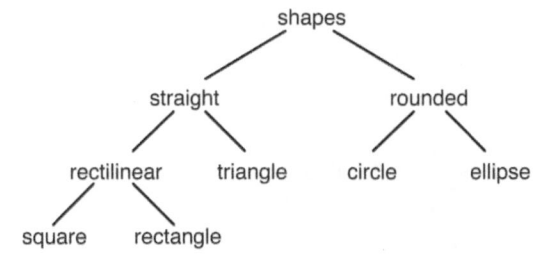

FIGURE 5.6 Shape taxonomy.

These are not only difficult to learn but fairly useless – rote learning again. This problem is called overfitting, when the rules seem to reflect accidental aspects of the training data rather than generalisable features.

A learning algorithm must have some bias – a tendency to choose certain types of rules rather than others. This reduces the set of possible rules, and in so doing makes the learning task both tractable and useful. Restricting the rules in the version-space method to conjunctions introduced just such a bias and so enabled the algorithm to learn. However, the downside of a bias is that it means that some sorts of rule cannot be learnt. In this case, we would not be able to learn rules of the form

if shape = triangle **or** colour = blue

Noise and Other Problems

The version-space method has several problems. It is very sensitive to noise – if any wrong examples are given, the algorithm will fail completely. It also demands a complete set of examples, in the sense that there must be exactly one rule that classifies them all. Finally, it is not well suited to multi-way classification (e.g. sorting animals into fish/bird/mammal). One must effectively treat these as several yes/no distinctions.

5.5.2 Decision Trees

Decision trees are another way of representing rules. For example, Figure 5.7 shows a decision tree for selecting all blue triangles. Imagine a tile coming in at the top of the tree. If it satisfies the condition at the top node, it passes down the yes (Y) branch; if it doesn't, it passes down the no (N) branch. It is passed down node by node until it comes to one of the leaves, which then classifies the tile.

Several algorithms learn by building decision trees in a top-down fashion. The most well known is the ID3 [226]

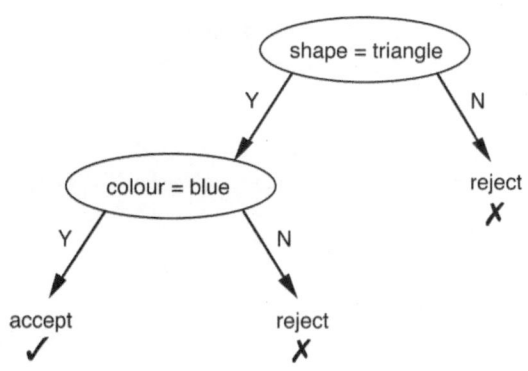

FIGURE 5.7   Decision tree.

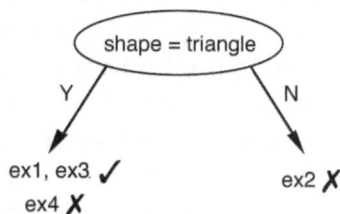

FIGURE 5.8   Starting to build a decision tree.

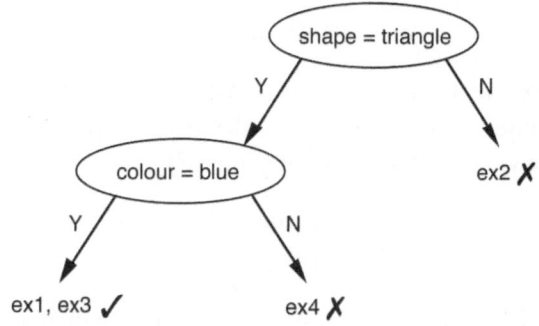

FIGURE 5.9   Completed tree.

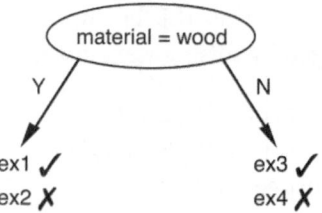

FIGURE 5.10   Starting a different tree.

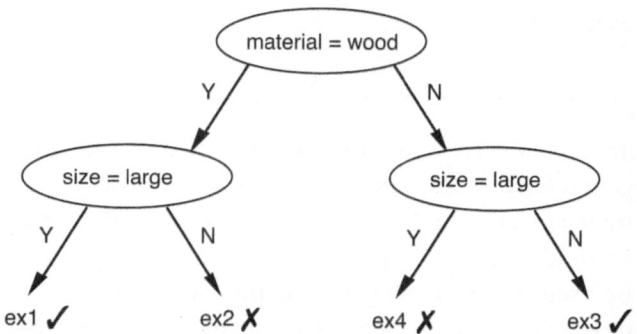

FIGURE 5.11   A different decision tree.

and its successor C4.5 [228], which use information entropy measures to choose the best decision at each stage. The example that follows uses the ID3 process.

5.5.2.1 Building a Binary Tree

Consider again the tiles in Table 5.2. We start off with the four examples and choose some condition to be the root of the tree, say "shape = triangle". Three of the tiles (ex1, ex3 and ex4) satisfy this, and one doesn't (ex2). The N branch has all negative examples, and so no further action is taken on that branch. The Y branch has a mixture of positive and negative examples, and so the same procedure is taken recursively (Figure 5.8).

We now choose another condition for this branch, say "colour = blue". The three examples are sorted by this condition and now both branches have examples of one type. At this point we stop and label the leaves in the obvious manner (Figure 5.9).

A different choice of condition at the root would lead to a different tree. For example, if we had instead chosen "material = wood", we would get to the stage in Figure 5.10. This time both branches have mixed examples, and we must build subtrees at each.

If we chose the same condition "size = large" for each branch, we would end up with the decision tree in Figure 5.11.

Note that this not only is a different tree from Figure 5.9 but also represents a completely different rule:

> **if**   material = wood **and** size = large
> **or** material ≠ wood **and** size ≠ large

as opposed to the original rule

> **if**   shape = triangle **and** colour = blue

How do we choose between these?

FIGURE 5.12 Contingency tables for different choices.

Well, one way would be to find the smallest tree (or at least one of the smallest). Unfortunately, the number of trees is huge and so an exhaustive search would be impractical. Instead, the algorithms are careful about the choice of condition at each node and use a condition that looks as though it will lead to a good tree (but might not). This decision is usually based on the numbers of positive and negative examples that are sent to the Y and N branches.

In Figure 5.12 these are tabulated for the two top-level conditions "shape = triangle" and "material = wood". In the first table, we see that the Y branch has two positive examples and one negative example giving three in total. The N branch has no positive examples and one negative example. In comparison the "material = wood" condition is very even handed with one positive and one negative example down each branch.

Of the two, the first is a more promising candidate as it makes the branches more uneven. Unevenness is important because we want the final tree to be very uneven – leaves must be either totally positive or totally negative. Indeed, one would expect a totally irrelevant attribute to give rise to an even split, as in the second table. Algorithms use different measures of this unevenness and use this to choose which condition to use at the node.

ID3 uses an entropy-based measure. The entropy of a collection of probabilities p_i is given by

$$\text{entropy} = -\sum p_i \log_2(p_i)$$

We calculate the entropy of each branch and then the average entropy (weighted by the number of examples sent down each branch). For example, take the "shape = triangle" table. The Y branch has entropy

$$-[\,2/3 \times \log_2(2/3) + 1/3 \times \log_2(1/3)\,] = 0.918$$

The N branch has entropy

$$-[\,0 \times \log_2(0) + 1 \times \log_2(1)\,] = 0$$

The average entropy is thus

$$3/4 \times 0.918 + 1/4 \times 0 = 0.689$$

(NB: When calculating entropy one assumes that $0 \times \log_2(0) = 0$. This usually has to be treated as a special case to avoid an overflow error when calculating $\log_2(0)$.)

In contrast, the entropy of the "material = wood" decision is:

$$2/4 \times -[0.5 \times \log_2(0.5) + 0.5 \times \log_2(0.5)]$$
$$+ 2/4 \times -[0.5 \times \log_2(0.5) + 0.5 \times \log_2(0.5)]$$
$$= -\log_2(0.5) = 1$$

Small values of entropy correspond to greatest disorder; hence the first decision would be chosen.

5.5.2.2 More Complex Trees

The original ID3 algorithm did not use simple yes/no conditions at nodes; instead it chose an attribute and generated a branch for each possible value of the attribute. However, it was discovered that the entropy measure has a bias towards attributes with large numbers of values. Because of this, some subsequent systems used binary conditions at the nodes (as in the above examples). However, it is also possible to modify the entropy measure to reduce the bias. Other systems use completely different measures of unevenness similar to the χ^2 statistical test. In fact, the performance of decision tree inductive learning has been found to be remarkably independent of the actual choice of measure.

As with the version-space method, decision tree building is susceptible to noise. If wrongly classified examples are given in training, then the tree will have spurious branches and leaves to classify these. Two methods have been proposed to deal with this. The first is to stop the tree growing when no condition yields a suitable level of unevenness. The alternative is to grow a large tree that completely classifies the training set, and then to prune the tree, removing nodes that appear to be spurious. The second option has several advantages, as it allows one to use properties of the whole tree to assess a suitable cut-off point, and is the preferred option in most modern tree-building systems.

The original ID3 algorithm only allowed splits based on attribute values. Subsequent algorithms have used a variety of conditions at the nodes, including tests of numerical attributes and set membership tests for attribute values. However, as the number of possible conditions increases, one again begins to hit computational problems in choosing even a single node condition. Set membership tests are particularly bad, as an attribute with m values gives rise to 2^{m-1} different possible set tests! The Query-by-Browsing example later in this chapter allows comparisons between attributes, which again increases the number of potential conditions.

While Quinlan's C4.5 algorithm and its variants are widely used, there are also alternative ways to create decision trees. For example, the decision tree can be constructed using a genetic algorithm rather than top-down, especially useful if the choices at each stage become complex. An increasingly popular alternative is forms of random decision forests [31, 134] where large numbers of trees are constructed using random subsets of attributes and the results combined.

5.5.3 Rule Induction and Credit Assignment

In both the version-space method and decision tree induction, the rules that are learnt are of the form "if condition then classify". The training can see whether a rule works simply by seeing whether the response it gives matches the desired response – that is, it classifies correctly. However, in more complicated domains it is not so easy to see whether a particular rule is correct. A classic example is pole balancing (Figure 5.13). The task is to move the railway carriage so that the upright pole does not fall over and so that the carriage stays between the buffers. At each moment, the system must choose whether to move the carriage to the right or left depending on its position and the position and velocity of the pole. However, if the pole falls over, which rule is held "responsible" – the last rule applied? In fact, in such tasks the mistake often happened much earlier, and subsequent rules might be good ones.

This problem is called the credit assignment problem. It arises in many domains. For example, in computer chess – if the computer won, which moves were the good ones? If it lost, which should be blamed? A human expert might be needed at this stage to analyse the game in order to tell the computer what went wrong.

There is no simple solution to this problem. The human expert will be useful in some circumstances, but often the nature of the problem makes this undesirable or impractical – for example a human expert would find it

FIGURE 5.13 Pole balancing.

hard to assign credit in the pole-balancing problem. If the problem domain is internal to the computer, it may be able to backtrack to each decision point and try alternatives. However, this approach will often be computationally infeasible. Sometimes there are special solutions dependent on the domain. For example, LEX, a theorem-proving program, searches for minimal proofs of mathematical propositions. All the heuristics that give rise to a minimal proof are deemed "good" – LEX assigns credit uniformly.

Humans can be equally bad at this, and one source of superstition is when we link an irrelevant thing we have done to a subsequent good or bad outcome. However, we do have means that help us including the way that events that are unusual or have obvious connection to the outcomes are more salient and more likely to be included in both conscious or unconscious learning. We will return to the lessons of human cognition for AI in Chapter 22 including the way regret can help focus learning.

5.6 EXPLANATION-BASED LEARNING

Algorithms for inductive learning usually require a very large number of examples in order to ensure that the rules learnt are reliable. Explanation-based learning addresses this problem by taking a single example and attempting to use detailed domain knowledge in order to explain the example. Those attributes which are required in the explanation are thus taken as defining the concept.

Imagine you are shown a hammer for the first time. You notice that it has a long wooden handle with a heavy metal bit at the end. The metal end has one flat surface and one round one. You are told that the purpose of a hammer is to knock in nails. You explain the example as follows. The handle is there so that it can be held in the hand. It is long so that the head can be swung at speed to hit the nail. One surface of the hammer must be flat to hit the nail with. So, the essential features extracted are: a long handle of a substance that is easy to hold, and a head with at least one flat surface, made of a substance hard enough to hit nails without damage. A couple of years ago, one of the authors bought a tool in Finland. It was made of steel with rubber covering the handle. The head had one flat surface and one flat sharp edge (for cutting wood, a form of adze). Despite the strange shape and not having a wooden handle it is recognisably a hammer.

Notice how explanation-based learning makes up for the small number (one!) of examples by using extensive domain knowledge: how people hold things; the hardness of nails; the way long handles can allow one to swing the end at speed. If the explanation is complete, then one can guarantee that the description is correct (or at least not overinclusive). Of course, with all that domain knowledge, a machine could, in theory, generate a design for a tool to knock in nails without ever seeing an example of a hammer. However, this suffers both from the search cost problem and because the concepts deduced in isolation may not correspond to those used by people (but it might be an interesting tool!).

In addition, explanations may use reasoning steps that are not sound. Where gaps are found in the explanation an EBL system may use abduction or induction to fill them. Both forms of reasoning are made more reliable by being part of an explanation.

Consider abduction first. Imagine one knows that hitting a nail with a large object will knock it into wood. If we have not been shown the hammer in use, merely told its function, we will have to use an abductive step to reason that the heavy metal head is used to knock in the nail. However, the match between features of the example and the possible cause makes it far more likely that the abductive step is correct than if we looked at causes in general (e.g. that the nail is driven into the wood by drilling a hole and then pushing it gently home).

Similarly, the inductive steps can be made with greater certainty if they are part of an explanation. Often several examples with very different attributes require the same assumption in order to explain them. One may thus make the inductive inference that this assumption is true in general.

Even if no non-deductive steps are made, explanation-based learning gives an important boost to deductive learning – it suggests useful things to learn. This is especially true if the explanation is based on a low-level, perhaps physical, model. The process of looking at examples of phenomena and then explaining them can turn this physical knowledge into higher-level heuristics. For example, given the example of someone slipping on ice, an explanation based on physical knowledge could deduce that the pressure of the person melted the ice and that the presence of the resulting thin layer of water allowed the foot to move relative to the ice. An analysis of this explanation would reveal, among other things, that thin layers of fluid allow things to move more easily – the principle of lubrication.

5.7 EXAMPLE: QUERY-BY-BROWSING

As an example of the use of machine learning techniques we will look briefly at Query-by-Browsing (QbB). This is an experimental "intelligent" interface for database that uses an extension of ID3 to generate queries for the user. This means that the user need only be able to recognise the right query, not actually produce it.

5.7.1 What the User Sees

Initially Query-by-Browsing shows the user a list of all the records in the database. The user browses through the list, marking each record either with a tick if it is wanted or a cross if it is not (see Figure 5.14). After a while the system guesses what sort of records the user wants, highlights them and generates a query (in SQL or an appropriate query method). The query is shown in a separate window so that the user can use the combination of the selected records and the textual form of the query to decide whether it is right (Figure 5.15).

Whereas so-called Query-by-Example works by making the user design a sort of answer template, Query-by-Browsing is really "by example" – the user works from examples of the desired output.

5.7.2 How It Works

The form of examples used by ID3, attribute–value tuples, is almost exactly the same as that of the records found in a relational database. It is thus an easy job to take the positive and negative examples of records selected by the user and feed them into the ID3 algorithm. The output of ID3, a decision tree, is also reasonably easy to translate into a standard database query.

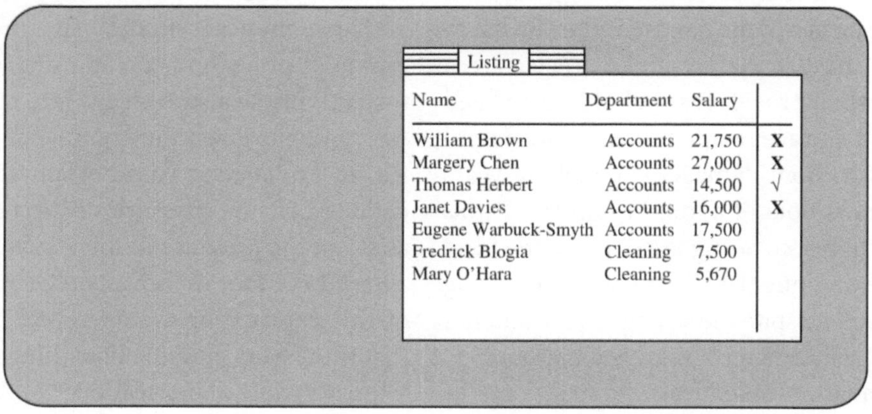

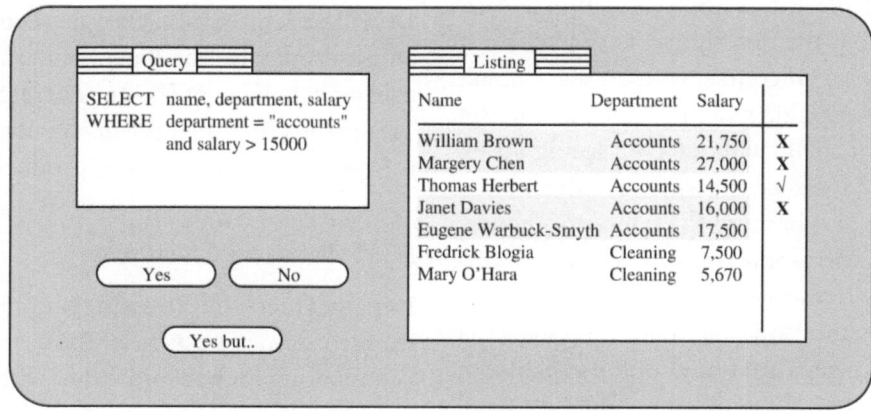

FIGURE 5.14   Query-by-Browsing – user ticks interesting records.

FIGURE 5.15   Query-by-Browsing – system highlights inferred selection.

In fact, QbB uses a variant of the standard ID3 algorithm in that it also allows branches based on cross-attribute tests (e.g. "overdraft > overdraft-limit") as these are deemed important for effective queries. Otherwise the implementation of the basic system is really as simple as it sounds.

5.7.3 Problems

Even a very simple implementation of QbB works very well – when the system gets it right. When it doesn't, things are rather more complicated. First of all the algorithm produces some decision tree which correctly classifies the records. However, there are typically many such trees. Sometimes the system produces a "sensible" answer, sometimes not. Although the answers are always "correct", they are not always the sort a reasonable human expert would produce. When QbB gets the wrong answer, the user can tell it and give more examples to help clarify the desired result. At some point the system generates a new query. However, the algorithm used starts from scratch each time, and so there may be no obvious relationship between the first attempt and subsequent guesses. Although the earlier queries were wrong, the resulting behaviour can appear odd and reduce one's confidence in the system.

The above problems can be tackled by modifying the algorithm in various ways, but the lesson they give us is that applications of machine learning must do more than work, they must work in a way that is comprehensible to those who use them. Sometimes the machine intelligence can be hidden away in a "black-box", where the mechanisms are invisible and hence don't matter, but more often than not someone will have to understand what is going on. This is a point we shall return to in Chapter 21.

5.8 SUMMARY

In this chapter, we have discussed the importance of machine learning, its general pattern and some of the issues that arise. Several specific machine learning methods have been described, including deductive learning, inductive learning and explanation-based learning. In particular we have examined two inductive learning algorithms: the version-space method and ID3. We ended the chapter with a discussion of an experimental system that uses machine learning in an intelligent database interface.

5.1   Apply the version spaces algorithm in Section 5.5.1 to the 'Fish and non-fish' training data in Table 5.1.

5.2   Consider the following vignette.

"The cook book said to use a whisk to beat egg white in a bowl until it is fluffy. I didn't have whisk, but the book had a picture of one alongside other tools. I could see it was similar in length to table cutlery. One end is tightly wrapped in wire, and looked easiest to hold. The other has wire loops with gaps. I guessed that you hold the tight end and then use your wrist to move the end with the loops quickly so that it mixes the egg white and gets air into it. Although I hadn't got a whisk, a fork looks a similar size and has gaps between the prongs, so I used that instead."

Identify the explanation-based learning applied in the vignette and the sources of background knowledge being applied.

5.3   In this exercise you will build decision tables using the 'Fish and non-fish' training data in Table 5.1.

   a.   First use the different column criteria in the order they appear: 'swims' for the top level decision, 'has fins' for the next, etc.

   b.   Now calculate the contingency table for each column and use this to choose the top-level decision.

   c.   Which gives the better tree?

FURTHER READING

J. D. Kelleher, B. Mac Namee, and A. D'arcy. Fundamentals of machine learning for predictive data analytics: Algorithms, worked examples, and case studies. Cambridge, MA: MIT Press, 2020.

A structured overview of machine learning techniques, illustrated throughout with concrete examples and cases studies.

Y. Anzai. *Pattern recognition and machine learning.* Academic Press San Diego, 1992.

Classic book that deals in detail with many of the algorithms discussed here including concept learning and decision trees.

V. Maini and S. Sabri. *Machine learning for humans,* 2017. https://everythingcomputerscience.com/books/Machine%20Learning%20for%20Humans.pdf

A short (approx. 100 pages) overview of machine learning. It is not as detailed as the other ML books but is very accessible and free to download.

Neural Networks

6.1 OVERVIEW

The earliest forms of AI focused on higher level human thinking, such as problem solving and tasks such as chess playing or mathematical proofs, which were thought to embody the highest forms of human intelligence. Some would still reserve the term AI for this kind of system. However, from the late 1980s and early 1990s a new form of AI took shape, based on a connectionist model of cognition and known as *neural networks*. These are now so common that for many people they are the first thought when the term "AI" is mentioned.

Rather than attempting to model the cognitive architecture and processing of the human mind, these systems use the physical architecture of the brain as inspiration. Supporters of this approach argue that we do not understand enough about cognitive processes to model them effectively. However, we do know how the brain operates at this lower, physical level. The idea is that if we can simulate the way the human brain operates, we may achieve some of its power in complex problem solving. Whether or not they faithfully emulate the way the brain actually works, neural nets have certainly proved effective in practical applications.

There are many kinds of neural network. In this chapter we will consider four early models: the multi-layer perceptron, associative memory, Boltzmann machines and Kohonen's self-organising network. These form the basic units of more complex networks. In Chapter 8 we will look at deep learning, which has, in large part, been

the technology which has established the idea of AI as an almost universal solution.

6.2 WHY USE NEURAL NETWORKS?

The brain consists of billions of small, basic processing units, called neurones. Each is connected to thousands of others, forming a rich network. The basic operations performed by each neurone are simple: summing the inputs received in some way and "reacting" if these exceed a certain level. The connections between neurons can adapt to reinforce those that are successful and to degrade those that are not. The power of the brain, therefore, is not in complex processing units but in the parallel operation of billions of simple units and the ability to adapt the configuration of these. Neural networks attempt to model this brain architecture, although current networks comprise hundreds rather than billions of neurons.

Connectionist models account for aspects of human thinking such as parallelism, the ability to do more than one thing at once, and graceful degradation, where the mind is able to operate even if impaired by fatigue or damage. Like the brain, a neural network consists of a network of simple processing units, all interconnected. Learning occurs through changes in the connections, and the configuration of connections constitutes the knowledge of the system. Because this knowledge is distributed among the units, the network is fault tolerant and performance degrades only gradually with damage.

6.3 THE PERCEPTRON

The roots of neural networks date back to 1960s when early neuroscience was beginning to piece together the behaviour of neurons and realise that they had parallels with electronic computers [112]. This lead to the idea of a perceptron (a single artificial neuron)

DOI: 10.1201/9781003082880-8

as in Figure 6.1 [198]. Each input is multiplied by the weight on its connection, which is set randomly to start with. The weighted inputs are summed by the neuron and compared with a threshold value. The simplest thresholding function is the step function where the response is "on" if the threshold is exceeded and "off" otherwise. The perceptron learns by adjusting the weights to reinforce a correct decision and discourage an incorrect one.

However, the single perceptron has major limitations and can only solve very simple problems. To illustrate this, imagine you have a group of dogs, some of which work as rescue dogs, some as sheep dogs. Your job is to assign them to the correct "class". Thinking about the characteristics of these two groups, you may decide that weight and speed are suitable measures to distinguish them, given that rescue dogs, often St Bernards and Newfoundlands, tend to be larger and slower than sheep dogs (assume that you do not have information about the dogs' breeds). You could plot the weight and speed of each dog on a graph as in Figure 6.2.

Looking at this graph, you can see two definite clusters (which you assume represent your two classes). You can in fact draw a straight line between these clusters and say that any point on one side of the line represents a sheep dog and every point on the other a rescue dog (as in Figure 6.3). The problem is *linearly separable* (see also Chap 7).

Unfortunately life is rarely as simple as that. In reality our pattern space is unlikely to be so neat and ordered. We may find when we ask our dogs' owners to identify their dogs' occupations that some of those we identified as sheep dogs are in fact rescue dogs (perhaps search and rescue collies). Similarly, some we thought were rescue dogs may in fact work as flock protection sheep dogs and so be on the larger side. So our graph may really look like Figure 6.4.

It is no longer possible to draw a straight line between the two groups; the problem has become linearly inseparable. Our simple perceptron can solve problems that are linearly separable but not those which are linearly inseparable, by far the more significant group.

6.3.1 The XOR Problem

One problem that is linearly inseparable is the exclusive OR or XOR function. Given two inputs (X and Y), which can be true or false, the XOR function returns true if either of the inputs is true, but false if both are true or both are false.

X	Y	Output
True	True	False
True	False	True
False	True	True
False	False	False

If we plot this on a graph, we will quickly see that it is linearly inseparable. In the graph in Figure 6.5 we represent true as 1 and false as 0.

Because this captures such a basic limitation of the perceptron, the so-called 'XOR problem' stalled further work on artificial neurons as a computational mechanism for many years.

6.4 THE MULTI-LAYER PERCEPTRON

The renaissance of neural networks came more than 20 years later in the 1980s with the development of the multi-layer perceptron [186] and the backpropagation learning algorithm.

The solution is two-fold: (i) link perceptrons together in layers so that different units can solve small parts of the problem, and (ii) combine the results and use a non-linear thresholding function where the neuron's value is not just 1 or 0 but can take values within a range. The resulting model is the multi-layer perceptron.

The standard multi-layer perceptron model has three layers: an input layer, an output layer and a single hidden layer that is not directly connected to inputs or outputs (see Figure 6.6). In the simplest case every input node is connected to every hidden node, and every hidden node is connected to every output node (fully connected), but there are alternatives with sparse connections.

The output and hidden layer units act like perceptrons (but with a new thresholding function); the input layer distributes the inputs through the network and so does not threshold. This implementation is able to solve linearly inseparable problems.

Crucially a multi-layer perceptron can solve the XOR problem. A simple network which does this is shown in Figure 6.7. It has two input units (for the two inputs X and Y) and one output unit (the output is either 0 or 1). The network also has two hidden units. Work through the network by hand and convince yourself that it does indeed solve the XOR problem (in this case the weights are multiplicative and the threshold function is a simple

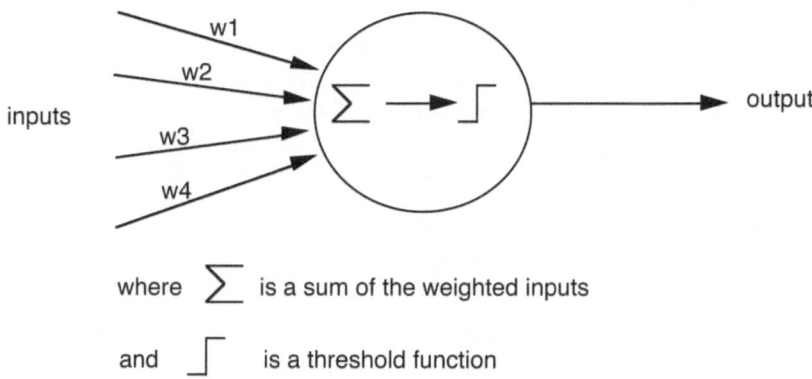

FIGURE 6.1   A single perceptron.

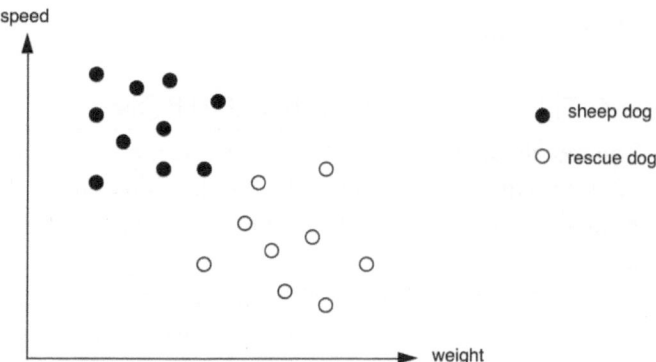

FIGURE 6.2   Sheep dogs or rescue dogs?

step function – if the sum is greater than the threshold output 1, otherwise output 0).

An early application was NETtalk [251] which used a multi-layer perceptron to pronounce English text. The network had 203 input units, 80 hidden units and 26 output units. The output units represented phonemes, the basic sound unit of the language. The network was presented with text in blocks of seven letters and learnt to pronounce the middle letter. It used the surrounding letters as context to distinguish between different sounds for the same letter. During the training phase the system appeared to mimic the speech sounds of young children. When the weights are random, the sounds are meaningless babble. As the network learnt it first produced the main sounds of English, finally producing intelligible speech. Listening to a tape recording of NETtalk in training is not unlike listening to a child learning to talk – speeded up of course!

In some ways modern language models, such as OpenAI's GPT-3 [34], are merely bigger versions of this,

albeit vastly larger both in numbers of network units (175 billion parameters for GPT-3 compared with about 20,000 for NETtalk). However, the crucial difference is that language models now include many hidden layers (96 for GPT-3 compared with a single hidden layer); that is they are deep neural networks.

6.5 BACKPROPAGATION

We said there were two aspects that enabled the multi-layer perceptron to work effectively. The first was the hidden layer(s) that enabled *more complex problems* to be solved. The other was the non-linear threshold function, this enabled the creation of an *effective learning algorithm* – backpropagation (often abbreviated as backprop).

Figure 6.8 shows on the left a simple step threshold function; if the input is below a certain value, it is zero, if it is higher, it is one. On the right is a sigmoid activation function that rises smoothly from zero to one, more steeply towards the centre, but without any discontinuous steps. The term 'sigmoid' comes because it is vaguely like a squashed 'S'.

This particular sigmoid is a logistic function, given by the equation:

$$S(x) = \frac{1}{1 + e^{-x}}$$

However, there are other variants. They are usually rotationally symmetric $S(x) = -S(-x)$ but critically have the following properties:

monotonic – rise from a lower to an upper bound, for neural networks usually zero to one.

asymptotic – become flat for large positive or negative inputs. This means that even quite substantial

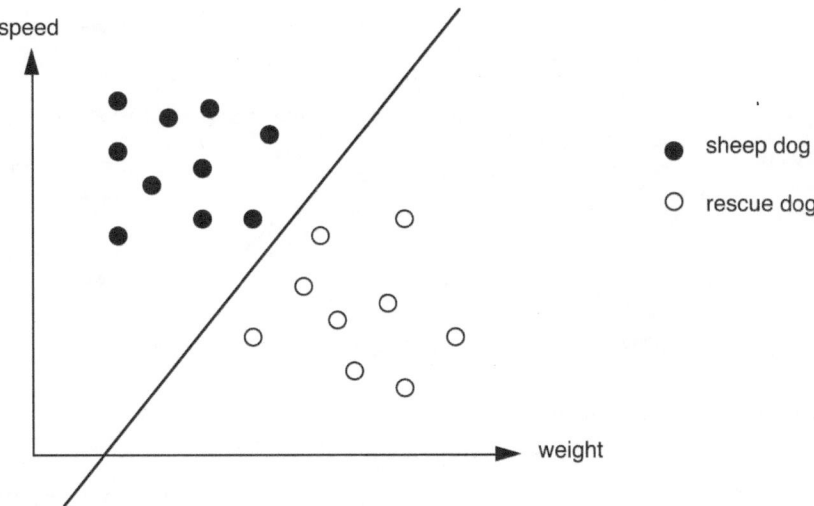

FIGURE 6.3   A linearly separable problem.

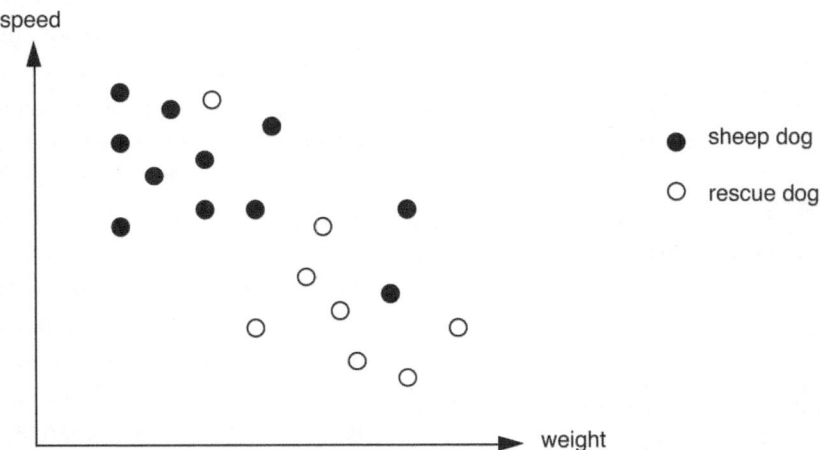

FIGURE 6.4   The actual pattern space.

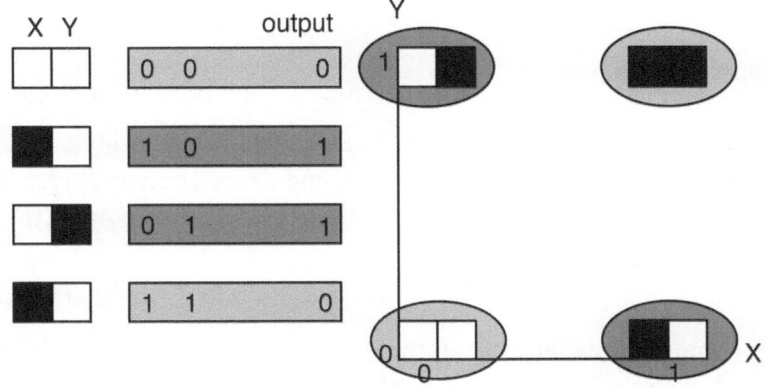

FIGURE 6.5   XOR problem (diagram after Beale & Jackson (1990)).

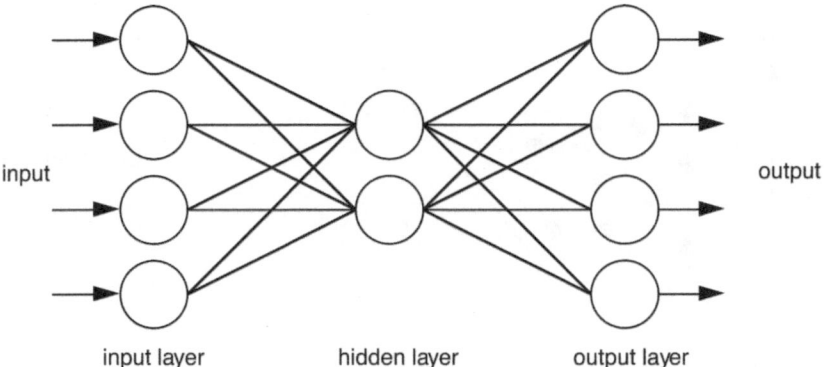

FIGURE 6.6   A multi-layer perceptron architecture.

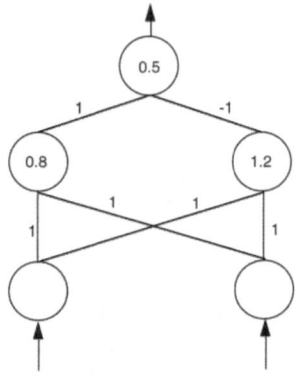

FIGURE 6.7   A simple multi-layer perceptron to solve the XOR problem.

changes in the inputs make little difference when they are at one or other extreme.

linear centre – have a steep but near-linear region towards the centre, creating a 'soft' threshold. In this region changes in the input make proportionate differences to the output.

continuity – have no sharp step changes.

Backpropagation is a supervised learning algorithm, which uses these properties to train the network given examples of input and desired outputs.

6.5.1 Basic Principle

Initially the weights are randomly assigned and the network is trained through repeated presentations of expected input and output. It learns by adapting the weights on the inputs to reinforce connections that

result in the correct output, until all the outputs are correct. The weights then remain stable, and the network is able to work on unseen input.

Each time an example input–output pair is presented to the network, the algorithm compares the actual output of the network from the given inputs with the desired output and calculates the difference (the error or delta). This says how much we'd have liked each output node to have been different and thus gives the information to start to change the weights. This is fairly straightforward for the final output layer but more complicated for hidden layers. To make this tractable backpropagation first works out the weight changes needed at the output layer but also calculates how the outputs of the layer below should change. The process is then repeated for each layer, at each stage propagating the error and desired change backward through the network.

Note that backpropagation is a form of hill climbing algorithm where one is trying to make small adjustments in the best direction to improve the results (get higher on the hill).

6.5.2 Backprop for a Single Layer Network

Let's look first at a single node with weights W_k, inputs I_k and desired output *Target*. The difference between the desired and actual output (the error or delta) is

$$\delta = Target - S\left(\sum W_k I_k\right)$$

Ideally we would like δ to be zero, and in order to make it smaller we adjust each of the weights a small amount $\hat{W}_k$ to reduce it. However, given the sigmoid activation function S is flatter at the extremes, we need to make

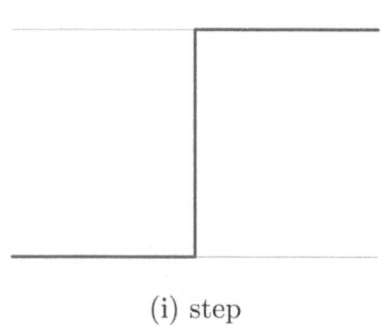

(i) step   (ii) sigmoid

$$S(x) = \frac{1}{1 + e^{-x}}$$

FIGURE 6.8   Different threshold functions.

bigger adjustments at those ends. To do this we calculate the slope (in mathematical terms differential) of S and then use this to work out how much each weight should change:

$$\hat{W}_k = \frac{\lambda \times \delta}{slope \times I_k}$$

Note that the change is bigger if δ is larger, because we want to make a bigger change in the output. However, it is smaller when the slope is larger, that is when we are close to the threshold and smaller changes make a greater difference. The change in the weight is also smaller when its associated input, I_k, is larger as this again means small differences in the weight have a larger impact.

The value λ is known as a relaxation term. It is usually quite small and determines how fast we change the weights. If we change them too slowly, we need to present the same examples many, many times before the network learns. However, if we set it too large, there is a danger that the weights may overshoot and the network bounce around and never settle down (converge) to a stable set of weights.

For a single-layer neural network this can be applied to every node.

6.5.3 Backprop for hidden layers

The above method works for the output layer, but what about the hidden layer?

Note that the single-layer step worked by comparing the desired output with the actual output to give a desired change. The outputs for the hidden layer are the inputs to the output layer.

We can use the same method that we used for calculating the change in the weights to see how much we'd like the inputs to the output layer to change in order to reduce the final error.

$$\hat{I}_k = \frac{\delta}{slope \times W_k}$$

Of course a single hidden layer node may be the input to many (or all) of the output layers, some of which might like it larger, some smaller. The above values are summed for the node to give an overall desired direction of travel (using the hill-climbing analogy) for the hidden layer's output. This is then used as the 'delta' value for the hidden nodes.

If there are multiple hidden layers, this process can be repeated again. However, for early networks it was rare to have more than one hidden layer, unless the other layers used different forms of algorithm for training. One reason for this is obvious, each layer adds more nodes and more weights, thus increasing computational cost. More critically, as the layers get 'further' from the input and output, they are typically less stable and less likely to converge to a final value. This is partly because the hidden layers are often underdetermined, there are many different ways in which the hidden layer weights could be assigned that give equally good answers; in particular for a fully connected network any shuffling of the hidden layer nodes is as good as any other. One can think of this a bit like moving furniture with a friend. When you move the bookshelf, it may be heavy but is easy to manoeuvre, but when moving the mattress, it is all floppy in the middle.

It is possible to adjust for this by using a very small value of the relaxation constant, or vary this dependent on the layer or dynamically during training. However, this then means you need many more presentations of the training data or very big data sets.

6.6 ASSOCIATIVE MEMORIES

Association of ideas is a familiar concept to us. We may associate a particular piece of music with a person or event, or we may associate a person with an activity. There are many examples in everyday life where we use association to remember things. Indeed, it is fundamental to models of human memory such as semantic networks.

An associative memory is a neural network that models the associative nature of human memory, by which a particular stimulus triggers a particular response [4, 155]. In the associative memory model, an input is stored with the required output, in such a way that when this input (or an incomplete version of it) is presented to the memory, the appropriate output is recovered.

There are two types of associative memory: heteroassociative memory, where input is associated with a different output pattern; and autoassociative memory, where the input is associated with itself.

In a heteroassociative memory, the input pattern is associated with a different output pattern, for example a class identifier. In this case, when the input pattern is encountered again, the class with which it is associated is returned, allowing the network to perform effectively in classification problems.

An autoassociative memory can be used to filter and "clean up" distortion in images, the latter for classification problems. In an autoassociative memory, the network is trained with the same pattern as both input and expected output. When this pattern (or a partial version of it) is presented to the memory, the stored pattern is retrieved. This allows the memory to deal with noise and distortion in patterns, as in Figure 6.9 below.

In more recent literature, especially concerning deep neural networks, an autoassociative memory is also called an autoencoder and these have become a central part of several practical technologies. If the autoencoder has some sort of layer or set of neurons that is smaller than the input space, then it can be used as a form of compressed representation of the input. This is often useful as an input to other layers of a network.

The ability to regenerate images from partial images is also a key part of deep fakes, both in the negative sense of fraudulent or exploitative imagery (see Chapter 20), but also legitimate uses in entertainment to allow dead actors to be 'brought back to life' in new films.

6.6.1 Boltzmann Machines

Boltzmann machines are an early form of neural network that is often used as an autoencoder [1]. They are important analytically because of their strong mathematical basis related to the spin glass models in physics, theoretically because of their mapping to plausible brain mechanisms, and practically because they are often used as a first stage of multi-layer neural networks.

In a Boltzmann machine nodes are normally binary valued (0/1) and are split into two classes:

visible – this includes both inputs and outputs, or in autoencoder mode, an image to be remembered.

hidden – used as part of the process of reconstructing the visible nodes during recall.

In the simplest Boltzmann machine, the visible and hidden nodes are all fully connected to each other (see Figure 6.10), not arranged in layers like a multi-layer neural network. The connections each carry a weight, and this is effectively the memory of the neural network. These are initially randomly assigned.

Training consists of two phases.

In the first phase the visible nodes are clamped to an example input and the hidden nodes modified so that the network settles into a 'lowest energy' state. After this phase, when a visible node and a hidden node are both active (value = 1), this is treated as a positive association between the nodes.

In the second phase both visible and hidden nodes are allowed to alter, again settling into a 'lowest energy' state. This can be repeated with different random initial values for the nodes. After this phase if a visible and hidden node are both active, this is regarded as a negative association.

The 'lowest energy' state is related to the analogy with physical systems including the way magnetic poles partially align when a ferrous metal is cooled quickly; however, there is also a neural analogy. If the sum of weighted inputs to a neuron is higher than a certain threshold, then it (is likely to) change to an active state, thus influencing the inputs of others. With some additional mechanisms to ensure convergence, the system settles into a steady state.

After these phases have been applied to one or more examples, the weights are updated, incrementing them where there is positive association and decrementing them where there is negative association. The former

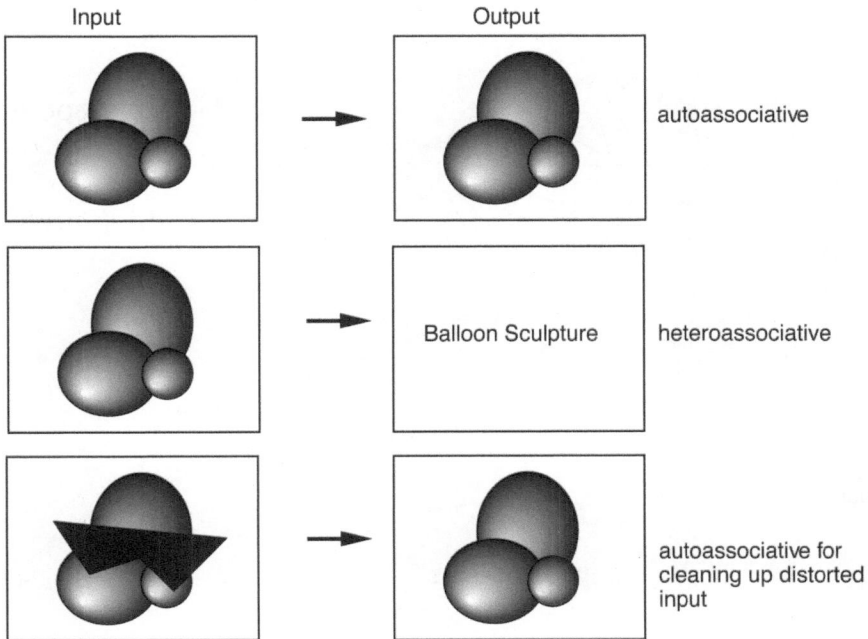

Input   Output

autoassociative

Balloon Sculpture   heteroassociative

autoassociative for
cleaning up distorted
input

FIGURE 6.9   Associative memory.

has an obvious biological analogy to Pavlovian learning – neurons that fire together stay together. The latter is often regarded as a kind of sleeping, where the mind is free to meander without sensory input [2, 225]. If nodes are associating based on random inputs, they are not storing 'useful' information, so their relationship should be weakened.

After the Boltzmann machine has been trained, it can be used to reconstruct partial inputs (often images) by clamping the known visible units to the known part of the input and letting the remaining visible units settle into reconstructed values using the same energy minimisation process as used in training.

If there are fewer hidden nodes than visible nodes, the values of the hidden nodes when the visible nodes are clamped can be regarded as a compressed or more abstract representation of the visible nodes.

A restricted Boltzmann machine (RBM) is identical except that the nodes are not fully connected [132, 133]. Instead, just the visible nodes are fully connected to the hidden nodes (see Figure 6.11), but there are no internal connections within the visible nodes, nor between the hidden nodes. This at first appears similar to a multi-layer perceptron, but in the RBM, the connections are two way. This does not matter during the phase when the visible nodes are clamped to the inputs but makes a difference during the sleeping phase or during reconstruction.

The layered structure means that RBMs can be stacked with the hidden nodes of one layer forming the visible nodes of the next layer. These can be used to create layers of abstraction. However, an RBM can also be used as the input layer to other forms of neural network or machine learning, notably as the first stage of a deep neural network (Chap. 8).

6.6.2 Kohonen Self-organising Networks

The previous methods we have encountered are all forms of supervised learning, because they are trained with an input and its desired output. In contrast, Kohonen networks are a form of unsupervised learning. Kohonen networks cluster the inputs into classes, according to common features [155] without any need to pre-label the inputs. This is often called a self-organising map or self-organising network. The idea is to emulate the ability of the human mind to make sense of unknown situations.

In a Kohonen network, neurons are not arranged in layers but in a flat grid (Figure 6.12), and all inputs are effectively connected to all nodes. The aim is that areas of the network form local neighbourhoods that act as feature classifiers (clusters or classifications) for the input

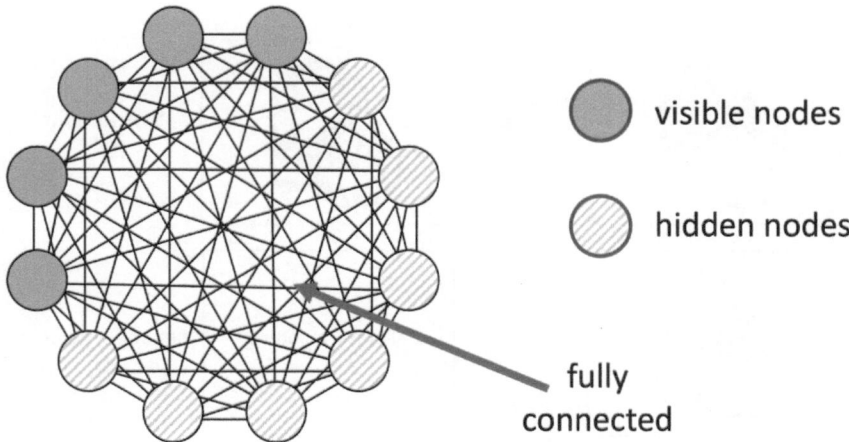

FIGURE 6.10 Boltzmann machine.

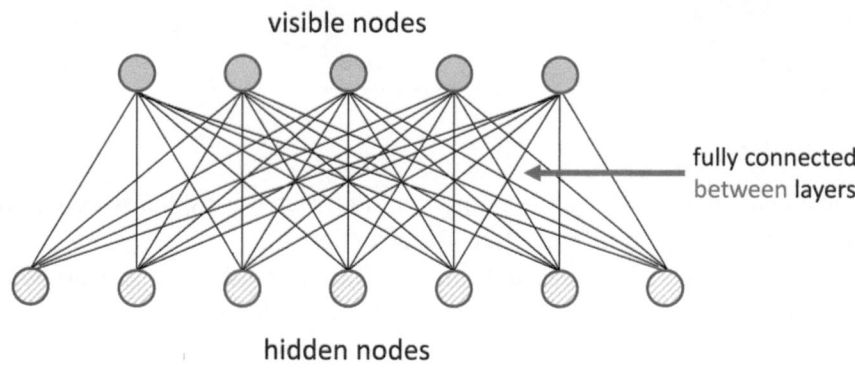

FIGURE 6.11 Restricted Boltzmann machine.

data and that close classes tend to end up close on the 2D grid.

At the start of training the network is initialised, usually with random weight vectors. During training each training example is compared with the weight vector at each of the nodes and the closest match 'wins'. A 'match' here can be that the vectors are close to one another or it may use an indirect measure such as a similarity matrix. The weight vector at the winning node is moved closer to the training example, and also neighbours are moved closer but by a lesser amount. The exact spread among neighbouring nodes may have a classic Mexican hat shape, growing smaller further away, and may sometimes have negative weights, that is close but not immediate neighbours may be moved further away to help reinforce distinctions. As training progresses the diameter of the neighbourhood may also be reduced so that early learning is quite vague and fuzzy but becomes more precise as the training progresses.

An early application of Kohonen's self-organising network was to perform speech recognition, in the form of a phonetic typewriter, a typewriter that could produce text from dictation. The network was used to cluster the phonemes into similar sounds, which could then be manually labelled. The phonetic typewriter is an example of an application that uses both neural and more conventional knowledge-based techniques, the neural network being used to preprocess the input to facilitate use of the knowledge base. This demonstrates an important point about connectionist models: although they were proposed as models of cognition, like production systems they can also be used for practical AI problem solving.

6.7 LOWER-LEVEL MODELS

Neural networks take inspiration from the function of human and other brains but typically simplify or ignore certain details. For example, it was found that to

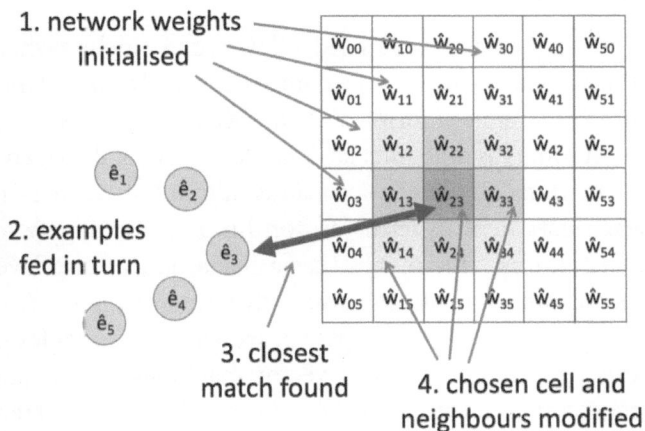

1. network weights initialised

2. examples fed in turn

3. closest match found

4. chosen cell and neighbours modified

FIGURE 6.12 Kohonen self-organising network.

model the function of a *single* cortical neurone a deep neural network was required with 5–8 layers and many thousands of artificial neurons – clearly a real neurone is more complicated than the simple sum-then-threshold perceptron.

Those interested in modelling actual brain function will often employ models that do involve greater levels of neural fidelity, and also these can sometimes be used to inspire variations in practical neural networks. We'll take a quick look at a few of these features.

6.7.1 Cortical Layers

When you look at the surface of the brain, it has a deeply folded structure. This is because the outer part, the cerebral cortex, is effectively a large sheet of neurons, a bit like a deflated balloon. The sheet of neurons is arranged in six rough layers, with different kinds of neurons in each layer. The majority of the neurone cell bodies in the brain (grey matter) are found within these layers.

At first this biological arrangement seems to justify multi-layer neural models and especially deep neural networks, which we'll see more of in Chapter 8. However, the patterns of connection are both more structured and more complicated than those usually found in artificial neural networks.

Within the cortex the cells have a column-like structure, where the connections are relatively local either up and down the columns or side-to-side, but the former are not limited to connecting to the 'next' layer. Furthermore as well as connections within this grey matter, there is a mass of nerve fibres deeper inside the brain, the white matter, which enables long distance connectivity between regions of the brain.

There are ongoing attempts to map and understand these rich patterns of interconnections and fMRI scans are revealing some of the complexity of cortical layering [160]. There are also attempts to model this computationally, in particular the large EU-funded Human Brain Project [183]. Crucially as the second edition of this book is being written, the first exascale computers have come online. This is the computational power that *in principle* would allow the real-time modelling of a complete human brain.

At a smaller scale, others have attempted to emulate the column-like structure both as a general purpose neural-network architecture [26] and for specific purposes including visual object recognition [125].

6.7.2 Inhibition

In most multi-layer neural networks the weights between layers can be negative, leading to a level of inhibition, where the firing of one neuron prevents the firing of another. The (fully connected) Boltzmann machine has potentially negatively weighted connections within a layer which allows a level of lateral inhibition as does the shape of the reinforcement function in a Kohonen network. However, by and large, rich connections within layers in general and lateral inhibition within a layer in particular are rare in many artificial neural networks, largely because of the complexity of learning.

However, it has long been known in neural science that inhibition structures are critical for many aspects of human motor control and perception [113]. Crucially patterns of neurons with mutual inhibition can give rise to spontaneous oscillations that are crucial for internal functions such as heart beat and external activity including locomotion and muscle control. Lateral inhibition is also central in efforts to use neural networks to understand the human visual system [39].

If we think back to the network reasoning structures in Chapter 3, there are negative associations between concepts. Furthermore, when we are attempting to disambiguate meanings of words or identify an object, it is important that alternative meanings compete with one another. That is, for the structures we need for semantic meaning mutual inhibition is key.

6.7.3 Spiking Neural Networks

The behaviour of a single perceptron and the connections between neurons on neural networks emulate the way that if a neuron has sufficient stimulus, it 'fires', triggering or inhibiting other neurons. In most artificial networks this firing is in lock-step with all neurons effectively firing simultaneously, and furthermore the firing is given a continuous value representing its strength.

The reality is a lot more messy! When a neuron is sufficiently excited, it does start to fire across a synapse to another neuron, but this is not a single coordinated value, rather a series of bursts of ionic activity, or spikes. It is the rate of these spikes which determines the amount of activation being transferred from one neuron to another. Furthermore there are often temporal dynamics, for instance hysteresis effects whereby if a spike is delivered across a synapse for a while it is easier for the next burst of activation to transfer, and then over slightly longer timescales chemicals deplete so it becomes harder to transmit for a while.

Spiking neural networks attempt to capture some of this complexity by emulating the dynamic spiking activity between neurons. The area was particularly active in the mid-2000s in the hiatus before deep neural networks began to deliver results, and so alternative forms of neural networks were being investigated. However, work in the area has continued at a lower but more sustained level. We will see in Chapter 22 that spiking neural networks alongside mutual inhibition may hold promise for disambiguation in rich semantic networks.

6.8 HYBRID ARCHITECTURES

Sometimes a single form of machine learning is used for a problem, but more often for practical uses a combination of techniques are used. Some parts may involve machine learning, others may be rule-based. Most often these hybrid architectures (or hybrid systems) involve some form of neural network, though not necessarily so.

Note the term hybrid simply means a combination of two things and so has other uses even within AI. Crucially data that includes a combination of discrete and continuous values may also be referred to as hybrid. However, the full term hybrid architecture should be unambiguous within AI.

6.8.1 Hybrid Layers

Often systems are organised as layers or a pipeline where the outputs of one layer feed into the next. The layers may use very different forms of AI or other algorithms.

In some cases the first layer acts as a form of data transformation. The lower layer adds additional richness to each data item in terms of non-linear combinations of the features. This means that a machine learning layer above can be relatively simple in terms of the *kinds* of things it can do. In Chapter 7, we will see examples of this, where both support vector machines (SVM) and reservoir computing use two layers with very different computational properties; in these cases, it is fundamental to the respective paradigms, as the lower layer does not in itself learn.

Another frequent form of hybrid architecture is to use some form of self-organising map (SOM), clustering or unsupervised learning at a lower level that acts as a kind of data reduction. This is particularly useful for datasets where each item is very large, such as an image or, as in the example below, a finely sampled time series. The resulting classes (e.g. in the case of clustering) or vector of values (e.g. for Kohonen nets) are then fed into an algorithm that works well with smaller numbers of features, such as a decision tree.

We saw how an early example of the use of Kohonen nets was to reduce sound sequences to phonemes. Figure 6.13 shows the main stages of processing and data at each level. There are a few things to note from this example that are common in hybrid architectures:

1. The input to the Kohonen network was not raw sound but transformed into a form of frequency space (using a variant of Fourier analysis) – that is there may be multiple layers of different complexity.

2. The phoneme labelling for the Kohonen network was added by hand, so that the vector output was reduced to a classification. This need not be the case and the whole process may be automatic, but this mixed machine–human training process (hybrid in a different sense of the word) is very common.

3. The low-level transformation (into frequency space) and the high-level recognition (phoneme sequences to words) are also hand-crafted rather

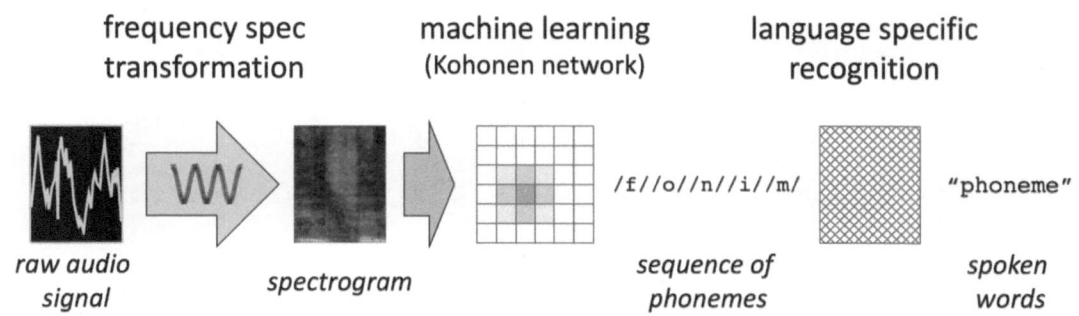

frequency spec transformation

machine learning (Kohonen network)

language specific recognition

raw audio signal

spectrogram

/f//o//n//i//m/

"phoneme"

sequence of phonemes

spoken words

FIGURE 6.13 Example of a hybrid architecture.

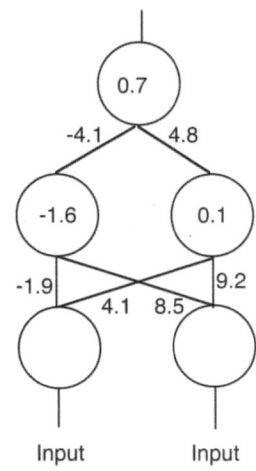

Input   Input

FIGURE 6.14 Does this network solve the XOR problem?

than using machine learning. The final system is fully automated, but many of the choices in the learning are human.

Even if the process is fully automated at the level of learning (no additional human labelling of intermediate representations), it will often be the case that the lowest level of transformation is predetermined (e.g. SVM, frequency space), the mid level is generating some sort of Gestalt understanding of the input as a whole, rather like the way we see a scene and recognise people's faces without consciously being aware of it, and the higher levels are using methods that are more comprehensible (e.g. decision trees, simple rule-systems or linear discriminant).

6.8.2 Neurosymbolic AI

Neurosymbolic AI is the general term used when symbolic and sub-symbolic aspects are combined in the same system. This may include systems, such as those described above, where there are layers or modules interacting loosely, but usually suggests much deeper integration. There are several styles including:

data transformation – In the example of hybrid layers above, a neural net can be used as a form of data reduction layer for symbolic AI.

sub-symbolic heuristic – We have seen in Chapter 4 how heuristics can guide search. A neural net can be trained to do this, for example to help a mathematical theorem prover to choose which axioms or lemmas to use next in generating a proof. We will also see in Chapter 11 that heuristics are very important for game playing and how AlphaGo [260] used neural networks as heuristics to guide Monte Carlo search.

symbolic guide – The symbolic AI may be used to guide the training of a neural network, for example an expert system to help decide on network parameters such as number or kinds of layers. In Chapter 22 we will see how a computational form of regret can use high-level counterfactual reasoning in order to guide the training of a sub-symbolic 'emotion' module.

symbolic learning – Sub-symbolic methods are also being used to reproduce aspects of reasoning that would normally be associated with symbolic AI, for example logical implication. This may use a symbolic representation, for example using genetic programming or similar techniques to create logical expression, but may also involve 'vanilla' or specially designed neural networks trained on lots of logical formulae.

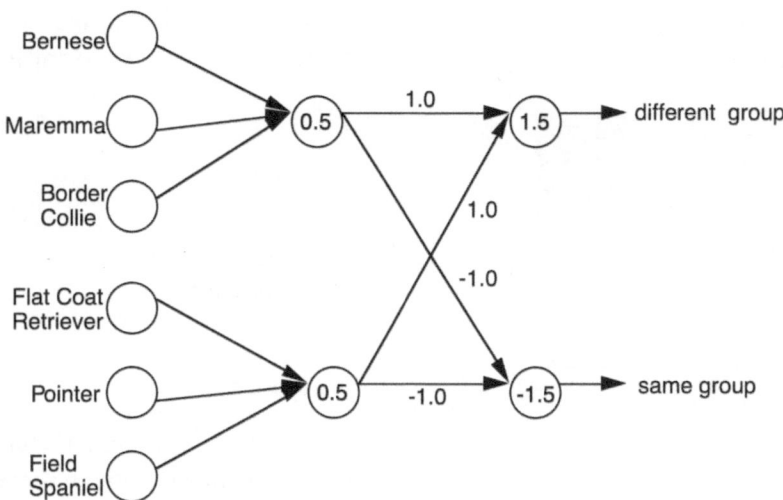

FIGURE 6.15   Does this network work?

It is likely that this area will grow substantially in the coming years, and it is a crucial next step for AI if it is to address the next level of challenging problems.

6.9 SUMMARY

In this chapter we have seen how neural networks emulate the physical properties of the brain to provide parallelism and fault tolerance. We have seen several types of neural network, which in many ways are similar with nodes and connections between those nodes. However, they differ (i) architecturally in terms of the pattern of connections and (ii) algorithmically in terms of the ways in which node values and weights on connections are updated. These apparently subtle differences affect the kinds of applications for which they are suitable. Different kinds of neural network can be combined together or with other forms of AI in hybrid architectures.

Simple neural networks are particularly suited to classification and pattern recognition tasks. They have also been used in many applications where their ability to learn from examples and generalise to new cases is beneficial. However, their disadvantages should also be acknowledged: they can take a long time to learn, be difficult to update quickly, and, perhaps most important, there is no explicit representation of the decision mechanism and therefore no explanation facility. The former can be addressed by using greater computing power (albeit with financial and environmental costs), and this has made possible the very large and deep networks that have transformed AI and will be discussed in Chapter 8. The latter, the need for explanation and interpretation, is a more fundamental problem which we shall return to in Chapter 21.

6.1   Does the network in Figure 6.14 solve the XOR problem? Show the outputs of the network to support your answer.

6.2   In the UK dogs are classified into six groups according to breed. Each group contains a number of different breeds. Given two breeds of dog the neural network in Figure 6.15 is required to indicate whether or not they belong to the same group. The two breeds of interest at any time are indicated by setting their input nodes to 1. All other input nodes are set to 0. Note that Maremma, Bernese and Border Collie are all members of the working group; Flat Coat Retriever, Pointer and Field Spaniel all belong to the gundog group.

a.   Does the network classify the dogs correctly?

b.   If not, how might you fix it?

c.   How do weights, thresholds and hidden units operate in this problem solution?

6.3   A student decides to create flash cards to help revise their knowledge of neural networks. They make a board with areas labelled by different kinds of neural network:

- perceptron
- multi-layer perceptron
- Boltzmann machine
- restricted Boltzmann machine
- Kohonen self-organising map

They then make cards for each area listing the qualities of each network, so that they can test themselves by placing the cards into the areas. Unfortunately they shuffle the cards before making a note of which card goes with which network. Here is what is written on each card:

- 2D layout
- auto-associative
- backpropagation training
- bi-directional connections between layers
- can solve XOR problem
- can't solve XOR problem
- form of clustering
- fully connected
- hard threshold
- hidden layer
- inspired by physics of spin glass
- modelled on single human neurone
- no connections within layers
- no connections within layers (two cards)
- one-way connections between layers
- sleeping phase
- two layers
- uses sigmoid
- winner takes all during training

Can you match the cards to the networks?

FURTHER READING

C. Aggarwal. *Neural networks and deep learning.* Springer, Cham, 2018.

Most recent version of classic textbook on neural networks.

R. Beale and T. Jackson. *An introduction to neural computing.* Adam Hilger Bristol, 1990.

An early book, but still in print. A readable introduction to neural networks which provides details and algorithms for most major classes of connectionist models.

Statistical and Numerical Techniques

7.1 OVERVIEW

By its nature advanced AI often includes aspects of a numerical or mathematical nature. In this book, we have tried to minimise more mathematical aspects as we are aware many find this difficult or even frightening. However, some techniques that are either used on their own or in conjunction with other forms of AI or ML have a statistical or mathematical nature, so this chapter attempts to introduce them without assuming a mathematical background.

7.2 LINEAR REGRESSION

One of the most common techniques for simple data analysis is linear regression. At its simplest this is about drawing a best fit line between points. For example, Figure 7.1 shows the time taken for afternoon walks of different lengths as recorded on a fitness tracker. Each point represents a single walk, for example the point marked A denotes a walk of 2.7 miles that took 71 minutes. The dots are quite scattered with walks of different lengths. The line through the middle is at 25 minutes per mile. Some walks are above the line, taking longer than this, perhaps more strenuous countryside, or just taken more leisurely, some are below representing faster walks. The solid line is a 'best fit' and can allow us to predict how long it might take for, say, a 3-mile walk (1 hour and 15 mins) or how many miles one might walk in 2 hours (4.8

miles). By looking at the typical spread above and below the line, one can also obtain a measure of uncertainty.

Note that one point lies well above the line and far away from the rest of the data. These extreme values are called outliers and can often skew the best fit line depending on the method used to calculate it. In this case it could represent a 'true' value where the walk was simply very difficult and slow, but might be where the user forgot to tell the fitness tracker the walk was over.

Mathematically the regression line is of the form:

$$y = mx + c$$

where (by convention) the y axis is the vertical axis, in this case time taken, and x is the horizontal axis, distance walked. The number m is the slope of the line, in this case miles per minute, and c is the intercept where the line crosses the y axis, the value of y when x is zero. Here the intercept, c, represents the time taken to walk no miles at all, which feels as though it should be zero, but perhaps represents the time it takes after the fitness tracker is turned on to put on boots, lock the house or car, etc.

The normal way to calculate this best fit is using the formula:

$$m = \frac{\sum (y_i - \bar{y}) \times (x_i - \bar{x})}{\sum (x_i - \bar{x})^2}$$

$$c = \bar{y} - m\bar{x}$$

where $\bar{x}$ and $\bar{y}$ are the average value of the x and y coordinates respectively and the sums are taken over all of the points.

This formula minimises the sum of the squares of the residuals, that is the distances between the points and the

DOI: 10.1201/9781003082880-9

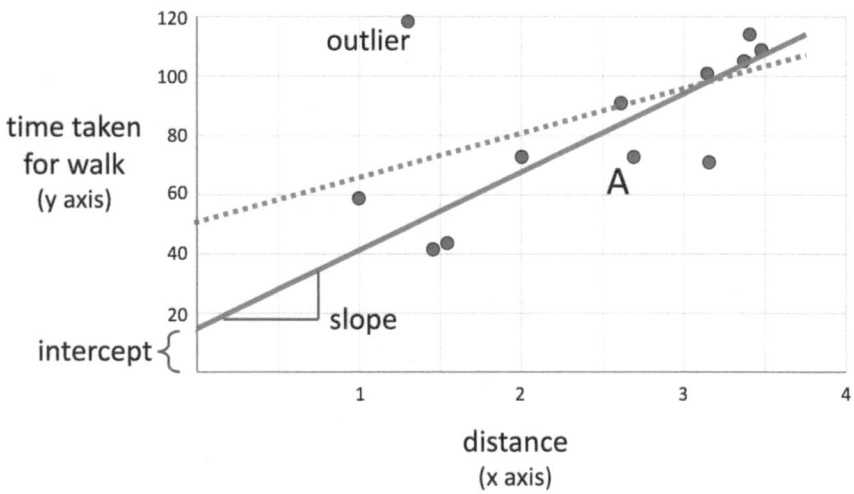

FIGURE 7.1   Linear regression for short walks: solid line ignoring the outlier, dotted line including all data.

line. In general, these least-squares methods are used frequently in various forms of statistical data analysis and model fitting. They work particularly well with data that fits the Normal distribution but also tend to have easy mathematical treatments. However, they tend to be sensitive to outliers and so sometimes extreme values are removed from the data prior to applying linear regression. Note that in Figure 7.1, the solid line is the regression having removed the outlier. The dotted line shows the regression line including the outlier. See how this distorts the line, meaning it no longer fits the rest of the data very well at all. Ideally outlier removal should be based on predetermined rules, to avoid simply removing elements until the data fits one's preconceived ideas.

This is a simple case where there is a single independent variable, the thing you know (in this case the length of the walk) and a single dependent variable, the thing that is measured or to be predicted (in this case how long it takes).

In general, you may want to predict several things, that is several dependent variables, perhaps both time taken and average heart rate while walking. This is a simple extension, you just create a separate best fit line for each thing you want to predict.

A more complex case is when you have more than one independent variable, for example you know both the distance and average gradient of a route and use both of these to predict the time taken. That is you attempt to obtain a prediction equation of the form:

$$y = mx + nz + c$$

where, in this case, x is the time taken and z the gradient. Note that this is a plane in 3D space that for any x, z values (distance, gradient) gives the time taken. This is called multi-linear regression, and it is possible to work out the general formula for this for any number of independent variables, but we omit this here (see *web resources* for more details).

Figure 7.1 was for short walks only. Figure 7.2 is extended to also include longer walks. However, now note that the single best fit line (solid) is a poor fit for the overall data. The two dashed lines are separate regressions, one for the shorter walks and a second one for the longer walks. Clearly the two short lines fit the data better than the single long one.

Thinking about actual walks, the reason is obvious. For longer walks one often stops for a lunch break, and indeed the short and long walks have nearly the same slope, but the long-walk intercept is larger, corresponding to around a one-hour lunch! In practice this is quite common, many phenomena are locally linear, that is linear models fit well so long as you restrict yourself to a small area, or a single cluster. One can often create piecewise linear models that fit the data well, even for curves if the patches are chosen small enough.

Note too the importance of visualising data. It would be easy to take the numerical data from Figure 7.2 and fit the single regression line, without ever looking at it. The single line would have been a predictor of the data but clearly misses a critical aspect of the data that allows the far better piecewise linear fit.

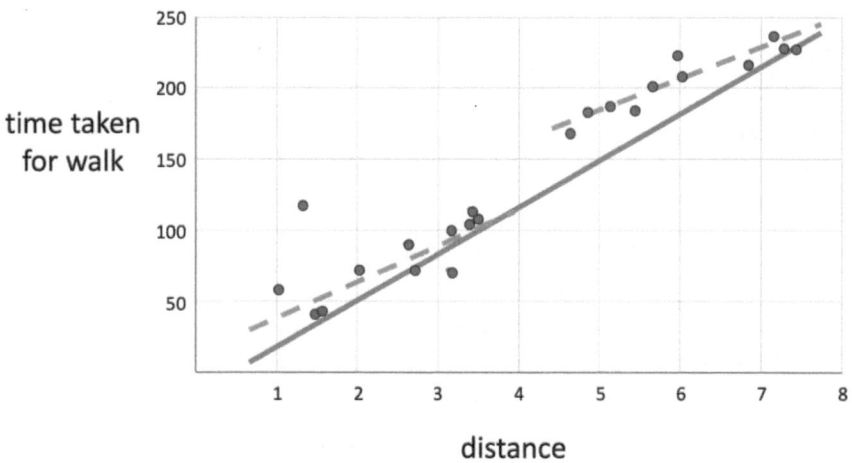

FIGURE 7.2   Piecewise-linear regression – longer walks.

7.3 VECTORS AND MATRICES

In physics, vectors represent things that have both a value and a direction, for example a moving car has both a speed and a direction it is moving in. Vectors in 2D can be represented as a direction angle and length (called radial coordinates), but more generally as a series of two or three numbers representing components in each direction, for example a northwards and eastwards speed for a car. The same representation is used for Cartesian coordinates in space, such as latitude/longitude on a map (ignoring for the moment the curvature of the earth). Both can be two-dimensional (as in the lat/long, or car's velocity) or three-dimensional (aircraft lat/long+height). In computing the word 'vector' is often used simply for a list or array of numbers. For example, in a recommender system for a media streaming platform, we might have an array MT where MT[person_id] is the number of movies each person has watched on a streaming platform. This corresponds roughly to the generalisation mathematicians make when talking about vectors with very large numbers of dimensions, not just two or three. In the movie example, the dimension of this vector is the number of people, which could be enormous.

Sometimes it can be useful to think about the geometric analogy between data arrays and vectors in physics, and we will use diagrams that follow this analogy. However, there are limitations to this as both the diagrams and our ability to conceptualise vectors are usually limited to three dimensions, and there are also sometimes crucial differences in behaviour for larger dimensions.

If you do not find this physical analogy helpful, you can think of these structures purely algorithmically.

Of course, a movie platform has many people (say 26952) and many movies (say 13575), and we may be interested in data about each movie. For this we might use a data structure M[person_id][movie_id] (see Figure 7.3). Mathematically this is regarded as a 26952×13575 matrix.

If you encountered matrices in school, they are likely to have been 2×2 or 3×3 matrices representing transformations of coordinates in 2D or 3D space. For example, the following matrix represents a clockwise rotation of about 37 degrees.

$$\begin{pmatrix} 0.8 & -0.6 \\ 0.6 & 0.8 \end{pmatrix}$$

Given an (x,y) coordinate pair each row of the matrix is multiplied by the corresponding coordinate values and added up to give new coordinates. In this case the new x is $0.8x - 0.6y$, and the new y is $0.6x + 0.8y$, so that (3,1) is transformed to (1.8,2.6). These 2×2 matrices can be added, subtracted and multiplied (and in some cases divided), somewhat like ordinary numbers. The same is true of more general N×M matrices (like our movies–people one), and many complex algorithms that involve embedded loops can be represented more concisely in this 'matrix algebra' form, albeit with care to make dimensions match up. Furthermore, we can exploit lots of known mathematical properties of matrices in order to both create algorithms and analyse their properties.

		movie id (13575 movies in total)					
		1	2	3	...	13574	13575
	1	0	2	0	...	1	0
	2	1	0	0	...	0	0
person id	3	0	1	0	...	0	0
(26952 in total)	...	...	...	...	...	...	...
	26952	1	0	0	...	0	1
	26953	0	0	3	...	0	0

FIGURE 7.3   Matrix representing number of times a specific movie has been seen by each person.

Note too that in 2 × 2 rotation matrix, both rows and columns correspond in a sense to the same x–y space. In contrast in the people–movies matrix, the rows and columns are indexed by different things. Matrices that have the same number of dimensions in both directions are called square matrices (for obvious reasons). Square matrices (say people × people, or movie × movie) have particularly powerful mathematical properties, and so it is quite common to manipulate data to create square matrices even if the original data is not square. For example, we might start with a people × movies matrix, L[person_id,movie_id] that has a positive number if the person has watched and liked it, a negative number if the person watched and disliked it, and zero if the person hasn't watched it at all. We might then transform this into a measure of similarity between people S[person_id1,person_id2] by summing the product of likes and dislikes for shared movies:

```
for p_id1 = 1 to nos_people
  for p_id2 = 1 to nos_people
    sum = 0
    for m_id = 1 to nos_movies
      sum = sum + L[P_id1][m_id]
                    * L[P_id2][m_id]
    S[p_id1][p_id1] = S[p_id1][p_id1] + sum
```

Note that if both users dislike the same film, the product is positive, so adding to their similarity. If one likes it and the other doesn't, it is negative hence reducing their similarity (maybe making it negative, a dissimilarity). If either of the people has not watched the film, then the product is zero, so has no effect on the similarity measure. In matrix algebra terms this can be written as:

$$S = L \times L^T$$

where L^T is the transpose of L, the same as L with rows and columns swapped and '×' is matrix multiplication. See how much more succinct it is! Note that this similarity matrix is symmetric, that is L[p1][p2] = L[p2][p1]. Symmetric matrices have yet more useful properties.

7.4 EIGENVALUES AND PRINCIPAL COMPONENTS

In the case of 2 × 2 rotation matrices, every vector (location as pair of coordinates), except (0,0), is moved to a new direction. However, for some forms of transformations vectors retain their direction and simply get longer or shorter. For example, the following matrix represents a shearing and stretching of 2D shapes.

$$\begin{pmatrix} 3 & 1 \\ 2 & 2 \end{pmatrix}$$

Consider the vector (1,1), it is transformed to (4,4), similarly any multiple of (1,1) is transformed to a vector precisely four times as big (e.g. (3,3) becomes (12,12)). If we now look at the vector (1,-2), it is transformed to itself, as are multiples of (1,-2) such as (7,-14).

These vectors (1,1) and (1,-2) that preserve their direction are called eigenvectors, and the multipliers (4 and 1) are called the corresponding eigenvalues. Crucially it is possible (with certain conditions) to transform the original data and represent it in terms of these eigenvectors. For example, (1,0) is $\frac{2}{3}(1,1) + \frac{1}{3}(1,-2)$.

For symmetric matrices these eigenvectors are also orthogonal (at 90 degrees) to one another, which makes it particularly easy to re-represent the original data items in terms of eigenvectors.

The 2 × 2 matrix had two eigenvectors and in general (with some caveats) an N × N matrix has N eigenvectors, although occasionally some of the eigenvalues for these may be identical. For matrices that represent some form of correlation or similarity measure, the large eigenvalues represent directions where there is a lot of change or difference. If one selects the eigenvectors corresponding to the largest few eigenvalues, then these in some way represent the aspects of the data that are potentially most important (as in vary a lot). These are called principal components.

In statistical analysis, principal components analysis is based on the correlation matrix between features/columns. The validity depends on aspects of the data, notably the mean of each data item is deducted and only residuals multiplied to give the correlation. Also if the level of variation of features is not consistent, some sort of normalisation may be performed.

For example, for the matrix of people × movie likes, some people may have watched a lot more movies, some may tend to be very positive and some very negative. We might pre-whiten the data by, for example, subtracting the mean score for the person from each rating and then dividing by their average or total ratings, so that everyone ends up with scores in a similar range. For very large datasets this kind of thing may sometimes be omitted as things 'average out', but it is worth checking whether this seems valid for a particular dataset, either analytically or by trying out normalised and unnormalised data analysis on test datasets.

The principal components can simply be used to help think about the data. In the movies and people example, a principal component might represent aspects of the dataset such as "people who love/hate thrillers". However, not all principal components can be easily described.

In addition, if you choose the top P principal components, these can be used to reduce the people dimension of your dataset so that instead of having a like score for every person for a given movie, you end up with P numbers for each movie, for example, the extent to which people on the "loves/hates thrillers" dimension like this film. This reduced data can then be used in other machine learning techniques, such as clustering (to create emergent movie genres) or neural networks (see Chap. 8 for more details).

Note we could have instead created a movies × movies similarity matrix and then used that to create reduced representations for each user and then, for example, created clusters of users which could be used for recommen-dations or even suggested new contacts in a movie social network.

7.5 CLUSTERING AND K-MEANS

Look at Figure 7.5. It is easy to see that there are two clusters of points. Furthermore it is possible to simply draw a line that separates them (more generally in two dimensions this might be a plane and in higher dimensions a hyperplane). Where this is the case, there are various algorithms to find the lines that separate clusters. However, if we have principal components, then it is often the case that a hyperplane orthogonal to the first principal component (the one with the largest eigenvalue) is a good separation between clusters. In general, finding such a separating line is called linear discriminant analysis.

However, things are rarely that simple! Figure 7.6 shows a more complex set of points, which again we can see by eye are in a number of clusters, but where no simple straight line can be used to make the distinction (they are not linearly separable) and there are even places where clear clusters slightly overlap. We have seen these issues of linear separability before when we discussed the move from perceptrons to multi-layer neural networks (Chap. 6).

There are more complex variants of linear separation using multiple hyperplanes, but also other techniques, that have fewer assumptions about the statistical properties of the data and can deal with non-numeric data.

One of the most common forms of clustering is to use variants of the k-means algorithm. The idea is fairly simple.

1 (humanly) choose how many clusters you'd like to identify; that is the value for 'k'.

2 (algorithmically) choose k of the data points at random, to act as 'seeds', and for each:

    2.1 create an initial cluster with its 'centre' at the random data point

3 For each data point d

    3.1 find the closest cluster to d, call it c

    3.2 add d to c

    3.3 recalculate the centre of c – for numeric data this is the average point (centroid), but for

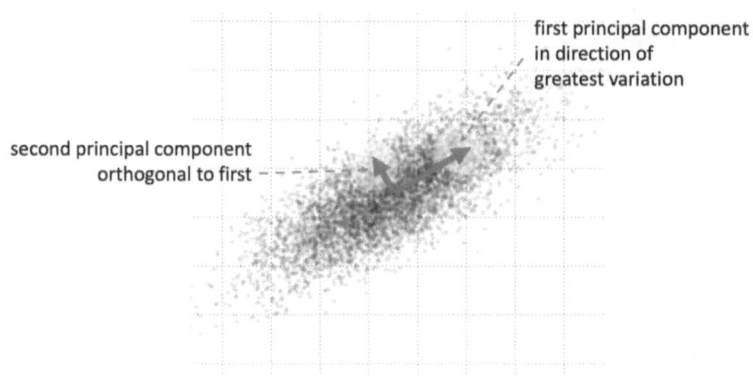

FIGURE 7.4 Principal components showing directions of maximum variation in the dataset. Adapted from Nicoguaro, CC BY 4.0, via Wikimedia Commons.

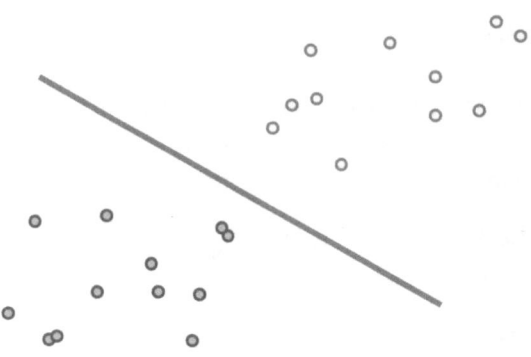

FIGURE 7.5 Linearly separable clusters.

non-numeric data alternatives can be used such as the data element that is most central in the cluster

4 Potentially, iterate 3 retaining the centre from the previous round, but re-allocating the data points.

Choosing k (step 1) can be critical. Sometimes there is an obvious value based on the details of the situation, but more commonly it is something we need to find out. If k is too large, we may end with lots of small and irrelevant clusters; if it is too small, we may not make important distinctions. Often the process is repeated with different values of k and the best chosen based on the tightness of the clusters. Of course, this choice process can itself be automated.

An alternative way to more inductively determine sensible numbers of clusters is to operate hierarchically. A clustering algorithm, such as k-means or linear separation, is used to find a small number of large initial clusters (possibly a binary split into two clusters). The data points allocated to each cluster are then clustered themselves, creating a tree of smaller and smaller clusters. The recursive process stops when clusters fail to be sufficiently distinct.

7.6 RANDOMNESS

In data analysis we often have to deal with data that has random or statistical properties, including noise. In addition, many algorithms also explicitly or implicitly use randomness to make them work.

7.6.1 Simple Statistics

Advanced statistical techniques are used in a number of areas, especially in machine learning. However, it is also common to see quite simple methods used, especially in early data preparation and in evaluation of models. You will almost certainly have encountered measures such as the mean, median and standard deviation. Here we'll just recap these common measures of centrality and spread.

The term 'average' is often seen in both media and technical reports but has two principal meanings, which are often substantially different:

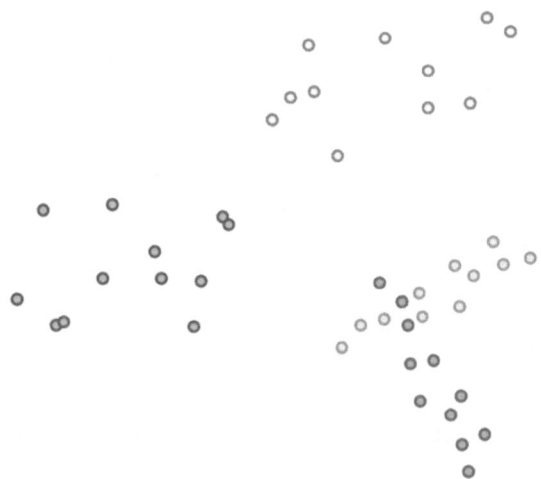

FIGURE 7.6   More complex clusters.

mean – the sum of the data items divided by the number of items. This is sometimes referred to as the arithmetic mean and often written as μ (Greek mu).

median – the 'half way' value where 50% of the items are smaller than the value and 50% bigger.

In general, if you hear the term 'average' be alert to the two meanings. Note too that, despite the ambiguity, if you use a spreadsheet, the function called AVERAGE is usually the arithmetic mean.

For some kinds of data, for example heights of people, the two are effectively the same, but for other kinds of data, especially 'long tail' data (see below), there can be a substantial difference. Most commonly when they differ, the mean is bigger than the median, but not always.

Figure 7.7 shows the distribution of weekly incomes in the UK for the financial year 2011/12. The mean income for the year is £528 per week, but the median income is £427. Usually when incomes are mentioned in the press or government statistics, the 'average income' is the median as this makes more sense practically – half of people earn more, half earn less.

Note also that the peak (smoothing out the bumpiness of the distribution) is around £300 per week. This value is called the mode, but distributions can have more than one peak, especially if there is a mix of two sources of data with different distributions – imagine the incomes of people working in a Wall Street or City of London office; there will be one peak at the lower end for cleaners and security staff and another at the upper end for the bankers.

The median is usually a more stable measure and better to use, but the mean has better mathematical properties. The means 'add up' in the sense that if you have several groups of data items and know the mean and number of items in each (μ_g, n_g), you can work out the overall mean as the 'weighted average':

$$\mu_{overall} = \frac{\sum_g \mu_g \times n_g}{\sum_g n_g}$$

There is no equivalent easy way to obtain an overall median from the parts.

As well as hearing about 'average' income, you will often hear about the top 5% or bottom 10%. These are called percentiles and a special case are the quarter way points, the bottom and top 25%, which are called the lower and upper quartile, respectively.

In addition, you will encounter the term residual in data analysis. Indeed, we have already seen this used when discussing linear regression earlier in this chapter. In general the residual is the difference between a data item and some sort of model or fitted value. In the case of linear regression this was the difference between the data points and the line, but it can be used more widely. Often as a first stage of data analysis one works out the mean and then subtracts this from every data item leaving 'residuals'.

As well as these measures of the middle, or 'central tendency', there are equivalent measures of spread or variability:

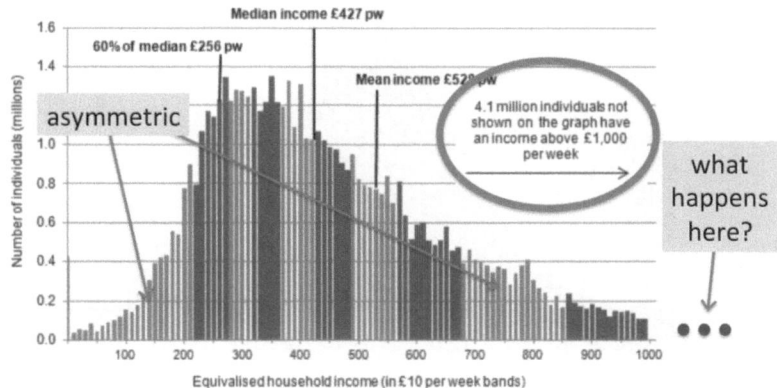

FIGURE 7.7   UK income distribution 2011/12. Source: Office of National Statistics [216].

standard deviation – This is the square root of the variance (below) and is typically written as σ (Greek sigma).

variance – The arithmetic average of the sum of the squares of the residuals, often written as μ^2.

interquartile range – This is the difference between the upper and lower quartile, that is the range from the bottom 25% to top 25%.

In more detail for the variance, one first of all works out the difference between each data item and the mean, then squares those values and then works out the arithmetic mean of those squares. If there are N data items x_i, the formula is

$$\sigma^2 = \frac{\sum_i (x_i - \mu)^2}{N}$$

As with mean and median, the interquartile range is the more stable measure, but the variance 'adds up'. As with the means, if we have several groups of data items with mean, variance and number of items in each group (μ_g, σ_g, n_g), the overall variance is given by:

$$\sigma^2_{overall} = \frac{\sum_g \sigma^2_g \times n_g}{\sum_g n_g}$$

Although it is the variances that add up, we most commonly quote the standard deviation as a measure as the variance is in square units. For example, if we look at males or females (not the mixed distribution), the standard deviation of each is around 6cm; however, the variance is 36 square cm – not very meaningful! Of course it is easy to move back and forth between the two.

7.6.2 Distributions and Long-tail Data

Most readers will be familiar with the bell shape of the Normal distribution (also called the Gaussian distribution), which is common in many natural phenomena, such as human heights, and often also areas such as exam marks. It shows that values are clustered around the mean value (μ) but spread symmetrically in both directions, with about 70% of the data within a single standard deviation (σ) either side and more than 95% of the values within two standard deviations.

The Normal distribution is a continuous distribution, in that it shows the probability (strictly probability density) for any value (Figure 7.8). You may also encounter discrete distributions where the variable you are considering only takes on discrete values. For example, the distribution of number of goals scored by a football team: in what proportion of games does it score nil, one, two, etc.

However, even discrete distributions start to look like the Normal distribution, especially if they are based on the sum or average of lots of small things. In particular, if you look at the arithmetic means of many kinds of things, they will be nearly Normal. Indeed, the Normal distribution is so ubiquitous, at least for means, that a large proportion of statistical techniques assume an approximately Normal distribution, and you will see many machine learning algorithms described in terms of Gaussian assumptions.

However, not every distribution is like this. Recall the income distribution by in Figure 7.7. It is clearly not Normal itself, as it is asymmetric, but worse than that even if you take the mean income of large groups of people, the result is not approximately Normal either. This is because

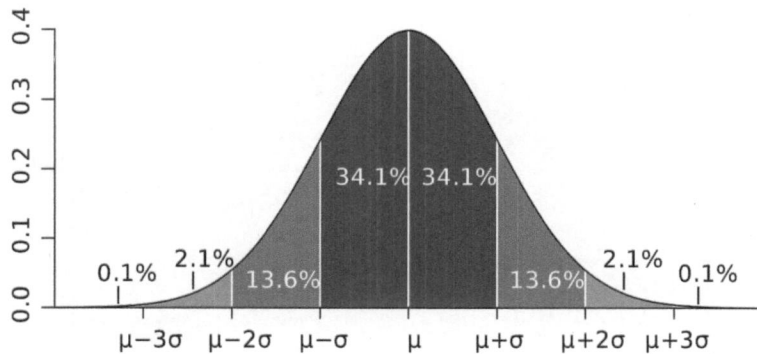

FIGURE 7.8 Normal distribution. Source: By Ainali–Own work, CC BY-SA 3.0, https://commons.wikimedia.org/w/index.php?curid=3141713.

it is a long-tail distribution. While the Normal distribution does allow values very far from the mean, these are so exceedingly unlikely they have little impact on most calculations. In contrast, the small number of extremely large incomes do have an impact on incomes.

Figure 7.9 expands the income distribution by looking at larger and more extreme examples. Each line zooms into the graph by an order of magnitude (×10). In 2011/12 the income of the average company director was around £100,000 per annum, the Prime Minister £142,500 per annum (£2000 and £2800 per week respectively). A few years after (when UK salaries were still relatively similar) there was considerable publicity about the salary of the Vice Chancellor of the University of Bath, which was around three times the Prime Minister's salary, and in the bottom line (300 times expanded from the original Figure 7.7), we see the highest paid footballer at the time. Of course, this graph doesn't include hedge-fund managers, CEOs of large companies, let alone the Royal Family.

These small numbers of large incomes can make a significant difference when we sample, leading to apparent outliers, but in particular make a huge difference to the variance, indeed for many practical purposes income distributions have no well defined variance.

This is not an isolated example, as many phenomena where there are complex feedback effects end up with long-tail distributions. This includes natural phenomena such as the sizes of earthquakes but also network phenomena such as friendship groups in social media or page links on the web [35, 202].

This is really important if the technique you are planning to use (e.g. linear regression or principal components) assumes a Normal or otherwise 'well-behaved' distribution. It is possible to transform the data to make it more Normal-like, for example simply capping large values or transforming data into percentile values. It is crucial, however, that one is aware that some adaptation may be necessary whenever long-tail data is encountered.

7.6.3 Least Squares

We've already mentioned that linear regression is a least-squares method; that is it minimises the sum of the squares of the difference between the actual values and the fitted values. Many algorithms either explicitly or implicitly follow this rule, for example the variant of gradient descent that is embodied in the backpropagation learning algorithm for neural networks is effectively minimising the average square difference during training between the current outputs of the net and the outputs expected in the training set.

Least squares methods are common because:

- just like the arithmetic mean and variance, they have nice mathematical properties, which make them easy to calculate and analyse

- they are closely related to the Normal distribution

- they often arise almost accidentally, as with backpropagation

Sometimes there can be more than one way to calculate least squares depending on what you take to be the gap between the fitted value and actual value. For linear regression, it is assumed that the x values are precise and all the noise or error is in the y value. It then minimises

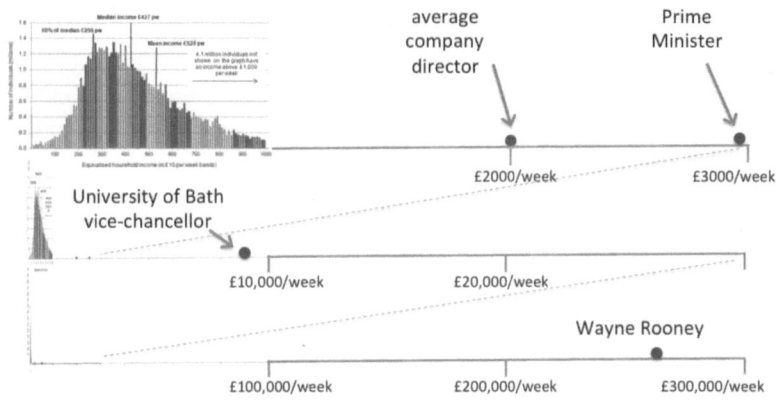

FIGURE 7.9   Long tail of large incomes. Source: Statistics for HCI [83], adapted from Office of National Statistics [216].

the sum of squares of the *y* error. You can do it the other way round, working out the regression line for:

$$x = m'x + c'$$

This minimises the sum of squares of the *x* distances and so gives a slightly different fit line (see Figure 7.10), which can be confusing as both are in a sense 'best fits'. The difference is due to what you count as 'error'.

We noted when discussing principal components that often the top P components can be used to reduce the dimensionality of the data (see also Chap. 8). The hyperspace spanned by these is the best in the sense that if points are projected into the hyperplane the square of the component orthogonal to the hyperplane is minimised. If P-dimensional hyperplanes are a little hard to imagine, consider the case of the single top principal component in 2D space (a line). The first principal component is based on the shortest distance between the points and the line (rather than *x* or *y* difference alone), and it typically lies between the two regression lines in Figure 7.10.

7.6.4 Monte Carlo Techniques

In the description of k-means we need to choose random data points to start with. This is because data often comes to you part-ordered, so that if you chose the first *k* data points, they might all naturally belong in the same cluster. In fact for k-means and many machine learning algorithms, it is a good idea to mix up data items by randomising the order before presenting them to the algorithms.

Various random features such as this are often essential to ensure that algorithms do not encounter Byzan-

tine conditions, that is particular orders of data items that cause poor behaviour. Of course, a random order of data items might just, by sheer chance, have all of the items that belong to a particular cluster or are similar in some other way together, but it is exceedingly unlikely. In contrast, the natural data order is far more likely to exhibit patterns that can be problematic.

Randomness can be used even more centrally in AI algorithms, either to make them more resilient, generalisable or efficient. For example, simulated annealing (Chap. 5) depends on making semi-random choices at each step, effectively a sort of drunkards walk across the space of possible solutions. Also in certain circumstances random vectors have been shown to perform nearly as well as principal components for data reduction.

At an extreme, we can have an algorithm that simply chooses values at random until one 'works'. For example, if you are trying to find the set of parameters that give the highest value for a function *F*, you simply choose lots of random parameters, *p*, work out *F(p)* for each one and keep track of the best so far. If the search space is relatively small, or F is very complex, this may even be an efficient algorithm, but it is usually combined with other algorithms.

We saw an example of this with hill climbing in Chapter 4. A key problem of hill climbing is that it gets stuck at local maxima, small peaks rather than the overall maximum. One way to address this is hill-climbing with random start points. For this, you start at an initial point and hill climb from there; you then choose another start point at random and redo the hill climb, and so on. At each step you record the best result so far and then stop when you seem to be making little further progress or

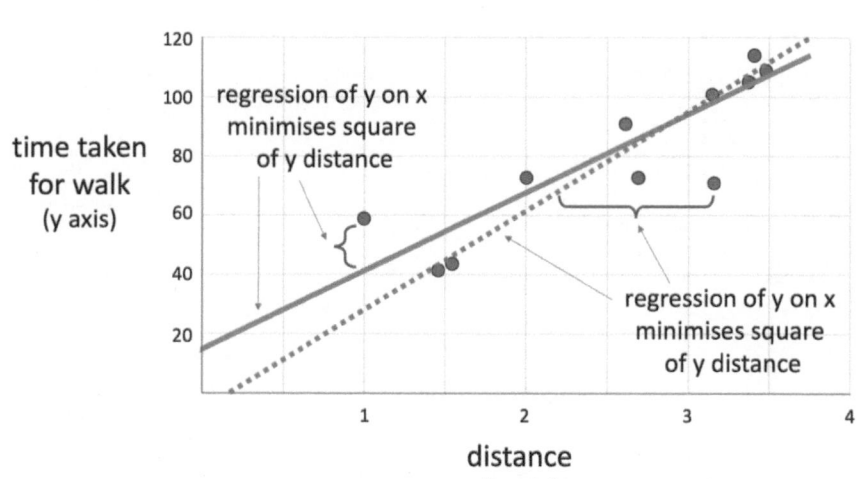

FIGURE 7.10   Linear regression as least squares – different ways to do it.

when you run out of time. If you think of the geographic analogy, your first start point may be in the middle of a large plain and so you end up on a tiny hill, but sooner or later a random start point will put you somewhere in the Himalayas and you'll get to the top of a big mountain.

Another application is in Monte Carlo tree search [55], which has been used especially for board games including Go. The search methods described in Chapter 4 and those we will see for games in Chapter 11 are deterministic, following fixed paths through the tree, guided potentially by heuristics. However, if the breadth of the tree is very large, it becomes impossible to follow more than a few paths. Monte Carlo tree search efficiently chooses lots of random paths and then uses this to create a probabilistic heuristic.

7.7 NON-LINEAR FUNCTIONS FOR MACHINE LEARNING

K-means and other forms of clustering explicitly break the data space into parts, but often we want less discrete non-linear behaviour. For example, the curve in Figure 7.11 has been approximated by linear patches, but we might wish for methods that can account for the curvature more directly.

To some extent deep neural networks (Chap. 8) can create quite complex non-linearities by simply having sufficient layers, but each stage of complexity requires more training data, so other methods are often used, sometimes as part of hybrid architectures.

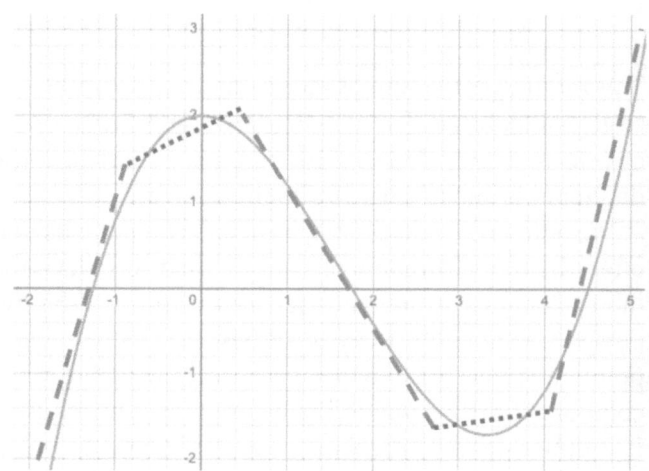

FIGURE 7.11   Piecewise linear fit, where non-linear function would be better.

7.7.1 Support Vector Machines

Support vector machines (SVM) achieve non-linear learning by applying a large number of fixed non-linear functions to the original inputs and then using these as the input to other forms of simpler machine learning. The idea is that with sufficient non-linearity to start with, any actual function between input and output can be approximated by the system.

Thinking back to the linear regression earlier in the chapter, one can imagine that instead of simply fitting $y = mx + x$, one could include x^2, x^3 terms, etc.

However, just as with piecewise linear fitting, more patch-like matching works better. For example, for image data, the patches are often based on 'Mexican hat'-like functions, rather like wavelets, that create a weighted sum with a lot at the centre and with smaller and smaller values moving outwards, rather like viewing the raw image, but also blurred with different scales of blurring.

The family of non-linear patches are often chosen so that they form what is known as a kernel, which allows detailed mathematical analysis and sometimes proofs of optimality under certain conditions.

More pragmatically, the crucial thing is to add lots of derived data based on many non-linear transformations of selections of the features. The formal use of an SVM kernel ensures a good spread and avoids redundant derived features, but other methods can also work especially where the non-linear derived data is in some way cheap to evaluate. For example, some early machine learning algorithms for black-and-white image analysis applied address decoders to random selections of pixels, basically turning 011 into 00001000 (8 bits with bit 3 turned on). As this could be done in hardware with decoders and FPGAs, it was very efficient compared to more exact methods.

7.7.2 Reservoir Computing

The idea of using hardware-derived non-linearity has resurfaced in the form of reservoir computing.

The fundamental idea is quite simple. You choose some sort of biological or physical system (the reservoir) where the relationship between input and output is complex and non-linear. The reservoir itself is then surrounded by relatively simple input and output transformation layers, the former is usually a minimal transformation of the overall inputs into a form suitable for the reservoir (e.g. maybe applying multiple thresholds if the reservoir requires binary input). The output transformation (the readout) employs some sort of machine learning, for example a simple 2- or 3-layer neural network. It is trained so that when a training input value is applied, the raw reservoir output is transformed into desired training set output.

The hope is that if the reservoir's internal processes are sufficiently rich, among the outputs are combinations of values that can relatively simply match any desired non-linear function. Virtual reservoirs, that is random mappings implemented in software, can be used, but for many the goal is to use hardware or biological processes that can rapidly react in a massively parallel way faster, cheaper or more energy efficiently than software alone.

One example of a hardware reservoir uses tiny nanoparticles deposited on a wafer [148]. After a while the nanoparticles start to connect and create patches or routes across the medium. If the deposition continues too long, then the wafer simply becomes a conductor, but at a critical point, just before that, there are no fully connected paths, but instead lots of near-connections. At this critical point a combination of quantum tunnelling and diffusion of atoms due to potential differences can create complex patterns of partial conductance.

These nanoparticle-based systems are not only non-linear but also have a level of time dependence as areas where there has been conductance in the past can be more likely to conduct in the future, a form of memristor [50]. Where this is the case the readout transformation needs to be a form of machine learning that can work on time series data (Chap. 14).

7.7.3 Kolmogorov-Arnold Networks

Kolmogorov-Arnold Networks (KAN) can be thought of as a variation of a multi-layer perceptron. However, instead of a fixed sigmoid activation function and trained weights, the KAN adjusts and learns the shape of the activation function itself (which is typically not a sigmoid). It is based on the Kolmogorov-Arnold representation theorem which says that any non-linear function of multiple variables (including the output of any neural network no matter how deep) can be constructed using two layers of sums of non-linear functions of single variables. Furthermore the hidden layer needs at most $2N + 1$ nodes where N is the number of inputs.

While theoretically interesting, the actual non-linear functions can be discontinuous and so complex that they are effectively unlearnable; so for many years this was merely an interesting theoretical result. However, work being published as this edition was being completed has changed this picture, proposing the KAN as a potentially practical machine learning method [172]. This is partly by using more than the absolute minimum number of nodes, and further by using B-splines, a restricted class of non-linear functions. B-splines glue together several polynomials to create an approximation of a continuous non-linear function; rather like an generalisation of

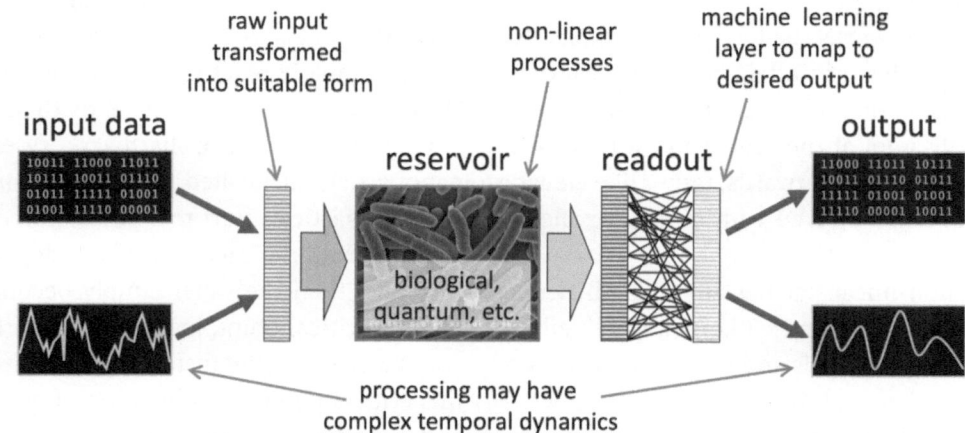

FIGURE 7.12   Reservoir computing – main stages.

a piecewise linear approximation. B-splines are used extensively in graphics. Because they are parameterised, they can be trained as part of a KAN.

At the time of writing, Kolmogorov-Arnold Networks have proved successful on smaller data, but struggle with scale. However, research is very active and may have changed considerably by the time you read this.

7.8 SUMMARY

This chapter has covered several statistical techniques that you are likely to encounter in AI algorithms. Linear regression, principal components and eigenvectors are particularly likely to occur in the description of how other algorithms work, emphasising the need for a basic understanding of matrices. Variants of the k-means algorithm are widely used, but also it is a good introduction to the concepts in other more complex clustering algorithms. We have discussed probabilities in earlier chapters of the book, and some level of uncertainty is often inherent in training data or in the environment in which AI is deployed. Data corresponding to the Normal distribution is especially common, leading in particular to the popularity of least-squares methods. But it is also important to be aware that many forms of computational data, including social network connections and web page links, have long-tail distributions, where a small number of very extreme values can skew results unless care is taken. However, we can also make use of randomness; many of the algorithms we use in AI have some stochastic elements and even purely Monte Carlo techniques can be surprisingly effective. We can also 'mix up' input data

using non-linear functions including carefully crafted mathematical functions in support vector machines; near-random physical or biological effects in reservoir computing; or B-splines in Kolmogorov-Arnold Networks.

7.1 Consider the following, treating all three columns as a single x,y dataset.

x	y	x	y	x	y
0.434	12.093	5.427	13.621	15.927	4.927
0.933	8.339	6.631	16.523	16.111	28.064
1.525	6.206	7.818	10.234	17.948	38.146
1.933	10.257	9.197	16.603	20.392	28.586
2.718	6.38	10.727	15.408	22.977	26.639
3.684	15.637	12.14	20.464	25.868	37.821
4.294	24.493	13.998	17.66		

a. Plot the data (by hand or digitally)

b. Using the formulae in Section 7.2, calculate m and c for the regression line $y = mx + c$

c. Draw the line on your data plot. Are there any outliers?

d. Repeat steps (b) and (c), but this time computing the regression the other way round, that is swapping the x,y values. Note that when you plot the line, you will need to be careful to plot the right x and y coordinates!

7.2 Consider the following matrix:

$$\begin{pmatrix} 9 & 2 \\ -2 & 6 \end{pmatrix}$$

a. Try to find an *eigenvector* by hand (you can look up methods for this, or just use trial and error).

b. What is its *eigenvalue*?

c. As this is a *symmetric matrix*, the other eigenvector is *orthogonal* to the one you have found, so if your eigenvector is (a,b), the other eigenvector is (b,−a). What is the *eigenvalue* of this second eigenvector?

d. Which is the principal eigenvector?

7.3 Consider the following matrix.

$$\begin{pmatrix} 3 & 1 \\ 2 & 2 \end{pmatrix}$$

You already know from Section 7.4 that its eigenvectors are (1,1) and (1,-2), but use the following *iterative method* to find the first eigenvector. You can do this by hand or with a calculator, but it will be easier if you either code it or use a spreadsheet with successive calculations flowing from row to row.

```
1. Start with any seed vector
2. Multiply the vector by the matrix
3. Normalise the resulting vector (x,y)
   by dividing it by sqrt( x*x + y*y )
4. Repeat 2 and 3 until the vector
   doesn't change much (or use a fixed
   number of iterations, such as 10)
```

If you are doing this by hand, you might find it easier to normalise by dividing by $|x| + |y|$ at step (3); this will mean you'll end up with a slightly different multiple of the same vector as the algorithm.

a. Try this with the following seed vectors: (1,0), (1,0), (1,1), (1,-2), (1.1,1), (1.1,-2). Document your results.

b. Does this always give the principal eigenvector?

FURTHER READING

Alan Dix. *Statistics for HCI: Making sense of quantitative data.* Morgan & Claypool, 2020. https://alandix.com/statistics/book/

This book deals with the things that aren't in a standard statistical textbook. It focuses on what statistical concepts and terms mean rather than how to calculate them.

Christopher M. Bishop. *Pattern recognition and machine learning.* Springer, New York, 2006. ISBN:0-387-31073-8

Mathematical treatment of statistical machine learning algorithms coming from an engineering perspective.

G. James, D. Witten, T. Hastie, and R. Tibshirani. *An introduction to statistical learning.* New York, NY: Springer STS 112, 2013.

As well as offering a detailed overview of statistical machine learning techniques from regression to SVM, the book has code examples with variants of the book for R and Python. The book's website https://www.statlearning.com/ includes links to downloadable PDF versions of the books.

William H. Press, Saul A. Teukolsky, William T. Vetterling, and Brian P. Flannery. *Numerical recipes 3rd edition: The art of scientific computing.* Cambridge University Press, Cambridge, 2007.

The Numerical Recipes series is the definitive reference for implementing numeric methods such as matrix algorithms.

Going Large: Deep Learning and Big Data

8.1 OVERVIEW

As we discussed in the introduction, the inflection point, during the 2010s, in the widespread adoption of AI was due in large part to the growing availability of large datasets (big data) and cloud computation. This has allowed various forms of large-scale processing, including the recommender systems that are widespread in internet shopping and social media. However, perhaps most well-known has been deep learning, which uses vast amounts of computational power (leading to environmental concerns) in order to train networks with billions of connections.

Deep learning requires lots of data, so sometimes it is necessary to grow the dataset; this can include generating data or using adversarial techniques so that networks learn from each other. More often we actually have too much data and so need various forms of data reduction, either in terms of the size of each data item or the number of data items we choose to process.

Even though computation is fast, often large numbers of processors are needed, and so it is important to be able to process data in parallel across lots of machines. Some forms of data are particularly difficult to process in volume; this includes data that has some sort of internal structure, notably graphs and temporal data, and also real-time data where the volume of sensed data

may mean that it is impossible to store all of the data for later processing.

8.2 DEEP LEARNING

One of the drivers of the resurgence of AI since the late 2010s has been deep learning. It is a term that is not just known in academic and technical circles but in the popular press too. Crucially, one of the most well-known early deep learning systems, AlphaGo, achieved one of the milestones of AI in 2016 when it defeated the Go world champion Lee Sedol. In some ways this paralleled the success of IBM's Deep Blue in 1996 when it beat Garry Kasparov at chess. However, while much of Deep Blue's success came through brute force, evaluating vast numbers of possible moves, this was impossible for Go. A game of Go may take hundreds of moves each of which has hundreds of possibilities, the search space is enormous.

While AlphaGo brought deep learning to popular attention, fundamentally deep learning was not a new technological breakthrough in the sense that backpropagation allowed the move from single-layer perceptrons to multi-layer neural networks. Indeed, all deep learning means is a neural network with many layers.

In principle this was possible for many years as backpropagation algorithms do not have a fundamental limit on the number of layers. However, in practice computational power limited the number of layers that was possible, especially for larger problems such as image recognition.

Things have changed, the combination of faster processors and cloud computing have transformed this

DOI: 10.1201/9781003082880-10

picture, allowing networks with many layers. Figure 8.1 shows an example of a five-layer network taking image pixels as input and producing some form of classification as output.

As is evident even from this image, there are many choices in terms of the number of layers, number of nodes at each layer and the way in which weights between the nodes are arranged. However, before looking at these choices, it is worth considering why deep networks were so computationally expensive in the first place, as this helps guide architectural choices and potential use.

8.2.1 Why Are Many Layers so Difficult?

To some extent it is obvious that if there are more layers, there are more nodes, so it is harder.

Suppose you have the simplest three-layer network with input–hidden–output. If all the layers have N nodes and the network is fully connected, then there are $N(N - 1)/2$ weights for the first layer and $N(N - 1)/2$ weights for the second layer, $N(N - 1)$ weights in total. However, if there are M layers, then there are $(M - 1)N(N - 1)/2$ weights in total. This sounds as if it only grows linearly with M, but in practice if you have twice as many nodes, algorithms take more than twice as long, so it may get worse far faster than the number of layers at first suggests.

In fact, there are more fundamental problems than simple number of weights, as we saw when we first encountered multi-layer perceptrons in Chapter 6.

First, the weights for the inner layers are under-determined, that is there are many configurations of weights that will give equivalent results. This is true even of a three-layer network, any permutation of the inner nodes will be equally good. However, the range of possibilities grows with the depth, not just the number.

This can lead to over-fitting; more weights mean more likelihood you will just be matching 'accidents' of the training dataset rather than creating generalisable solutions. This means that more training data is needed, that takes longer to process.

Even more problematic the inner layers of deep networks are poorly constrained. During training in a three-layer network, the input and output layers have known values, so each middle node is only one step away from a known fixed or target value. In contrast in a four-layer

network, the inner layers only have one known side, the other is being learnt, and for deeper networks the inner-most layers are trying to train themselves when the layers, either side are themselves in flux.

If the rate of training is too great, this will lead to instabilities, the inner layers 'flopping about', rather like carrying a flexible mattress. This means that the training rate parameters have to be set very low leading to slower learning and more iterations of the training data.

These problems could potentially be solved by throwing more computation and data at the training; however, the greater capabilities of deep learning also mean that more complex problems become tractable: more complex games such as Go, larger images or video. Understanding the fundamental computational issues behind deep learning can help guide more effective solutions to these tough challenges.

8.2.2 Architecture of the Layers

Let's look again at Figure 8.1. The very first layer is connected to the inputs; in the case of an image, this will be the pixels, but for other examples, such as the Go board position, some encoding of the data is needed. The output may be some form of single classification or score (e.g. the emotion of a face), a slightly more complex decision or move (as in Go), or even a complete image. The last is a special case, so we'll assume initially that the output layer is relatively small.

There are many choices in a deep learning network including:

the number of layers – Usually relatively small, although there are 'very deep' networks with tens to hundreds of layers.

the number of nodes in each layer – This is typically not uniform, with some layers having fewer than others; these pinch points can be critical architecturally.

the connections between layers – The simplest choice is to fully connect, but other choices such as a smaller number of random connections or local connections may be used (see Figure 8.2).

the learning rule – The simplest choice is to use back-propagation everywhere, but other forms are possible, including forms of unsupervised learning.

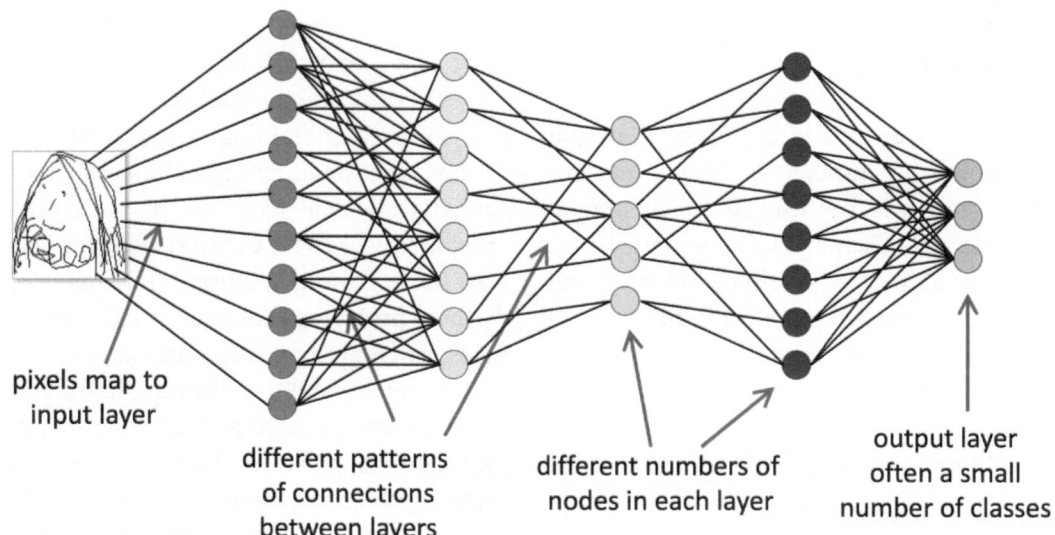

FIGURE 8.1 Deep learning architecture – multiple layers, with varying connection topologies.

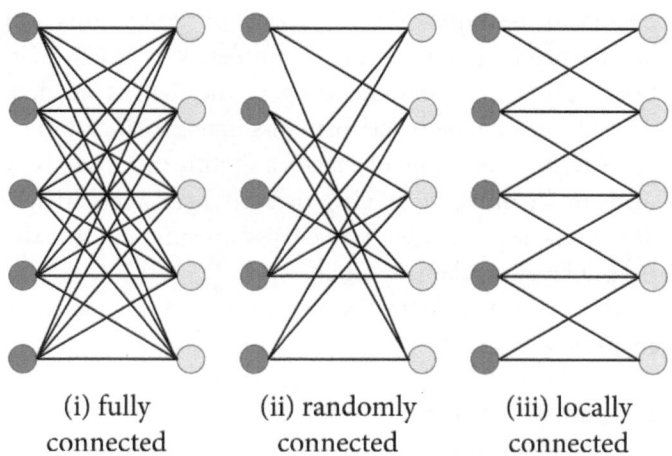

(i) fully connected

(ii) randomly connected

(iii) locally connected

FIGURE 8.2 Different kinds of connection patterns between layers.

the order of learning – Again the simple choice is simply to start it all randomly and let rip, but where there are unsupervised learning layers, these may be trained first, or the network may be built up successively adding layers/nodes.

Together these choices comprise the architecture of the network. The choice of these can be quite principled, or more a matter of trial and error. Those working with a particular type of data or in a particular domain can become expert at choosing an appropriate architecture but may not be able to fully justify every decision. Sometimes another layer of AI, such as a genetic algorithm, is used to choose the best parameters for the network.

In the end this is a bit of an art, not an exact science. However, there are some general principles that can help understand existing networks and also help you design your own. Even when the layers are all using the same learning rules, the different layers often achieve different parts of the machine learning task. This is very like our brains, where individual neurons are relatively similar, but the way they are interconnected leads to parts of the brain having specialised purposes.

Early Layers – Data Transformation

When the input is quite large, such as an image, the first few layers are often performing a level of dimension reduction. In particular the first two layers often form a restricted Boltzmann machine (Chap. 6). This has the property that it can be run backwards to reconstruct (closely) the input, so that the second layer is effectively a compressed coding of the input. This can reduce the number of nodes without substantially losing information.

These early layers may also implicitly create non-linear combinations of inputs, rather like support vector machines (Chap. 7). Alternatively, some form of kernel such as radial basis functions or recoding such as a wavelet transformation (Chap. 14) may be explicitly included.

In some ways the early layers perform a similar function to hand-coded data preparation and transformation, and it can be possible to solve problems by just 'adding more layers'. However, sometimes more bespoke pre-processing can be helpful. This may use non-linear transformations (Chap. 7), dimension reduction (Section 8.4.1), or the media specific methods we'll see in Part III.

Middle Layers – Feature Identification

The middle layer in Figure 8.1 is shown with a smaller number of nodes. This hourglass shape is in fact quite a common choice. These intermediate nodes often embody a form of feature space. For example, in images of landscapes are they sunny days or cloudy, mountains or seascapes, have people in or none? It is not that individual nodes represent these features, the representation is likely to be distributed across the nodes, but more that the important features of the images are being captured.

If the pinch-point is too small, there may not be enough dimensionality to encode all of the information needed to produce the output. This will typically lead to poor learning, especially for smaller classes. On the other hand, if there are too many nodes at this point, the opposite may occur with too much 'freedom' leading to over-training and poor generalisation.

Later Layers – Feature Combination and Higher-Level Features

The later layers between the pinch point and the output are then sufficient to create a final decision based on non-linear combinations of the features. This is again very like a support vector machine with the penultimate layer creating the non-linear combinations of features and the final layer the discrimination surfaces.

Of course when we look at a picture or consider a problem, we may build concepts on top of concepts: see eyes and mouths to make faces, see many faces to see a crowd.

For these deeper networks it is common to see a sort of rippling shape, with a few wider layers punctuated with pinch points. Sometimes the strict layers may be broken by copying the lower level 'features' further along the pipeline (see Figure 8.3).

But Really Just Lots of Nodes

Thinking of the deep network in terms of functions at different levels can be helpful, in particular this can help one decide where to vary the number of nodes in a layer. However, if the deep net were simply performing known predetermined functions at each layer, it would probably be best to swap the layer for one that is specifically designed for the purpose. One of the strengths of deep networks is that they 'sort of' do the jobs above but also can do the unexpected. This is particularly the case with deeper networks.

8.3 GROWING THE DATA

Deep learning has lots of weights to learn, that is many degrees of freedom. In order to avoid overfitting and ensure generalisation it therefore needs lots of data on which to train.

Sometimes this is not a problem, for example social media companies have massive numbers of profile pictures and other shared pictures, many with tags to say who is who. However, there is not always enough real data. Furthermore as trained models get bigger and bigger there is a worry that even all the world's texts will not be sufficient for adequate training [289].

Happily, it may be possible to fill gaps in solid data by generating synthetic data. There are different ways to do this, although all require a substantial amount of domain knowledge and carry the risk of training a system on unreal data.

8.3.1 Modifying Real Data

Sometimes we can work out ways in which data could have been a little different, small tweaks we can make to real data that are plausible but different.

The simplest case is to add a little noise, for example, given numerical data, we might simply add a small amount to each reading. For images we can take a high-resolution high-quality image and then add noise in different ways, perhaps blurring it or adding small optical distortions. It is important to understand the algorithm you are using when adding noise; for example, adding noise to the training data of many kinds of neural network can improve generalisation, but adding linear noise before linear regression reduces the accuracy without any gain in generalisation.

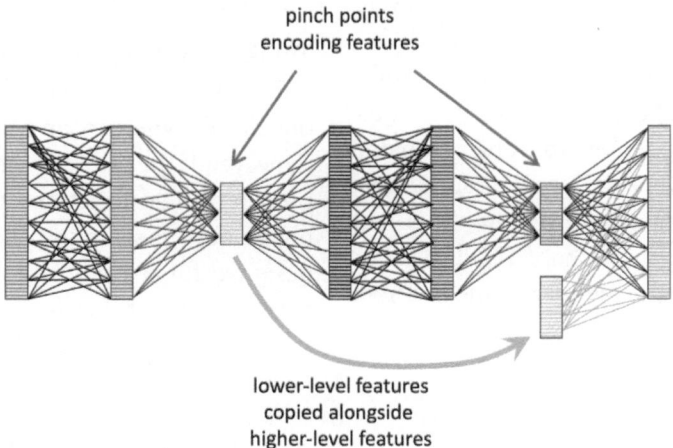

FIGURE 8.3 Combining high- and low-level features in deep learning.

We might also be able to use knowledge about the domain to create slight variants. For example, with ECG data (heart monitoring), we might detect the gaps between beats and then slightly shift the beats, or slightly stretch a single beat. For images, we might crop the image in slightly different ways or slightly magnify or shrink it. In all these cases, there is a danger that the modifications might affect the underlying data in ways that interfere with the things for which we are trying to train the network. Moving the image is fine if the aim is to be able to identify the person, as shifting an image doesn't change who the person is. However, if a heart condition is marked by irregular rhythms, shifting the beats may interfere with training.

8.3.2 Virtual Worlds

We might be able to go further and create data. For example, if we have a model of how the heart works, we can simulate different kinds of anomalies and then generate ECG traces from the model.

This technique has been used for training robots and also self-driving cars. Gathering data for either takes a long time and also can run safety risks to the equipment or the environment in which they are being trained. Furthermore, we may want to train for situations that are rare and hazardous.

For example, for the autonomous car, we would like to train the car how to behave if a child runs into the road. Happily this does not occur often, but if it does we do not want to use this as a 'training opportunity'. Training and testing the guidance software in a virtual world means we can try out these difficult situations.

In addition, when we run a simulation, we can generate simulated video and run recognition or guidance software with that as input. However, we also know 'ground truth', what is 'really' in the simulated video, hence we can more easily use supervised learning.

8.3.3 Self-Learning

When the designers of AlphaGo wanted to train their deep learning network, there was only a limited amount of training data available. It could be fed all the records of major Go tournaments, and for a human reader this would seem like a lot of data (around 30 million moves), but for the deep network, it was not nearly enough.

To supplement the training from real games, versions of AlphaGo were pitted against each other (see Chap. 11). That is the same underlying code but trained with slightly different random perturbations in their training. Each variant tried to outwit the other, training themselves on the games they played. Effectively the ML was in an arms race against itself. Each variant getting better but against an opponent that was also improving. There can be far more computer–computer games than have ever been played in human tournaments.

While this appears to be only suited for games, in fact the same principle of adversarial learning is used elsewhere. In particular, image generation networks are trained by having one network attempt to create realistic images and a second, 'playing against' it, that is trying to distinguish the fake images from real images. This internal arms race between generation and discrimination networks, called generative adversarial networks (GAN), leads to better and better image generation [116].

8.4 DATA REDUCTION

The great thing about big data is that there is a lot of it, so it is easy to use it to find patterns and learn rules. The difficult thing about big data is that there is a lot of it, so it is hard to run algorithms over it all. One of the first stages of any big data analysis is often data reduction, that is in various ways reducing the total size of the dataset we are going to deal with. This might reduce it to a size that can be managed on a single computer. However, even if we still need to break the problem up so that it can be executed in parallel on lots of machines, at least there will be less of it.

Stated like this, data reduction sounds like a necessary evil; however, sometimes it can actually make things better. Having smaller data for machine learning can sometimes lead to more comprehensible and more generalisable rules. Precise methods of data reduction vary for different kinds of data such as graph data, event streams and image sets. However, general rules apply reducing the number of items you need to consider by some form of selection or abstraction and reducing the complexity of the data about each item.

We'll primarily consider a large table with N records/rows (perhaps records representing users or web pages) and C columns of features/properties of them. Both are typically large, but we'll assume N, the number of rows/records, is extremely large, several millions or even billions, and C, the number of columns, is at least in the thousands.

There are a number of options for this:

1. reduce the complexity of the information considered for each item (reduce C) – often called dimensionality reduction

2. reduce the total number of data items we consider (reduce N)

3. reduce the number of data items considered by one processor at one time

Even when we have massive computation available, we may still use the two strategies, especially (1), as this makes each computation more effective. Also note that while option (3) is primarily used to allow parallelism (working with lots of processors at the same time), it can also be used sequentially on a single machine as a means to ensure that it can have all data in memory.

We'll look at a few options for each in turn.

8.4.1 Dimension Reduction

When data is numeric, the C data features can be thought of as coordinates in a C dimensional space $\mathbb{R}^C$. The term dimension reduction is about trying to find a smaller set of features, so that the space has a smaller number of dimensions. However, the term may also be used when the features are non-numeric.

8.4.1.1 Vector Space Techniques

When the features are numeric, we can apply statistical techniques to find a smaller dimensional space that retains as much as possible of the variation of the original data. Commonly one looks for a collection of B 'basis vectors' b_1, b_2, ..., b_B, in the same C-dimensional vector space as the item features, where B is a lot smaller than C, the original number of dimensions. For each data item d, we construct B new features $f_1, f_2, ..., f_B$ using the vector dot product:

$$f_i = b_i \bullet d$$

The features f_i are then used as the new B-dimensional representation of the data item. Mathematically this is projecting the C-dimensional space into the smaller space that is spanned by the vectors b_i.

One way of choosing the basis vectors (and hence the C-dimensional subspace) is by choosing the first B principal components. These are the directions in which the data varies most (see Chap. 7). They can be derived by calculating the eigenvectors and eigenvalues of the correlation matrix between the original C features. This gives the optimal space in the sense that the data loss is smallest in terms of 'least squares', the sum of the squares of the distance of the points from the chosen subspace.

Thinking in terms of B-dimensional subspaces of C-dimensional feature spaces when C may be many thousands and B still dozens of dimensions can be a bit mind blowing, so Figure 8.4 shows this in two dimensions, with a 1-dimensional subspace being chosen. See how the first

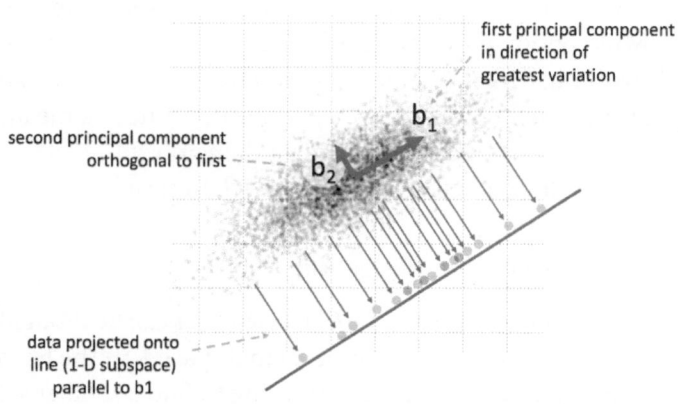

FIGURE 8.4 Projecting into subspace defined by principal components. Adapted from Nicoguaro – CC BY 4.0, https://commons.wikimedia.org/wiki/File:GaussianScatterPCA.svg.

principal component points along the direction where the data varies most.

Calculating principal components when the vector space is very large can be computationally expensive as the correlation matrix is $C \times C$, so typically contains many millions of elements. Although the principal components are optimal, choosing a few more 'good enough' vectors can work nearly as well. Indeed, even completely random vectors may be sufficient.

Another possibility is to calculate principal components for smaller collections of features that are in some way linked, for example working on the engine characteristics of a car separately to the appearance and internal comfort features. Because the algorithms use scale with powers of C, doing, say, 3 lots of $C/3$ features typically is substantially less time consuming than dealing with all C together.

Finally, the calculation of principal components relies on each individual feature having roughly similar variability. Sometimes the library functions you use will deal with this, but if not, you may need to scale features before performing the analysis.

8.4.1.2 Non-numeric Features

Some forms of non-numerical data can be used in vector-based approaches by taking simple 0/1 indicator variables for categorical data or other transformations. However, this is not always possible. In such cases one can take subsets of features and use some form of unsupervised clustering or similar technique to reduce several features into one or more categories.

If you did this for all of the data at the same time, you would hit the same learning problems that you are trying to avoid by using data reduction, but operating on small groups of features can make this manageable. In addition, some techniques, notably multi-dimensional scaling (MDS) can transform similarity data into two-dimensional (or other) spaces preserving distance. Using an algorithm such as this can mean a group of non-numerical features can be transformed into two or more numeric fields.

8.4.2 Reduce Total Number of Data Items

8.4.2.1 Sampling

A simple way to reduce the number of data items is to sample. When the data does not start digital, this is common practice: except for an occasional census or tax return, one rarely gathers information about every person in a country, but academic or market researchers will interview a sample of people and then use these to estimate figures for the overall population.

When the data is digital, it almost feels wrong to ignore some of it, but you can use a similar process. For example, you could randomly choose 20,000 records from the complete dataset and perform a machine learning algorithm on these. So long as the sample size is large and the sampling does not create any bias (don't just pick the first 20,000!), this may give good results, certainly for more general characteristics.

The downside of this approach is that it may not work so well if there are relatively small parts of the dataset with special characteristics. For example, if the data is based on users of a sporting website, sampling may give good results for popular pastimes such as football, but less so for more niche ones such as cockroach racing. One way to deal with this problem is to use a form of boosting. You first train on a (relatively) small sample, then check this against a larger sample. You then select some of the examples on which it behaves less well, add these to the training set and try again (see Figure 8.5).

Alternatively the records that are poorly matched could be taken as the basis of a completely different machine learning phase, training rules specifically for them. This is a form of segmentation, which we'll look at in more detail below.

```
Given a data set D consisting of N records
                        (N extremely large)
1. Choose a random sample S out of D
   where size(S) << size(D).
   use this as initial training set
2. Run machine learning algorithm on training
   set to generate rules R
3. Apply R to all or a large sample of D
   checking how well it works on each record
4. Generate X the set of exceptions where R
   performs badly
5. Add a sample of X to the training set
6. Repeat from step 2
```

FIGURE 8.5 Pseudocode for boosting niche record sets.

8.4.2.2 Aggregation

Rather than ignoring some records, we can collect together records that have some form of common characteristic and create one or more summary records. This can be done using some pre-existing criterion, or example geographic region for people or genre for music, or it can be derived from the data, such as the poorly matched records above, or a clustering algorithm.

For numeric data the summary for the aggregate may be the sum or average value (mean or median), maybe with some measure of spread such as min/max quartiles or standard deviation. For example, we may have global geographic data with data for many small regions which we are collecting to give a single record for each country. The summary might have the sum of the region population figures, but the average rainfall (maybe weighted by region area).

For non-numeric data it is often less clear how to form a summary. In such cases the many individual records could be replaced by a small number of records chosen to be 'typical' of the group as a whole. This is often a good choice for text or image data where, for example, an 'average' image may just be a grey smudge.

8.4.3 Segmentation

One way to reduce the number of records is through segmentation, that is simply dividing the large dataset D into a number of smaller datasets and then working on them all, but separately. This does not reduce the total amount of data but means that it can more easily be processed in parallel. In addition some of the techniques make it possible to incrementally add batches of data (time of availability acting as the segmentation rule) rather than completely re-training when new data becomes available.

As with sampling, some of the segmentation methods rely on each segment of the data being in some way representative of the dataset as a whole. Where this is the case, random segmentation may be sufficient, but with the same limitations we saw for sampling when the dataset is unbalanced.

8.4.3.1 Class Segmentation

One way to divide the data is if it falls into natural classes. As with aggregation, this can use a pre-existing criterion, or be derived from the data, through some form of unsupervised learning. Machine learning can then be applied to each of the segments yielding rules for each. When we want to deal with a fresh input, the same criterion is applied to choose the appropriate rule set (see Figure 8.6).

8.4.3.2 Result Recombination

Some algorithms can process parts of the data and then combine the results. This is particularly the case for several kinds of statistical processing including calculating minima, maxima, sums, and averages (below), and a similar technique can be used for more complex calculations creating correlations between fields and n-grams (multi-word word frequencies in text).

```
1. split dataset D into N segments D_1 ... D_N
2. for each D_i
```

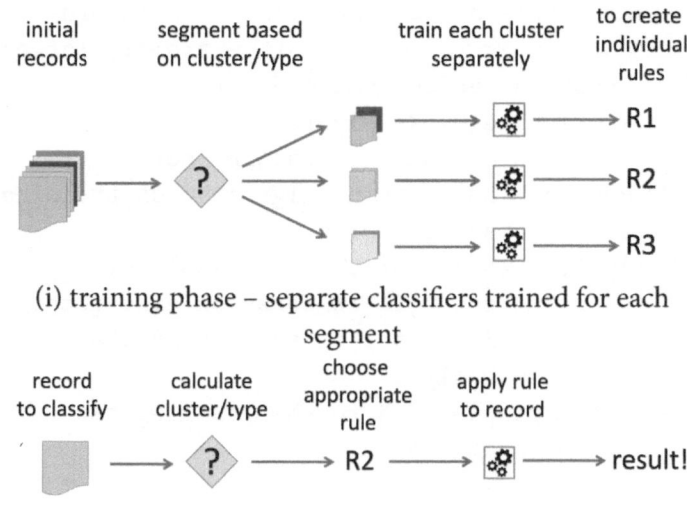

(i) training phase – separate classifiers trained for each segment

(ii) recognition phase – the appropriate classifier chosen for each input

FIGURE 8.6   Segmentation of training set into classes to allow dedicated classifiers for each.

```
2.1  calculate the minimum, maximum,
     sum and count of items
2.2  store these as mn_i, mx_i, s_i
     and ct_i respectively
3. take the stored results and calculate:
   3.1  D_min = minimum of the mn_i
   3.2  D_max = minimum of the mx_i
   3.3  D_sum = sum of the s_i
   3.4  D_ct  = sum of the ct_i
   3.5  D_avg = D_sum / D_ct
```

Some ensemble algorithms including random forests deliberately segment the dataset in order to create diversity (see Chap. 16). Here the combination rule is part of the fundamental algorithm. For example, for random forests each segment creates a single decision tree, and these are simply gathered together into a forest.

Both this and class-based segmentation are particularly appropriate for processing by MapReduce (see Section 8.5.3 below).

8.4.3.3 Weakly Communicating Partial Analysis

Other algorithms lend themselves to parallel execution on parts of the dataset but with some low-volume interactions. A good example is genetic algorithms. Subsets of data, perhaps selected randomly, are sent to different processors which individually use a GA to work out rules for their sample of the data. Occasionally the processors share a few of their top ranked rules with each other and swap some of their data sample for fresh data.

You can think of this a bit like animals breeding in small valleys, with natural selection happening locally in each. Occasionally a few of the stronger individuals make it over into neighbouring valleys and so spread good genes in the global gene pool.

8.5 PROCESSING BIG DATA

8.5.1 Why It Is Hard – Distributed Storage and Computation

When we looked at recommender systems in Chapter 7, we discussed a simple way of constructing product–product scores from the matrix M of user–product engagement scores (films liked by each user). This was effectively computed by multiplying an $N \times P$ matrix by its own transpose, where N is the number of users and P is the number of products. The pseudocode for this is:

```
score(product A, product B)
  sum = 0
  foreach user u
    sum = sum + M(u,A) * M(u,B)
  return sum
```

When we are looking at web data, N, the number of users could be anything from a few million to billions, and P, the number of products, could be from tens of thousands to millions. That is the matrix M may range from 10s of billions to more than a quadrillion (1,000,000,000,000,000) entries; that is from about a hundred gigabytes to more than a petabyte of data. The simple algorithm above would take around $P \times P \times N$ iterations of its inner loop to calculate every product–product entry, that is 100s of trillions of times even for the smaller end.

It is clear that

1. The data needs to be stored in ways that take advantage that it is sparse (most entries are zero)

2. Even when reduced, the data still needs to be split over many data stores (disks/memory) – often called sharding

3. The computation needs to be divided among many processors

4. The computation needs to be organised so that it accesses the data in efficient ways

If each user engages with on average E products, maybe a few hundred, then a sparse representation of M (point 1) could take space proportional to $N \times E$, that is 'only' billions to trillions of entries. At the lower end, we might just fit into a single machine, but if not, (2) is essential.

Storage for the sparse matrix will tend to either group all the product entries for a single user together on a single disk, or all the user entries for a given product. If they are stored the first way round, it is easy to do something to all the product entries for a given user but hard to find all users that have engaged with a particular product. Whichever way round it is done, the code above will be very slow and involve a lot of network traffic shunting small packets of data around.

If there are more users than products, it is most likely that the entries for a single user are clumped together. Given this, it is more efficient to organise the code by first of all looping over all users:

```
S = PxP matrix    // product-product score
foreach user u
  foreach pair of product A, B
    add  M(u,A) * M(u,B)  to  S(A,B)
```

If each user's entries are together on the same disk, then this is efficient grabbing all the entries for a single user and working on them at once. That is the code is local – accessing data that is stored physically close to one another.

Notice that this code does something to lots of entries about the same user and then adds up results. Furthermore the order of adding up doesn't matter, making it easy to combine results of smaller calculations. That is we can create a separate product–product score for each shard (portion of the dataset stored on a single device), and then add up the S(A,B) value from each individual shard to give an overall product–product score.

This is a fairly common pattern. For example, n-gram calculation can be performed by computing all the n-gram frequencies for a group of web pages and then pooling the results.

This pattern has led to a particular style of cloud computing called MapReduce [71].

8.5.2 Principles behind MapReduce

MapReduce is based on two concepts that themselves have their origins in the AI language LISP and later functional programming languages (see Figure 8.7).

In LISP map is a higher-order function, it takes a function and a list of values and applies the function to each item in the list returning a new list:

```
map( f, [ a, b, c, d, ..., z] )
   = [ f(a),  f(b),  f(c),  f(d), ..., f(z) ]
```

For example:

```
function square(x)  return  x * x;

map ( square, [1, 2, 7, 42, 6 ] )
   =  [ 1, 4, 49, 1764, 36 ]
```

Notice how this 'does the same thing' to lots of different entries, just as we needed to process each user's entries.

The other part of MapReduce is based on another higher-order function, reduce, which is a generalisation of operations such as 'sum' that add up all the entries in a list. It takes a function that can combine two values and then applies it successively to create a sort of running total:

```
reduce( g, init, [ a, b, c, d, ..., z] )
   = g( z, g(...
               g(d,g(c,g(b,g( a,init))))
               ...) )
```

```
map( f, entries ) =
  match entries
    case []:
      return  []
    case [a, rest...]:
      return prepend( f(a),  map( f, rest ) )

reduce( g, running, entries ) =
  match entries
    case []:
      return  running
    case [a, rest...]:
      return reduce ( g, g(a,running), rest )
```

FIGURE 8.7   Pseudocode for map and reduce in functional programming.

For example:

```
reduce ( '+', 0, [1, 2, 7, 42, 6 ] )
    =  6 + 42 + 7 + 2 + 1 + 0
```

Many operations can be accomplished by combining the two, for example to compute the sum of squares:

```
sum_of_squares( entries)
    = reduce ( '+', 0, map( square, entries) )
```

8.5.3 MapReduce for the Cloud

The versions of map and reduce for cloud computing are different but borrow loosely from this pattern. The key difference is that instead of operating in sequences or lists, they operate using hashes. The hash of a value is a way of mapping values to near unique, shorter versions that are effectively randomly spread. One example for a string of letters would be to take their numeric values, add them together and then take the last two digits. This would map every text string, no matter how long, onto a number between 0 and 99. Real hashes 'shuffle' the values up a little more.

The reason for using hashes in large-scale data processing is that if, for example, you use the actual values and these have some pattern (perhaps lots of names start with 'A'), then the data is spread unevenly leading to bottlenecks. Let's assume that the data in M is initially stored collected in a field called pscore in user records and these records are stored in lots of different data stores. We first define a function (Figure 8.8.i) that

is executed on whatever processor is convenient, possibly the user records are sent in batches to processors, or perhaps the code is sent to a processor that is attached to the data store of the user records. Crucially this code can run in parallel by different processors on different user records.

The output consists of a hash code plus a data packet. The hash codes are allocated arbitrarily to a number of processing units for the reduce stage. Say we have decided to use 37 processors for this stage, we would simply send all the packets with hash code h to processor h mod 37. Because the hash function is designed to mix up the values, we can be confident that this will create a fairly even balance between processors.

The data packets with the same hash code are then collected together and passed to a task-specific reduce function (Figure 8.8.ii), which combines all of the individual user results for a particular product pair. Because each data packet for a pair A, B goes to the same hash, they will all end up in the same processor, and so there will be only one processor generating a particular A–B value as output.

The outputs from this stage can be used as input for further MapReduce phases, or can be collected together, as they would in this example to generate the product–product scores matrix.

8.5.4 If It Can Go Wrong – Resilience for Big Processing

Figure 8.9 summarises the main steps in MapReduce. There are two innovative and essential aspects of

```
my_mapper ( user_record )
    foreach pair of products A and B in user_record.pscore
        score_ab  = user_record.pscore(A) * user_record.pscore(B)
        output( hash(A,B), { A, B, score_ab } )
```

(i) map function that operates on a single user's product engagements

```
my_reducer ( hash, collection_of_data_packets )
    sum = 0
    foreach data packet `pkt' in collection_of_data_packets
        sum = sum + pkt.score_ab
    output( hash, { first.A, first.B, sum } )
        // note A and B are the same for every item with the same hash
```

(ii) reduce function applied to all outputs from map stage with same hash code. Note that, in practice, packets with the same hash are delivered in a piecemeal fashion rather than all at the same time.

FIGURE 8.8 Pseudocode for map and reduce in cloud computing framework.

MapReduce cloud computing framework over and above the Lisp/functional programming foundations in Section 8.5.2. The first, which we've already discussed, is the use of the hash to enable balanced distribution over processors. The second is that both Google's own implementation and alternatives, such as the Apache Hadoop framework, are built to be robust to failure.

When you write code, you probably assume the computer will do what you ask it to. You will of course have bugs or parts of your code with behaviour you don't quite understand, but you will normally assume that the computer works reliably.

Indeed, if your computer does develop a fault, it is most likely to show up in the failure of one of the other programs that are executing such as the operating system or web browser, before you notice a problem in your own code. If something does go wrong with your computer, you send it for repairs and restart a failed computation from scratch (when it is fixed). In other words you treat your computer as though it is always perfectly correct, or completely broken, and happily the latter is rare and exceptional.

However, a data centre may have tens of thousands of computers and a single computation may be executing on large numbers of them. At this scale, failure isn't an exception but normal. It will typically be the case that some processors in the data centre have a fault.

Happily MapReduce lends itself well to fault-tolerant computing. There is no communication between processors within each stage, and so if a processor fails, its calculations can simply be repeated. If the partial results have already been distributed, the framework keeps track of this and re-executes any knock-on reduce or gathering stages as necessary.

Dealing with this kind of failure is complex, and one of the reasons for the success of MapReduce is that it deals with all of this for you. Of course you have to be able to transform your chosen algorithm into an equivalent MapReduce form, and this might mean modifying it slightly. However, where possible it can be a relatively rapid way to create systems that can be deployed at scale.

8.6 DATA AND ALGORITHMS AT SCALE

8.6.1 Big Graphs

A lot of web data is in the form of graphs: the links between pages, friendship connections in social media, the triples of RDF and the Semantic Web (Chap. 17). Most web graphs, and indeed many large graphs, have two key properties:

Long–tail distribution – Some web pages have vast numbers of links inwards, or outwards, but the vast majority have few; similarly in a social network

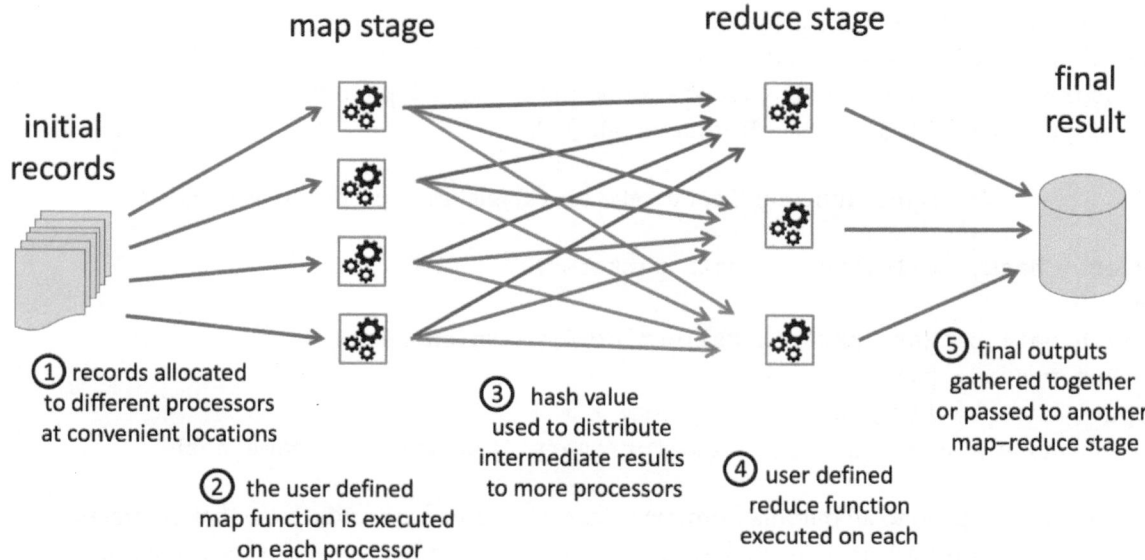

FIGURE 8.9 MapReduce distributed computation pipeline.

a small number of people have vast numbers of contacts, where most have few.

Small world – You may have heard of 'six degrees of separation', everyone you meet in the world is typically linked by a chain of six or more friends of friends; social networks naturally then follow this pattern. Web pages are the same with any pair of pages connected by a relatively small number of link hops.

The first of these means that you have to be very careful of algorithms that work with averages or samples, as the small number of exceptional pages/people can skew results. It also means that one has to be careful that algorithms do not fail when they hit these highly connected individuals, especially when thinking about load balancing over processors.

The second means that any form of 'local crawl', starting at an individual and then looking at friends, friends-of-friends, etc. will grow very rapidly. Effectively they have poor locality, with links rapidly cross-cutting between regions of the web.

In addition, all forms of graph processing have an additional problem:

Combinatorial explosion – The number of possible links between nodes increases with the square of the size of the graph, the number of paths between nodes, even faster. For example, a graph of size 10

has 90 possible directed links, while one of size 1000 has nearly a million.

Because of this, graph theorists in mathematics joke (yes mathematicians do have a sense of humour!) that problems on graphs with less than four nodes are trivial, those with four or five nodes are challenging and those with six or more impossible. Of course, we may be looking at graphs with millions or billions of nodes!

Processing big graphs therefore poses equally big challenges. MapReduce allows massive-scale processing because the data can be broken up and processed separately – it exploits locality, but because graphs have links everywhere, some problems we'd like to address are inherently non-local.

Happily, there are exceptions and important ones, notably the PageRank algorithm [32] used by Google has a MapReduce implementation. Indeed this was one of the reasons for developing MapReduce in the first place.

For smaller graphs there are special systems for storing the data, such as triple stores, discussed in Chapter 17. Typically these make it easy to do 'link chasing' algorithms, for example finding all the friends of a person, and some support forms of reasoning, such as understanding that 'descendant of' is transitive so that if A is a descendant of B and B is a descendant of C, then A is also a descendant of C.

Truly scalable graph processing frameworks are harder as they inevitably have to deal with non-locality, but can help, for example with fault tolerance, and packaging up messages between vertices. Some are built as specialised systems from the ground up, others built over existing big-data frameworks such as MapReduce.

Sometimes the uneven, clumpy nature of graphs can be an advantage and used to create an abstraction of the overall graph. It may be possible to cluster nodes and then look at the graph of clusters, for example looking at the graph of websites rather than web pages. These clusters may be based on predetermined features, such as the domain name of a URL or geographic location of a person, but clustering algorithms can be used to detect cliques. Alternatively in a social network it may be possible to focus on the highly connected individuals that arise because of the long tail nature of networks.

Local patterns can be used as features for non-graph algorithms. For example, the number of friends (connectedness), how many of those friends are friends of each other (cliquiness). Of course, these properties themselves can recursively be used, for example is someone friends with lots of highly connected people. Although more sophisticated than this, social networks largely use this form of local processing as it is impossible to perform complete analysis of graphs of billions of nodes.

8.6.2 Time Series and Event Streams

There are special issues and algorithms when dealing with time series or event stream data (see Chap. 14). Some techniques are based on windowing (working on sections of the data stream) and so are naturally parallel algorithms, but others depend on processing data serially and are thus hard to deal with using multiple processors.

8.6.2.1 Multi-scale with Mega-windows

In some cases, even if the underlying algorithms need to process the data in a serial fashion, it may be possible to break the data into large windows, process each serially and then bring the different parts together in some way. In particular, parallel algorithms can be used as the first part of a multi-stage/scale algorithm (see Chap. 14). For example, millisecond sample data can be split into large windows, say an hour for each window. Each hour's data is used by an unsupervised algorithm to build classifiers for short, say one minute, sections. It is usually easier to test a classifier than build it, so the resulting rules can

be shared with other processors and the most successful overall used by each processor to classify their data into one-minute regions. The classified one-minute chunks then form a much smaller sequence that can be collected together and processed using small-data techniques.

8.6.2.2 Untangling Streams

Event streams often have some natural form of source or topic that can be used to disentangle the single stream into multiple smaller streams. For example, separating a Twitter (X) stream based on people or tags. Alternatively, some form of classifier can be used to classify each event into streams. Note that this fits well into a MapReduce framework. The input event stream is split into chunks or randomly routed to multiple servers. Each performs a map operation (the classification) assigning a hash based on the class or topic. The reduce stage collects together the events for a single topic, sorts them based on time of arrival and then processes the event stream using standard techniques.

8.6.2.3 Real-time Processing

A pinch point in gathering any form of time-series big-data is the initial arrival of data. Sometimes this will mean some very fast data reduction simply to make it of a scale that can be stored. For example, the detectors in the particle accelerator at CERN have to deal with 600 million events per second, with substantial data for each event [42]. Dedicated processors use fast hand-written algorithms to discard uninteresting events and reduce this to about 100,000 events per second. Second stage algorithms then further process this and reduce this to a few hundred events per second. It is only then that data is stored and passed on for higher-level processing.

8.7 SUMMARY

We have seen that while in some ways deep learning is just about neural networks with lots of nodes and lots of layers, it is also evident that the behaviour of these networks has a distinct nature and feels as though it is not just *more* but also *different*. We can see how the various layers in the network may embody levels of abstraction and that we can tune this for various applications by making appropriate architectural choices as to sizes, connectivity and learning rules of each layer.

We saw a similar story for other forms of big data analysis, at some point a change in quantity can lead to a change in the qualitative nature of results produced including apparently intelligent behaviour from simple statistical techniques.

Sometimes the data is insufficient for certain forms of deep learning, so we need to use ways to grow the dataset. Adversarial techniques have proved particularly useful in a number of applications, not least game playing. More often we need to look for ways to reduce the volume of data that we have to consider in total or by any single processor. We have seen in particular that MapReduce has proved powerful in allowing robust large-scale computation of data that is far too big for a single computer.

Some forms of big data have particular problems; notably graphs, such as social network data, do not have good locality making it hard to divide over processors. Time-based data can also be complicated, especially large quantities of real-time data that may need to perform data reduction close to the source.

8.1 A large dataset contains school records for every pupil in the country over several years. As well as basic demographics and subject-by-subject exam results, it also includes social data such as lists of friends. The data is proving too large to process as a unit, so several strategies are being considered:

   (i) Take all the pupils in a school class and average the values for exam results and other appropriate fields, resulting in one record per class.

   (ii) Process the data from each region of the country separately.

   (iii) choose 5000 pupils at random and perform the analysis only on their records.

      a. For each of the above identify the kind of data reduction being employed.

      b. Can you think of advantages or disadvantages to any of these options, maybe for particular kinds of analysis?

8.2 School records include exam data for each pupil in the form subject:score in the first and second semesters:

```
semester_1: { maths:53, history: 67,
              geography: 63, ... }
semester_2: { maths:82, history: 71,
              geography: 59, ... }
```

Following the pattern in Figure 8.8, write MapReduce pseudocode to calculate the average improvement for each subject between semesters 1 and 2 over all pupils. (That is the semester 2 score minus the semester 1 score for the subject.)

8.3 It is hypothesised that friends will tend to improve in the same subjects. Consider how you might process the very large school dataset to investigate this. Which aspects are easy and which are difficult? If you think of a suitable strategy, write pseudocode for your solution.

Note, this may be a good exercise to work on in a pair or small group.

FURTHER READING

I. Goodfellow, Y. Bengio, and A. Courville. *Deep learning. Adaptive computation and machine learning*. MIT Press, 2017.

This includes both more statistical approaches and also deep learning including aspects such as convolutional neural networks, which we'll encounter in later chapters.

J. Luengo, D. García-Gil, S. Ramírez-Gallego, S. García, and F. Herrera. *Big data preprocessing*. Springer, Cham, 2020.

This book covers in depth many of the issues in this chapter including data reduction and the use of processing frameworks such as MapReduce and Hadoop.

M. Nielsen. *Neural networks and deep learning, 2019.* http://neuralnetworksanddeeplearning.com/

A short free online book that offers an accessible overview of neural networks and deep learning including examples in Python.

J. Dean and S. Ghemawat. MapReduce: Simplified data processing on large clusters. *Communications of the ACM*, 51(1):07–113, 2008.

This is the definitive paper on MapReduce. The paper not only explains the framework but is written by developers who had extensive experience building distributed algorithms and therefore can attest to the value of the infrastructure.

Making Sense of Machine Learning

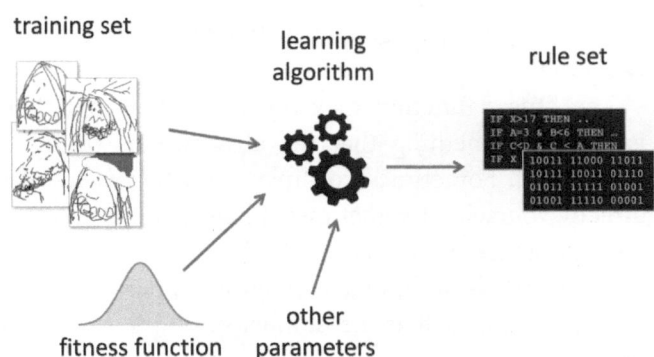

FIGURE 9.1   Training phase of machine learning.

9.1 OVERVIEW

We've seen various examples of machine learning techniques including classic knowledge-rich methods (Chap. 5), neural networks (Chap. 6) and statistical techniques (Chap. 7). In this chapter we'll reflect on some of the broader issues that cut across many of these techniques. The chapter starts with a recap of the main stages of machine learning: training, validation and application, looking at general principles and issues for each. It then looks at properties of the fitness landscape and how understanding this can both help choose an appropriate algorithm and then guide its application. Finally we will look at some of the potential complexities and pitfalls you may encounter in using machine learning.

9.2 THE MACHINE LEARNING PROCESS

In Chapter 5 we saw that machine learning has three main phases (Figure 5.3):

training – Building some sort of collection rules or other representation based on a training set of examples.

validation – In some way check that the rules generated are operating satisfactorily.

application – Using the rules to address new situations or data beyond those in the training set.

We'll first look at these in more detail before exploring other issues.

9.2.1 Training Phase

The training phase (Figure 9.1) usually starts on some set of existing data, the training set. In Chapter 5, we saw that there were two main classes of algorithm, supervised and unsupervised learning, depending on whether the data has some sort of pre-existing label. We have seen examples of each:

supervised learning – When the training set comes pre-labelled with some form of classification or expected response. Examples we've seen include version spaces, decision trees and, of course, many kinds of neural networks.

unsupervised learning – When the data is unlabelled and the algorithm creates its own labels or structure. Examples we've seen include Kohonen nets, principal components and k-means.

DOI: 10.1201/9781003082880-11

The learning algorithm also usually needs some form of fitness function (sometimes called an objective function), that says how well the algorithm is doing. For supervised learning this is usually how close the algorithm is to giving the right answer. For a binary yes/no categorisation this may simply be a count of how many times the algorithm gets the right answer. However, in other cases the fitness function needs to capture a measure of "*how close am I?*". For unsupervised learning the fitness is more about the coherence of the grouping or classification, for example, if there is some measure of similarity between items in the training set, this might compare the average similarity of items put in the same category to the average similarity of those in different categories.

The fitness function may not be explicit, but there is usually something that the algorithm is trying to be good at. Sometimes you may need to provide this directly yourself, or sometimes things that contribute to it (e.g. a similarity measure). In addition, you may often need to provide additional configuration parameters for the algorithm, such as the number of nodes in a neural network.

The output of the learning phase is usually some form of rule set. These may be very clearly 'rules', for example IF–THEN rules, or a decision tree, but may also be represented more abstractly, for example as weights in a neural network.

9.2.2 Application Phase

During the application phase (Figure 9.2) the rules are used to classify or process unseen examples, for example images to be classified, or board positions in the middle of a game. A recognition algorithm takes the rule set and the unseen example and allocates the class label or decides the next move.

In the case of supervised learning, this will be from the original set of labels used during training, in the case of unsupervised learning the categories, or other representation created by the learning algorithm itself.

In some cases, for example in many neural networks, the training and recognition algorithms look very similar, but this need not be the case. For example, a genetic algorithm may be used to create a set of IF–THEN rules, which are later converted into raw code to run on a target platform. The learning algorithm must in some sense

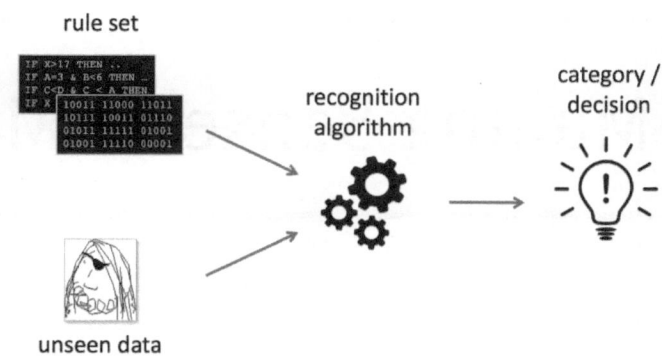

FIGURE 9.2   Application phase of machine learning.

know about the eventual recognition algorithm as it is creating the rules for it, but this can be relatively indirect.

9.2.3 Validation Phase

In Chapter 5, we noted that this stage may be omitted or not explicitly present. This is especially the case when further learning is taking place after deployment, for example the speech recognition in a home assistant that is constantly adapting to the voices of the home occupants. There are two different reasons one may want to validate a system:

Evaluation – checking how successful the learning has been

Interpretation – making sense of the rules generated

We will look at evaluation in more detail below (Section 9.3).

In the case of supervised learning, we might simply be happy with the evaluation, it does what it should. However, sometimes we want to understand in more detail. This can be trivial, for example a small set of IF–THEN rules, or apparently impossible, for example a billion weights in a large neural network. Even if we understand the rules, making sense of how they came about, the learning process, may be important. This issue of explainable AI is a topic in itself and we will return to it in Chapter 21.

For unsupervised learning, there are three broad ways the output may be used (see Figure 9.3).

Visualisation and understanding – Here the end point is to help a user make sense of the data. There may never be a further application phase.

The techniques for this are similar to those for explainable AI when the output is also used in an application (see Chap. 21).

Expert labelling – In some cases, the machine learning algorithm is used to allocate the data into groups/-clusters or perhaps create a representation in terms of a small number of features. An expert then looks at these and labels them (see also Chap 18). Later, when a new data item is seen it can be allocated the expert label. For example, suppose patients have been grouped by ML based on symptoms, a doctor then looks at the groups and labels each group by potential ailments. Later when a new patient arrives, the recognition algorithm allocates the patient to one of the learnt groups, and then this is used to associate them with the relevant ailment.

Pre-processing – The second way unsupervised machine learning is used is as a pre-processor for data that is then passed into another supervised machine learning algorithm. Revisiting the medical example, we may actually have a diagnosis for each patient in the training set, but initially ignore this to perform unsupervised learning. The label for each patient is then re-attached to the reduced representation created by the unsupervised algorithm and this is fed into another learning algorithm. This sounds a little indirect, but the unsupervised algorithm is effectively simplifying the data.

9.3 EVALUATION

For supervised learning we want to know whether the machine learning has been successful. There are two separate questions:

1. How well does it do on the training set?

2. How well does it do on unseen examples?

Sometimes you will see both figures quoted. However, we would normally expect better response on the training set than on unseen examples, so it is the latter that is most critical as this tells us how well our learnt rules generalise.

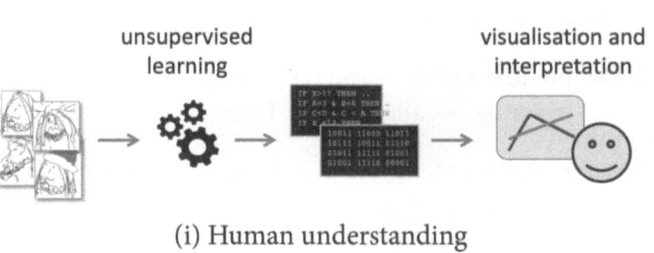

(i) Human understanding

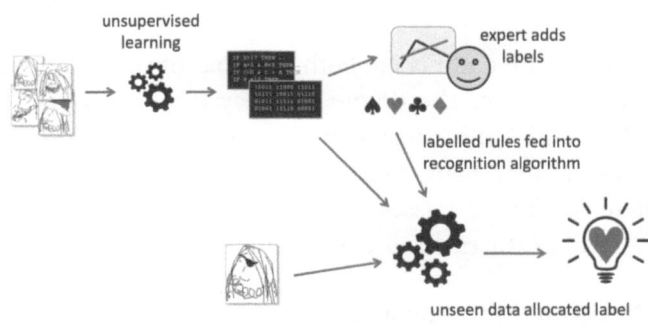

(ii) Expert labelling

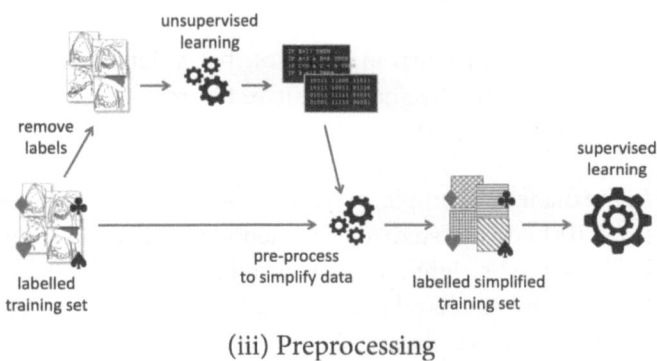

(iii) Preprocessing

FIGURE 9.3   Three ways of using the output of unsupervised learning.

9.3.1 Measures of Effectiveness

The most obvious measure is accuracy, how often the predicted classification or label is correct. However, this may be a very poor measure, particularly if the base rate of the thing you are interested in is low. For example, just under 10% of people worldwide have diabetes (the base rate). If a diagnostic test for diabetes always says "No" it will be 90% accurate ...but utterly useless. For Fabry disease affecting about 1 in 40,000 the simple "No" answer will be 99.975% accurate, so whenever you see an accuracy figure on its own, think base rate!

For a simple binary classification such as "Has diabetes Yes/No?", there are four numbers which between them define the effectiveness of the learning.

True positives (TP) – the test says "yes" and this is right

False positives (FP) – the test says yes, but it is wrong (in statistics called a Type I error)

False negatives (FN) – the test says "no", but it should have said "yes" (in statistics called a Type II error)

True negatives (TN) – the test says "no" and it is right!

Ideally FP and FN are both zero, but of course that is rarely the case. The simple accuracy measure is $(TP + TN)/Total$, but as we saw this is a poor measure if there is a large discrepancy between the number of people/data items in the two classes.

Two measures that are often quoted are:

Precision – the proportion of times that a data item with a positive test result really is positive
= TP/(TP+FP)

Recall – the proportion of times that a data item that really is positive has a positive test result
= TP/(TP+FN)

In the diabetes example, precision is addressing the question "if the ML system says someone has diabetes, how likely is it to be right", whereas recall addresses the question "if someone has diabetes, how likely is it the ML system will diagnose them?".

Sometimes these are combined into a single measure, the F score:

$$F = 2PR/(P + R)$$

This F score can be useful as a quick summary, but while this is better than the simplistic accuracy measure, do always look at the details, not just the single score.

9.3.2 Precision–Recall Trade-off

Often you can make choices that trade-off precision against recall. The simplest example is when the learning system outputs an "evidence for diabetes" score rather than a simple "yes/no". This can be converted into a "yes/no" by using a threshold. If the threshold is high, you reduce false positives, so precision is high, but also increase false negatives, so recall goes down. You may get similar effects by varying parameters for the learning system, such as the number of nodes in hidden layers of a neural network.

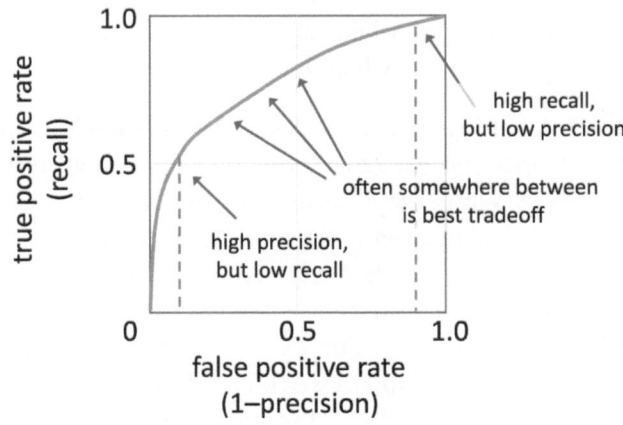

FIGURE 9.4 ROC curve – trade-off between false positive and false negative rates.

As you vary the threshold, or other parameter, you can plot precision vs recall on a graph and see a curve, called the Receiver Operating Characteristic (ROC) curve, where increasing precision reduces recall and vice versa. Figure 9.4 shows an example ROC curve. Note that the axes in a ROC curve are usually shown as true positive rate (same as recall) on the y-axis and 'false negative rate' (one minus the precision) on the x-axis. That is a small value on the x-axis is best and a large value on the y-axis is best.

Ideally we would have 100% precision (no false positives) and 100% recall, but that is rarely the case! The ROC curve can be used to answer questions such as "if I want precision to be at least 90% what is the best recall", or to make cost–benefit trade-offs. The curve in Figure 9.4 is typical, choosing a high value for precision means recall is low and vice versa. Often we choose a point somewhere on the middle of the ROC curve.

The ROC curve can also be used to compare different machine learning algorithms, say A and B. If you just test with a single threshold for each, it can be hard to compare the different precision and recall as neither may be better on both, but if you plot the ROC curve for them (Figure 9.5) and the A curve lies above the B one, you know that whatever parameter values you chose for B, there will be a choice of A parameter that will beat it on both criteria. As A is uniformly better than B, it would be the best choice between the two. Often things are not so easy. Suppose we want to choose between classifiers B and C; B is better than C if we require high recall, but it is worse when we want high precision. In addition, other considerations need to be taken into account, such

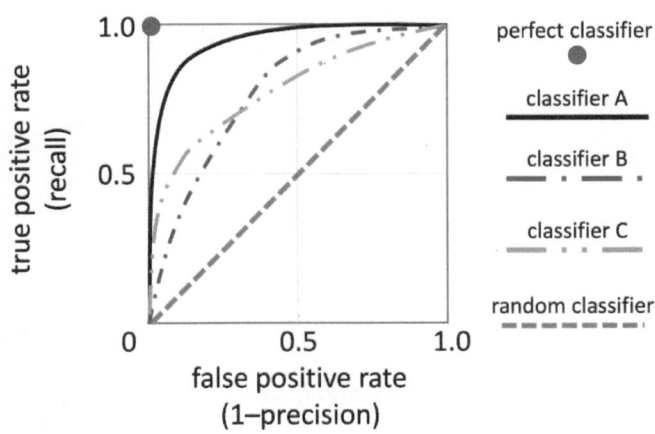

FIGURE 9.5   ROC curve – choosing between classifiers.

as computational cost of the classifiers and the relative harms of false positives and false negatives.

Sometimes the area under the curve on a ROC diagram is used as an overall measure of how well an algorithm is performing. Note that a completely random classifier that tossed a coin or rolled a dice to choose between the options would have an equal true positive and true negative rate. By choosing a different probability we can shift recall and precision, but they are exactly trading off one another. This random classifier has an area under the curve of a half, so an algorithm should be a lot better than that.

9.3.3 Data for Evaluation

In order to evaluate the output of a machine learning system, we need something to evaluate it against, that is data with a known label. It is tempting to use all your labelled data on training, especially if your dataset is not large. However, this means you have nothing left for evaluation and you can't tell if the algorithm can generalise to unseen examples.

To address this you normally 'hold out' some data for evaluation. A typical value is 10% of your data. You then train on the remaining 90% and evaluate on the hold out data items.

A slightly more complex variation on this is cross-validation. You split the data into ten random subsets, D1, D2, ..., D10 (or in general k subsets, known as k-fold cross-validation). You start by holding out D1 and training on the other 90% of the data (D2, D3, ..., D10) and evaluate the learnt model against D1. However, you then do the same for the other nine subsets in turn, holding

out D2, and training on (D1, D3, ..., D10); then holding out D3 and so on. By the time you have finished you have run the algorithm 10 times against different subsets of the data and evaluated it against every data item.

Of course with cross-validation you end up with 10 different rule sets. It is really more about evaluating the algorithm itself rather than a specific rule set.

9.3.4 Multi-stage Evaluation

A little extra care needs to be taken when dealing with 'meta' algorithms, high-level algorithms that apply multiple techniques, or the same technique with different tuning parameters. For example, in a multi-layer neural network, one might apply a higher level algorithm to choose the number of nodes at each layer. Similarly, a random forest algorithm might evaluate each of the individual decision trees in the forest to work out a best weighting between them.

The high-level algorithm is likely to use some variant of k-fold cross-validation to do its internal optimisation, and so it can be tempting to think this is good enough. However, this can be susceptible to the same issues of overfitting and lack of generalisation as the lower-level base algorithms on which it is working. You must apply the same principles to the high-level meta algorithm, holding back a portion of the training set for evaluation, even though you know the algorithm will do the same to the reduced set when it works with the lower-level algorithms.

9.4 THE FITNESS LANDSCAPE

The fitness function says how well the learning algorithm is doing. It is usually the sum or average of the individual fit of each data item, that is how close the algorithm is to correctly predicting the class label. Sometimes this is arranged so that larger values are better, but sometimes the other way round particularly when some sort of error is measured so that zero is the perfect score.

This is often visualised in terms of a landscape, either an energy landscape when the goal is to minimise error/energy or a fitness landscape when the goal is to maximise fitness. The coordinates of the landscape are the parameter values that are being learnt (e.g. weights in a neural network) and the height/depth is the fitness or error. When the parameters are two numbers, this looks just like a physical landscape with high mountains (optimal values) and deep valleys (low fitness). In reality there are typically large numbers of parameters, and these may be discrete, binary or even structured like a decision tree, so the image breaks down at that point, but if you think of the physical landscape you can obtain some useful insights.

In these terms learning is simply an optimisation task, find the parameters (weights, rules) so that the fitness is maximised (or error minimised). That is, find the highest point in the fitness landscape.

All algorithms make some implicit assumptions about the structure of the landscape. If there were no structure, then no algorithm could perform better than a Monte Carlo search, simply choosing parameters at random and selecting the best guess. Many algorithms assume some sort of locality, if one choice of parameters is good, then small changes in the parameters are also likely to be good; that is they assume that the landscape is relatively smooth, with few discontinuities

9.4.1 Hill-Climbing and Gradient Descent/Ascent

When you examine the internal mechanism of many algorithms, they are in part doing a form of 'hill climbing', looking for small changes from the current location that are better. You can explicitly choose to use hill climbing, but often this is implicit, for example mutation in genetic algorithms or the backpropagation rules in neural networks.

We need to work out the best direction to take. On a hill this is the direction that gives the most rapid height gain. Sometimes this direction is approximated by evaluating a number of possible small steps and choosing the one that gives the best result – imagine feeling with your foot in dense fog to work out which direction is best. However, if the fitness function has suitable properties, it can be calculated exactly by differentiating (finding the slope of) the fitness function with respect to the various parameters. Using this optimal direction is called gradient descent (or gradient ascent for maximising).

Backpropagation is just such a gradient-descent algorithm that is seeking to minimise the sum of squares of the differences between the actual and desired outputs. The multi-layer neural network can be seen as a series of function applications where each layer is a function of the previous layer and the weights of connections. Differentiating this gives precisely the chains of backpropagated errors in the algorithm. Strictly the exact gradient descent would work out the small weight changes due

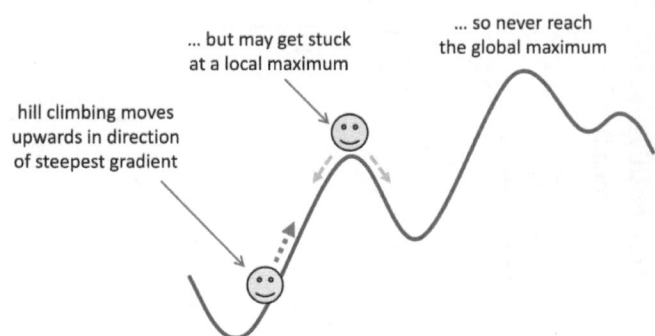

FIGURE 9.6 Hill climbing may get stuck at local maximum.

to the entire training set and apply them together. The incremental algorithm that updates the weights for each training example can be seen as an approximation to the optimal direction if the changes are all very small.

9.4.2 Local Maxima and Minima

A common problem in any optimisation is getting stuck in local minima/maxima as we saw when we first discussed hill climbing in Chapter 4. You want to get to the top of the highest mountain on a foggy day, you keep climbing upward until you get to the summit of what appears to be the highest point, but it is really just a small hill in the middle of a plain (Figure 9.6). If you look at a map, or the fog clears, it is obvious that the real mountain top is on the horizon. Sometimes to get higher you have to initially go downhill, but locally it is impossible to tell.

The learning algorithm can have exactly the same problem. It finds a selection of parameter values, a location in the fitness landscape, that performs better than any close location. If it is working by small increments, then nothing it tries is better, so it thinks it has got the best solution. You can think of the search problem as being in two main parts:

global search – finding a good overall region (Himalayas rather than the Netherlands);

local search – finding the best point within the region.

Algorithms get around this in different ways. For example, some have lots of attempts from different random starting points. In some cases this needs to be explicitly considered and parameters carefully chosen, but in some algorithms this happens almost as an accident of the nature of the algorithm.

In particular, the *overall average* effect of all the backpropagation steps in neural net training is a pure hill climb, but the fact that these are performed incrementally for each training item means the algorithm is taking a slightly wiggly path rather than directly downhill (to minimise error). In fact, this slightly wiggly path can be beneficial as it is adding an element of noise and makes it less likely that the network gets stuck in a local minima.

One algorithm that makes explicit use of these effects is simulated annealing (Chap. 4), which effectively tries random steps, and probabilistically decides whether to proceed based on how much better or worse the new position is; it is more likely to choose the better position but may also choose a worse one. The algorithm has a 'temperature' setting that determines how likely it is to take a step even if the new position is worse. The system starts off hot, so it is more likely to move near randomly around the landscape, but spending more time in the 'better' parts (global search). Over time the temperature is reduced, meaning it is less likely to take these speculative moves and instead ends up taking small steps more like a simple hill-climb (local search).

Genetic algorithms can similarly be seen as a form of gradient descent. Mutations effectively take small steps and the best individuals survive, but rather than following a single route, they are effectively trying multiple paths in parallel and with a lot of random choices. In addition, the inter-breeding in genetic algorithms allow sharing of partial knowledge of the fitness landscape (see Section 9.4.4).

9.4.3 Plateau and Ridge Effects

A related problem is plateaux, large areas of the fitness landscape where there is little if any improvement. That is, configurations of the parameters being learnt where no small local changes make any improvement at all.

A plateau often happens at poor points where certain combinations of parameters are so bad they make the whole solution poor. For example, a particular set of hat, coat, scarf and gloves where every colour clashes with every other colour, no small change makes it any better! There can also be ridges, where some small changes do make things much worse (stepping sideways off the ridge), but among the changes that don't make it worse, there is no clear direction. This may happen when the model's degrees of freedom are too large, so that lots of different model parameters all lead to reasonably good solutions.

In both cases, the problem is there is no clear direction. It might be that you simply end up with one of many 'good enough' solutions, but in fact it might be that beyond the plateau or at one or other end of the ridge, there are higher regions; it is just impossible to tell how to get there. The impact of the plateau or ridge in the fitness landscape is a corresponding plateau in the learning rate; you will see a long period of time with no overall improvement.

Furthermore it is possible that given the lack of clear global direction, the algorithm locally optimises to some irrelevant pattern (a local maxima); such patterns can often be a spurious feature of the training sets (overfitting).

It may be possible to automatically identify these issues arising within an algorithm, for example if the 'best so far' doesn't improve for a while, it may be possible to 'kick' the algorithm, increase mutation rate in a genetic algorithm or temperature for simulated annealing. Alternatively you might simply restart the algorithm entirely. The algorithm can always keep track of the 'best seen', so that if the radical move does not make any improvement, it can still return the solution in the plateau.

In addition, as this can be a sign of overfitting, it can be a clue that the architecture of the model needs to be changed, perhaps reducing the number of nodes in a layer of a neural network.

9.4.4 Local Structure

Genetic algorithms in addition make use of the fact that the 'shape' of the landscape is similar in different places so that if a subset of parameters are optimised for particular values of the rest of the parameters, they may well still be a good choice if one or more of the other parameters changes. This typically occurs when some of the choices are only weakly dependent on others.

For example, imagine trying to work out the best set of clothes to wear. If a particular choice of hat, scarf and gloves work together with one pair of boots, then even if you change your mind and wear a different pair of boots, the effort of working out that the hat, scarf and gloves look good together is not wasted. Of course, it might sometimes be that the boot colour clashes with the scarf, so you have to rethink, but for many boot choices, the rest of the outfit works.

9.4.5 Approximating the Landscape

Many algorithms do not use this local structure, in which case it may not matter if the landscape is in a sense slightly smoothed or fuzzy. When you choose your mountain climbing holiday, you may initially use a small-scale map, where the whole mountain region simply appears as a clump, but as you start to climb a specific mountain you will use a large-scale map that shows individual peaks and paths.

One variant of this is to use a small sample from the training dataset in early stages of learning and then use more of the data later for fine tuning – here using fewer training data items initially is giving a less precise fitness landscape. Some image processing algorithms start off using lower resolution copies of the image and gradually move to higher resolution – here it is the accuracy of the individual training data items that is being manipulated.

If the training data is provided by a simulation, as in some Industry 4.0 applications, then there may be choice in the simulation that can make it more or less accurate. For example, many simulations are themselves iterative, so taking fewer iterations gives a less precise answer, which can save time in the early stages of learning.

In each case, the smoothing of the fitness landscape due to the poorer quality of the early training data not only makes this faster and more efficient but may also, paradoxically, make the global search stage better, more likely to end up with robust generalisable solutions.

9.4.6 Forms of Fitness Function

As noted, most commonly the fitness function can be seen as the sum or average of the accuracy of each data item:

$$fitness(params) = \sum_{(d,v) \in Training\ set} diff(v, alg_{params}(d))$$

Here the difference function, *diff*, may take various forms depending on the algorithm and data. In the case of classification algorithms, it may simply be a binary 1 for matches, 0 for doesn't match. For more numeric outputs, it could be the absolute difference or, very commonly, the squared distance.

The fact that the fitness function is effectively the sum of lots of little per-data-item costs is crucial for many machine learning algorithms that present the training set item by item and modify the parameters slightly for each

item. The net effect is similar to a single gradient descent over the landscape as a whole.

However, the fitness function does not always split into lots of per-data-item costs but may also involve aspects that relate to the algorithm or dataset as a whole:

algorithm metrics – Often there are measures of complexity of the rules or structures generated by the training. We may want to penalise more complex cases (e.g. a very large decision tree) in order to improve generalisation and explanation.

cohort metrics – In some cases we may want to take into account the joint outputs of multiple data points. This is particularly the case for unsupervised algorithms, for example, the coherence of the clusters. Alternatively we may want to ensure that the algorithm behaves fairly for different classes of people, for example working equally well for women and men.

Some learning algorithms, for example k-means, effectively build this into their normal operations. However, others may need modification. For example, in the ID3 algorithm it is hard to modify the entropy-based learning steps for the decision tree, so additional pruning stages are often added after initial training to simplify the tree. Alternatively one might more radically change the learning mechanism, for example retain decision trees as the outcome, but use genetic programming to create suitable trees as it is easy to add extra factors in the genetic algorithms fitness function.

9.5 DEALING WITH COMPLEXITY

Many factors can increase the complexity of the machine learning process. Some are specific to particular algorithms, but we'll look at a few of the more common factors here.

9.5.1 Degrees of Freedom and Dimension Reduction

Do you remember school geometry? Euclid's first postulate is that a straight line can be drawn between any two points.

Imagine you have been collecting data on butterfly numbers every day for three years. You have over a thousand points, and they run close to a straight line; you can reasonably believe that you have found a pattern. Even if you collect three or four days' data and they lie exactly on a line, you might feel this is suggestive. However, if you have just two days' data, it is not interesting at all that they are connected by a line; it is bound to be the case.

Looking at this in terms of numbers, if you have two observations on day 1 and day 2, that is just two numbers. The equation of a line is $y = mx + a$, it has two parameters: m – the slope and a – the intercept. You have two numbers in your data and two parameters to adjust, hardly surprising you can make them fit.

Roughly speaking, if you have N parameters you can exactly fit data with N independent numbers. However, this rarely represents anything interesting, an extreme form of overfitting. An interesting generalisable pattern needs to have a lot more data than the number of parameters being fitted.

In statistics the term for the number of independent things that are being fitted in the data is the degrees of freedom. If this is not substantially larger than the degrees of freedom in the model being fitted, then you are likely to get overfitting. This is when the model is not creating a general pattern but simply matching potentially arbitrary aspects of the particular training set.

For example, if we have N items in the training dataset $d_1, d_2, ..., d_N$ with a classification $c_1, c_2, ..., c_N$, respectively, then we could create the rule set:

```
Classify(x)
    IF  x = d1   THEN   c1
    IF  x = d2   THEN   c2
    ...
    IF  x = dN   THEN   cN
```

Totally accurate on the training dataset but unlikely to be useful for later use.

For machine learning models it is very easy to have lots of parameters being fitted. For example, if you have a fully connected neural net with 3 layers: 10 inputs, 10 outputs and 20 nodes in a hidden layer, this will have 400 weights to be fitted. In this case, and often, the number of weights grows with the square of the size of the input. This means that typically you need to have a lot more training items than the number of fields or columns in each data item.

At first this may not seem too daunting, but it is easy to exceed these numbers. Imagine you are doing interviews for medical research – you may have dozens of questions

for each person and interview several hundred people. That sounds fine. However, maybe only 10% of people have ailments that you are interested in with perhaps 5%, 4% and 1% with three different variants of the disease. The data looks good for distinguishing ill/well, but once you look at the finer distinctions between the variants of the illnesses you have too few people to avoid overfitting.

This is even worse when data is gathered automatically. In the UK, the Met Office has 200 weather stations gathering data including "*air temperature; atmospheric pressure; rainfall; wind speed and direction, humidity; cloud height and visibility*". That is at least 1600 readings. These are gathered every minute, but there will be a high correlation between subsequent readings; so, in terms of new data, maybe an equivalent of about 50 independent readings a day. At least a month's data is needed before one can start to have any confidence in fitting of the data.

If you are dealing with image data, then there are millions of pixels in a medium resolution image and 8K TV at 60Hz frame rate is about 3.5 billion RGB channel values in a minute of video. You would need a lot of training data for that! In practice there are ways to reduce this, crucially in those 3.5 billion RGB values neighbouring pixels and successive frames won't be so different. We looked at ways to harness this through dimension reduction in more detail in Chapter 8. However, this does explain why unsupervised machine learning is sometimes used as a pre-processing stage to simplify data.

9.5.2 Constraints and Dependent Features

A complication we have seen in some applications is that there are constraints – not all parameter values are possible. For example, imagine you are using a genetic algorithm to design a kitchen, and each appliance is allocated a position in the kitchen. We might need the sink to be within a metre of the water connections, and an electric cooker to be at least a certain distance from the sink.

We've seen that machine learning can be seen as an optimisation problem, finding the highest point on the fitness landscape. In this context one often talks about solutions in terms of feasibility and optimality:

Feasibility – Does the proposed solution satisfy the constraints?

Optimality – Is the proposed solution the best among those that satisfy the constraints?

In some application areas, such as timetabling, the core requirement is finding any feasible solution. Indeed, constraint satisfaction is a whole area of study with its own specialised algorithms (see Chap. 4). However, it is also a potential application of some forms of AI: recall that genetic algorithms are able to make use of repeated or similar structures within the problem space, which is precisely a feature of constraint satisfaction problems.

Assuming you have found a feasible solution, an obvious approach is to look for close solutions that are also feasible. However, this suffers a similar problem to local maxima. If one thinks of the fitness landscape, but with areas covered with water representing the infeasible solutions, the areas of feasible solutions may well form separate islands. If you only perform local search within these islands, but the best solution is in a different island, you are stuck. The author is writing this in Wales, and the highest point reachable from here without crossing the sea is Ben Nevis in Scotland at a height of 1,345 metres, somewhat short of Mount Everest at 8,849 metres.

There are various ways around this problem.

Sometimes the constraints can be removed by reparameterising the problem. For example, a financial planning application might include the constraint that a bank balance always exceeds the overdraft limits:

```
balance >= limit
```

This is a constraint between the two variables.

However, we could instead represent these as a limit plus available cash, that is:

```
available = balance - limit
```

The original limit can be recreated if needed (balance = limit + available), so no information is lost, but this reparameterisation no longer has any constraints between parameters, available can be any non-negative number. Another way is to treat solutions that break constraints as bad but not completely off limits. We do this by incorporating the constraints into the fitness function. Imagine the original fitness function is $fit(x)$, this takes a potential solution x and says how good it is. However, in addition there are N constraints $c_1(x)$, $c_2(x)$, ..., $c_N(x)$ that all have to be true. We create a modified fitness function:

$$fit_{new}(x) = sig(fit(x)) + \text{number of } c_i(x) \text{ that are true}$$

where $sig(z)$ is a sigmoid function that maps the original fitness function to the range (0,1).

This modified function has the properties that

1. reducing the number of broken constraints always increases $fit_{new}(x)$, irrespective of $fit(x)$

2. for the same number of violated constraints, improving $fit(x)$ improves $fit_{new}(x)$.

The first means that an optimal solution always satisfies as many constraints as possible. The second means that it gets the best possible value of $fit(x)$. So if this modified fitness function is given to a machine learning algorithm, it will seek a solution that is both feasible and optimal.

A special kind of constraint is a dependent feature, where some field of a data record only exists, or is only relevant, when other fields have particular values. For example, one project used various forms of constraint solving and optimisation for early submarine design. The overall layout of the submarine depends critically on the chosen fuel source. If it is diesel, then there need to be fuel and air tanks and the means to exhaust gasses. If nuclear, then there needs to be extensive shielding between the reactor and crew quarters. If the data-representation of each data record is flat, then some of the fields are irrelevant, rather like junk DNA; this is in one sense wasteful, especially if the volume of data is large, but for certain algorithms can be a positive feature.

Various algorithms were used including a commercial constraint solving system, modified hill-climbing and (most successful) a genetic algorithm. In the latter the dependent fields for diesel (tank size, placement, etc.) were present even when the nuclear option was active and similarly fields pertaining to nuclear power (shielding thickness, placement) were in the representation when diesel was selected. This meant that work optimising the dependent options for one fuel type was not 'lost' from the gene pool even if the fuel type switched and so was still there if the reverse switch happened later.

Note that for a genetic algorithm both constraints and dependent features can be used as a guide to have more efficient breeding rules. Often when two individuals are combined in a genetic algorithm, the new artificial genes are randomly chosen from the 'parents'. However, in real genetics the chromosomes break and recombine at a few specific points (cross-over). Over many millennia genes that code for highly related features have migrated to be close on the chromosome and thus are more likely to be inherited as a single unit. Genes that work together breed together.

In the case of the submarine example, if two individuals have the same fuel type, then the fuel specific features can be inter-mingled, but all of the 'junk' fields for the unselected fuel type should be taken as a unit from one parent or the other. If they differ in the fuel type, then the dependent fields of the relevant parent would normally be copied intact. For constraints a similar principle can be applied. Fields that are closely related via constraints can be biased so that they are most often inherited as a unit from one or other parent, with only occasional mixing.

9.5.3 Continuity and Learning

Note how encoding constraints into the fitness function changes a binary "feasible vs not feasible", into a softer "more or fewer constraints". This can make it easier for machine learning algorithms to work, incrementally improving initially infeasible solutions towards ones where all the constraints are satisfied.

It is rather like the child's game of "hunt the thimble". The parent hides a thimble and the child starts to look: "cold", says the adult as the thimble is far away, and then as the child hunts further "warmer", "warmer", "colder", "warmer", "hot", "you've found it!". Imagine instead that the adult simply says, "no", "no", "no" unless the child is exactly where the thimble is hidden. It would not be much of a fun game and also it would be a lot harder to play. In general, whether in human or machine learning, continuity helps, if there are shades between "not right" and "got it", it is easier to find the right direction to improve.

Another very successful example of this is the sigmoid activation functions used in backpropagation in neural networks (see Figure 6.8). Early perceptrons simply had a hard threshold. If the inputs exceeded the threshold the artificial neuron 'fired', otherwise nothing happened. There was no difference except at the exact point when the input passed the threshold. The sigmoid function still has a roughly similar shape, but slightly softened. If the inputs are nearly at the threshold, the node fires a little, if they are only just over, it fires slightly less than when it is fully on. When the backpropagation learning step works out the differential (slope) of the sigmoid function, it is

effectively saying "what is the impact of a small change", and this is used to shift the weights towards a better solution.

Look out for these effects in different algorithms. Sometimes hard edges are softened by the shape of functions, as in the sigmoid, sometimes by adding probabilities: each toss of a coin is either a head or a tail, but over time the probabilities can shift continuously. If you are creating your own algorithms, think about whether it is possible to deliberately introduce these soft boundaries.

If you are using a pre-packaged algorithm, you may not be able to change the algorithm's internal behaviour, but, as with the constraint satisfaction example, you may be able to choose a fitness function that is better for learning.

If we revisit the fit_{new} function for constraint satisfaction, there are still quite hard boundaries at the point when each constraint changes from not satisfied to satisfied. In some cases this can be further softened by adding levels of 'nearly satisfied'. For example, the sink has to be within 1 metre of the water, so we give a score of 1 for this, but maybe we can give a score of 0.5 if it is 1.1 metres away, and possibly even give it slightly less than a full 1 out of 1, if it is just at the limit, say 99cm away. In the end we want all of the constraints to be fully satisfied, but by softening the constraints it is easier to find the solution.

Note that if soft constraints like these are used, the first of the properties is no longer true for fit_{new} as given, but there are ways of dealing with this, the simplest is simply to make the overall fitness function a weighted sum of the constraint satisfaction score (css(x)) of each item x:

```
overall fitness(x)
        = fit(x) + penalty * css(x)
```

The penalty can be gradually increased during learning, so that the algorithm is initially quite relaxed about a few broken constraints but gets more strict as it gets closer to a chosen solution.

9.5.4 Multi-objective Optimisation

The problem we faced with the precision–recall trade-off is that both are important. Depending on where we set the threshold, we might make one better but the other worse. In fact this is also often an issue with the problems we are solving with AI algorithms. Consider a company

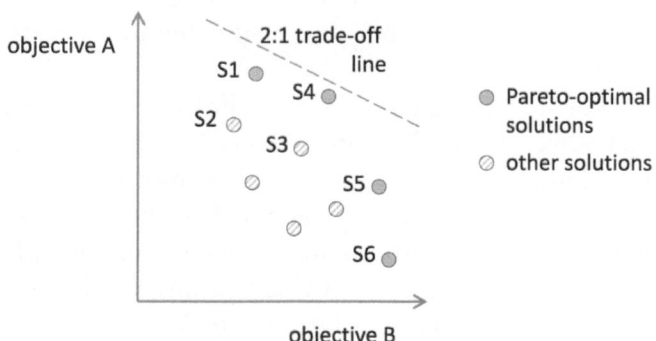

FIGURE 9.7   Potential solutions in multi-objective optimisation.

trying to optimise productivity while minimising environmental impact, or a country trying to offer the best healthcare and education while keeping taxes low.

These are called multi-objective optimisation problems. Instead of a single fitness function, we have several, one for each aspect we want to optimise. Typically improving one may reduce another. We have already seen a special case of this in the ROC curve where precision and recall are two competing optimisation criteria.

Figure 9.7 shows multiple potential solutions of a multi-objective optimisation problem with two objectives A and B. Solution S1 is better than S2 on all objectives, so there is no problem choosing between them. Similarly, S4 is better than S3 on every objective, so we would not choose S3. When choosing between S1 and S4 (or between S2 and S3), things are far less clear. S1 is better than S4 on objective A, but worse than S4 on objective B.

Looking more closely there is no solution that is uniformly better than S1 and similarly no solution that is uniformly better than S4. Solutions S5 and S6 also share this property. We call such solutions Pareto-optimal after the Italian economist Vilfredo Pareto who first identified the issue in the 19th century. It is clear that we would prefer to have a Pareto-optimal solution as any of the others can be improved upon for *all* objectives. However, the choice *between* Pareto-optimal solutions is far less clear.

Sometimes there is a clear hierarchy, for example in Figure 9.7 we may decide objective A is most important in which case S1 is the chosen solution. In the case of constraints, we needed to ensure that as many constraints as possible are satisfied before worrying about optimising the target metric. The function $fit_{new}(x)$ captured this, but

this is only possible because the number of constraints satisfied is discrete. If the primary objective is continuous, then there may be never-ending tiny improvements one can make to the primary objective, so that one never worries about the other at all. For example, if objective B were the primary criterion in Figure 9.7, then we would choose solution S6; yet it makes a very small improvement for objective B compared with S5, while being only half as good at objective A – is this really what we want?

Another approach is to add a weight to each objective and create the overall fitness function as a weighted sum of the fitness of each one. This effectively gives a value to the level of preference between the objectives. The dashed line in Figure 9.7 shows a 2:1 trade-off where objective A is given twice the weight of objective B. With this weighting S4 comes out best.

There are also algorithms that seek to find all Pareto-optimal solutions (the Pareto frontier), especially if they can then be visualised as an aid to human decision making.

9.5.5 Partially Labelled Data

We have seen that algorithms are normally divided into supervised and unsupervised learning, depending on whether the training data is or is not labelled with classes or intended outputs. However, sometimes we may need to perform semi-supervised learning when the dataset is only partially labelled.

If there are only a few unlabelled items, then we may simply discard these and use a fully supervised technique. Similarly if the labelled data is too small, we might simply ignore the labels and use a full unsupervised technique. However, where we have some form of partial labelling it may seem wasteful either to discard the unlabelled data or to discard the labels from the rest. Ideally we use all the available data.

There are two forms of partially labelled data:

partial outcome labelling – Here we have the desired output labelled for some of the data. For example, we might have some ECG traces that have been labelled by a heart specialist as in danger of imminent heart attack, or as less critical, but also have a large bank of unlabelled ECG traces from elsewhere.

intermediate value labelling – Here we may have partial or complete labelling of some derived features that are deemed to be of interest for the final outcome as well as (complete or incomplete) labelling of outcomes. For example, some or all of the ECG traces might be labelled as having different forms of arrhythmia.

In the former case, partially labelled outcomes, we can adopt a two-phase approach. First use an unsupervised approach on the large unlabelled dataset in order to create some form of data reduction, perhaps clusters or principal components. Then use this to reduce the dimensionality of the labelled data provided as input to a supervised algorithm.

Alternatively, we may be able to modify the algorithm itself. For example, with k-means we can initially seed the clusters with both an initial central data item and a classification. Unlabelled data items are simply added to the nearest cluster. For a labelled item, if the closest cluster has the right label, it is also simply added in. However, if the closest cluster has the wrong label, its centroid is 'pushed away' from the data item and the item is added instead to the closest matching cluster.

The case of intermediate labels can likewise be approached in different ways.

One approach is two-stage. First some form of supervised or semi-supervised learning builds a classifier for the intermediate feature based on the raw inputs. Then a second classifier is built for the final outcome where the intermediate feature is added as an additional input (maybe more heavily weighted).

Alternatively, we may seek to embed the intermediate labelling into an existing algorithm. For example, assume we are using some form of deep learning with a pinch-point layer. We can choose an arbitrary node in that layer and 'clamp' it to the value of the label (or for more complicated values choose a small collection of nodes). That is, downstream (between pinch point and output) it is treated as though that was the value of the node(s) for generating the output and subsequent backpropagation, and upstream (between pinch point and input) it is treated as the target output for the node(s) for backpropagation. When the label is not present the node functions as normal in the backpropagation learning process.

9.6 SUMMARY

In this chapter we've taken a high-level view of machine learning, complementing the more specific approaches discussed in earlier chapters and drawing out general issues. Some of this, such as the supervised–unsupervised distinction, has acted as a principled recap of material presented previously in a piecemeal fashion. Other material has been new including the in-depth analysis of accuracy metrics and trade-offs using the ROC curve. The fitness landscape is a crucial part of the conceptual understanding of machine learning. It can be used as a vocabulary to discuss issues arising during practical application, as a way to suggest criteria for algorithm selection and as an inspiration for the development of new techniques. We have also looked at a number of issues that can be particularly difficult in applying machine learning including dealing with excess dimensionality, constraints, continuity, multiple optimisation criteria and partially labelled data.

9.1 A new early test has been developed for tantilitis, a condition which is not usually diagnosed until symptoms appear. In order to validate the test, 10,000 volunteers were administered the new test. Of these one hundred tested positive and of these 80% went on to develop the condition. Of those who tested negative, the vast majority (95%) were indeed disease free. Calculate:

a. The numbers who were true positive (TP), false positive (FP), false negative (FN), and true negative (TN).

b. The Precision and Recall.

c. The F score.

9.2 Eight employees are being considered for a new post of head of design. They have each been evaluated against three criteria: diligence, efficiency and creativity. The scores are as follows:

	Diligence	Efficiency	Creativity
Anderson	90	60	50
Brown	70	70	40
Clark	40	30	50
Davies	20	90	20
Evans	80	60	20
Fraser	30	30	90
Gordon	10	60	90
Hughes	60	50	40

a. Which employees would be Pareto-optimal choices for the position?

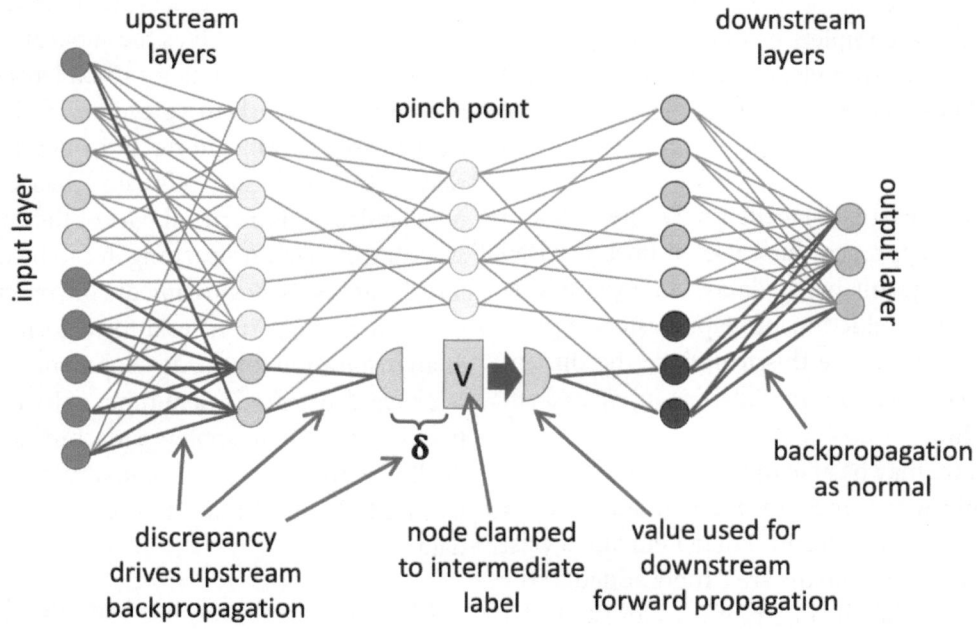

FIGURE 9.8   Pinch-point node clamped to intermediate label in semi-supervised learning.

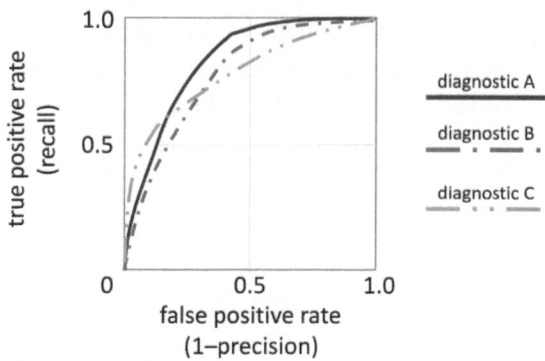

FIGURE 9.9   ROC curve for Exercise 9.3.

b. To resolve the choice one suggestion is to simply sum the scores. Given this, which candidate is optimal?

c. Because of the nature of the job, there is an argument to weight the criteria in the ratio 1:2:3. Which is the optimal choice given this weighting?

9.3   Three diagnostic tests (A, B and C) are being considered for early screening for tantilitis. Each is capable of giving a strength reading, so can be adjusted for sensitivity. Figure 9.9 shows the ROC curves for each diagnostic test.

a. Your colleague suggests diagnostic B as it seems more consistent – what do you think?

b. Another colleague suggests using area under the curve. If you follow this advice, which test would you choose?

c. Assume that the vast majority of people do not have tantilitis, and the follow-up tests to verify a potential diagnosis are very unpleasant and costly. Given this which test would you choose and how would you adjust its sensitivity threshold?

d. Assume instead that follow-up tests are cheap and easy but that the health consequence of missing a diagnosis is very severe. Given this which test would you choose, and how would you adjust its sensitivity threshold?

Although the concepts introduced in this chapter are familiar vocabulary for those working in machine learning and neural networks, it is hard to find detailed works focused on these topics.

FURTHER READING

S. Sinai and E. Kelsic. A primer on model-guided exploration of fitness landscapes for biological sequence design. *arXiv:2010.10614*, 2020.

Although targeted at those working in biological science, this is an accessible article that has lessons for anyone wanting to apply machine learning in practice. It systematically covers ways to make informed choices about appropriate methods using a fitness-landscape-oriented approach.

K. Traoré A. Camero and X. Zhu. Fitness landscape footprint: A framework to compare neural architecture search problems. *arXiv:2111.01584*, 2021.

A more mathematical treatment of the fitness landscape.

Data Preparation

10.1 OVERVIEW

In this book we normally use examples crafted for the particular representation or algorithm being presented. This is of course designed to be helpful for you to learn about each, but the real world is often somewhat more complicated. Data is rarely in exactly the right place or form that is needed for a particular algorithm, there is almost always some form of manipulation required, sometimes fairly systematic, others more 'hacking'. This manipulation is often called data wrangling.

In this chapter we'll look at some of the typical stages and processes used.

10.2 STAGES OF DATA PREPARATION

Figure 10.1 shows some typical stages in preparing data for machine learning or another form of AI algorithm. However, each dataset has its own quirks, and so the processes needed and the order in which they apply will vary substantially. In some ways it is better to think about a data preparation toolkit, a set of tools, techniques and heuristics to use at various points and not necessarily in the same order for any particular datasets.

Raw data is found in various places: sometimes it is relatively well processed for algorithmic manipulation in existing data files or feeds, but often it is more raw or unstructured, gathered from web pages or data streams.

Once the data is extracted from its sources and gathered together there will be various stages of normalisation, transformation and data cleaning, some

dealing with errors or omissions in the data and some changing it into formats and types that are suitable for subsequent processing.

Problems with the data may not be apparent at once, and so there is often a degree of iteration, where sanity checks or data validation at a later stage may highlight changes to earlier processing. This may highlight problems or properties of the initial data, but it is also important to make sure that data transformations are appropriate and correct. It is too easy to apply automated transformations without fully understanding their impact.

10.3 CREATING A DATASET

Sometimes you are given a well-described and well-managed dataset to analyse, or you may have control of the initial collection yourself. However, often you need to bring the data together from multiple sources, of varying quality.

10.3.1 Extraction and Gathering of Data

If you are fortunate, data will be in a semi-structured form such as CSV, tab-separated files or even a database. Care may still be needed, for example to convert dates to a standard format. Names and addresses also often have very different storage formats – for example a single field vs. separate given and family name fields. You may find examples where several things are encoded in the same field, perhaps a list of hobbies delimited by commas or semicolons.

Often some form of data cleaning is necessary as data may have been mis-entered or even entered using different character encodings. Some may be impossible to detect except by going back to original sources; however, often it is possible to create validation rules. This can include simple face-validity or sanity checks, such as

DOI: 10.1201/9781003082880-12

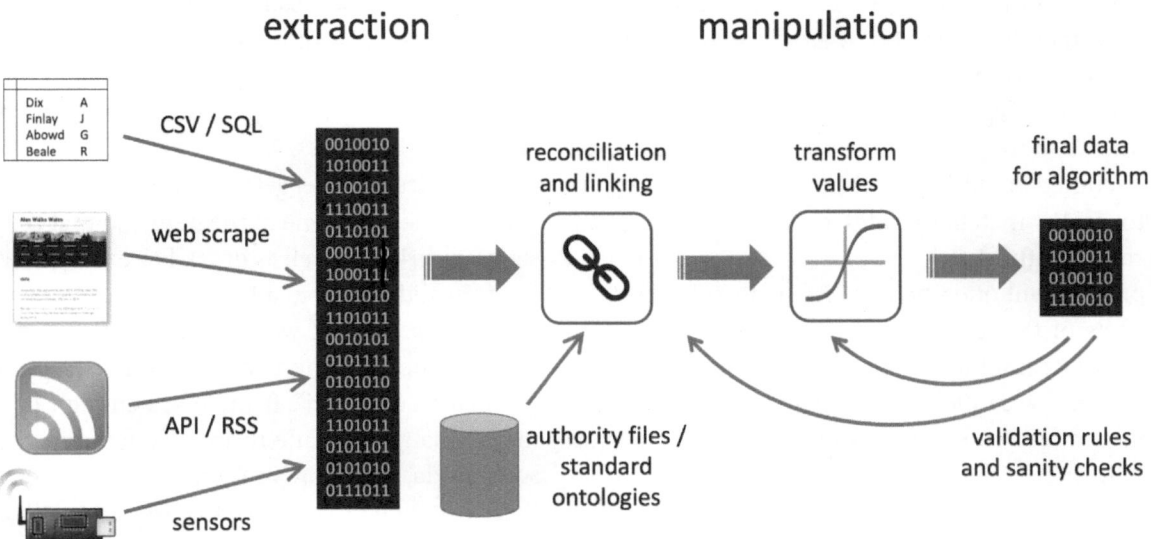

FIGURE 10.1   Typical data preparation stages.

checking that the date is in the right format, but may also include deeper semantic integrity, for example in historical records checking that the date of birth precedes the death date and that the age is not more than 150 years (to catch century digit errors).

In other cases the raw data is in free text or semi-structured text such as web pages. The experienced data wrangler becomes expert at a range of Python or similar scripts to process text looking for patterns; these typically include a combination of:

Known structure – for example, CSS classes on web pages, or row/column location on an old 80x25 screen-based system.

Keyword matching – for example, "Bibliography" to mark the beginning of the references section of an academic paper, or "ISBN" before the 10 or 13 digit ISBN of a book.

Regular expressions – for example, the following regular expression to find possible names:
/[A-Z][a-z]+ [A-Z][a-z]+/

Hand crafted parsers – for example, looking for a line containing a number and a street word such as 'Road', 'Street' or 'Avenue'.

The semi-structured data may be obtained from human-readable output of a legacy data-based system targeted at the web or old-style terminals. In such cases one usually refers to this process as web scraping or screen scraping.

This process may also itself use AI or machine learning, for example there are parsers for academic citation lists based on string-learning models.

Special care is needed when a single output dataset is composed from several different sources as there may be differences in collection style. For example, historic climate data is based on a variety of instruments, and so care is needed to ensure that this does not give rise to anomalies. In some cases there is overlap between datasets, which can be used for automatic cross-calibration, but in other cases this will need to be hand-coded.

10.3.2 Entity Reconciliation and Linking

If data comes from a single source and has been carefully constructed, there will be some form of unique identifier for each object in the data. However, real data may come from a variety of sources each of which may refer to the same thing in a different way. For example, the author is @alanjohndix on X/Twitter, "Alan Dix" on most academic papers and "Alan John Dix" on his birth certificate. In addition, at the time of writing, there is an Alan Dix who is a governor of the Leeds Playhouse, so the same apparent identifier in two different datasources may refer to different people or things. Matching entities between different datasources is thus a critical part of many data gathering exercises.

Having the same entity linked across datasources is an important first step for many large-scale AI or ML processes. However, AI may also be used as part of the process. Typically, there is a combination of coded rules, and weighted similarity metrics. These may be hand-crafted or use a machine learning algorithm either to create matching rules or choose weights of hand-crafted features.

The process will often be iterative, with rules used to match parts of the dataset, which are then checked by hand either by sampling or focusing on those where the algorithm has low confidence. This will identify misidentification and lead to either fresh hand-crafted rules or feedback to an ML algorithm.

Case Study: SAIL Databank

Swansea University is home to the SAIL Databank, which stores archives of health histories of nearly every person in Wales linked via a pseudonymised unique identifier for each person [104]. The databank is heavily used (with extensive privacy preserving protocols) for health and social research and policy purposes. This data has been collected over a period of more than 15 years using a wide variety of health-related datasets from different health services and government sources across Wales. That is, it is coming from heterogeneous sources and thus requires substantial work to connect the records.

Each individual has a unique health service number and National Insurance number, but few people know these in the UK; so if someone visits a hospital, the record will have a combination of name, address and their normal doctor's surgery ... again if known. Of course addresses change, postcodes may be entered incorrectly, and in general linking these records to the same person is not trivial. The expertise and algorithms to do this linking have become one of the core capabilities of the centre and one of the reasons it has become a world leader in the area [180].

10.3.3 Exception Sets

Crucial to many early data preparation tasks are exception sets. These are sets of specific examples where the rules do not apply. For example, matching datasets of 18th- and 19th-century concert venues in London was complex because Almack's Assembly Room had changed its name to Willis' Rooms [85, 86]. No amount of AI or human intelligence could guess this, so it is entered as a special case. Other examples would be whether a name has been misspelt in a record, or unusual formats such as the postcode for the former Girobank in the UK "GIR 0AA", which doesn't obey the normal rules for a UK postcode.

If the exception set gets too large, one might seek more generic rules for some of the items, so that the exception set may shrink as well as grow. For example, when looking for valid names in a UK/US context, one might start looking for pairs of words, each starting with a capital. After a time, one might start to build exceptions such as "Leonardo da Vinci". If there are enough of them, one might change the rule to include family name prefixes such as 'da', 'de la' or 'van'. The exception set entries for these can then be pruned. However, later still exceptions such as "Henry VIII" might start to proliferate and a new rule for Queen/King names might be added.

10.4 MANIPULATION AND TRANSFORMATION OF DATA

Often the values that you have for features of your data are not of the kind needed by an algorithm, so various forms of transformation are needed. You may also need to make decisions about missing values (where there is simply no value collected/stored for the feature) or outliers (data values that seem in some way extreme).

10.4.1 Types of Data Value

First of all, get to know the data you are working with, do not just 'throw' it at an algorithm and hope for the best.

Typical kinds of data you might encounter include:

Binary – Where there are exactly two possible values; e.g. true/false, 0/1, adult/child;

Categorical/enumerated – Where the data comes from a finite set of known values; e.g. blood type O/A/B/AB, class of employment;

Integer – Whole numbers: e.g. 0, 1, 2, 3, -42;

Continuous – Where the data takes on any numeric value: e.g. 3.142;

String – Where the data is alphanumeric, but relatively short: e.g. names;

Text – Longer alphanumeric data, sometimes in known meaningful language; e.g. the text of this section;

Rich text – HTML, RTF or other formats include fonts, character styles (such as bold) and possibly sections which can be used to help parse the text;

Hierarchical categorisation – Some form of class/subclass, part-whole, or taxonomic categorisation, for example the Dewey decimal code in a library, or a file path;

Images/media – These may be static such as a picture or time varying such as audio or video;

Temporal data – For example an ECG trace, climate data or sampled audio (see Chap. 14);

External links – Identifiers of other data items; e.g. URIs in linked data, external keys in a relational database;

Structured data – A single value might contain structured data such as XML or JSON encoded as a string;

Special formats – For example medical equipment or geographic information systems have their own data formats.

In addition, numeric data may be bounded, for example positive or in a particular range such as [0,1] for probabilities, and string data may be in particular formats, such as dates or ISBNs. Also data may be coded numerically but may really be categorical (e.g. multiple choice options).

This looks like a relatively easy distinction to make, but an integer might represent a number, time stamp or database identifier. Consult documentation (if there is any!) or talk to the people who collected the data or gave it to you so that you understand how the values you are seeing relate to the real-world things they encode.

For numeric data plot the values. Are there large numbers of zeros? If so, does this really mean zero, or a missing value indicator? If the numbers are integers and they are exactly or almost continuous over a range, then this is a good clue that they represent a database id or sequence number. For string fields, try sorting by the field and/or calculating counts for each value. This may help you see whether they are unique (and maybe an identifier field), only take on a small number of values (in which case they might be categorical) or have a small number of very frequent values, which may be a special code (e.g. 'unknown' in an address field).

10.4.2 Transforming to the Right Kind of Data

The particular algorithm you want to use, or the implementation of it that is easily available, may require a particular kind of data. For example, some genetic algorithms need binary data, statistical analysis usually requires numeric data. Typically, you will find yourself making data that is in some ways smaller or more constrained, for example turning string data into categorical data, or continuous data into binary.

Sometimes the algorithm you are using will perform transformations itself. For example, decision trees (Section 5.5.2) are usually based on binary decisions at each branch point. The algorithm will accept categorical or continuous data but create binary decisions based on these (e.g. "salary $> 5,000$"). If this is the case, it is still helpful to understand these processes in order to make sense of outputs (e.g. sharp changes in behaviour at a salary of 5,000). Other times you will need to perform data conversions yourself. Here are some common things you may need to do:

Numeric to binary – Choose a threshold T and use $x \leq T$ vs $x > T$

Integer to categorical – If the integer values are bounded (say 1 to 7), simply treat each number as a category label: $1 \rightarrow$ cat_1, $4 \rightarrow$ cat_4, etc. If the values are unbounded, then you may need to have catch-all categories for those bigger than some maximum value (another threshold choice).

Continuous to integer – Round to the closest whole number or use 'floor' to get the integer part of the value, for example $3.142 \rightarrow 3$. More generally, choose multiple thresholds that split the data into a small number ranges; e.g. if the thresholds are 0,1,2,4,8,16,32,64..., then $42 \rightarrow 6$.

String, text or link to categorical – Use some sort of clustering algorithm on the values or a pre-existing classification (e.g. language of text).

Categorical to binary – One option is to generate indicator variables: an "is it in category A" variable for each category A. This technique is used by many traditional statistical packages. Another option is subset-based where the variables are "is it one of {A,B,C}" for some subset of categories.

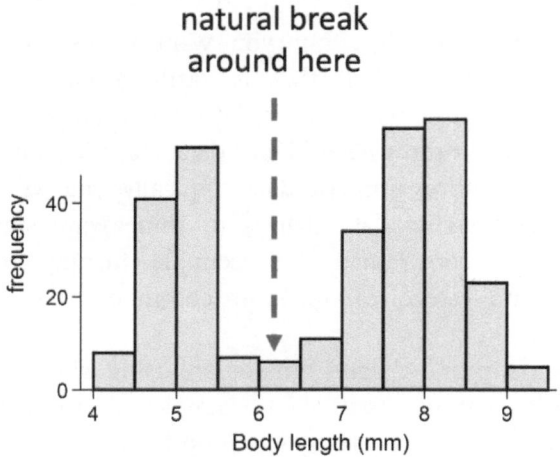

FIGURE 10.2 Bimodal distribution with natural break point (adapted from Qwfp at English Wikipedia, CC BY-SA 3.0, ht tps://commons.wikimedia.org/wiki/File:BimodalAnts.png).

Sometimes there may be domain knowledge that can help, for example in weather data known temperature ranges that are better or worse for plant growth. Alternately if you examine the data, there may be clear patterns. For example, Figure 10.2 shows a bimodal distribution; if we need binary data, it seems sensible to choose a threshold in the gap between the peaks. Effectively we are doing clustering by eye.

Similarly, when you are transforming categorical data into binary using subsets, it is always possible to create a minimal representation of a set of n items as $log_2 n$ subset indicator variables. However, not all subsets are equally meaningful. You may have some idea that some categories are more similar than others, so it is more sensible to group these into a single variable. Even if you are going to push this into a very general machine learning algorithm, it does no harm to 'help' the algorithm, although you might need to be careful about building in your own prejudices (see Chapter 20). Furthermore, if the initial coding is more meaningful, it is more likely that the eventual output of the algorithm will also be easier to interpret.

10.5 NUMERICAL TRANSFORMATIONS

10.5.1 Information

Many data transformations lose information. This is common, indeed the essence of much of learning is precisely discarding the irrelevant or unnecessary in order to concentrate on the important features. Of

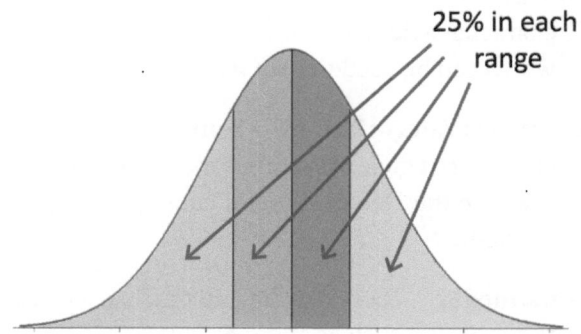

FIGURE 10.3 Normally distributed data split into four equally probable categories at upper and lower quartiles and median.

course, the danger is that some of the information you lose is precisely the relevant parts! This is where it is important to know your data. Often algorithms work better when there are roughly similar amounts of data in each category, or 50:50 for binary data. This can be used as a guide for choosing thresholds. If you choose the median, then 50% of the data is below and 50% above. Similarly, if you want N categories, you choose thresholds so that approximately $1/N$ of the data is in each range; for example Figure 10.3 shows the Normal distribution split into four equally probable categories.

This equal split also minimises information loss as measured by entropy. Entropy, as an information measure, is defined as:

$$-\sum p_i \times log_2 p_i$$

This is at its largest (most information) when all the p_i are equal.

A similar technique can be used for categorical data that is part of a hierarchy or tree, for example taxonomic categories of animals, or files on a disk. If you need a flat set of categories for the algorithm, an obvious choice is to chop off at some level, either just the top-level categorisation or some fixed depth. However, a better choice may be to keep on subdividing the larger categories, so that you end up with approximately equal sized bins as shown in Figure 10.4.

10.5.2 Normalising Data

The use of balanced thresholds is a form of pre-whitening, transforming the data so that it is in a form that is statistically better for future processing steps. It is especially important for data with a few extreme

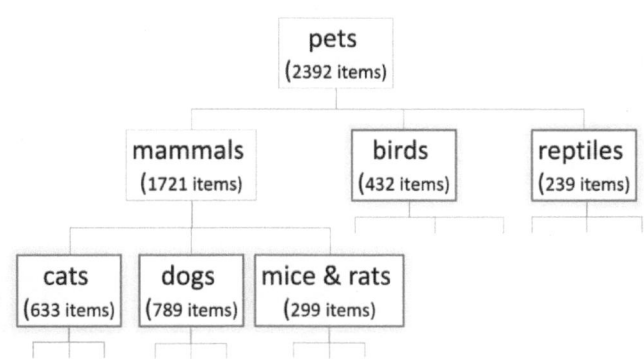

FIGURE 10.4 Taxonomy of animals in pet shop with cut-off nodes chosen to have (very) roughly similar sizes.

but important values such as power-law data, or where the data is clustered very tightly, as a simple choice of thresholds or even passing the raw data into an algorithm may lead to poor results. The equally spaced thresholds effectively make the data more like a uniform distribution (one where the probability of all outcomes is equal), but you may also want to transform the data to make it look more like a Normal distribution, especially where statistical algorithms are going to be used; for example a logarithmic transform is often used for data such as delay times as this is often closer to a Normal distribution.

For numerical data you may also want to scale the data. Imagine if you have one field representing monthly income and another representing height in metres; the income figure will typically be hundreds of times bigger than the height figure. For some algorithms this may not matter, but for others this may effectively make the income figure count much more than height in learning rules. This is particularly true for algorithms that use statistical methods, such as principal components analysis.

For data that is reasonably well spread over a finite range, you can simply scale each value so that the range of each is the same, say [0,1]. Figure 10.5 shows this, first calculating the minimum and maximum value of each column/facet and then using this to translate and scale the data values X[i,j]. For more spread data, such as Normally distributed values, you can translate the data to make the average zero and fixed variance. Figure 10.6 shows this, at first calculating the mean and standard deviation of each data facet/column and then using this to transform the data.

10.5.3 Missing Values – Filling the Gaps

Missing values can occur when data is collected, but not recorded, because of a fault (e.g. a sensor failed), or because in an interview or similar process someone forgets to ask a question. We can think of these as true missing values. They may also occur because the data is being re-used and the original process did not need the particular feature. For example, the record of one hospital visit may include the value of a particular blood test, whereas for other visits this was not necessary. Similarly one might only have French exam results for pupils who sat the French exam. These are perhaps better regarded as optional values, rather than true missing values, but for many purposes behave the same: (i) they are equally not available for processing, but (ii) there is probably a potential value that we simply do not know – what the blood test would have been, what the person would have got in the French exam.

It is important to distinguish these missing values from a zero or null value. A user's profile might have an empty list of interests meaning they have none, or they may simply never have filled in the field (say if they are a new user). The first is definitive knowledge about lack of interests, the latter is lack of knowledge about interests. Ideally "no interests" should be recorded differently from "not filled in", or perhaps even "don't want to say", but often they are all represented as a blank field. In these situations, there is no easy answer, so it is important that early data cleaning or reading does not accidentally conflate these.

When you encounter a missing value, there are three main alternatives:

1. Leave it as a missing value if the algorithm you want to use deals with them itself.

2. Skip the record entirely, especially during learning phases.

3. Attempt to fill in the value using the average or default value for the feature, or attempt to fill it in from the others using a statistical or ML process.

The first of these is the preferred option where possible, but unfortunately many algorithms cannot deal with data that is at all 'messy'. Some machine learning algorithms may not be able to cope with missing values

```
Input unnormalised data X[i,j]  - i data item/row j facet/column
Output normalised data Y[i,j] in the range 0 <= Y[i,j] <= 1
Foreach j
    Lo[j]  =  min over i  X[i,j]
    Hi[j]  =  max over i  X[i,j]
Foreach i,j
    Y[i,j]  =  (X[i,j] - Lo[j]) / (Hi[j] - Lo[j])
```

FIGURE 10.5   Normalising data to a fixed range.

```
Input unnormalised data X[i,j]  - i data item/row j facet/column
Output normalised data Y[i,j] with mean=0 and sd=1
Foreach j
    M[j]  = average X[i,j]   // average over rows varying i
    V[j]  = variance X[i,j]  // again ranging over rows
    SD[j] = sqrt( V[j] )     // standard deviation
Foreach i,j
    Y[i,j]  =  (X[i,j] - M[j]) / SD[j]
```

FIGURE 10.6   Normalising data to a fixed variance.

during learning but are able to use them when the rules are applied.

The last of these feels like a bit of a kludge ... and it is, but sometimes it is all that is possible. For some types of algorithms, it is possible to use this in a bootstrap fashion. For example, some auto-associative neural networks, in particular restricted Boltzmann machines, can be used to recreate inputs from partial outputs (Chap. 6). In these one can use a simple filling in, such as a default value, for a first pass of the learning algorithm, then use the rules produced to fill in values and use this revised data for a second pass, and so on. For some statistical algorithms this iterative algorithm turns out to be the least squares estimate of the missing value, so is theoretically satisfying as well as practically useful.

Note that the implicit assumption underlying most ways of dealing with missing values is that the fact that a value is missing is not correlated in any way with items of interest. However, this may not always be the case. For example, medical records of rough sleepers may well have more missing values than those of professional workers. That is missing values can lead to a form of sampling bias. The counter to this is that in such cases it may be possible to use the fact that a value is missing as a feature in its own right; for example, the fact that a blood test has not been ordered indicates that a doctor had deemed it unnecessary. That is the presence or absence of the test is an implicit record of a clinical judgement.

Missing values are often viewed as exceptional cases that can largely be ignored, and many algorithms will only work with complete data. However, there are many domains, not least medical records, when missing values may be the norm rather than the exception. Indeed, for medical records, you may not even know that an event is missing, for example if someone visits a hospital while abroad on holiday and this does not get entered in their local doctor's records.

So, if you are ever creating your own novel algorithms for AI or ML, do think about whether it is possible to account for missing values.

10.5.4 Outliers – Dealing with Extremes

Outliers are values which are in some way unusual or extreme, such as the very slow walk in Figure 7.1 when we discussed linear regression in Chapter 7. If you are eyeballing data, this may simply be something that sticks out from the rest of the values, but in automated algorithms some rule is used, for example a data value that is more than three standard deviations away from the mean.

Outliers are normally associated with numeric data, but it is also possible to have extreme forms of other data. For example, a black and white image among colour ones, a person's name that isn't in two parts such as 'Madonna', or a place name that is very long such as "Llanfairpwllgwyngyllgogerychwyrndrobwllllantysiliogogogoch".

Outliers occur for two main reasons:

1. Some form of fault or failure, for example a sensor that is misbehaving or an overflow in numerical processing.

2. A real value that is just unusual, such as Llanfairpwll station above or Robert Wadlow who was 2.72 metres tall.

The former clearly need dealing with as they are false data. Sometimes they are particularly obvious, especially when due to numerical overflow as they may then be orders of magnitude larger than real data. However, it is important to realise that not all faults show up in this way; if a sensor is stuck or sluggish, the values may look normal but still be wrong.

The latter, unusual real values are typically less extreme and do represent the true data. However, often one has to manipulate these both because they may not be distinguishable from faulty data and because some numerical algorithms behave badly when there are extreme values.

Algorithms that use some form of averaging can react particularly badly especially when given completely erroneous data. For example, many time-series algorithms use some form of moving average to smooth data prior to other forms of processing, these effectively average the data over a time period before (see Chap. 14). If there is a really massive spike, this can affect processing for a long time afterwards, known as infinite impulse response.

It is also easy to make assumptions about the sizes of inputs, especially when optimising code for constrained computation such as on IoT devices or mobile phones. The impact of unexpectedly large values can be dramatic. In 1996 the Ariadne 5 rocket exploded less than a minute after take-off. The cause turned out to be a horizontal velocity that was too big for a 16 bit integer [171].

Extreme values can also be problematic as part of a learning set, even if they are true values. Many algorithms implicitly treat the training set as though it

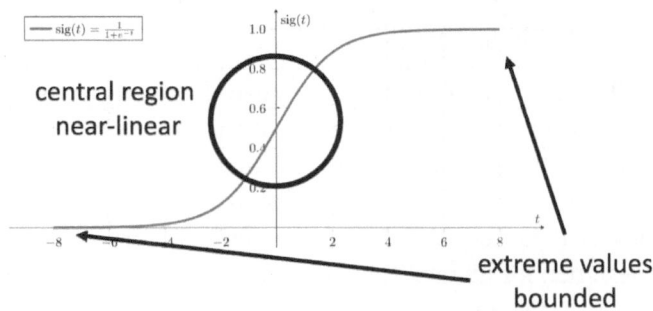

FIGURE 10.7 Sigmoid function smoothly limits extreme values.

is a typical sample, so a single extreme value could skew the rules, which are then applied in perpetuity.

There are several ways of dealing with extreme values:

1. Treat it as a missing value, this is particularly appropriate when the missing value is assumed to be a fault.

2. Cap the value in some way. For example, for numerical data with an acceptable range [-500,500] a value of -3792 is turned into -500.

3. Use a bounding function that retains difference but restricts the range. For example the sigmoid functions that are heavily used in neural networks are linear in their middle ranges, but cap the size of very large values in a smooth fashion (see Figure 10.7).

4. Leave the value as it is, but mark it in some way for the algorithm. For example, it may be possible to allocate it a lower weighting.

In some kinds of data extreme values can be normal, in particular long-tail distributions as found in social networks (Chap. 7). In such cases the problem for machine learning may not be having unwanted extreme values but that a random sample may have none. This is not just a web issue but happens in medical data. Most people are well or have common ailments, rare diseases may each only affect a very small number of people, even though between them they are very important. In such cases it may be important to deliberately look for data items with extreme values, but then to be able to tell the algorithm their expected prevalence. Not all algorithms allow this, especially generic implementations found in machine learning libraries.

10.6 NON-NUMERIC TRANSFORMATIONS

10.6.1 Media Data

When dealing with image or video data you may need to transform the colour space. Although monitors use RGB, a transformation to HSV may often be more effective as shape information is often available in the grey levels. If you are dealing with any sort of library functions designed for images, they may deal with this themselves or may suggest the best format to use.

The image may be just a single feature among many others, for example the profile picture among other elements on a social media profile such as interests, number of friends or age. In such cases treating the image pixels as values alongside the rest of the features will not work well as the image data will swamp the rest, and algorithms or neural networks for images tend to be specialised. In such cases it may be best to process the images first, perhaps using some form of clustering or other unsupervised learning mechanisms, to reduce the image to a small number of category labels and then use this smaller feature set as part of the overall profile data.

Images of printed text can also be turned into text using optical character recognition (OCR). However, this may need additional human processing to deal with difficult cases. Often forms of crowdsourcing or human computation are used for this, that is large numbers of people perform relatively small parts of the overall task (see Chapter 17 for more details). Handwriting can also sometimes be recognised and turned into text, especially if there is substantial writing by the same hand, or if the writer is deliberately trying to be clear, for example digits on cheques. Historical handwriting and old fonts can add to challenges, especially as spellings and grammar may be different from modern text. Furthermore, for languages with fewer speakers and especially those with non-Latin scripts, OCR can be poor or non-existent.

Audio data can also be processed either as an audio stream, as discussed in more detail in Chapter 14, or through automated or human transcription.

Note for both audio and image transcription it is not always important that recognition is 100% accurate. If there is sufficient data, it may be more effective to use imperfect automated transcription for vast quantities of data compared to relatively small volumes of hand-transcribed data.

10.6.2 Text

Some algorithms treat text as a raw character stream, but for others the text has to be reduced to something with more structure, or even reduced to numeric values. Sometimes this may involve bespoke pattern matching as described in Section 10.3, or natural language processing, which we'll deal with in more detail in Chapter 13.

Often string or text data may need some form of normalisation, for example if different character encodings are used or if one data file uses all-caps for names whereas another uses mixed case (e.g. 'ALAN DIX' vs. 'Alan Dix'). Ideally one should retain the maximum information (e.g. initial caps tell us more than all upper or all lower case), but often it is easier to simply reduce everything to lower case for subsequent processing. Special care is often needed if text has been prepared for human reading, for example the text may include ligatures such as 'ff' as a single character that needs to be broken into separate characters.

Care may also be needed if text has had line breaks inserted. For example, a frequent early tidying operation is to remove leading and trailing spaces from strings, but in RTF trailing spaces at the end of lines are significant as they denote a word break, whereas lines without a trailing space need to be joined without a space. Hyphens at line ends can also be problematic as sometimes it denotes hyphenation inserted to break a long word at the line end, whereas at other times it may be a real hyphen that should be retained.

Many techniques for reducing text to some sort of signature or metric require comparison of words and phrases in the text with typical frequencies in the corpus in order to identify the more unusual ones. This may even include applying some sort of statistical or neural network to a corpus to create a reduced representation. If there is sufficient data, this can be done relative to the text in the dataset itself. This is because, say, a phrase such as 'principal components' will be less unusual in texts about machine learning than in day-to-day language. However, this does require substantial quantities of text, and so often comparisons have to be made largely against generic corpora, or using pre-produced lists of word frequencies.

A special case of this form of processing is simply removing the most common words, known as stop words, from the text before further processing. Examples of typical stop words include 'a', 'the', 'and' and 'or', although these of course depend on the language of the text.

Sometimes reduced forms of the text are used directly by subsequent algorithms, but often text fields are used to create similarity measures between corresponding text fields in different data records. Two common ways to do this are Jaccard similarity and cosine similarity.

Jaccard similarity uses the bag of words in each text, that is the set of words ignoring their frequency (call them *words1* and *words2* respectively). It then looks at the relative proportion of shared words by dividing the size of the intersection (number of common words) by the size of the union (total words in either text):

$$Jaccard\_similarity(doc1, doc2) = \frac{|\,words1 \cap words2\,|}{|\,words1 \cup words2\,|}$$

Note that in set theory $|S|$ means cardinality of the set S, that is the number of items in the set.

If the documents A and B have no words in common, the Jaccard similarity is zero. In contrast if they have exactly the same words, it is one, so intuitively this is a sensible measure. Note however that by random chance larger documents end up with larger Jaccard similarity. Also, as it ignores the frequency, two short documents that both mention an obscure word such as 'accipiter' multiple times would be treated as no more similar due to it than two large documents that each used it once.

Cosine similarity treats the frequency of words within the two texts as if it were a very large dimensional vector (call them *f1* and *f2* respectively) and then calculates the cosine of the angle between the vectors using the dot product (•).

$$cosine\_similarity(doc1, doc2) = \frac{\sum_i f1_i \bullet f2_i}{|f1| \bullet |f2|}$$

Here $|v|$ is used in its vector theory sense of the length of the vector:

$$|v| = \sqrt{\sum_i v_i^2}$$

Note that both set theory and vector theory use $|A|$ to mean the 'size' of A, but in different senses.

Arguably while Jaccard similarity does not take multiple occurrences of words into account, cosine similarity may be swayed too strongly by a few very frequent words. Sometimes variants are used, such as the cosine metric applied to logarithms of the counts rather than the raw frequencies (typically $log(1 + count)$ to avoid $log\,0$). This means that two occurrences of a word count more than one, but less than twice as much.

Frequently both metrics may be modified by removing stop words first and/or weighting by overall corpus frequencies. Choosing these metrics is a bit of an art rather than an exact science as what appears to be the 'optimal' mathematical metrics often end up slightly fragile in practice.

10.6.3 Structure Transformation

As well as transforming individual fields, it may be necessary to look at transformations involving multiple fields in the same data record, or even more radical transformations of the structure of the dataset.

An example of the former, which we've already discussed, is where two fields are the given name and family name of the same person and so need to be treated as a unit for name lookups. Similarly latitude–longitude may be two fields but represent a single entity. We may also want to introduce additional fields based on calculations, such as the available cash as the difference between bank balance and overdraft limit in Chapter 9. This might just be a boolean indicator such as is_overdrawn. While some algorithms may be able to find these things out for themselves given sufficient data, others may need a little help.

More radical transformations may look across data items. For example, the text similarity measures mentioned above could be used to transform table-like data into a network with similarity measures on each network connection. This can then be used in similar ways to those discussed for recommender systems in Chapter 8. Alternatively given a network, it is possible to work out local metrics on the network nodes, for example the number of friends in a social network, or, for a web page, the words used as the anchor text on hyperlinks pointing to it. This can then be used to create a more tabular representation which is more suitable for large-scale cloud processing (Chap. 8).

10.7 AUTOMATION AND DOCUMENTATION

It is easy to get lost in the process of data analysis and forget what one has done to get to the cleaned data. However, it is important to keep track of this for two reasons.

documentation – You may need to tell others what you have done, whether in an internal report or for external publication. This may have important legal consequences if, for example, the data is later used as part of personal information processing (see also Chap. 20).

repeatability – You may need to re-run the analysis on the new data or re-run the processing on the same data.

The last of these, the need to repeat steps on the same data, is less obvious than new data or documentation for reporting. However, it is very common.

We already discussed validation rules and sanity checks for early data gathering and extraction from raw sources. However, there may be additional checks that may need to be carried out that were impossible earlier as they depend on the analysis and transformations.

For example, as part of processing the OCR of an old gazetteer, one stage identified the names of places starting each entry, based partly on capitalisation. After this stage the alphabetic order of the entries was checked. Sometimes this was wrong because the original editor has misordered the entries, sometimes because there was an error in the OCR of the entry and sometimes because some capitalisation or OCR errors in the middle of an entry made the entry-detection algorithm think a new entry was starting. The different forms of misorder required different kinds of changes; some, such as correcting the OCR, entailed re-running the entire processing pipeline.

In addition, processing may accidentally introduce errors. A classic form of this is when global substitutions capture unintended strings. For example, in a historic text one might encounter the name 'Henry the FifthIII' – not because it was in the original text but because a global substitution to transform 'Henry V' to 'Henry the Fifth' accidentally matched 'Henry VIII'. The substitution rule can easily be corrected to only begin and end at word breaks, but if this is only noticed later in the process, the analysis pipeline may need to be re-run from when the substitution was performed.

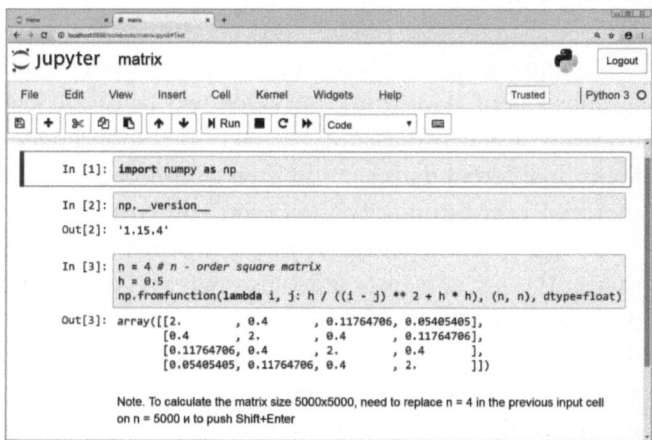

FIGURE 10.8 Jupyter Notebook showing replayable Python code (Image: Andrey Yakimchik – CC BY 4.0, https://commons.wikimedia.org/w/index.php?curid=97158987).

In some cases tools you are using can help this process. You may be able to use a notebook-style interface, such as Jupyter (Figure 10.8), which automatically records your actions and allows you to edit and replay [154]. Alternatively, if you are using some form of command-line shell (Unix or DOS), you can make sure you save the transcript after every session. Interactive tools can make this more difficult as many do not have easy ways to record the actions you have performed let alone to replay them.

Even when you have recorded and can replay by hand or automatically what you have done, it is important to ensure you are able to recreate the environment. If the datafiles or online data have changed since you originally processed the data, then the outputs will be different. This may not matter, if you simply want the most up-to-date results, but can be a problem if you want to be able to reproduce previous analysis. Where possible it can be useful to make time-stamped copies of critical datafiles, or make zip archives of folders of data, although this may not be possible for very large datafiles and various forms of deltas (files recording differences) may be needed.

When heavy computation is needed, it may also be worth making copies of intermediate files to avoid completely re-running processing pipelines but with care to recompute them if anything has changed on earlier stages of processing. If you are not using an environment that supports this for you, you should create your own processes, file naming conventions, etc. and document the main steps you are following.

If you create bespoke analysis code, you should consider using a version control system such as GitHub

TABLE 10.1   Data Used in Exercise 10.2.

mountain	height	source
Everest	8,850 meters	https://education.nationalgeographic.org/resource/mount-everest/
Yr Wyddfa	3,560 feet	https://www.britannica.com/place/Snowdon
Agiocochook	6,288.3 ft	https://en.wikipedia.org/wiki/Mount_Washington
Kilimanjaro	5895 m	https://www.statista.com/statistics/1237791/highest-mountains-in-africa/

so that changes made to the code are recorded, and maybe factoring the code into a core execution engine (that changes rarely) plus rules (that change more often). If you are making an interactive tool, then make sure there is some way to save the actions and/or configuration of the system so that it can be replayed!

File systems are good at storing datafiles, but poor at the meta-information about the files. Do make sure you use a system to record data documentation such as the format of different datafiles and what the columns mean in a CSV file, and also the provenance including which program created it, what versions of input data were used, and parameters supplied. This all seems obvious at the time but can be obscure even a few weeks later.

10.8 SUMMARY

This has been a chapter about the messy side of AI, with lots of practical advice as well as more overarching methods and processes. We have looked at the main stages of preparing data for use in AI and machine learning. This started with extracting and gathering the data, potentially from separate sources including cleaning and validating it. This is then followed by various forms of data transformation and manipulation to put it in a form suitable for the chosen algorithm. This can be relatively straightforward, for example transforming data values from continuous to discrete values, but may require substantial structural changes to the dataset. We saw various places where AI may itself be used during this data preparation, including entity recognition and data reduction of media resources.

10.1 Place the following into the data type categories in Section 10.4.1:

    a. "Hello World!"
    b. a photo of Everest
    c. 42
    d. {mountain:"Everest";height:8849}
    e. false

    f. https://alandix.com/

10.2 In the data fragment { mountain:"Everest"; height:8849 } the height is given in metres. How might you preprocess data with mixed units such as in Table 10.1.

10.3 Consider the following data – ignore the three columns: treat it as a single x–y dataset. This data is also available in the chapter web resources.

x	y	x	y	x	y
82.092	2.480	65.875	2.691	35.956	4.989
24.601	4.521	64.667	2.010	34.963	4.971
8.120	5.065	76.328	2.493	73.086	2.508
72.420	7.975	80.725	8.000	56.624	2.219
15.439	4.589	58.251	2.268	28.790	4.418
21.561	4.488	30.578	5.115	86.419	7.521
72.306	2.715	67.602	1.347	23.260	3.865
65.517	7.597	36.831	5.433	82.545	7.707

Using a spreadsheet, code or by hand:

a. Calculate the minimum, maximum, mean and standard deviation for x and y

b. Use the min-max to normalise by range

c. Use the mean-std'dev to normalise by standard deviation

d. Use quartiles to create an information-oriented classification of x and y into four categories (i.e. code 1 for lower quartile, 2 for lower quartile to median, 3 for median to upper quartile and 4 for larger than upper quartile)

e. Calculate contingency tables for x quartile vs y quartile

f. Plot a histogram for x and y and look for natural breakpoints in each

g. Using the natural breakpoint for x, create a Low-High class for the x values and create a contingency table for x-breakpoint vs y quartiles

h. Do you spot any patterns in (e) or (g)?

FURTHER READING

Although there is lots of scattered information on the web, it is hard to find a coherent text in this area.

R. Mitchell. *Web scraping with Python: Data extraction from the modern web.* O'Reilly, Sebastopol, CA, 2015.

Latest edition of this popular O'Reilly title, offering practical methods for dealing with web data.

A. Doan, A. Halevy, and Z. Ives. *Principles of data integration.* Elsevier, Waltham, MA, 2012.

The book is rooted in an organisational database context of how to integrate the mass of unstructured web data with structured databases. However, the methods and issues covered are just the same as those needed when preparing data for AI use.

J. D. Kelleher, B. Mac Namee, and A. D'arcy. *Fundamentals of machine learning for predictive data analytics: Algorithms, worked examples, and case studies.* MIT Press, Cambridge, MA, 2020.

This book is in part a general overview of machine learning techniques, but a particular strength is the practical focus through concrete examples and case studies including discussions of issues such as data quality and data preparation.

III

Specialised Areas

Game Playing

11.1 OVERVIEW

Game playing has been an important part of the history of AI. The techniques for game playing can also be applied to other situations where factors are unknown but will be discovered only after action is taken. This chapter will consider algorithms for playing standard games (non-probabilistic, open, two-person, turn-taking, zero-sum games). Such games include chess, draughts, tic-tac-toe and Go. In particular, we will look at minimax search techniques and alpha–beta pruning. This builds on the search techniques studied in Chapter 4. The chapter will also consider other types of game where co-operation is important, where players can take simultaneous moves and where random events happen (such as the throw of a die). We will see in Chapter 15 that acting in the presence of uncertainty is essential for robotics and other practical planning tasks, and this chapter will show how game-playing algorithms can be used to tackle such non-gaming problems. Neural networks have been successfully used alongside other game-playing algorithms to tackle some problems that were previously thought to be nearly impossible for machines.

11.2 INTRODUCTION

Game playing has always been an important part of AI. Indeed, the earliest attempts at game-playing computer programs predate the field. Even Babbage considered programming his Analytical Engine to play chess.

Games have been seen as a good testing ground for two reasons. First, because the mixture of reasoning and creative flair seems to epitomise the best of human intelligence. Secondly, because the constrained environment of play with clearly formulated rules is far more conducive to computation than the confused and open problems of everyday life. This advantage is also a weakness of game playing as a measure of intelligence. Instead human intelligence is regarded as being more thoroughly expressed in the complexity of open problems and the subtlety of social relationships. Arguably the brute force approaches that were so effective for chess are no longer mainstream AI.

This critique of game playing should not detract from its own successes and its enormous importance in the development of the field of AI. When chess programs were still struggling at club level, they were regarded as a challenge to AI; now they compete at grandmaster level. Game-playing programs have also led to the development of general purpose AI algorithms; for example, iterative deepening (discussed in Chap. 4) was first used in CHESS 4.5 [261]. Game playing has also been a fertile ground for experiments in machine learning, in particular adversarial learning which has been adopted across other areas of AI.

> The single problem that has received most attention from the artificial intelligence community is the playing of chess, a game whose whole attraction is that it runs to precise rules within which billions of games are possible. As Stephen Rose, the British brain biologist, says, getting a computer to do this is not too great a wonder. Get one to play a decent game of poker, he says, and he might be more impressed.
>
> Martin Ince, THES (1994)

Most interesting games defy pure brute force approaches because of the sheer size of their branching factor. In chess there are typically around 30 legal moves at any time (although only a few "sensible" ones), and it is estimated around 10^{75} legal chess games. We say "legal" games, as few would be sensible games. In order to deal with this enormous search space the computer player must be able to recognise which of the legal moves are sensible and which of the reachable board positions are desirable. Search must be heuristic driven, and the formulation of these heuristics means that the programs must capture, to some extent, the strategy of a game.

These factors are exemplified by the game of Go. Its branching factor is nearly 400, with as many moves. Furthermore, the tactics of the game involve both local and global assessment of the board position, making heuristics very difficult to formulate. However, effective heuristics are essential to the game. The moves made in the early part of the game are critical for the final stages; effectively one needs to plan for the end game, hundreds of moves later. But the huge branching factor clearly makes it impossible to plan for the precise end game; instead one makes moves to produce the right kind of end game.

Applying machine learning and neural networks to Go also encounters problems as the tactical advantage of a move is partly determined by its absolute position on the board (easy to match) but partly also by the local configuration of pieces. We will see in Chapter 12 that position independence is a major problem for pattern matching, and so this is not a parochial problem for game playing.

For these reasons the success of AlphaGo in 2016, already discussed in the introduction (Chap. 1), was not just surprising but shocking to many. Crucially, this success was not just about brute force. Although the deep learning networks were very large, they were not large enough to encode every game play but were clearly encoding some form of strategy and tactics, even though of a different form to a human player.

Perhaps even more surprising have been games that require a level of understanding of human players including poker [200] and web-based Diplomacy [140]. For the latter natural language models were combined with game playing as a critical part of the game is chat-based negotiations. The AI is even capable of levels of deceit, withholding critical information.

11.3 CHARACTERISTICS OF GAME PLAYING

Game playing has an obvious difference from the searches in Chapter 4: while you are doing your best to find the best solution, your adversary is trying to stop you! One consequence of this is that the distinction between *planning* and *acting* is stronger in game play. When working out how to fill out a magic square, one could always backtrack and choose a different solution path. However, once one has made a choice in a game there is no going back. Of course, you can look ahead, guessing what your opponent's moves will be and planning your responses, but it remains a guess until you have made your move and your opponent has responded – it is then too late to change your mind.

The above description of game playing is in fact only of a particular sort of game: a non-probabilistic, open, two-person, turn-taking, zero-sum game.

- *non-probabilistic* – no dice, cards or any other random effects.

- *open* – each player has complete knowledge of the current state of play, as opposed to games like "battleships" where different players have different knowledge.

- *two-person* – no third adversary and no team playing on your side, as opposed to say bridge or football.

- *turn-taking* – the players get alternate moves, as opposed to a game where they can take multiple moves, perhaps based on their speed of play.

- *zero-sum* – what one player wins, the other loses.

In addition, the games considered by AI are normally non-physical, but there are also game-playing robots, including an annual RoboCup for robotic football. With a bit of effort one can think of games that have alternatives to all the above, but the "standard" style of game has been most heavily studied, with the occasional addition of some randomness (e.g. backgammon).

As with deterministic search, we can organise the possible game states into trees or graphs, with the nodes linked by moves. However, we must also label the branches with the player who can make the choice between them. In a game tree alternate layers will be controlled by different players.

Like deterministic search problems, the game trees can be very big and typically have large branching factors. Indeed, if a game tree is not complex, the game is likely to be boring. Even a trivial game like noughts and crosses (tic-tac-toe) has a game tree far too big to demonstrate here. Because of the game tree's size it is usually only possible to examine a portion of the total space.

Two implications can be drawn from the complexity of game trees. First, *heuristics* are important – they are often the only way to judge whether a move is good or bad, as one cannot search as far as the actual winning or losing state. Secondly, the choice of which nodes to expand is critical. A human chess player only examines a small number of the many possible moves but is able to identify those moves that are "interesting". This process of choosing directions to search is knowledge rich and therefore expensive. More time spent examining each node means fewer nodes examined – in fact, the most successful chess programs have relatively simple heuristics but examine vast numbers of moves. They attain grandmaster level and are clearly "intelligent", but the intelligence is certainly "artificial". In contrast, AlphaGo, which leans more heavily on complex learnt heuristics, is perhaps more human-like, although still plays very differently from a human player.

11.4 STANDARD GAMES

11.4.1 A Simple Game Tree

In order to demonstrate a complete game tree, we consider the (rather boring) game of "placing dominoes". Take a squared board such as a chess board. Each player in turn places a domino that covers exactly two squares. One player always places pieces right to left, the other always places them top to bottom. The player who cannot place a piece loses. The complete game tree for this when played on a 3 × 3 board is shown in Figure 11.1. In fact, even this tree has been simplified to fit it onto the page, and some states that are equivalent to others have not been drawn. For example, there are two states similar to b and four similar to c.

The adversaries are called Alison and Brian. Alison plays first and places her pieces left to right. Consider board position j. This is a win for Alison, as it is Brian's turn to play, and there is no way to play a piece top to bottom. On the other hand, position s is a win for Brian, as although neither player can place a piece, it is Alison's turn to play.

We can see some of the important features of game search by looking at this tree. The leaves of the tree are given scores of +1 (win for Alison) or −1 (win for Brian – Alison loses). This scoring would of course be replaced by a heuristic value where the search is incomplete. The left-hand branch is quite simple – if Alison makes this move, Brian has only one move (apart from equivalent ones) and from there anything Alison does will win. The right branch is rather more interesting. Consider node m: Brian has only one possible move, but this leads to a win for him (and a loss for Alison). Thus position m should be regarded as a win for Brian and could be labelled "−1". So, from position e Alison has two choices, either to play to l – a win – or to play to m – a loss. If Alison is sensible, she will play to l. Using this sort of argument, we can move up the tree marking nodes as win or lose for Alison.

In a win–lose game either there will be a way that the first player can always win, or alternatively the second player will always be able to force a win. This game is a first-player win game; Alison is a winner! If draws are also allowed, then there is the third alternative that two good players should always be able to prevent each other from winning – all games are draws. This is the case for noughts and crosses, and it is suspected that the same is true in chess. The reason that chess is more interesting to play than noughts and crosses is that no one knows, and even if it were true that in theory the first player would always win, the limited ability to look ahead means that this does not happen in practice.

11.4.2 Heuristics and Minimax Search

In the dominoes game we were able to assign each leaf node as a definite win either for Alison or for Brian. By tracing back we were able to assign a similar value for each intermediate board position. As we have discussed, we will not usually have this complete information and will have to rely instead on heuristic evaluation. As with deterministic search, the form of this will depend on the problem. Examples are

- *chess* – One can use the standard scoring system where a pawn counts as 1, a knight as 3 and so on.

- *noughts and crosses* – One can use a sum based on the value of each square where the middle counts most, the corners less and the sides least of all. You add up the squares under the crosses and subtract those under the noughts.

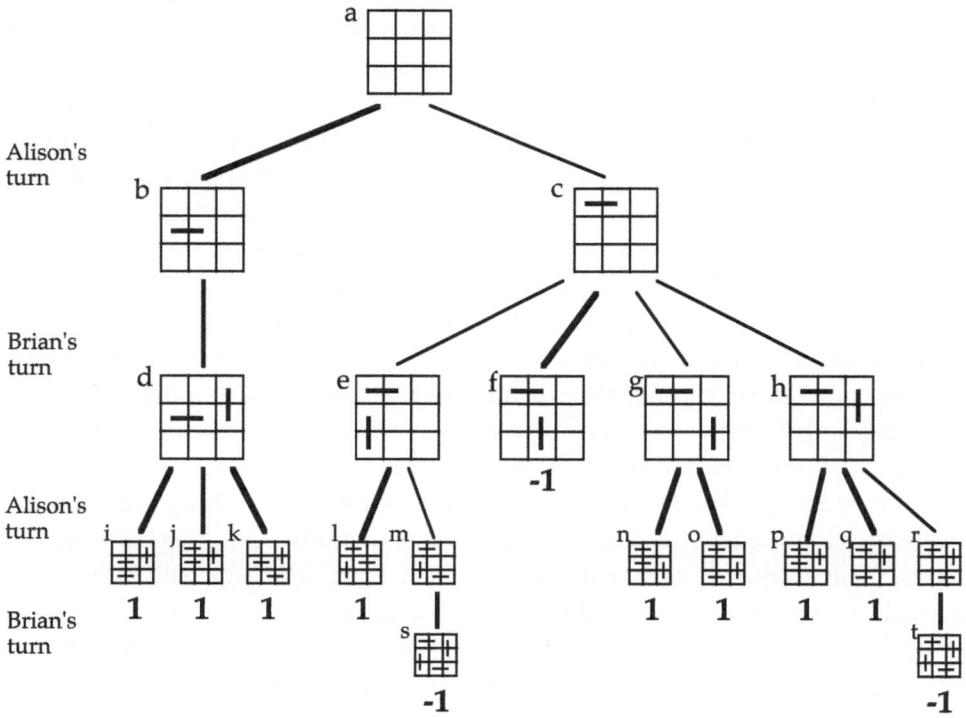

FIGURE 11.1   Game tree for "placing dominoes".

Note that these heuristics may give values outside the range 1 to −1, so one must either suitably scale them or choose large enough values to represent winning and losing positions.

Figure 11.2 shows an example termgame tree with heuristic values for each position. The heuristic values are the unbracketed numbers (ignore those in brackets for the moment). Alison's moves are shown as solid lines, and Brian's moves are dashed. This is not the whole game tree, which would extend beyond the nodes shown. We will also ignore for now the difficult issue of how we decided to search this far in the tree and not, for example, to look at the children of node k. The portion of the tree that we have examined is called the *search horizon*.

It is Alison's move. There are obviously some good positions for her (with scores 5 and 7) and some very bad ones (−10). But she cannot just decide to take the path to the best position, node j, as some of the decisions are not hers to make. If she moves to position c, then Brian might choose to move to position g rather than to f. How can she predict what Brian will do and also make her own decision?

We can proceed up the tree rather as we did with the dominoes game. Consider position i. It is Brian's move, and he will obviously move to the best position for him, that is the child with the *minimum* score, n. Thus, although the heuristic value at node i was 2, by looking ahead at Brian's move we can predict that the actual score resulting from that move will be −3. This number is shown in brackets. Look next at node d. It is Alison's move. If she has predicted Brian's move (using the argument above), her two possible moves are to h with score −2 or to i with score −3. She will want the best move for her, that is the *maximum* score. Thus the move made would be to h and position d can be given the revised score of −2. This process has been repeated for the whole tree. The numbers in brackets show the revised scores for each node, and the solid lines show the chosen moves from each position.

With this process one alternately chooses the minimum (for the adversary's move) and the maximum (for one's own move). The procedure is thus called minimax search. Pseudocode for minimax is shown in Figure 11.3.

Note that the numbers on the positions are the worst score that you can get assuming you always take the indicated decisions. Of course you may do better if your adversary makes a mistake. For example, if Alison moves to c and Brian moves to f, Alison will be able to respond with a move to j, giving a score of 7 rather than the worst

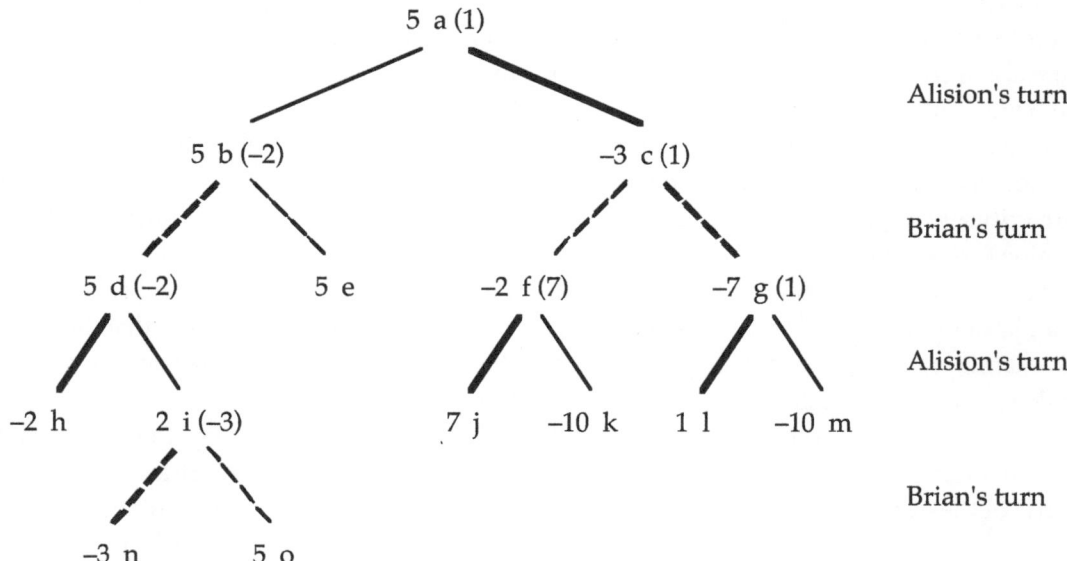

FIGURE 11.2   Minimax search on a game tree.

```
to find minimax score of n
   find minimax score of each child of n
   if it is Alison's turn
      score of n is the maximum of the children's scores
   if it is Brian's turn
      score of n is the minimum of the children's scores
```

FIGURE 11.3   Minimax pseudocode.

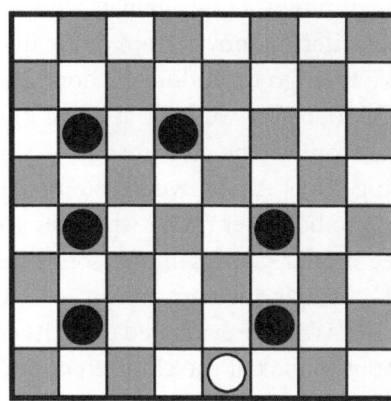

FIGURE 11.4   Horizon effect – simple heuristics can be wrong!

case score of 1. However, if you don't take the indicated moves, a good opponent will fight down your score to below the minimax figure. Minimax is thus a risk averse search.

11.4.3 Horizon Problems

It is important to remember that the portion of tree examined in determining the next move is *not* the whole tree. So although minimax gives the worst case score *given the nodes that have been examined*, the actual score may be better or worse as the game proceeds, and one gets to previously unconsidered positions. For example, imagine that Alison looks ahead only two moves, to the level d–g. A minimax search at this level gives scores of 5 to b and −7 to c, so Alison will move to b, whereas by looking further ahead we know that c would be better.

Looking even further ahead, our choice might change again. These rapid changes in fortune are a constant problem in determining when to stop in examining the game tree. Figure 11.4 shows a particularly dramatic example. The white draught is crowned, so it can jump in any direction, and it is white's move. A simple heuristic would suggest that black is unassailable, but looking one move further we find that white jumps all black's draughts and wins the game!

Look again at Figure 11.2. Positions a, b, d and e all have the same heuristic score. That is, they form a plateau rather like we saw in hill climbing. While we only look at the positions within a plateau, minimax can tell us nothing. In the example tree, the search horizon went beyond the plateau, and so we were able to get a better estimate of the score for each position. In fact, if you examine the suggested chess heuristic, this only changes when a piece is taken. There are likely to be long play sequences with no takes, and hence plateaux in the game tree.

Plateaux cause two problems. First, as already noted, minimax cannot give us a good score. Secondly, and perhaps more critically, it gives us no clue as to which nodes to examine further. If we have no other knowledge to guide our search, the best we can do is examine the tree around a plateau in a breadth first manner. In fact, one rule for examining nodes is to look precisely at those where there is a lot of change – that is, ignore the plateaux. This is based on the observation that rapid changes in the heuristic evaluation function represent interesting parts of the game.

11.4.4 Alpha–beta Pruning

The minimax search can be speeded up by using branch and bound techniques. Look again at Figure 11.2. Imagine we are considering moves from d. We find that h has score −2. We then go on to look at node i – its child n has score −3. So, *before* we look at o, we know that the minimax score for i will be *no more than* −3, as Brian will be minimising. Thus Alison would be foolish to choose i, as h is going to be better than i whatever score o has.

We can see similar savings on the dominoes game tree (Figure 11.1). Imagine we are trying to find the move from position c. We have evaluated e and its children and f, and are about to look at the children of nodes g and h. From Brian's point of view (minimisation), f is best so far. Now as soon as we look at node n we can see that the minimax score for g will be at least 1 (as Alison will play to maximise), so there is no reason to examine node o. Similarly, having seen node f, nodes p and q can be skipped. In fact, if we look a bit further up, we can see that even less search is required. Position b has a minimax score of 1. As soon as we have seen that node f has score "−1" we know that Brian could choose this path and that the minimax score of c is at most −1. Thus nodes g and h can be ignored completely. This process is called alpha–beta pruning and depends on carrying around a best-so-far (α) value for Alison's choices and a worst-so-far (β) for Brian's choices.

11.4.5 The Imperfect Opponent

Minimax and alpha–beta search both assume that the opponent is a rational player using the same sort of reasoning as the algorithm. Imagine two computers, AYE and BEE, playing against one another. AYE is much more powerful than BEE and is to move first. There are two possible moves. If one move is taken, then a draw is inevitable. If the other move is taken, then, by looking ahead 20-ply, AYE can see that BEE can force a win. However, all other paths lead to a win for AYE. If AYE knows that BEE can only look ahead 10-ply, then AYE should probably play the slightly risky move in the knowledge that BEE will not know the correct moves to make and so almost certainly lose.

For a computer to play the same trick on a human player is far more risky. Even though human players can consider nowhere near as many moves as computers, they may look very far ahead down promising lines of moves (actually computers do so too). Because AYE knew that BEE's search horizon was fixed, it could effectively use probabilistic reasoning. The problem with human opponents, or less predictable computer ones, is that they might pick exactly the right path. Assuming random moves from your opponent under such circumstances is clearly foolhardy, but minimax seems somewhat unadventurous. In preventing the worst, it throws away golden opportunities.

11.5 NON-ZERO-SUM GAMES AND SIMULTANEOUS PLAY

In this section we will relax some of the assumptions of the standard game. If we have a non-zero-sum game, there is no longer a single score for each position. Instead, we have two values representing how good the position is for each player. Depending on the rules of play, different players control different choice points, and they seek to maximise their own score. This formulation allows one to consider not only competitive but also co-operative situations, where the choices are made independently, but where the players' ideas of "good" agree with one another. This leads into the area of *distributed AI*, where one considers, for example shop-floor robots co-operating in the building of a motor car (see Chap. 16). However, there we will consider the opposite extreme, where all parties share a common goal. In this section we will consider the in-between stage when the players' goals need not agree but may do so. We will also examine simultaneous play, that is when both parties make a move in ignorance of each other's choice.

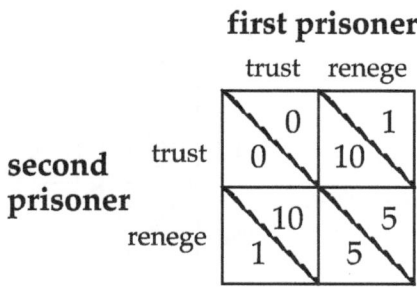

FIGURE 11.5 The prisoner's dilemma.

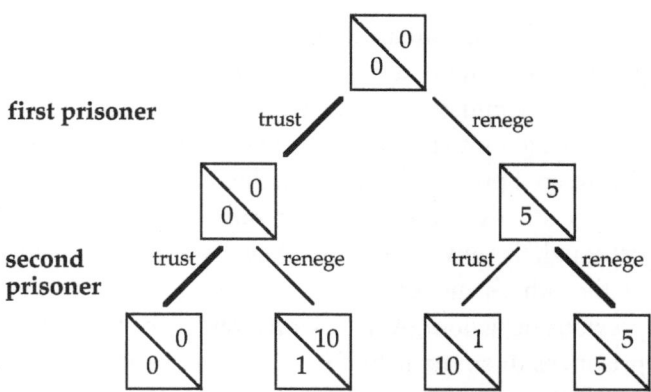

FIGURE 11.6 Game tree for prisoner's dilemma.

11.5.1 The Prisoner's Dilemma

A classic problem in game theory is the prisoner's dilemma. There are several versions of this. The one discussed in Section 11.5.4 is the most common, but we will deal with a more tractable version first! This comes in several guises, and the most common is as follows. Imagine two bank-robbers have been arrested by the police and are being questioned individually. The police have no evidence against them, and can only prosecute if one or the other decides to confess.

Before they were arrested, the criminals made a pact to say nothing. Each now has the choice either to remain silent – and trust their colleague will do the same – or to renege on their promise. Is there honour among thieves?

If neither confesses, then the police will eventually have to let them go. If both confess, then they will each get a long, five-year sentence. However, the longest sentence will be for a prisoner who doesn't confess when the other does. If the first prisoner confesses, then the other prisoner will get a ten-year sentence, whereas the first prisoner will only be given a short, one-year sentence. Similarly, if the second prisoner confesses and the first does not, the first will get the ten-year sentence. The situation is summarised in Figure 11.5. In each square the first prisoner's sentence is in the upper right and the second in the lower left.

Let's consider the first prisoner's options. If he trusts his colleague, but she reneges, then he will be in prison for ten years. However, if he confesses, reneging on his promise, then the worst that can happen to him is a five-year sentence. A minimax strategy would suggest reneging. The second prisoner will reason in exactly the same way – so both confess.

11.5.2 Searching the Game Tree

The above problem was drawn as a matrix rather than a tree, because neither prisoner knew the other's moves. If instead the two 'played' in turn, then the situation would be far better. In this case we can draw the prisoner's dilemma as a game tree (see Figure 11.6). At each terminal node we put the two values and use a minimax-like algorithm on the tree.

Imagine the first prisoner has decided not to confess, and the second prisoner knows this. Her options are then to remain silent also and stay out of prison, or to renege and have a one-year sentence. Her choice is clear. On the other hand, if the first prisoner has already reneged, then it is clear that she should also do so (honour aside!). Her choices are indicated by bold lines, and the middle nodes have been given pairs of scores based on her decisions.

Assuming the first prisoner can predict his partner's reasoning, he now knows the scores for each of his options. If he reneges, he gets five years; if he stays silent, he walks away free – no problem!

Notice that although this is like the minimax algorithm, it differs when we consider the second prisoner's moves. She does not seek to minimse the first prisoner's score, but to maximise her own. More of a maximax algorithm?

So, the game leads to a satisfactory conclusion (for the prisoners) if the moves are open, but not if they are secret (which is why the police question them separately). In real-life decision making, for example many business and diplomatic negotiations, some of the choices are secret. For example, the Cuban missile crisis can be cast in a similar form to the prisoner's dilemma. The "renege" option here would be to take pre-emptive nuclear action.

Happily, the range of options and the level of communication were substantially higher.

Although there are obvious differences, running computer simulations of such games can be used to give some insight into these complex real-world decisions. In the iterated prisoner's dilemma, the same pair of players are constantly faced with the same secret decisions. Although in any one game they have no knowledge of the other's moves, they can observe their partner's previous behaviour. A successful strategy for the iterated prisoner's dilemma is tit-for-tat, where the player "pays back" the other player for reneging. So long as there is some tendency for the players occasionally to take a risk, the play is likely to end up in extended periods of mutual trust.

11.5.3 No Alpha–Beta Pruning

Although the slightly modified version of the minimax algorithm works fine on non-zero-sum games, alpha–beta pruning cannot be used. Consider again the game tree in Figure 11.6. Imagine this time that you consider the nodes from right to left. That is, you consider each renege choice before the corresponding trust choice. The third and fourth terminal nodes are considered as before, and the node above them scored. Thus the first prisoner knows that reneging will result in five years in jail. We now move on to the second terminal node. It has a penalty of ten years for the first prisoner. If he applied alpha–beta pruning, he would see that this is worse than the reneging option and so not bother to consider the first node at all.

Why does alpha–beta fail? The reason is that it depends on the fact that in zero-sum games the best move for one player is the worst for the other. This holds in the right-hand branch of the game tree but not in the left-hand branch. When the first prisoner has kept silent, then the penalties for both are minimised when the second prisoner also remains silent. What's good for one is good for both.

11.5.4 Pareto-optimality

In the form of the prisoner's dilemma discussed above, the option when both remain silent was best for both. However, when there is more than one goal, it is not always possible to find a uniformly best alternative. Consider the form of the prisoner's dilemma in Figure 11.7. This might arise if the police have evidence

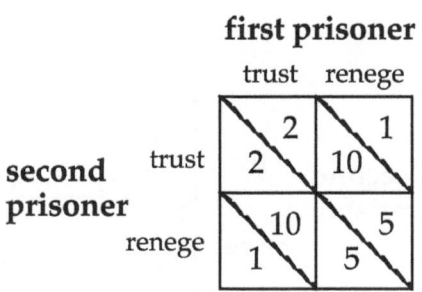

FIGURE 11.7   Modified prisoner's dilemma.

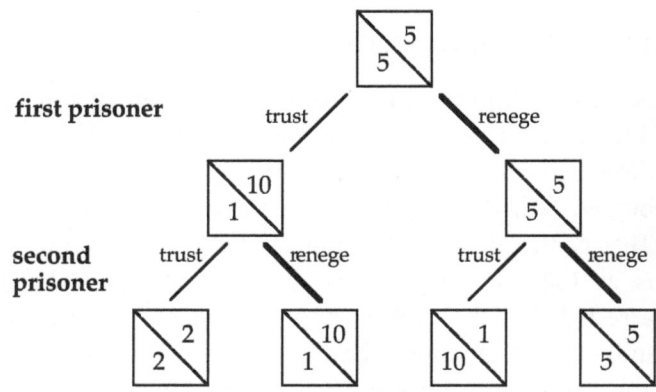

FIGURE 11.8   Non-Pareto-optimal solution.

of a lesser crime, perhaps possession of stolen goods, so that if neither prisoner confesses they will still both be imprisoned for two years. However, if only one confesses, that prisoner has been promised a lenient sentence on both charges.

This time, there is no uniformly optimal solution. Neither prisoner will like the renege–renege choice, and the trust–trust one is better for both. However, it is not best overall as each prisoner would prefer the situation when only they confess. The trust–trust situation is called Pareto-optimal. This means that there is no other situation that is uniformly better. In general, there may be several different Pareto-optimal situations favouring one or other party.

Now see what happens when the prisoners make their choices. The first prisoner wonders what the second prisoner might do. If she reneges, then he certainly ought to as well. But if she stays silent, it is still better for him to renege as this will reduce his sentence from two years to one. The second prisoner reasons similarly and so they end up in the renege–renege situation.

This time, having an open, turn-taking game does not help. Figure 11.8 shows the game tree for this version of the dilemma, which also leads to the renege–renege option. Even though both prisoners would prefer the Pareto-optimal trust–trust option to the renege–renege one, the latter is still chosen. Furthermore, if they both did decide to stay silent, but were later given the option of changing their decision, both would do so. The Pareto-optimal decision is, in this case, unstable.

The lesson is that, in order to get along, both computers and people have to negotiate and be able to trust one another. It is thus crucial for some applications that software agents (see Chap. 16) have an idea of trust.

11.5.5 Multi-party Competition and Co-operation

The above can easily be extended to the case of multiple players. Instead of two scores, one gets a tuple of scores, one for each player. The modified minimax algorithm can again be used. At each point, as we move up the tree, we assume each player will maximise their own part of the tuple. The same problems arise with secret moves and non-Pareto-optimal results.

11.6 THE ADVERSARY IS LIFE!

Game playing is similar to interacting with the physical environment – as you act, new knowledge is found, or circumstances change to help or hinder you. In such circumstances the minimax algorithm can be used where the adversary is replaced by events from the environment. This effectively assumes that the worst thing will always happen.

Consider the following coin-weighing problem:

King Alabonzo of Arbicora has nine golden coins. He knows that one is a counterfeit (but not which one). He also knows that fake coins are slightly lighter than true ones. The local magician Berzicaan has a large and accurate balance but demands payment in advance for each weighing required. How many weighings should the king ask for and how should he proceed?

Figure 11.9 shows the search space, expanded to one level. The numbers in bags represent the size of the pile that has the heavier coin in it. This starts off as size 9. The king can weigh two coins (one on each side of the balance), four, six or eight. If the balance is equal, the coin must be in the remaining pile; if unequal, he can confine his search to the heavier pile.

For example, imagine the king chose to weigh four coins. If the balance was unequal, he would know that the lighter side had the fake coin in it; hence the pile to

test would now consist of only two coins. If, on the other hand, the balance had been equal, the king would know that the fake coin was among the five unweighed coins. Thus if we look at the figure, the choice to weigh four coins has two branches, the "=" branch leading to a five-coin bag and the "≠" branch leading to a two-coin bag.

The balance acts as the adversary, and we assume it "chooses" to weigh equal or unequal to make things as bad as possible for King Alabonzo! Alabonzo wants the pile as small as possible, so he acts as minimiser, while the balance acts as maximiser. Based on this, the intermediate nodes have been marked with their minimax values. We can see that, from this level of look-ahead, the best option appears to be weighing six coins first. In fact, this is the best option, and, in this case, the number of coins remaining acts as a very good heuristic to guide us quickly to the shallowest solution.

11.7 PROBABILITY

Many games contain some element of randomness, perhaps the toss of a coin or the roll of a die. Some of the choice points are replaced by branches with probabilities attached. This may be done both for simple search trees and for game trees. There are various ways to proceed. The simplest is to take the expected value at each point and then continue much as before.

In the example of Alabonzo's coins we deliberately avoided probabilities by saying he had to pay in advance for the number of weighings, so only the worst case mattered. If instead he paid per weighing when required, he might choose to minimise the expected cost. This wouldn't necessarily give the same answer as minimax. Figure 11.10 shows part of the tree starting with five coins. The lower branches have been labelled with the probability that they will occur. For example, if two coins are weighed, then there is a probability of 2/5 that one of them will be the counterfeit and 3/5 that it will be one of the three remaining coins. At the bottom of the figure, the numbers in square brackets are the expected number of further weighings needed to find the coin. In the case of one coin remaining, that must be the counterfeit and so the number is zero. In the cases of two or three coins, one further weighing is sufficient.

With five coins the king can choose to weigh either two or four coins. The average number of weighings for each has been calculated. For example, when weighing two coins, there is that weighing, and if the scales are

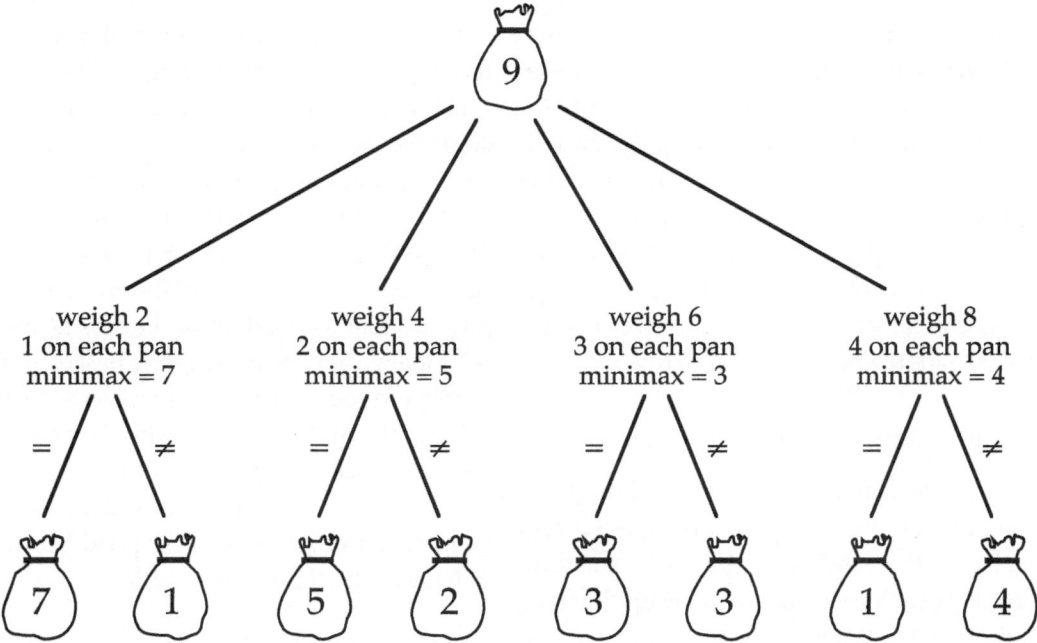

FIGURE 11.9   Minimax search for King Alabonzo's counterfeit coin.

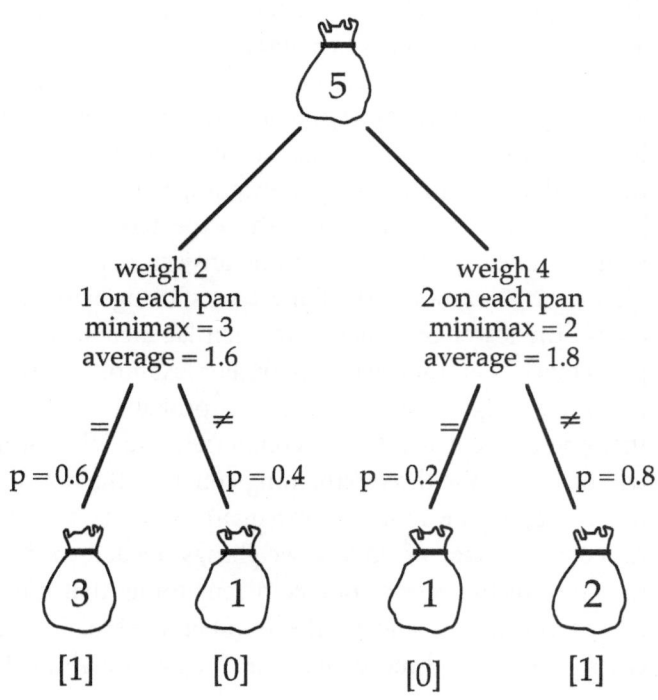

FIGURE 11.10   Probabilistic game tree for King Alabonzo's counterfeit coin.

equal (with a probability of 0.6), then a further weighing is required, giving an average number of 1.6.

See how the average number of weighings required is 1.6 when two coins are weighed and 1.8 when four are weighed. So, it is better to weigh two. However, if the number of coins in the piles is used as a heuristic, the minimax score is better for four weighings. In general the two methods will not give the same answer, as minimax will concentrate on the worst outcome no matter how unlikely its occurrence.

One problem with calculating the average pay-off is that it leads to a rapid increase in the search tree. For example, in a two-dice game, like backgammon, one has to investigate game situations for all 21 different pairs of die faces (or 13 sums). One way to control this is by using a probability-based cut-off for the search. It is not worth spending a lot of effort on something that is very unlikely to happen.

Averages are not the only way to proceed. One might prefer a choice with a lower average pay-off (or higher cost), if it has less variability – that is, a strategy of *risk avoidance*. On the other hand, a gambler might prefer a small chance of a big win. This may not be wise against a shrewd opponent with no randomness but may be perfectly reasonable where luck is involved.

Because of the problem with calculating probabilities, game-playing programs usually use complex heuristics rather than deep searches. So, a backgammon program will play more like a human than a chess program. However, there are some games where the calculation of probabilities can make a computer a far better player than a person. In casinos, the margin towards the house is quite narrow (otherwise people would lose their money too quickly!), so a little bit of knowledge can turn a slow loss into a steady win. In card games, the probability of particular cards occurring changes as the pack is used up. If you can remember every card that has been played, then you can take advantage of this and win. But don't try it! Counting (as this is called) is outlawed. If the management suspects, it will change the card packs, and anyone found in the casino using a pocket computer will find themselves in the local police station – if not wearing new shoes at the bottom of the river!

11.8 NEURAL NETWORKS FOR GAMES

Neural networks have been applied to games for many years [47, 223, 270]; however, as mentioned previously, it was the application of deep learning in AlphaGo that gave a new impetus not only to games applications but to deep learning in general. The success of AlphaGo was based on classic techniques, including game searches and Monte Carlo tree search, combined with multiple neural networks and adversarial techniques [260]

We will use AlphaGo as the running example, but the techniques are general.

11.8.1 Where to Use a Neural Network

The most obvious use of a neural network in games is to ask the question, "*what move do I make next*", or perhaps more broadly, given the board position, "*which moves seem most promising?*" However, game search algorithms effectively answer this question by looking ahead in the game tree. Effectively this means thinking about each potential move and then asking two subsequent questions "what move do I think my opponent will make?" and when one has reached a particular depth of lookahead "how good or bad is the board state".

AlphaGo tackled these questions by training two deep neural networks on different kinds of heuristics (Figure 11.11):

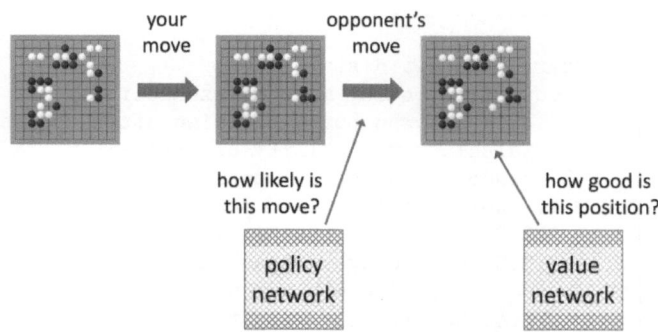

FIGURE 11.11 Policy and value networks to guide search in the game tree.

policy network – given a board position with the opponent to move, estimate the probability of what move the opponent will make next.

value network – given a board position and which player is next to move, estimate the probability that the game will end up as a win.

Given these two networks it is fairly straightforward to apply Monte Carlo tree search using minimax criteria, as shown in Figure 11.12. There are variants for this, for example one can apply the value network to the board position after step 2.1 to decide whether to just use the heuristic itself or to expand the search in step 2.2.

Note too that similar techniques can be used in other domains. For example, in a mathematical proof, the valid steps are well defined (like moves in a game), making it relatively easy to check a proof. Normally the human mathematician chooses which valid rule to attempt at any point in time; however, a neural network can be trained to generate heuristics in a similar way [174].

Note that both the policy network and the value network can be trained on examples of games played to completion so that the final outcome is known. However, the value network can also be trained inductively. If one has an existing heuristic (say the value network from a previous iteration), then for any board position a one-step or two-step minimax lookahead can be used to create a revised heuristic value for the board (in a similar way to Figure 11.12). This can then be used as training data for the next generation of the value network.

```
1. Start at a board state
2. Choose a potential move M (exhaustively, or in random order)
  2.1. Work out the board position after the move
  2.2. Calculate probability of opponent moves using policy network
  2.3. Randomly choose an opponent move O based on likelihood from step 2.2.
  2.3.1.  Work out the board position after the opponent move
      2.3.2. Calculate the heuristic using the value network for this position
      2.3.3  Depending on the heuristic and past ones, either treat this value as the board
             position minimax value, or ...
      2.3.4. Recursively repeat steps 1-3 from this board position to work out a minimax value.
  2.4. Potentially repeat 1.3 for more possible opponent moves
  2.5. Record the minimax for M as the minimum (least good) of the minimax value of each opponent
       move O tried, potentially weighted by probability
3. Record the overall minimax of the board position as the maximum of the minimax of each move M.
4. Make the move M with the largest minimax value.
```

FIGURE 11.12   Pseudocode for Monte Carlo tree search using policy and value networks.

11.8.2 Training Data and Self Play

Initially the deep neural networks that powered AlphaGo were bootstrapped using datasets of large numbers of human tournaments. However, there have been too few human games to train sufficiently large networks.

To deal with this, self play techniques were used to expand training data as described in Chapter 8. Variants of the AlphaGo network were played against one another, each learning from the matches they played and each getting better and better (Figure 11.13). With sufficient computer resources far more computer–computer tournaments can be generated than have ever been played in the history of human Go. In AlphaGo, this started based on human tournaments, but AlphaGo Zero did not use any human games but learnt entirely from the rules of Go and lots of adversarial learning.

Lee Sedol eventually retired as a Go player because AI "cannot be defeated" [19]. However, it is worth noting that Lee Sedol learnt his Go based on a tiny fraction of the games that AlphaGo played virtually against itself. While the AI defeated Lee Sedol, the way in which it did so is very different, and in many ways far less efficient, than the human grandmaster, an issue we will return to in Chapter 22.

11.9 SUMMARY

In this chapter we have looked at algorithms for playing standard games (non-probabilistic, open, two-person, turn-taking, zero-sum games). Such games include

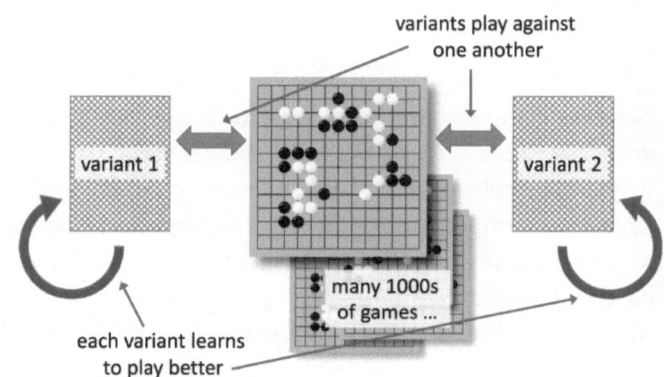

FIGURE 11.13   Self play.

chess, draughts, tic-tac-toe and Go. We considered minimax search techniques and alpha–beta pruning, which relate to the search techniques studied in Chapter 4. We also discussed games where co-operation is important, where players can take simultaneous moves and where random events happen (such as the throw of a die). Finally, we have seen how deep learning has transformed the level of computer game playing but still makes use of the fundamentals. We will see in Chapter 15 that acting in the presence of uncertainty is essential for robotics and other practical planning tasks, and that chapter will show how game-playing algorithms can be used to tackle such non-gaming problems.

11.1 Consider the alternatives to the "standard" game (the non-probabilistic, open, two-person,

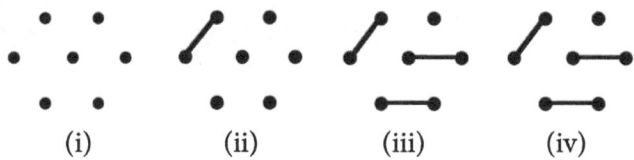

FIGURE 11.14   Hex-lines.

turn-taking, zero-sum game). Confining yourself to turn-taking games, consider all possible combinations of game types, and attempt to find a game to fit in each category. Only worry about the "zero-sum" property for two-person games; this means you should have 12 categories in all. For example, find a game that is probabilistic, open and not two person.

11.2   Consider the three-person game hex-lines, a variant of "placing dominoes". A piece of paper is marked with dots in a triangular pattern. Different sizes and shapes of playing area give rise to different games. Each person in turn connects two adjacent points. However, they are only allowed to use points that have not yet been used. The players each have a direction and are only allowed to draw lines parallel to their direction. We'll assume that the first player draws lines sloping up (/), the second horizontal (—) and the third sloping down (\). If players cannot draw their direction of line, then they are out of the game. When no player can draw a line, the lines for each player are counted, giving the final score.

Consider an example game on a small hexagonal playing area. The board positions through the game are shown in Figure 11.14. The initial configuration is (i).

a.   First player draws sloping up (ii).

b.   Second player draws horizontal (iii).

c.   Third player cannot play and is out.

d.   First player cannot play either and is out.

e.   Second player draws again giving (iv).

The final score is thus [1,2,0] (1 for the first player, 2 for the second and 0 for the third).

Taking the same initial configuration draw the complete game tree. Could the first player have done better?

11.3   In the game of Nim there are a number of piles of stones. Each player takes turns to choose a pile and take as many stones from the pile as they like. The winner is the person who takes the last stone.

a.   Draw a game tree for the game of Nim starting with three piles one with two stones in and the other two piles with just one stone each, You could write this "(2,1,1)". Think carefully about symmetries. This will reduce the size of your game tree considerably as was the case for 'placing dominoes' in Figure 11.1.

b.   Calculate the minimax values for each state, scoring +1 for a win by the first player (say A) and −1 for a win by the second player (say B).

c.   Is there a winning move for the first player?

d.   Repeat the minimax, but instead using the difference in the number of stones taken by each player as the score. That is, if player A has taken three stones and player B one stone, the score is 2 in favour of A.

e.   Does this change player A's strategy?

11.4   Repeat the above exercise, but this time assume the second player, player B, is entirely random in the sense that among all possible next moves, they choose randomly between them. Instead of minimax, use expected scores.

FURTHER READING

G. Yannakakis and J. Togelius. *Artificial intelligence and games*. Springer, Cham, 2018.

> *An entire book dedicated to the use of AI in games. The earlier parts cover broad algorithms including both classic algorithms such as minimax and stochastic tree search and also the use of genetic algorithms and neural networks. The book has a website https://gameaibook.org/ with additional resources.*

J. Pearl. *Heuristics: Intelligent search strategies for computer problem solving*. Addison-Wesley, Reading, MA, 1984.

> *Part 3 of this book concentrates on game-playing strategies and heuristics.*

D. Silver, A. Huang, C. Maddison, et al. Mastering the game of Go with deep neural networks and tree search. *Nature*, 529:484–489, 2016. DOI:10.1038/nature16961

The paper by the creators of AlphaGo that can be seen as marking the start of the current phase of deep-learning-based AI.

Computer Vision

12.1 OVERVIEW

Computer vision is one way for a computer system to reach beyond the data it is given and find out about the real world. There are many important applications, from robotics to airport security. However, it is a difficult process. This chapter starts with an overview of the typical phases of processing in computer vision. Subsequent sections (12.3–12.9) then follow through these phases in turn. At each point deeper knowledge is inferred from the raw image. Neural networks are used extensively in computer vision, augmenting or replacing some of these stages, although often using specialised networks based on understanding from the more algorithmic techniques. Finally, in Section 12.11, we look at the special problems and opportunities that arise when we have moving images or input from several cameras or moving images.

In this chapter we shall assume that the cameras are *passive* – we interpret what we are given. In Chapter 15 we shall look at active vision, where the camera can move or adjust itself to improve its understanding of a scene.

12.2 INTRODUCTION

12.2.1 Why Computer Vision Is Difficult

The human visual system makes scene interpretation seem easy. We can look out of a window and can make sense of what is in fact a very complex scene. This process is very difficult for a machine. As with natural language interpretation, it is a problem of ambiguity. The orientation and position of an object changes its appearance, as does different lighting or colour. In addition, objects are often partially hidden by other objects.

In order to interpret an image, we need both low-level information, such as texture and shading, and high-level information, such as context and world knowledge. The former allows us to identify the object, the latter to interpret it according to our expectations.

12.2.2 Phases of Computer Vision

Because of these multiple levels of conformation, most traditional computer vision is based on a hierarchy of processes, starting with the raw image and working towards a high-level model of the world. Each stage builds on the features extracted at the stage below. Typical stages are (see Figure 12.1):

- *digitisation*: either the image is captured digitally or an analogue image is converted into a digital image

- *signal processing*: low-level processing of the digital image in order to enhance significant features

- *edge and region detection*: finding low-level features in the digital image

- *three-dimensional or two-dimensional object recognition*: building lines and regions into objects

- *image understanding*: making sufficient sense of the image to use it

Note, however, that not all applications go through all the stages. The higher levels of processing are more complicated and time consuming. In any real situation one

DOI: 10.1201/9781003082880-15

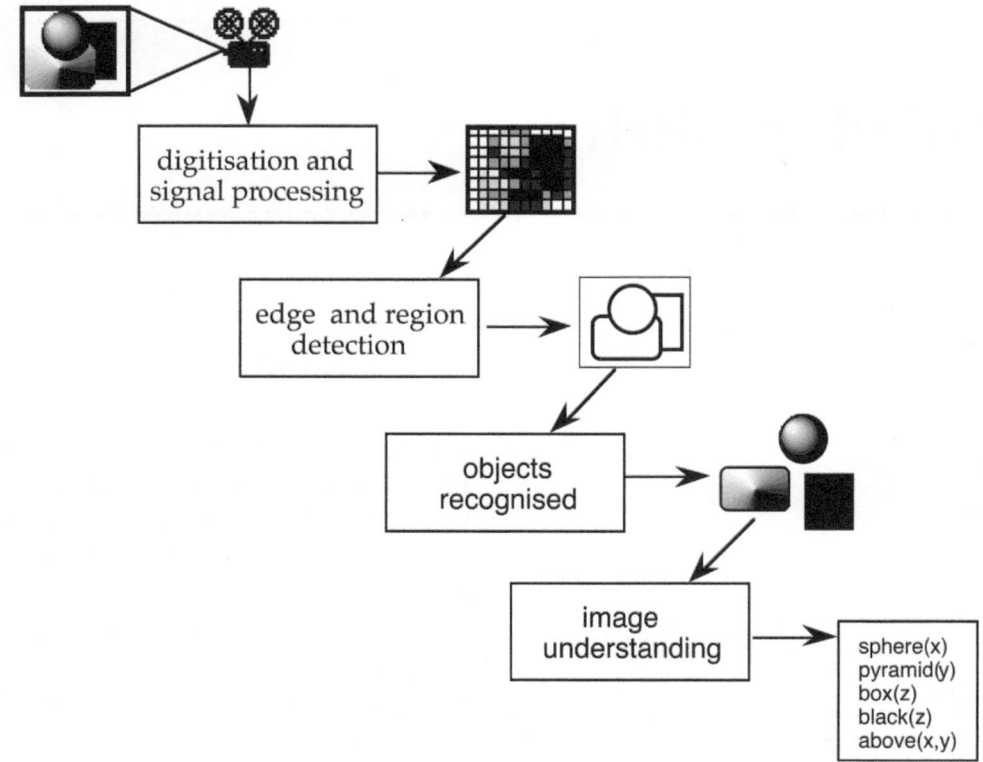

FIGURE 12.1   Phases of computer vision.

would want to get away with as low a level of processing as possible. Neural networks may be used for different parts of this pipeline.

The rest of this chapter will follow these levels of processing, and we will note where applications exist at each level.

12.3   DIGITISATION AND SIGNAL PROCESSING

The aim of computer vision is to understand some scene in the outside world. This may be captured using a video camera but may come from a scanner (e.g. optical character recognition). Indeed, for experimenting with computer vision it will be easier to use digital photographs than to work with real-time video. Also, it is not necessary that images come from visible light. For example, satellite data may use infrared sensing. For the purposes of exposition, we will assume that we are capturing a visible image with a video camera. This image will need to be digitised so that it can be processed by a computer and also "cleaned up" by signal processing software. The next section will discuss signal processing further in the context of edge detection.

12.3.1   Digitising Images

For use in computer vision, the image must be represented in a form that the machine can read. The analogue image is converted into a digital image. For born-digital media this may happen within the camera as the image is focused onto a CCD array; however, for older paper or film media, this may be performed separately by some form of scanner or video digitiser. The digital image is basically a stream of numbers, each corresponding to a small region of the image, a pixel. In the case of 'black and white' (really grey) images there is a single number for each pixel, which measures the light intensity at the pixel, the grey level. The range of possible grey levels is called a grey-scale (hence grey-scale images). If the grey scale consists of just two levels (really black or white), the image is a binary image.

Figure 12.2 shows an image (ii) and its digitised form (i). There are ten grey levels from 0–white to 9–black. More typically there will be 256 or more grey levels rather than ten and often 0 is black (no light). However, the digits 0–9 fit better into the picture. Also, in order to print it, the image (ii) is already digitised and we are simply looking at a coarser level of digitisation.

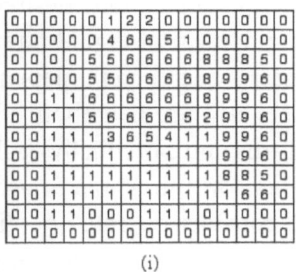

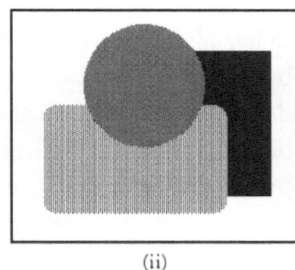

FIGURE 12.2   Digitised image.

Most of the algorithms used in computer vision work on simple grey-scale images. However, sometimes colour images are used. In this case, there are usually three or four values stored for each pixel, corresponding to either primary colours (red, blue and green) or some other colour representation system.

Look again at Figure 12.2. Notice how the right-hand edge of the black rectangle translates into a series of medium grey levels. This is because the pixels each include some of the black rectangle and some of the white background. What was a sharp edge has become fuzzy.

As well as this blurring of edges, other effects conspire to make the grey-scale image inaccurate. Some cameras may not generate parallel lines of pixels, the pixels may be rectangular rather than square (the aspect ratio) or the relationship between darkness and grey scale recorded may not be linear. However, the most persistent problem is noise: inaccurate readings of individual pixels due to electronic fluctuations, dust on the lens or even a foggy day!

12.3.2 Thresholding

Given a grey-scale image, the simplest thing we can do is to threshold it; that is, select all pixels whose greyness exceed some value. This may select key significant features from the image.

In Figure 12.3, we see an image (i) thresholded at three different levels of greyness. The first (ii) has the lowest threshold, accepting anything that is not pure white. The pixels of all the objects in the image are selected with this threshold. The next threshold (iii) accepts only the darker grey of the circle and the black of the rectangle. Finally, the highest threshold (iv) accepts only pure black

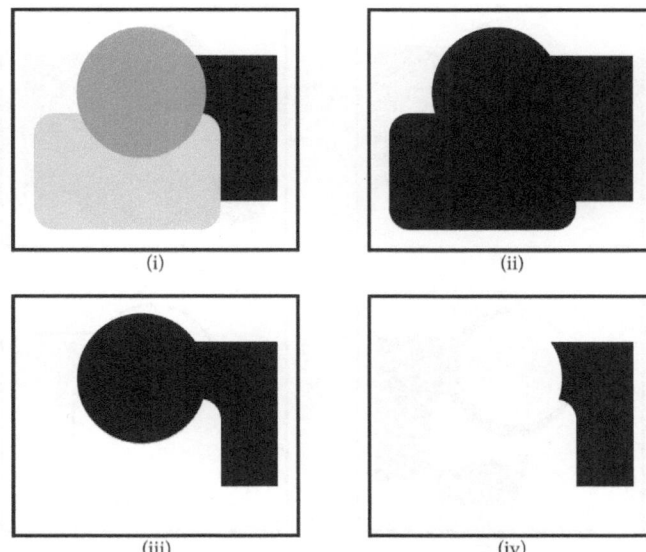

FIGURE 12.3   Thresholding.

pixels and hence only those of the obscured rectangle are selected.

This can be used as a simple way to recognise objects. For example, [175] shows how faults in electrical plugs can be detected using multiple threshold levels. At some levels the wires are selected, allowing one to check that the wiring is correct; at others the presence of the fuse can be verified. In an industrial setting one may be able to select lighting levels carefully in order to make this possible.

One can also use thresholding to obtain a simple edge detection. One simply follows round the edge of a thresholded image. One can do this without actually performing the thresholding as one can simply follow pixels where the grey changes from the desired value. This is called contour following.

However, more generally, images resist this level of interpretation. Consider Figure 12.4. To the human eye, this also consists of three objects. However, see what two levels of thresholding, (ii) and (iii), do to the image. The combination of light and shadows means that the regions picked out by thresholding show areas of individual objects instead of distinguishing the objects. Indeed, even to the human eye, the only way we know that the sphere is not connected to the black rectangular area is because of the intervening pyramid.

Contour following would give the boundary of one of these images – not really a good start for image understanding. The more robust approaches in the next section

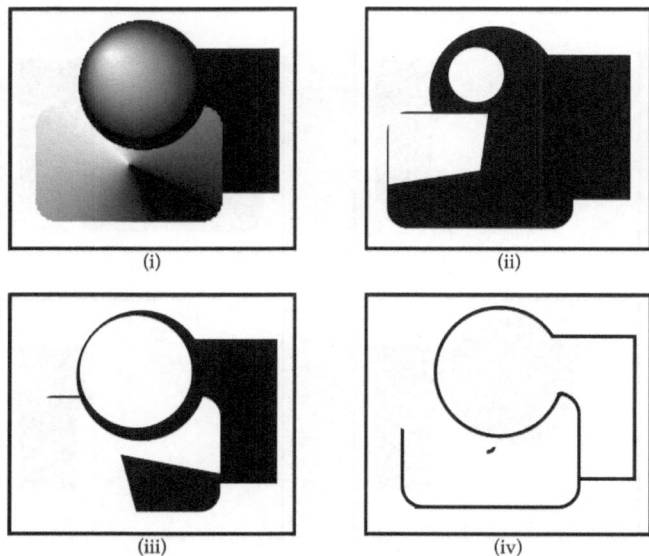

FIGURE 12.4   A difficult image to threshold.

will instead use the rate of change in intensity – slope rather than height – to detect edges. However, even that will struggle on this image. The last image (iv) in Figure 12.4 shows edges obtained by looking for sharp contrasts in greyness. See how the dark side of the sphere has merged into the black rectangle, and how the light shining on the pyramid has lost part of its boundary. There is even a little blob in the middle where the light side of the pyramid meets the dark at the point.

In fact, as a human rather than a machine, you will have inferred quite a lot from the image. You will see it as a three-dimensional image where the sphere is above the pyramid and both lie above a dark rectangle. You will recognise that the light is shining somewhere from the top left. You will also notice from the shape of the figures and the nature of the shading that this is no photograph, but a generated image. The algorithms we will discuss later in this chapter will get significantly beyond thresholding, but still nowhere near your level of sophistication!

12.3.3   Digital Filters

We have noted some of the problems of noise, blurring and lighting effects that make image interpretation difficult. Various signal processing techniques can be applied to the image in order to remove some of the effects of noise or enhance other features, such as edges. The application of such techniques is also called digital filter-

ing. This is by analogy with physical filters, which enable you to remove unwanted materials, or to find desired material. Thresholding is a simple form of digital filter, but whereas thresholding processes each pixel independently, more sophisticated filters also use neighbouring pixels. Some filters go beyond this and potentially each pixel's filtered value is dependent on the whole image. However, all the filters we will consider operate on a finite window – a fixed-size group of pixels surrounding the current pixel.

12.3.3.1 Linear Filters

Many filters are linear. These work by having a series of weights for each pixel in the window. For any point in the image, the surrounding pixels are multiplied by the relevant weights and added together to give the final filtered pixel value.

In Figure 12.5 we see the effect of applying a filter with a 3 × 3 window. The filter weights are shown at the top right. The initial image grey levels are at the top left. For a particular pixel the nine pixel values in the window are extracted. These are then multiplied by the corresponding weights, giving in this case the new value 1. This value is placed in the appropriate position in the new filtered image (bottom left).

The pixels around the edge of the filtered image have been left blank. This is because one cannot position a window of pixels 3 × 3 centred on the edge pixels. So, either the filtered image must be smaller than the initial image, or some special action is taken at the edges.

Notice also that some of the filtered pixels have negative values associated with them. Obviously this can only arise if some of the weights are negative. This is not a problem for subsequent computer processing, but the values after this particular filter cannot easily be interpreted as grey levels.

A related problem is that the values in the final image may be bigger than the original range of values. For example, with the above weights, a zero pixel surrounded by nines would give rise to a filtered value of 36. Again, this is not too great a problem, but if the result is too large or too small (negative), then it may be too large to store – an overflow problem. Usually, the weights will be scaled to avoid this. So, in the example above, the result of applying the filter would be divided by 8 in order to bring the output values within a similar range to the input grey

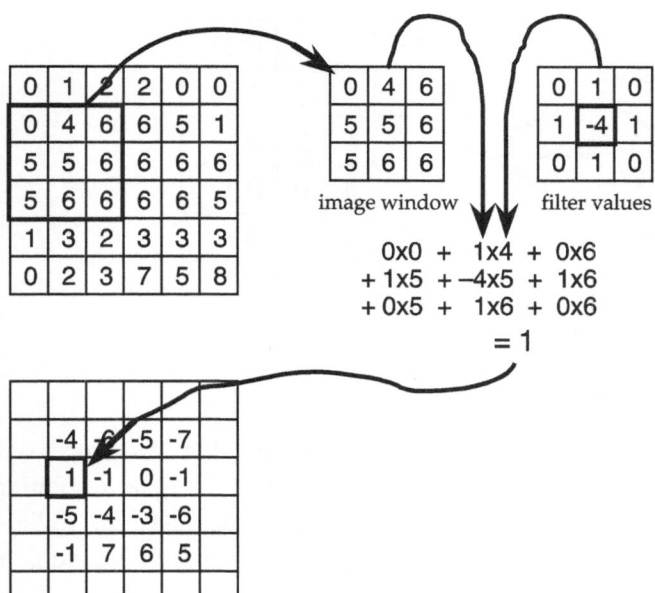

FIGURE 12.5 Applying a digital filter.

tance to pixels near the chosen one and less to those far away.

There are disadvantages to smoothing, especially when using large filters. Notice in Figure 12.6 that the boundary between the two regions has become blurred. There is a line of pixels that are at an average value between the high and low regions. Thus, the edge can become harder to trace. Furthermore, fine features such as thin lines may disappear altogether. There is no easy answer to this problem – the desire to remove noise is in conflict with the desire to retain sharp images. In the end, how do you distinguish a small but significant feature from noise?

12.3.3.3 Gaussian Filters

The Gaussian filter is a special smoothing filter based on the bell-shaped Gaussian curve, well known in statistics as the "Normal" distribution (Chap. 7). One imagines a window of infinite size, where the weight, $w(x, y)$, assigned to the pixel at position x, y from the centre is

$$w(x, y) = \frac{1}{2\pi\sigma^2} \exp[-(x^2 + y^2)/2\sigma^2]$$

The constant σ is a measure of the spread of the window – how much the image will be smeared by the filter. A small value of σ will mean that the weights in the filter will be small for distant pixels, whereas a large value allows more distant pixels to affect the new value of the current pixel. If noise affects groups of pixels together, then one would choose a large value of σ.

Although the window for a Gaussian filter is theoretically infinite, the weights become small rapidly, and so, depending on the value of σ, one can ignore those outside a certain area and so make a finite windowed version. For example, Figure 12.7 shows a Gaussian filter with a 5×5 window. Notice how it is symmetric and how the weights decrease towards the edge. This filter has weights totalling 256, but this took some effort! The theoretical weights are not integers, and the rounding errors mean that in general the sum of weights will not be a nice number.

One big advantage of Gaussian filters is that the parameter σ can be set to any value yielding finer or coarser smoothing. Simple smoothing methods tend only to have versions getting "bigger" at fixed intervals (3×3, 5×5, etc.). The Gaussian with $\sigma = 0.7$ would

scales. The coefficients are often chosen to add up to a power of 2, as dividing can then be achieved using bit shifts, which are far faster.

12.3.3.2 Smoothing

The simplest type of filter is for smoothing an image. That is, surrounding pixels are averaged to give the new value of a pixel. Figure 12.6 shows a simple 2×2 smoothing filter applied to an image. The filter window is drawn in the middle, and its pivot cell (the one which overlays the pixel to which the window is applied) is at the top left. The filter values are all ones, and so it simply adds the pixel and its three neighbours to the left and below and averages the four (note the ÷4). The image clearly consists of two regions, one to the left with high (7 or 8) grey-scale values and one to the right with low (0 or 1) values. However, the image also has some noise in it. Two of the pixels on the left have low values and one on the right a high value. Applying the filter has all but removed these anomalies, leaving the two regions far more uniform, and hence suitable for thresholding or other further analysis.

Because only a few pixels are averaged with the 2×2 filter, it is still susceptible to noise. Applying the filter would only reduce the magnitude by a factor of 4. Larger windows are used if there is more noise, or if later analysis requires a cleaner image. A larger filter will often have an uneven distribution of weights, giving more impor-

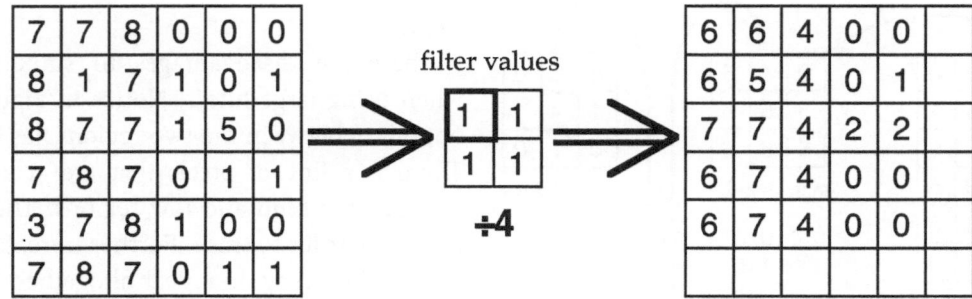

FIGURE 12.6   Applying a 2 × 2 smoothing filter.

0	1	3	1	0
1	13	30	13	1
3	30	64	30	3
1	13	30	13	1
0	1	3	1	0

FIGURE 12.7   Gaussian filter with $\sigma = 0.8$.

also fit on a 5 × 5 window but would be weighted more towards the centre (less smoothing).

12.3.3.4 Practical Considerations

We have already discussed problems of overflow when computing filtered images, and in general there are various computational factors that influence the choice of filter. Indeed, the cost of image processing can be so high that it is often better to use a simple method rather than an optimal one. It's no good an industrial robot recognising a nut ten seconds after it has passed by on the conveyor belt.

Images are large. Even a small 512 × 512 image with 256 grey levels consumes 256 kilobytes of memory. This is expensive in terms of storage, but also those 262 144 pixels take a long time to process one by one. A linear filter with a 2 × 2 window takes four multiplications per pixel, a 3 × 3 window takes nine and 5 × 5 takes 25! Also, a simple filter with coefficients of ±1 or powers of 2 can be calculated by simple adds and shifts, further reducing the cost. So, the simple 2 × 2 smoothing filter in Figure 12.6, although crude, only takes 1 million

additions, whereas the Gaussian filter in Figure 12.7 takes over 6 million multiplications. A higher resolution colour image, say 4K x 2K pixels, takes 24 million bytes to store and correspondingly more operations to process.

One solution is to use special hardware, DSP (Digital Signal Processing) chips or parallel processing, or to use the Graphics Processing Unit (GPU) found in most desktop computers. Indeed, your brain works in something like this fashion, with large areas committed to specific tasks such as line detection. It processes the whole image at once, rather than sequentially point by point. Whether or not this is available, care in the choice of processing method is essential.

The large amounts of storage required make it imperative that algorithms do not generate lots of intermediate images (unless you have masses of memory!). One way to achieve this is to overwrite the original image as it is filtered. But beware – look again at the 3 × 3 filter in Figure 12.5. If the image is processed from the top left downwards, then by the time a pixel is processed those pixels above and to the left of it will have been overwritten. With some very simple filters (such as the averaging filter in Figure 12.6) this is not a problem, but in general one must be careful to avoid overwriting pixels that will be needed. It is possible with care! An alternative way to avoid intermediate storage is to work out the effects of multiple steps and to compute them in one step. We see an example of this in the next section, in the calculation of the Laplacian-of-Gaussian filter.

12.4 EDGE DETECTION

Edge detection is central to most computer vision. There is also substantial evidence that edges form a key part of human visual understanding. An obvious example is the

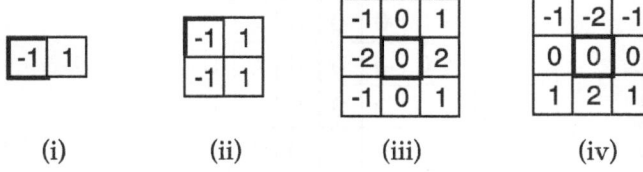

FIGURE 12.8   Different gradient filters.

ease with which people can recognise sketches and cartoons. A few lines are able to invoke the full two- or three-dimensional image. Edge detection consists of two sub-processes. First of all, potential edge pixels are identified by looking at their grey level compared with surrounding pixels. Then these individual edge pixels are traced to form the edge lines. Some of the edges may form closed curves, while others will terminate or form a junction with another edge. Some of the pixels detected by the first stage may not be able to join up with others to form true edges. These may correspond to features too small to recognise properly or simply be the result of noise.

12.4.1 Identifying Edge Pixels

The grey-level image is an array of numbers (grey levels) representing the intensity value of the pixels. It can be viewed as a description of a hilly landscape where the numbers are altitudes. So a high number represents a peak and a low number a valley. Edge detection involves identifying ridges, valleys and cliffs. These are the edges in the image. We can use gradient operators to perform edge detection by identifying areas with high gradients. A high gradient i.e. a sudden change in intensity) indicates an edge. There are a number of different gradient operators in use.

12.4.1.1 Gradient Operators

If you subtract a pixel's grey level from the one immediately to its right, you get a simple measure of the horizontal gradient of the image. This two-point filter is shown in Figure 12.8(i), together with two alternatives: a four-point filter (ii), which uses a 2×2 window, and a six-point filter (iii), which uses a 3×3 window. The vertical version of the six-point filter is also shown (iv).

The effects of the six-point filters are shown in Figure 12.9. The image shows the corner of a rectangular region in the bottom right-hand corner. Notice how the horizontal gradient operator picks out the left edge of the

region and the vertical operator picks out the upper edge. Both operators would detect a diagonal edge, but less efficiently than one in their preferred direction. So, in Figure 12.10, the pixel values are large, but the filtered values at the edge are smaller and more smeared.

These operators can be useful if edges at a particular orientation are important, in which case one can simply threshold the filtered image and treat pixels with large gradients as edges. However, neither operator on its own detects both horizontal and vertical edges.

12.4.1.2 Robert's Operator

Robert's operator uses a 2×2 window. For each position (x, y), a gradient function, $G(x, y)$, is calculated by

$$G(x, y) = |f(x, y) - f(x + 1, y - 1)|$$
$$+ |f(x + 1, y) - f(x, y - 1)|$$

where $f(x, y)$ is the intensity of the pixel at that position. Notice that this is not a simple linear filter, as it involves calculating the absolute value of the difference between diagonally opposite pixels. This is necessary in order to detect lines in all directions.

The results of the gradient function can be compared with a predetermined threshold to detect a local edge. Consider the various examples in Figure 12.11(i–iv):

(i) $G = |3 - 3| + |3 - 3|$
      $= |0| + |0|$ $= 0$
(ii) $G = |7 - 2| + |3 - 8|$
      $= |5| + |-5|$ $= 10$
(iii) $G = |5 - 6| + |7 - 1|$
      $= |-1| + |6|$ $= 7$
(iv) $G = |7 - 8| + |1 - 2|$
      $= |-1| + |-1|$ $= 2$

A threshold of 5 would detect (ii) and (iii) as edges, but not (i) or (iv). Let's look at each example. The first is a constant grey level: there are no edges, and none are detected whatever threshold is chosen. The second is a very clear edge running up the image, and it gets the highest gradient of the four examples. The third example also has quite a strong gradient. It appears to represent an edge running diagonally across the image. The final example has dramatic changes in intensity but a low gradient. This is because there is little overall slope in the image. It represents a sort of ridge going across the picture. This might be a line a single pixel wide but not an edge between regions.

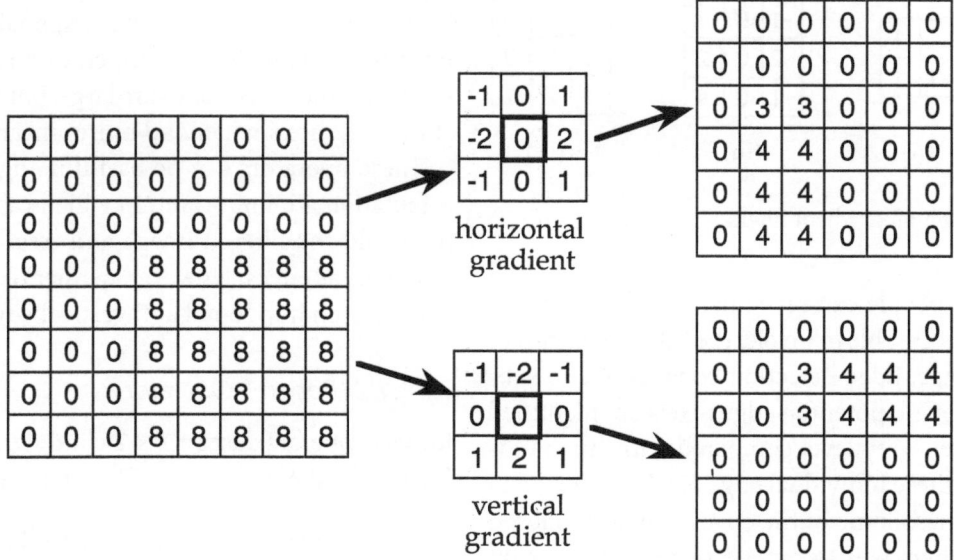

FIGURE 12.9   Applying gradient filters.

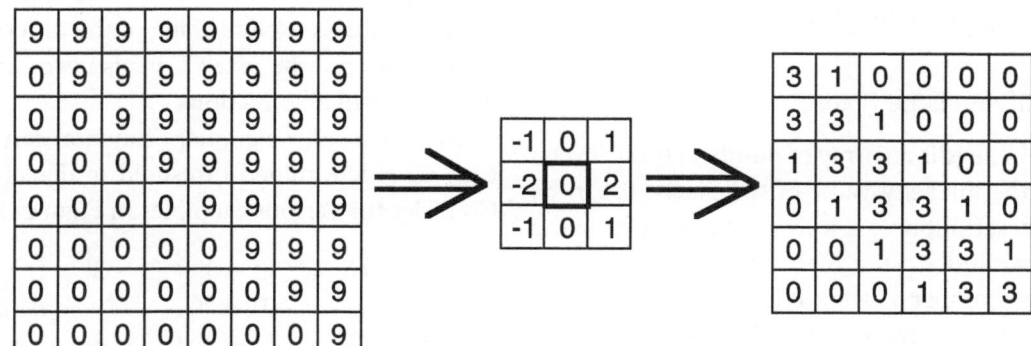

FIGURE 12.10   Gradient filter on a diagonal edge.

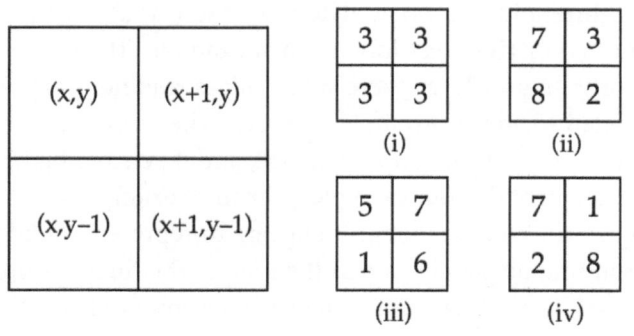

FIGURE 12.11   Robert's operator.

Robert's operator has the advantage of simplicity but suffers from being very localised and therefore easily affected by noise. For example, (ii) got a high gradient reading and would have been detected as a potential edge, but this is largely based on the bottom right pixel. If this one pixel were wrong, perhaps as a result of random noise, a spurious edge would be detected.

12.4.1.3 Sobel's Operator

Sobel's operator uses a slightly larger 3×3 window, which makes it somewhat less affected by noise. Figure 12.12 labels the grey levels of the nine pixels. The gradient function is calculated as

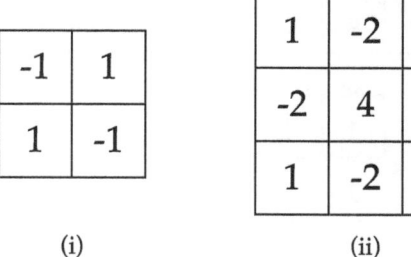

a	b	c
d	e	f
g	h	i

where
a = f(x–1,y+1)
b = f(x,y+1)
c = f(x+1,y+1)
d = f(x–1,y)
e = f(x,y)
f = f(x+1,y)
g = f(x–1,y–1)
h = f(x,y–1)
i = f(x+1,y–1)

FIGURE 12.12 Sobel's operator.

FIGURE 12.13 Approximations to the Laplacian.

(i)

(ii)

$$G = |(c + 2f + i) - (a + 2d + g)| \\ + |(g + 2h + i) - (a + 2b + c)|$$

Again, this can be thresholded to give potential edge points.

Notice that the grey level at the pixel itself, e, is not used: the surrounding pixels give all the information. We can see the operator as composed of two terms, a horizontal and a vertical gradient:

$$H = (c + 2f + i) - (a + 2d + g)$$
$$V = (g + 2h + i) - (a + 2b + c)$$
$$G = |H| + |V|$$

The first term, H, compares the three pixels to the right of e with those to the left. The second, V, compares those below the pixel with those above. In fact, if you look back at the six-point gradient filters in Figure 12.8, you will see that V and H are precisely the absolute values of the outputs of those filters. An edge running across the image will have a large value of V, one running up the image a large value of H. So, once we have decided that a pixel represents an edge point, we can give the edge an orientation using the ratio between H and V. Although we could follow edges simply by looking for adjacent edge pixels, it is better to use edge directions (as we shall see later).

Note that it is also possible to give an orientation with Robert's operator, as the two terms in it correspond to a northwesterly and northeasterly gradient respectively. However, this estimate of direction would be even more subject to noise.

Note also that Sobel's operator uses each pixel value twice, either multiplying it by two (the side pixels: f, d, h and b) or including it in both terms (the corner pixels: a, c, g and i). However, an error in one of the corner pixels might cancel out, whereas one in the side pixels would always affect the result. For this reason, some prefer a modified version of Sobel's operator:

$$G = |(c + f + i) - (a + d + g)| \\ + |(g + h + i) - (a + b + c)|$$

On the other hand, there are theoretical reasons for preferring the original operator, so the choice of operator in a particular application is rather a matter of taste!

12.4.1.4 Laplacian Operator

An alternative to measuring the gradient is to use the Laplacian operator. This is a mathematical measure (written ∇) of the change in gradient. Its mathematical definition is in terms of the second differential in the x and y direction (where the first differential is the gradient):

$$\nabla f = \frac{d^2f}{dy^2} + \frac{d^2}{dy^2}$$

However, for digital image processing, linear filters are used which approximate to the true Laplacian. Approximations are shown in Figure 12.13 for a 2 × 2 grid and a 3 × 3 grid.

To see how they work, we will use a one-dimensional equivalent to the Laplacian which filters a one-dimensional series of grey levels using the weights $(1, -2, 1)$. The effect of this is shown in Figure 12.14. Notice how the edge between the nines and ones is converted into little peaks and troughs. The actual edge detection then involves looking for zero crossings, places where the Laplacian's values change between positive and negative.

Notice that in Figure 12.14 the boundary between the nines and the ones is a 5. The one-dimensional image is slightly blurred. When Robert's or Sobel's operators encounter such an edge, they are likely to register several possible edge pixels either side of the actual edge. The Laplacian will register a single pixel in the middle of the blurred edge.

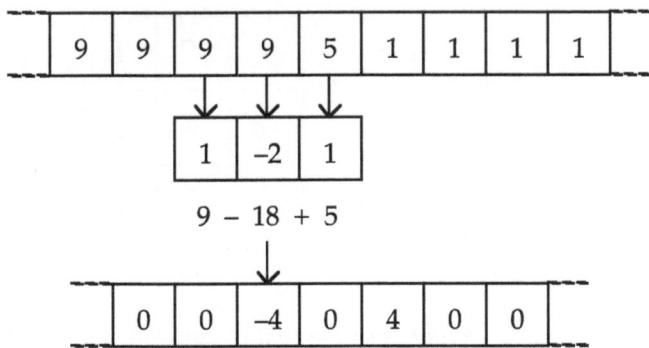

FIGURE 12.14 Using the Laplacian.

The Laplacian also has the advantage that it is a linear filter and can thus be easily manipulated with other filters. A frequent combination is to use a Gaussian filter to smooth the image, and then follow this with a Laplacian. Because both are linear filters, they can be combined into a single filter called the Laplacian-of-Gaussian (LOG) filter.

Note that the Laplacian does not give any indication of orientation. If this is required, then some additional method must be used once an edge has been detected.

12.4.1.5 Successive Refinement and Marr's Primal Sketch

We saw earlier that images are very large and hence calculations over the whole image take a long time. One way to avoid this is to operate initially on coarse versions of the image and then successively use more detailed images to examine potentially interesting features. For example, we could divide a 512 × 512 image into 8 × 8 cells and then calculate the average grey level over the cell. Treating each cell as a big "pixel", we get a much smaller 64×64 image. Edge detection is then applied to this image using one of the methods suggested above. If one of the cells is registered as an edge, then the pixels comprising it are investigated individually. Assuming that only a small proportion of the cells are potential edges then the savings in computation are enormous – the only time we have to visit all the pixels is when the cell averages are computed. This method of successive refinement can be applied to other parts of the image processing process, such as edge following and region detection (discussed later).

One representation of images, Marr's primal sketch [184], uses similar methods to detect features at different levels of detail, but for a very different reason. Instead of averaging over cells, Laplacian-of-Gaussian filters are used with different standard deviations, where small standard deviations correspond to fine detail. Recall that Gaussians use large windows, so this is definitely not a cost-cutting route to image processing! The concept of different levels of detail is central to the model. The primal sketch is divided into edges, terminations (ends of edges), bars (regions between parallel edges) and blobs (small isolated regions). In particular, blobs are regions of pixels that register as edges (zero crossings of the Laplacian) at fine resolution but disappear at high resolution. Look at the room and then screw up your eyes. If you can see it when your eyes are open, but not when they are screwed up, then it is a blob.

12.4.2 Edge Following

We have now identified pixels that may lie on the edges of objects. We are not there yet! The next step is to string those pixels together to make lines, that is to identify which groups of pixels make up particular edges. The basic rule of thumb is that if two suspected edges are connected, then they form a single line. However, this needs to be modified slightly for three reasons:

- because of noise, shadows and so on, some edges will contain gaps

- noise may cause spurious pixels to be candidate edges

- edges may end at junctions with other lines.

The first means that we may have to look more than one pixel ahead to find the next edge point. The other two mean that we have to use the edge orientation information in order to reject spurious edges or detect junctions.

A basic edge-following algorithm is then as follows:

1. Choose any suspected edge pixel that has not already been used.

2. Choose one direction to follow first.

3. Look for an adjoining pixel in the right general direction.

4. If the orientation of the pixel is not too different, then accept it.

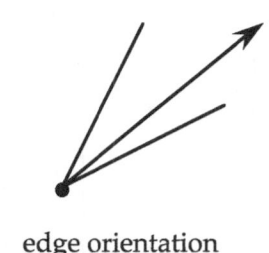

edge orientation
in this octant

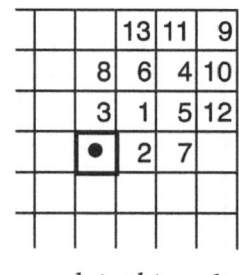

search in this order

FIGURE 12.15   Following edges.

5. If there is no adjoining pixel, scan those one or two pixels away.

6. If an acceptable pixel has been found, repeat from 3.

7. If no acceptable pixel is found, repeat the process for the other direction.

The pixels found during a pass of this algorithm are regarded as forming a single edge. The whole process is repeated until all edge pixels have been considered.

A few of the steps in this algorithm need unpacking slightly. First, in step 2, a line has two ends, so one has to choose which to follow first. As both will eventually be traced, the choice is unimportant and some default, say towards the right, can be chosen. Remember, though, that the orientation of the edge is at 90° to the line of maximum slope. At step 3, one only bothers to look for pixels that are in the general direction of the edge. For example, if the orientation is northeast, one would look at the pixels to the top right, right and top. Similarly at step 5, one only looks slightly further in the relevant directions. Figure 12.15 shows a typical order in which pixels are scanned. You have to look at quite a wide swath of pixels, as even a straight line is quite jagged when digitised and also the edge may bend. Note that the figure includes the additional pixels searched at step 5. The threshold used to decide whether two edge pixels have a "close enough" orientation will depend somewhat on the sort of images, noise levels and so on. However, a typical rule might be to accept if the orientations lie within 60° of one another.

The output of this algorithm is a collection of edges, each of which consists of a set of pixels. The end points

of each edge segment will also have been detected at step 7. If the end point is isolated, then it is a termination; if several lie together, or if it lies on another edge, then the end point is at a junction. This resulting set of edges and junctions will be used by Waltz's algorithm in the next section to infer three-dimensional properties of the image.

However, before passing these data on to more knowledge-rich parts of the process, some additional cleaning up is possible. For example, very short edges may be discarded as they are likely either to be noise or to be unimportant in the final image (e.g. texture effects). Also, one can look for edges that terminate close to one another. If they are collinear and there are no intervening edges, then one may join them up to form a longer edge. Also, if two edges with different orientation terminate close together, or an edge terminates near the middle of another edge, then this can be regarded as a junction. One problem with too much guessing at lower levels is that it may confuse higher levels (the source of optical illusions in humans). One solution is to annotate edges and junctions with certainty figures. Higher levels of processing can then use Bayesian-style inferencing and accept or reject these guesses depending on higher-level semantic information. However, for the purposes of exposition, we will assume that the output of this level of analysis is perfect.

12.5 REGION DETECTION

In the previous section we likened edge detection to understanding a cartoon. In contrast, an oil painting will not have lines drawn at the edges but will consist of areas of different colours. An alternative to edge detection is to concentrate on the regions composing the image. We considered this briefly when we discussed thresholding but will now look at a more sophisticated algorithm.

12.5.1 Region Growing

A region can be regarded as a connected group of pixels whose intensity is almost the same. Region detection (or segmentation) aims to identify the main shapes in an image. This can be done by identifying clusters of similar intensities. The main process is as follows:

- group identical pixels into regions

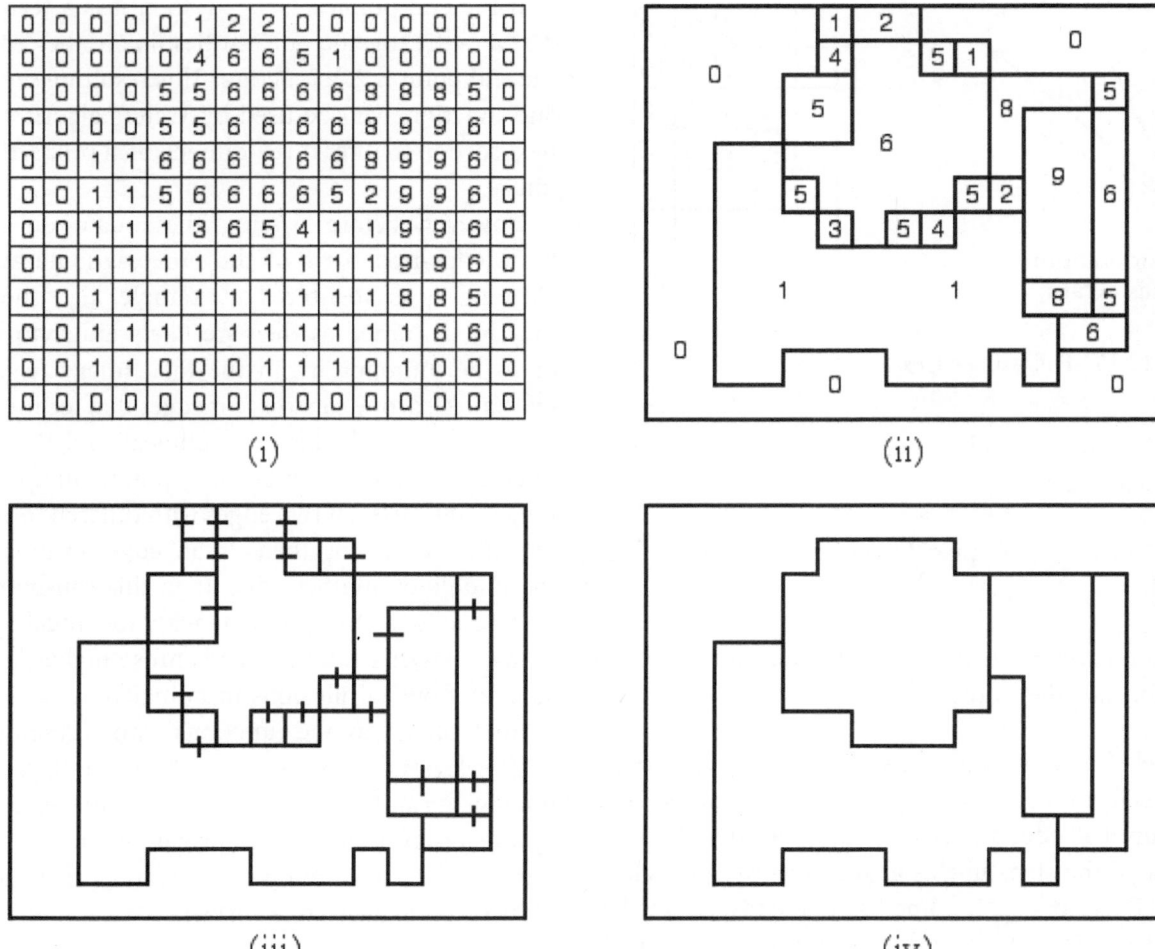

FIGURE 12.16   Region merging.

- examine the boundaries between these regions – if the difference is lower than a threshold, merge the regions.

The result is the main regions of the image.

This process is demonstrated in Figure 12.16. The first image (i) shows the original grey levels. Identical pixels are merged giving the initial regions in (ii). The boundaries between these are examined and in (iii) those where the intensity is less than 3 are marked for merging. The remainder, those where the difference in intensity is more than 2, are retained, giving the final regions in (i).

12.5.2 The Problem of Texture

Texture can cause problems with all types of image analysis, but region growing has some special problems.

If the image is unprocessed, then a textured surface will have pixels of many different intensities. This may lead to many small island regions within each large region. Alternatively, the texture may "feather" the edges of the regions so that different regions get merged. The obvious response is to smooth the image so that textures become greys. However, if the feathering is bad, then sufficient smoothing to remove the texture will also blur the edge sufficiently that the regions will be merged anyway. In a controlled environment, where lighting levels can be adjusted, one may be able to adjust the various parameters (level of smoothing, threshold for merging) so that recognition is possible, but where such control is not easily possible region merging may fail completely.

12.5.3 Representing Regions – Quadtrees

In the previous section we represented regions by simply drawing lines round them on the page. In a computer program it is not that straightforward! The simplest representation would be to keep a list of all the pixels in each region. However, this would take an enormous amount of storage. There are various alternatives to reduce this overhead. One popular representation is quadtrees. These make use of the fact that images often have large areas with the same value – precisely the case with regions. We will describe the algorithm in terms of storing a binary image and then show how it can be used for recording regions.

Start off with a square image where the width in pixels is some power of 2. Examine the image. Is it all black or white? If so, stop. If not, then divide the image into four quarters and look at each quarter. If any quarter is all black or white, then leave it alone, but if any quarter is mixed, then split it into quarters. This continues until either each region is of one colour, or else one gets to individual pixels – which must be one colour by definition. This process is illustrated in Figure 12.17. The first part (i) shows the original image, perhaps part of a black circle. This is then divided and subdivided into quarters in (ii). Finally, in (iii) we see how this can be stored in the computer as a tree data structure. See how the 64 pixels of the image are stored in five tree nodes. Of course the tree nodes are more complicated than simple bitmaps and so for this size of image a quadtree is a little over the top, but for larger images the saving can be enormous.

This can be used to record regions in two ways. Each region can be stored as a quadtree where a black means that the pixel is part of the region. Alternatively, one can use a multi-coloured version of a quadtree where each region is coded as a different colour. In either case, regions can easily be merged using the quadtree representation.

Variants of quadtrees are also used in geographic information systems (GIS) to store spatial information, and 3D equivalents, octrees, can be used for voxel data, for example to store the output of an ultrasound scan.

12.5.4 Computational Problems

Region growing is very computationally expensive, involving many passes over the digitised image. Operating on reduced representations such as quadtrees can reduce the number of operations, but at the expense of more complicated data structures. For this reason, [288] suggests that region growing is not generally applicable in industrial contexts. Instead, edge detection methods are preferred. The contrast is easy to see – a 100×100 pixel square has 10 000 interior pixels, but only 400 on the boundary! However, against this one should note that region growing is easily amenable to parallel processing and so the balance between different techniques may change.

12.6 RECONSTRUCTING OBJECTS

12.6.1 Inferring Three-Dimensional Features

Edge and region detection identify parts of an image. We need to establish the objects that the parts depict. We can use constraint satisfaction algorithms to determine what possible objects can be constructed from the lines given. First, we need to label the lines in the image to distinguish between concave edges, convex edges and obscuring edges. An obscuring edge occurs where a part of one object lies in front of another object or in front of a different part of the same object. The convention is to use a "+" to label a convex edge, a "−" for a concave edge and an arrow for an obscuring edge. The object that the edge is "attached" to lies to the right of the arrow; the obscured object lies to its left. Figure 12.18 shows an object with the lines in the image suitably labelled.

How do we decide which labels to use for each line? Lines meet each other at vertices. If we assume that certain degenerate cases do not occur, then we need only worry about trihedral vertices (in which exactly three lines meet at a vertex). There are four types of such vertices, called L, T, fork (or Y) and arrow (or W). There are 208 possible labellings using the four labels available, but happily only 18 of these are physically possible (see Figure 12.19). We can therefore use these to constrain our line labelling. Waltz proposed a method for line labelling using these constraints.

Waltz's Algorithm

Waltz's algorithm [294] basically starts at the outside edges of the objects and works inward using the constraints. The outside edges must always be obscuring edges (where it is the background that is obscured). Therefore, these can always be labelled with clockwise arrows. The algorithm has the following stages:

1. Label the lines at the boundary of the scene.

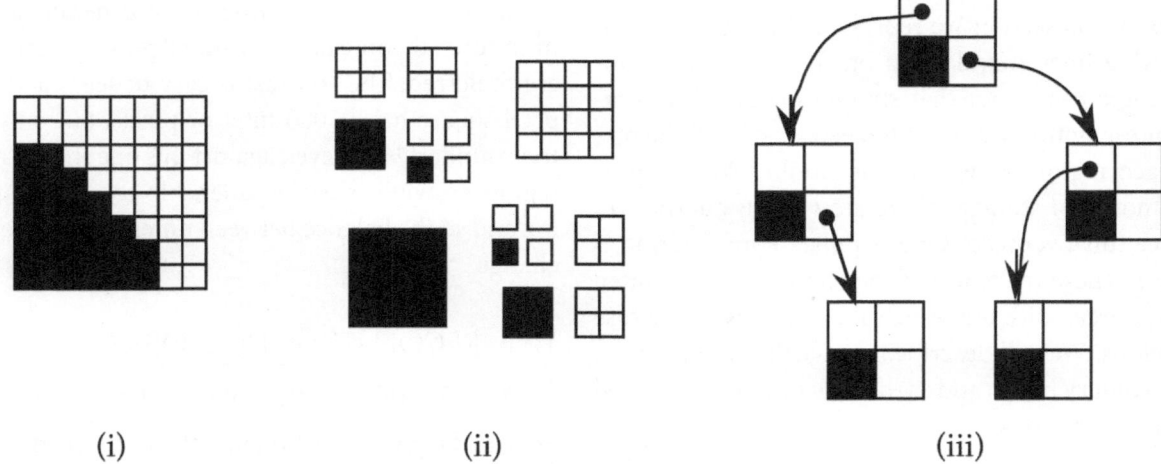

(i)   (ii)   (iii)

FIGURE 12.17   Quadtree representation of image.

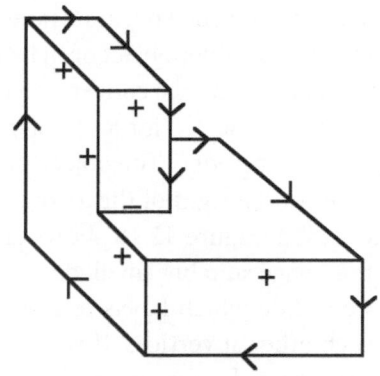

FIGURE 12.18   Scene with edges labelled.

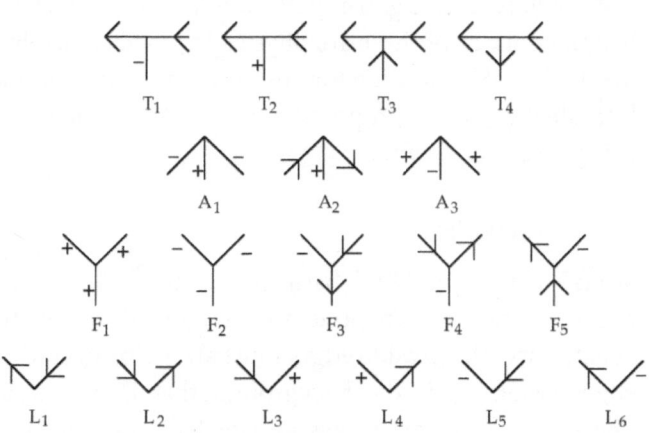

FIGURE 12.19   Possible trihedral vertices – T junctions (T_1–T_4), arrows (A_1–A_3), forks (F_1–F_5) and L junctions (L_1–L_6).

2. Find vertices where the currently labelled lines are sufficient to determine the type of the vertex.

3. Label the rest of the lines from those vertices accordingly.

Steps 2 and 3 are repeated either until there are no unlabelled lines (success) or until there are no remaining vertices which are completely determined (failure).

We will follow through the steps of this algorithm attempting to label the object in Figure 12.18. We start by naming the vertices and labelling the boundary lines. This gives the labelling in Figure 12.20(i).

We now perform the first pass of steps 2 and 3. Notice how a, c , f and h are arrow vertices with the two side arms labelled as boundaries (">"). Only type A_6 matches this, so the remaining line attached to each of these vertices must be convex ("+"). Similarly, the T vertex d must be of type T_4; hence the line d–k is a boundary. Vertices e and i are already fully labelled, so add no new information. The results of this pass are shown in (ii).

On the second pass of steps 2 and 3 we concentrate on vertices j, k and l. Unfortunately, vertex k is not determined yet; it might be of type L_1 or L_5 and we have to wait until we have more information. However, vertices j and l are more helpful: they are forks with one concave line. We see that if one line to a fork is concave, it must be of type F_1 and so all the lines from it are concave. These are marked in (iii).

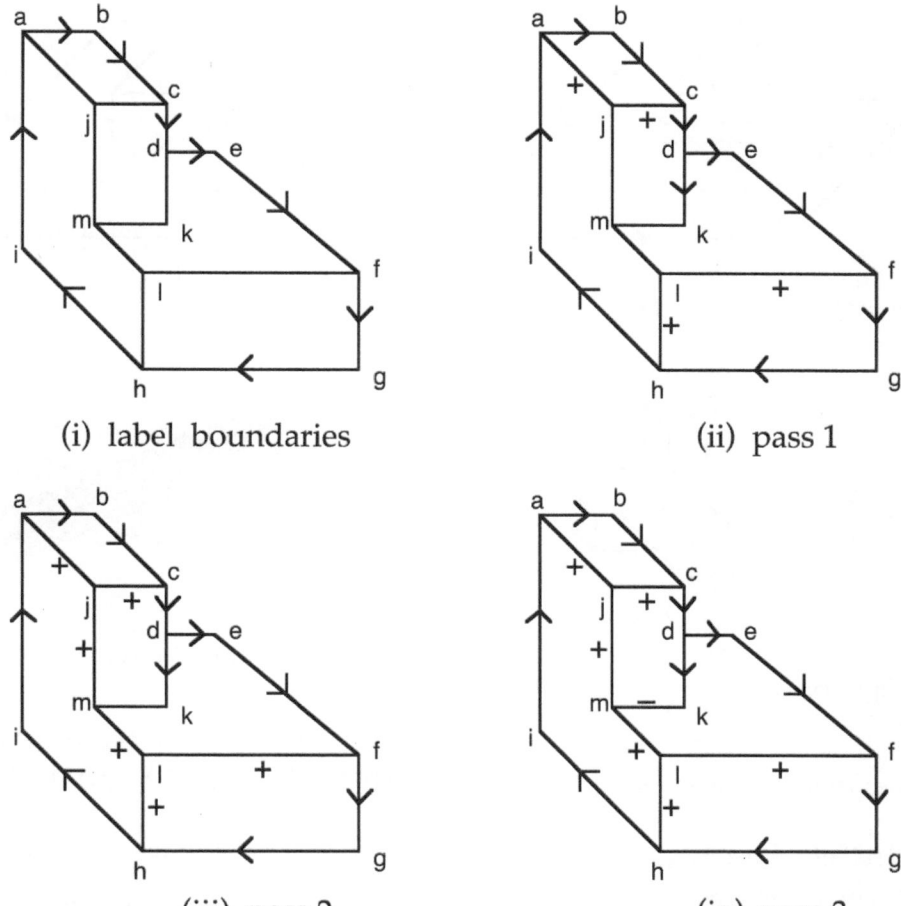

(i) label boundaries

(ii) pass 1

(iii) pass 2

(iv) pass 3

FIGURE 12.20   Applying Waltz's algorithm.

As we start the third pass, we see that k is still not determined, but m is an arrow with two concave arms. It is therefore of type A_3 and the remaining edge is concave. This also finally determines that k is of type L_5. The fully labelled object (iv) now agrees with the original labelling in Figure 12.18.

12.6.1.1 Problems with Labelling

Waltz's algorithm will always find the unique correct line labelling if one exists. However, there are scenes for which there are multiple labellings, or for which no labelling can be found. Figure 12.21 shows a scene with an ambiguous line labelling. The first labelling corresponds to the upper block being attached to the lower one. In the second labelling the upper block is "floating" above the lower one. If there were a third block between the other two, we would be able to distinguish the two, but with no further information we cannot do

so. With this scene, Waltz's algorithm would come to an impasse at stage 2, when it would have unlabelled vertices remaining, but none that are determined from the labelled edges. At this stage, you could make a guess about edge labelling, but whereas the straightforward algorithm never needs to backtrack, you might need to change your guesses as you search for a consistent labelling.

Figure 12.22(i) shows the other problem, a scene that cannot be labelled consistently. In this case Waltz's algorithm would get stuck at step 3. Two different vertices would each try to label the same edge differently. The problem edge is the central diagonal. Reasoning from the lower arm, the algorithm thinks it is convex, but reasoning from the other two arms it thinks it is concave. To be fair, the algorithm is having exactly the same problem as you have with this image. It is locally sensible,

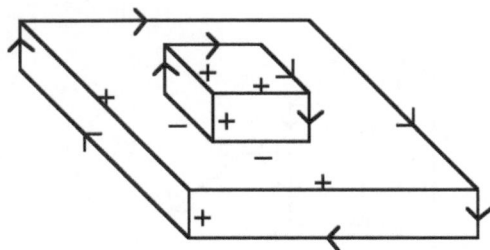

FIGURE 12.21   Scene with ambiguous labelling.

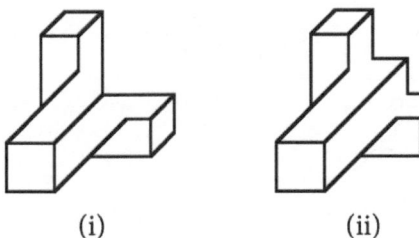

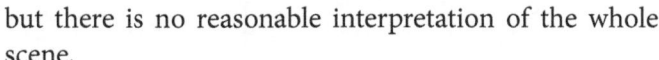

(i)   (ii)

FIGURE 12.22   Improper scene.

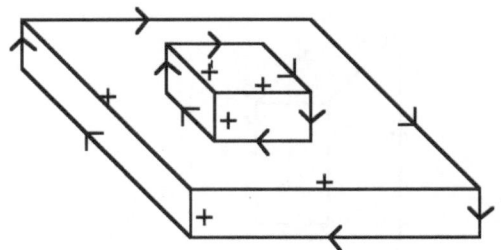

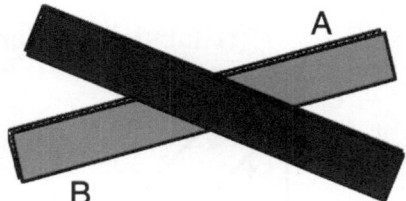

FIGURE 12.23   Two objects or three?

but there is no reasonable interpretation of the whole scene.

Given only the set of vertex labellings from Figure 12.19, there are also sensible scenes that cannot be labelled. A pyramid that has four faces meeting at the top cannot be labelled using trihedral vertices. Even worse, a piece of folded cloth may have a cusp, where a fold line disappears completely. These problems can be solved by extending the set of vertex types, but as one takes into account more complex vertices and edges, the number of cases does increase dramatically.

Note also that the algorithm starts with the premise that lines and vertices have been identified correctly. Given what you know about edge detection, you will see that this is not necessarily a very robust assumption. If the edge detection is not perfect, then one might need to use uncertain reasoning while building up objects. Consider Figure 12.22 (ii) – a valid scene that can be labelled consistently. However, if the image is slightly noisy at the top right vertex, it might be uncertain whether it is a T, an arrow or a Y vertex. If it chose the last of these, it would have the same problems as with the first, inconsistent Figure If the edge detection algorithm instead gave probabilities, one could use these with *Bayesian reasoning* to get the most likely line labelling.

However, the search process would be somewhat more complicated than Waltz's algorithm!

12.6.2 Using Properties of Regions

Edge detection simply uses lines of rapid change but discards the properties of the regions between the lines. However, there is a lot of information in these regions that can be used to understand the image or to identify objects in the image. For example, in Figure 12.23, it is likely that the regions labelled A and B are part of the same object partly obscured by the darker object. We might have guessed this from the alignment of the two regions, but the fact that they are the same colour reinforces this conclusion.

Also, the position and nature of highlights and shadows can help to determine the position and orientation of objects. If we have concluded that an edge joins two parts of the same object, then we can use the relative brightness of the two faces to determine which is facing the light. Of course, this depends on the assumption that the faces are all of similar colour and shade. Such heuristics are often right but can sometimes lead us to misinterpret an image – which is precisely why we can see a two-dimensional picture as if it had depth.

Once we know the position of the light source (or sources), we can work out which regions represent possible shadows of other objects and hence connect

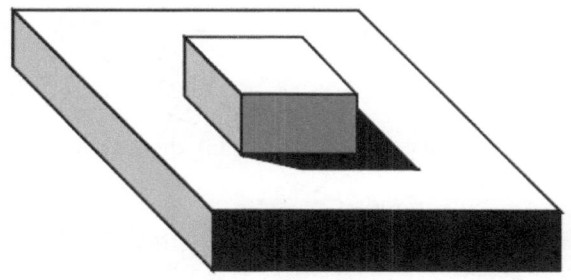

FIGURE 12.24   Shadows and highlights.

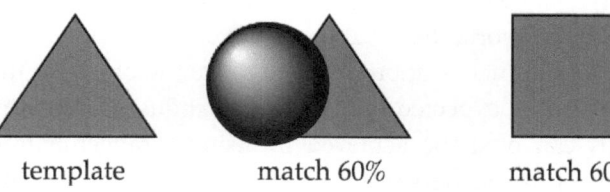

template   match 60%   match 60%

FIGURE 12.25   Simple template matching.

them to the face to which they belong. For example, in Figure 12.24, we can see from the different shadings that the light is coming from above, behind and slightly to the left. It is then obvious that the black region is the shadow of the upper box and so is part of the top face of the lower box.

Shadows and lighting can also help us to disambiguate images. If one object casts a shadow on another, then it must lie between that object and the light. Also, the shape of a shadow may be able to tell us about the distance of objects from one another and whether they are touching. Recall in Figure 12.21 how the edges had no unambiguous line labelling. However, looking at the shadow in Figure 12.24 it is clear that the upper box is in contact with the lower one.

Lighting effects can also help us to interpret curved objects. For example, in Figure 12.4 at the beginning of this chapter, the sphere gets darker and darker until it becomes indistinguishable from the black rectangle in the background. However, we have no trouble identifying it as a sphere as we infer a boundary based on the rate of change of colour. A similar rule can be built into an image analysis program.

12.7   IDENTIFYING OBJECTS

Finally, having extracted various features from an image, we need to establish what the various objects are. The output of this will be some sort of symbolic representation at the semantic level. We will discuss three ways of doing this that operate on different sorts of lower-level representation.

12.7.1 Using Bitmaps

The simplest form of object identification is just to take the bitmap, suitably thresholded, and match it against various templates of known objects. One can simply count the number of pixels that disagree and use this as a measure of fit. The best match is chosen, and so long as its match exceeds a certain threshold it is accepted.

This form of matching can work well where one can be sure that shapes are not occluded and where lighting levels can be chosen to ensure clean thresholded images. However, in many situations the match will be partial, either because of noise, or because the object is partly obscured by another object. Simply reducing the threshold for acceptability will not work. Consider the two images in Figure 12.25. They have a similar amount of pixels in common, but the first is clearly a triangle like the template whereas the latter is not.

If a neural network is trained using noisy as well as perfect images, it may be able to deal with noisy pattern matching. After training, when the network is presented with an image, it identifies the object it thinks it matches, sometimes with an indication of certainty. Neural networks can often give accurate results even when there is a large amount of noise, but without some of the unacceptable spurious matches from crude template matching. One reason for this is that many nets effectively match significant features (such as the corners and edges of the triangle). This is not because they have any particular knowledge built in but simply because of the low-level way that they learn. We'll look at the use of neural networks for image processing in more detail in Section 12.9.

One problem with both template matching and neural networks is that they are looking for the object at a particular place in the image. They have problems when the object is at a different location or orientation than the examples with which they are taught. One solution is to use lots of examples at different orientations. For template matching this increases the cost dramatically (one test for each orientation). For neural nets, the way in which the patterns are stored reduces this cost to some extent, but if too many patterns are taught without increasing the size of the network, the accuracy will eventually decay.

An alternative approach is to move the object so that it is in the expected location. In an industrial situation this can often be achieved by using arrangements of chutes and barriers that force the object into a particular position and orientation. Where this is not possible, an equivalent process can be carried out on the image. If one is able to identify which region of the image represents an object, then this can be moved so that it lies at the bottom left-hand corner of the image, and then matched in this standard position. This process is called normalisation. A few stray pixels at the bottom or left of the object can upset this process, but alternative normalisation methods are less susceptible to noise, for example moving the centre of gravity of the object to the centre of the image.

Similar methods can be used to standardise the orientation and size of the object (the size may be different if it is closer or farther away than the examples). The general idea is to find a co-ordinate system relative to the object and then use this to transform the object into the standard co-ordinate system used for the matching. A typical algorithm works like this:

1. Select a standard point on the object (say its centre of gravity).

2. Choose the direction in which the object is "widest"; make this the x-axis.

3. Take the axis orthogonal to the x-axis as the y-axis.

4. Scale the two axes so that the object "fits" within the unit square.

The definitions of "widest" and also "fits" from steps 2 and 4 can use the simple extent of the object, but are more often based on measures which are less noise sensitive. The process is illustrated in Figure 12.26. The resulting x and y axes are called an object-centred co-ordinate system. Obviously all the example images must be transformed in a similar fashion so that they match!

12.7.2 Using Summary Statistics

An even simpler approach than template matching is to use simple statistics about the objects in the image, such as the length and width of the object (possibly in the

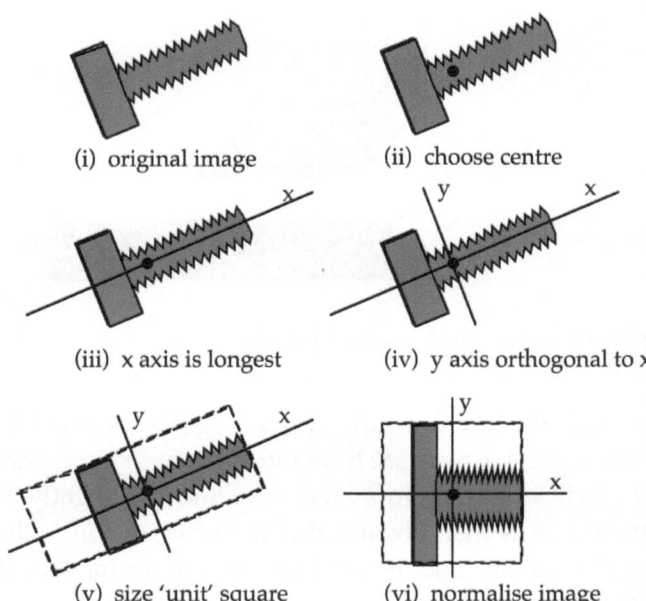

(i) original image

(ii) choose centre

(iii) x axis is longest

(iv) y axis orthogonal to x

(v) size 'unit' square

(vi) normalise image

FIGURE 12.26 Choosing an object-centred co-ordinate system.

object-centred system), the number of pixels with various values and so on. For example, if one were trying to separate nuts and bolts on a production line, then those objects with an aspect ratio (ratio of length to width) greater than some critical value would be classified as nuts. Another example would be a line producing washers where you are trying to reject those that have not had their centres properly removed. Those objects with too many pixels would be rejected as defects.

12.7.3 Using Outlines

We saw when discussing template matching that issues of location, orientation and size independence cause some problems. These get far worse when we have to consider three-dimensional rotations. At this point techniques using higher-level features, such as those generated by Waltz's algorithm, become very attractive. In essence one is still template matching, but now the templates are descriptions of the connectivity of various edges. Of course, the same object will have different edges visible depending on its orientation. However, one can generate a small set of representative orientations whereby any object matches one or other after a certain amount of deformation.

Figure 12.27 shows some of the representative orientations of a simple geometric object. The example set of all possible orientations can be generated by hand, or (ide-

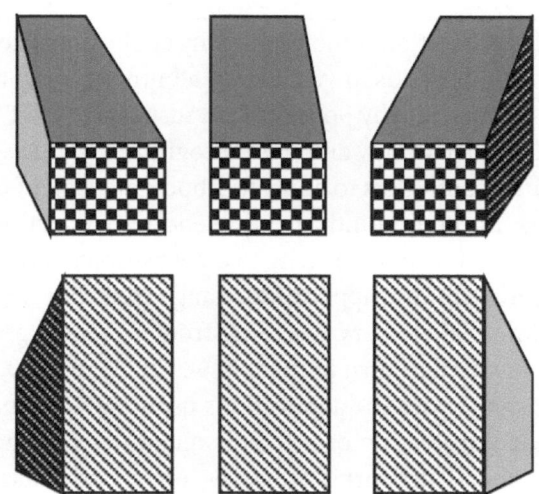

FIGURE 12.27 Different orientations of an object.

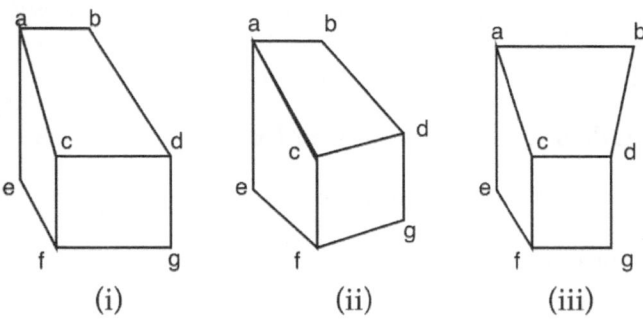

FIGURE 12.28 Matching an object.

ally) automatically using a three-dimensional geometric model of the object. The number of orientations can be reduced dramatically if one can make any assumptions, say that the object's base stays on the ground, or that the camera position is within certain bounds.

When an object is to be recognised, it is matched against the representatives of all known objects. Each vertex and edge in the image object is matched with a corresponding one in the example. If such a correspondence can be found, then the match succeeds. The exact positions of the vertices and edges don't matter, but the relative geometric constraints must match. For example, in Figure 12.28, image (ii) matches the template (i). However, (iii) doesn't because vertex d is a fork-type junction rather than an arrow.

The matching process can be more or less precise. As well as the types of junctions, it may use information such as whether certain lines are parallel or vertical. However, adding constraints tends to mean that there are more

cases to consider when producing the set of all representative orientations.

Note that this type of method can be used to match more complicated objects. If an object consists of various pieces, then the pieces can be individually identified by the above method and then a description of the connectivity of the pieces within the object can be matched against the known objects. This allows one to detect objects which change their shape, such as people.

12.7.4 Using Paths

Finally in this section we look at the special case of handwriting and gesture recognition. Reading human handwriting has been a long-term aim of AI (as well as many of the authors' friends) and is now commonplace, albeit not perfect. Pen-based systems recognise both characters (to enter data) and gestures (such as a scribble to mean "delete"). These systems either demand that the writer uses very stylised letters or that new writers spend some time training the system. Even when the system is trained, the writer must write each character individually. Reliable and flexible writer-independent recognition of connected writing is not yet with us.

One way to approach handwritten text is to take the bitmap generated by the path of the pen and then process it. Some applications demand this approach, for example if you want to interpret proof corrections written onto paper copies and then scan these in, or if you want to transcribe historic hand-written correspondence. However, for an interactive system this throws away too much useful information. If we trace the path of the pen, we not only have the lines already separated from the background (why bother to detect them again!), but we also know the direction of the strokes and the order in which they were written.

These path data differ from the grey-scale or bitmap images we have considered so far. Instead of a set of intensities at positions, we have a set of positions of the pen at various times. We can match the strokes in the image against those learnt for the particular writer. However, handwritten letters and gestures are never exactly the same and so we must accept some variation. There are various approaches to this.

One way is to look for characteristic shapes of strokes: lines, curves, circles and so on. Characters and gestures are then described using this "vocabulary". A letter "a" might be described as either "a circle with a line con-

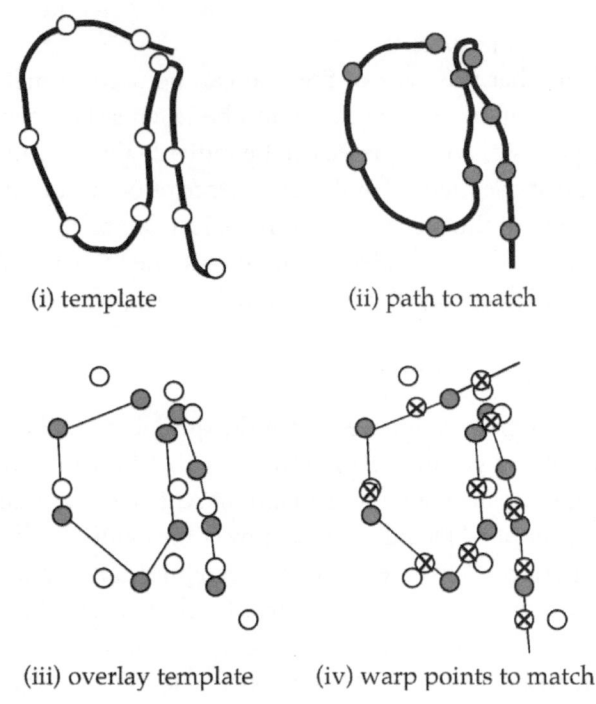

(i) template   (ii) path to match

(iii) overlay template   (iv) warp points to match

○ original sample   ● sample to match   ⊗ warped points

FIGURE 12.29 Handwriting recognition – warping the sample.

nected to the right" or "a semi-closed curve with a line closing it to the right".

An alternative is to try and match the strokes against examples stored during training. However, not only may the written characters vary but also the points at which the pen is sampled may differ between the training example and the one to be recognised. One therefore has to "warp" the points on the path and find intermediate points that match most closely the example. The idea is to choose points that were not in the sample, but might have been!

The process is illustrated in Figure 12.29. The sample points on the original template and the character that needs to be matched are shown in (i) and (ii). The template sample points are overlaid as closely as possible (iii), and then intermediate points (the warp points) are chosen on the lines connecting the sample points. These are chosen so as to be as close to the template points as possible. At the ends of the stroke the warp points must be chosen on the extrapolation of the last lines. Now the warped points are used rather than the originals in deciding whether or not the character really matches the template.

12.8 FACIAL AND BODY RECOGNITION

Facial recognition is found in many applications: to label family and friends in your photo album, as a way to unlock phones, and by police to find suspects in CCTV images. Similarly body and pose recognition is used widely in game controllers to allow full-body movement digital sports, in the film industry to blend human action and CGI and by airport security to detect suspicious activity. As is evident, the applications range from the frivolous to some that are very worrying from a privacy point of view – we will return to these issues in Chapter 23.

In some cases the applications use standard pipelines of video processing or use large quantities of images to train neural networks. However, often both facial and body pose recognition use some form of model of the human face or body in order to create more knowledge-informed systems.

Facial recognition can be used for various purposes:

identification – Whose face is this?

authentication – Is this your face?

attention – Where is this person looking?

emotion – What are they feeling?

The first two are about matching a face to a person, the last two about understanding the facial expression irrespective of who it is, something humans can do with strangers as well as friends.

For matching, the number of faces you want to match to makes the job easier or harder. The Chinese government's Skynet Project matches people from tens of thousands of CCTV cameras with a database of hundreds of thousands of police suspects – a small false positive rate of even a few per cent would lead to thousands of false identifications every day. In contrast you may have hundreds of friends to be matched against the photographs in your album, and just you for your mobile to authenticate and unlock the phone (or more often not!). Clearly the last is easiest.

Often face-based systems work by identifying key features: eyes, mouth, nose, cheeks, and then either morphing the image to match other images or using metrics such as nose-to-mouth or eye-to-eye distance. Identification and authentication applications simply use these to match known faces. Facial expression recognition will look in more detail at fine features such as micro-muscle movements. In fact exactly the things that facial recognition wants to ignore as it is still the same face whether happy or sad, looking left or right.

Oddly, computer emotion recognition can be better in some ways than humans as there are certain muscles that are involuntary, so a computer system may be able to distinguish a false smile (deliberately formed) from a real one. However, a level of complicit deceit is part of the normal patterns of human intercourse; it is often a bad idea if the person asking 'how do I look today' can tell if the answer is honest. So some emotions are best left undetected, even when detection is possible.

For film production actors may wear special suits with either reflective dots or miniature radio-location devices to record movements which can later be used to render false bodies. Similarly sports players may wear such devices in order to gather data for research or for coaches to help them improve their performance. This special equipment generates precise 3D coordinates for each limb position.

However, often an equivalent job is needed without the person wearing special clothes or devices, for example, when playing virtual sports games. For this vision systems use combinations of plain and depth cameras. Sometimes, structured light may also be used; that is patterns of infra-red light projected on the scene to allow easy 3D depth detection.

These systems detect key features such as the head or elbows. This is combined with models of the human body both to help eliminate impossible interpretations (the neck is not connected to the ankle), and, once disambiguated, to make an accurate 3D virtual skeleton including arm and leg positions and head angle. Often such systems use a hybrid architecture (Chap. 6) combining multiple types of AI approach; for example a system might use statistical signal processing at a low level, neural networks for key feature identification and model-based approaches with Bayesian reasoning to create the final skeleton.

12.9 NEURAL NETWORKS FOR IMAGES

Neural networks have been used for various forms of image categorisation or reconstruction since the earliest days and many of the neural net examples in this book involve images. The classical image processing techniques presented in this chapter are quite complex, but when we look at a scene or a face we just 'get it', we have some immediate impression or recognition. We don't feel as though we are going through many stages because the processing is unconscious and virtually instantaneous.

The knowledge-rich techniques described already are of course all performed unconsciously in our brains, and many are based on psychological experiments that seek to unpack these conscious processes. However, much of what is going on in our own brains is still not understood. Neural networks have therefore always been an obvious approach, especially for those hard to codify aspects – you just know a photo is a 1950s city scene rather than contemporary, but it may be hard to pin that down to lines, shapes or textures.

For many applications generic networks are simply applied to an image, setting the pixels as input and some classification as output. This may be preceded by some form of image-specific pre-processing such as thresholding, or a wavelet transform (Chap. 14), but otherwise is a straightforward application of a generic algorithm.

However, there are some algorithms that have been specifically designed for images or have been particularly closely connected. We will discuss two, convolutional neural networks and autoencoders.

12.9.1 Convolutional Neural Networks

Many of the techniques we've seen in this chapter involve doing the same thing across every pixel position in an image. Some of them look at a single pixel at a time, notably thresholding, but others look at a region around the pixel. These include simple smoothing and Gaussian filters, and also gradient filters and Sobel's operator for edge detection. In mathematical terms these are called convolutions and are also applied to linear data such as time series.

The strength of convolutions is that they operate uniformly, no particular area of an image is any different to any other. In contrast a 'vanilla' application of neural networks to an image regards every pixel as completely different. This may mean that a car is only detected if it lies in exactly the same spot as in an example image. There are some aspects of images where this may be appropriate, for example landscape pictures have sky at the top, passport photos have a face in the middle. However, typically we want more position independent recognition.

Convolutional neural networks achieve this by adopting the same techniques as convolutional filters. The image is divided up into potentially overlapping patches

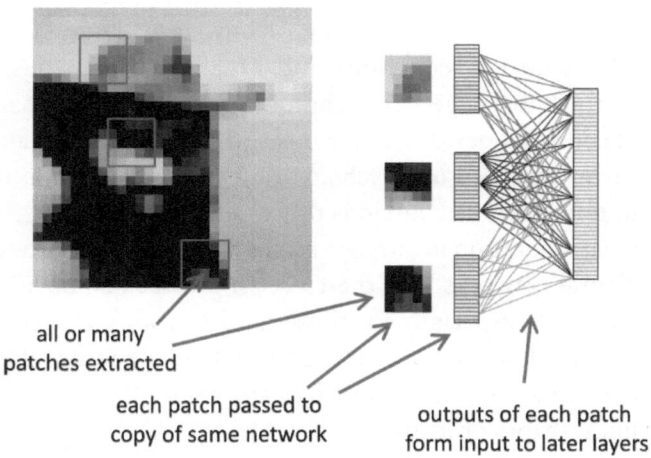

all or many
patches extracted

each patch passed to
copy of same network

outputs of each patch
form input to later layers

FIGURE 12.30   Convolutional neural network.

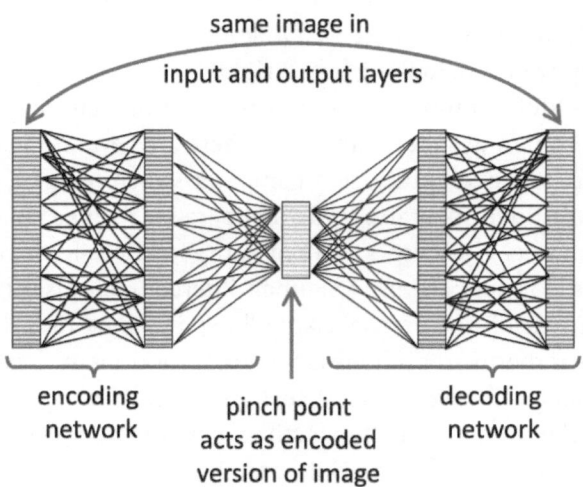

same image in
input and output layers

encoding
network

pinch point
acts as encoded
version of image

decoding
network

FIGURE 12.31   Autoencoder.

(called the receptive field) and the same (relatively small) neural network is applied to every patch. This is then used as the input layer of a deep neural network (see Figure 12.30).

The training of the convolutional part can still use the error values fed back from a standard backpropagation algorithm, but as the same weights are used in multiple places, the change in each weight is an average of all the changes generated at each patch independently.

In some implementations this is achieved by replicating the initial network, once for each patch, to create the first layer(s) of processing, but then 'clamping' the corresponding weights together. Here backpropagation would be applied as normal, but at the end of each cycle of learning each weight in the convolution layer would be set to the average of all its corresponding weights.

While this technique is especially used for images, it can also be applied to other forms of 2D data or indeed 1D data such as time series data or more complex 3D or forms of graph data that have some sort of regular structure.

12.9.2 Autoencoders

Recall from Chapter 6, an autoencoder is a form of autoassociative memory where the input and output are the same. Some of the early connectionist systems, in particular Hopfield networks and Boltzmann machines, functioned as autoassociative memory and have been used for image construction, but the term 'autoencoder' is most often used where there is a clear internal encoding that is much smaller than the original images, as can be the

case with a restricted Boltzmann machine when there are fewer hidden units than visible units.

As with all AI there are many variations, but typically an autoencoder is a deep neural network where:

1. The input and output layers are intended to be the same image (or other form of data),

2. There is a narrow hidden layer that acts as the encoding,

3. Where there are additional layers, the architecture is often symmetric.

Figure 12.31 shows these features. In the simplest case the network is trained by simply presenting the same example image in both input and output during standard training. However, one of the uses is to 'clean up' noisy data or to fill in where there is a gap in an image. To achieve this as well as training the net on exact matching pairs, the input may have noise added so that the network learns how to associate a noisy image with its clean equivalent and thus with enough training examples learns in general how to remove noise.

In its basic form this is unsupervised learning and so is especially useful when large numbers of untagged images are available, as is the case with many images on the web. However, after unsupervised training, the encoding layer can also be used as a reduced dimension version of the image for other kinds of learning.

For example, if a small proportion of the images have some form of human tagging, the encoded version of each together with the tag can be used as input–output pairs to train a classifier. As the encoded layer is a lot smaller than the original image, this is possible to do robustly with far fewer tagged examples than would be needed with the raw image data. Once trained in this way, the classifier can be applied to the encoded form of unseen images.

Other uses include using random values for the encoding or perturbing it and then letting the decoding part of the network generate realistic images. The ability of this and more sophisticated forms of autoencoder to generate or clean up images can also be potentially problematic in the generation of deep fakes. For example, if the face of a politician is placed over the body of someone doing something illegal and then this image is presented as input to an autoencoder, the join between face and body would be 'tidied up' so that it might look as though it were a real photograph of the politician.

12.10 GENERATIVE ADVERSARIAL NETWORKS

Autoencoders can generate very realistic images but also ones that are manifestly unreal, for example placing facial features in unnatural positions. One approach to deal with this is to use generative adversarial networks (GANs). This takes inspiration from game playing, in particular the idea of a zero-sum game where one player's gains are the other player's loss. For image generation one AI acts as generator and is trained to create artificial images that mimic as closely as possible the properties of real ones; another AI acts as critic and is trained to distinguish real images from those that are artificially generated. The AI generator 'wins' if it fools the AI critic and vice versa. The process of generation and testing leads to vast numbers of training examples, and so machine learning can be used to improve both the generator and the critic, a form of artificial arms race.

This is a general machine learning technique that can be used for different kinds of data and problems but is most widely known for its applications in image generation.

12.10.1 Generated Data

In Chapter 8 we saw that deep neural networks may need generated or virtual data in order to have sufficient examples from which to learn. This is particularly true for

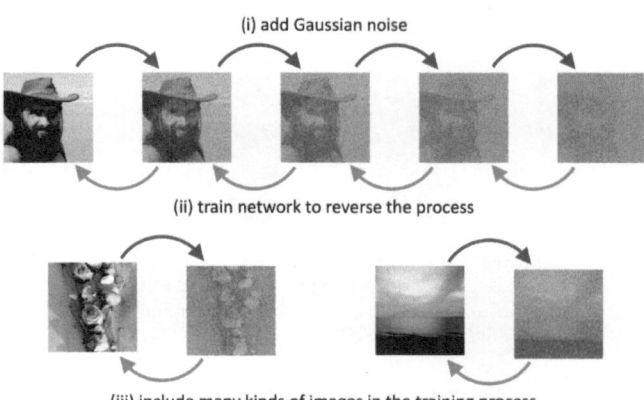

(i) add Gaussian noise

(ii) train network to reverse the process

(iii) include many kinds of images in the training process

FIGURE 12.32   Training a diffusion model.

machine learning on image or video data. The problem is typically not about getting sufficient data – for example Google have thousands of hours of street-mapping video, but about getting sufficient *labelled* data. Crowd-sourcing techniques can help, but often there are limits, particularly when dealing with rare or dangerous situations, such as potential accidents for autonomous cars.

We have already mentioned adding noise to data as a way to grow the labelled dataset – every time we have an image labelled 'Eiffel Tower' we can generate hundreds more: shifted, resized, rotated or with noise added.

In addition engines designed for virtual reality or immersive games create high fidelity images of their virtual worlds. These are based on models of the world, so that effectively they have ground truth. We know precisely where each person, building and car is positioned in the environment, and so can train vision systems to recreate the model world from the generated image.

12.10.2 Diffusion Models

Diffusion models use this technique as a primary mechanism [61]. Instead of simply using noise to expand the datasets, they are trained to gradually remove noise from images. Crucially they do not try to remove the noise in a single step, but rather reduce the noise step by step by modifying each pixel of the noisy image in a way that moves it in the direction most likely to be a real image (Figure 12.32). This is a form of gradient descent as discussed in Chapter 9 (Section 9.4.1).

The system can then be presented with an image consisting of nothing but noise and is asked to remove the noise first assuming very high amounts of noise, and then

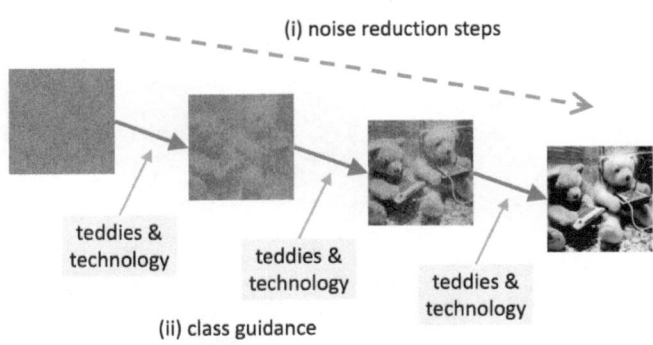

FIGURE 12.33 Using a diffusion model to generate images from prompts.

assuming smaller and smaller amounts of noise. In the early steps large-scale structures start to emerge, say a face or tree, and then in later steps these get refined (details of eyes or leaves).

This kind of image can be trained on a particular class of image, say landscapes, and then run to produce completely new images that appear to come from the original class. However, their full power is exploited when the diffusion model is trained on all possible images, class-free, so that it is a general purpose de-noising process. Left to its own devices, this would produce plausible images but they could be of anything. Instead, the step-wise process of image re-construction can be guided using a classifier nudging towards images that have that particular classification (Figure 12.33). Furthermore, the classifier doing the guiding can use descriptions that combine several simple classes that are maybe never found together in a training image, such as 'teddies' and 'technology'.

It is this process that caught public attention in systems such as OpenAI's Dall-E, which are able to create realistic images based on text descriptions, even in the style of a particular artist [229, 230].

12.10.3 Bottom-up and Top-down Processing

Note that while this chapter is principally about processing and understanding visual images as *input*, diffusion models create images as *output*. In traditional computing terms this is perhaps more the domain of graphics, but in AI the two start to come together. This is not altogether surprising as this is exactly what happens in the human visual system when we try to imagine something or when we dream. 'Seeing' in a dream or in our mind's eye is not the same as seeing with our real eyes, indeed many people cannot conjure up a clear mental image at will, but

brain scans do show that our visual cortex is activated in a very similar way when imagining a scene as when we actually see it. These top-down processes are important also when we are awake and alert as they help our perception to rapidly make sense of otherwise ambiguous or partial senses.

Most of the techniques we have described are bottom-up, moving from sensation (raw sensory input) to perception (meaning attached to images). However, one of the general lessons of both human perception and AI, in many areas including text and speech as well as vision, is that bottom-up processes are usually mixed with top-down processes ... we already know what we expect to see or read and this guides our lower-level processes.

12.11 MULTIPLE IMAGES

So far, we have looked at single images. This may be all we have to work on, for example a single photograph of a scene. However, in some circumstances we have several images, which together can be used to interpret a scene. On the one hand, this can make life more difficult (lots of images to process!). On the other hand, we may be able to extract information from the combined images that is not in any single image alone. These multiple images may arise from various sources:

1. Different sensors may be viewing the same scene.

2. Two cameras may be used simultaneously to give stereo vision.

3. We may have continuous video of a changing scene.

4. A fixed camera may be panning (and possibly zooming) over a scene.

5. The camera may be on a moving vehicle or mounting.

The first of these, the combination of different sorts of data (e.g. infrared and normal cameras), is called data fusion. It is especially important for remote sensing applications, such as reconstructing images from satellite data. Different sensors may show up different features; hence edges and regions in the two images may not correspond in a one-to-one fashion. If the registration between the

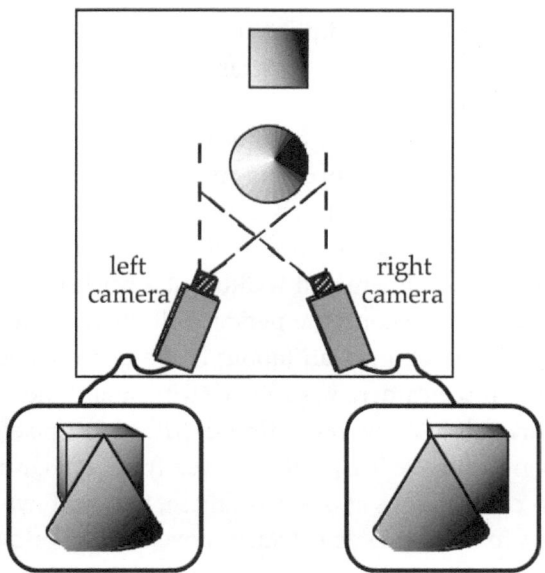

FIGURE 12.34 Stereo vision.

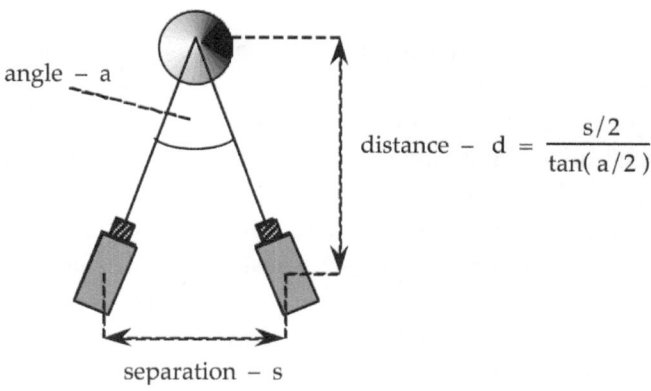

$$\text{distance} - d = \frac{s/2}{\tan(a/2)}$$

FIGURE 12.35 Triangulation.

sensors is known (i.e. one knows how they overlap), then the images can simply be overlaid and the information from each combined. Often this registration process is the most difficult part and the high-level data may be used to aid this process. For example, terrain-following cruise missiles rely on the matching of ground features with digital maps to calculate their course and position.

The last three sources of multiple images have somewhat different characteristics but are similar enough to discuss together. We will therefore look at stereo vision (Figure 12.34) and moving images in more detail.

12.11.1 Stereo Vision

Look out into the room and hold a finger in front of your face. Now close each eye in turn. Your finger appears to move back and forth across the room. Because your eyes are at different positions they see slightly different views of the world. This is especially important in determining depth. If you have not tried it before, here's another simple experiment. Hold a pencil in one hand and try to touch the tip of the finger of your other hand with the point of the pencil. No problem? Try it with one eye closed. The properties of stereo vision are one of the clues our eyes use to determine how far away things are.

One way to determine depth is to use triangulation in a similar way to a surveyor. Assuming you have been able to identify the same feature in both images, you can work out the angle between the two and hence the distance

from the camera (see Figure 12.35). To use this method to give exact distances, you need very accurate calibration of the cameras. However, even without such accuracy, one can use this method to obtain relative distances (which is probably what your eye is doing with the pen and finger).

In fact, it is not necessary to do any explicit calculations in order to obtain qualitative estimates of relative distance. Notice in Figure 12.34 how the cone moves back and forth relative to the cube. This effect is called parallax. If the amount of movement between the two views is great, then we know there is a considerable distance between the two objects.

So far we have assumed that we know which objects are the same in each image. However, this matching of objects between images is a difficult problem in itself. One can attack it at various levels. On the one hand, we can simply look for patterns of pixels that match one another in the grey-scale image. To do this, we work out the correlation (a measure of similarity) between groups of pixels in the two images at small offsets from one another. Where the correlation is large, we assume that there is some feature in common. The size of the offset then tells us the disparity in angle between the two images. Note that this will usually highlight the boundaries of objects, as the faces often have near constant intensity. Alternatively, one can wait until objects have been identified in each image and then match the objects. The low-level approach has the advantage that the information from both images can be used for subsequent analysis. For example, parallax can allow us to label lines ready for Waltz's algorithm and, indeed, is a very good edge indicator in its own right.

12.11.2 Moving Pictures

Recall that we listed three types of movement: objects may move in the scene, the camera may pan or zoom, and the camera may be mounted on a moving vehicle. These all lead to similar but slightly different effects. For example, an object moving towards the camera will have a similar effect to zooming the camera, or moving the camera closer to the object. Of course, several or all of the above effects may occur and we may even have stereo cameras and multiple sensors! To simplify the discussion, we will consider principally the case of a single stationary camera.

One special advantage of a stationary camera is that it may be possible to calibrate the camera when the scene is "empty". For example, if the camera is used for surveillance in an airport departure lounge, we can take an image when the lounge is empty. This will contain the fixed furniture, pillars and so on. Then, when we look at an image of the lounge in use, we will be able to match it with the fixed image and so identify the additional objects. In fact, it is not quite so easy! Changes in lighting levels, or indeed automatic light level controls in the camera, mean that one has to perform some adjustment to remove the fixed background.

Whether or not one has removed part of the image, some parts of the image change more rapidly than others. It is these regions of change that correspond to the moving objects. As with stereo vision, one can use local correlation to determine where groups of pixels in the image correspond to the same feature. This optical flow can be performed at a high level, matching whole objects, or at a low level, similar to edge detection, yielding a pixel-level flow pattern. With stereo vision we need only look for change in one direction, parallel to the separation of the "eyes"; however, in contrast, the objects in video may move in any direction. Furthermore, when an object gets closer or further away, the edges of the object move in different directions as the image of the object expands or contracts. Note also that we can usually only calculate the direction of movement orthogonal to the edge. Any movement parallel to the edge is (at least locally) invisible. Again, in the stereo case, this is only a problem when a long flat object is being viewed.

This all sounds quite complicated. Happily some things are easier! Because we have many images in sequence we can trace known objects. That is, once we have identified an object moving in a particular direction, we have a pretty good idea where to find it in the next image. Furthermore, the optical flow field can be used as an additional level of input for other bottom-up algorithms such as edge detection, or as input to neural networks.

It is worth noting for both moving images and stereo vision the magnitude of change that is likely between images. Imagine we are tracking someone walking across the airport lounge. Assume that the person is 10 metres from the camera and walking at a brisk 1.5 m/s. At 15 frames per second the person will move through an angle of 0.01 of a radian (about half a degree) between frames. If the camera has a 60° viewing angle and we are capturing it at a low resolution of 512×512 pixels, the person will move five pixels between frames. So, we have to do comparisons at one, two, three, four and five pixel offsets to be able to detect such movements – calculated all over the image 15 times per second! Even then, what about someone moving closer to the camera or a high-resolution image? Clearly, one has to design the algorithms carefully in order to save some of this work.

12.12 SUMMARY

The processing in a typical computer vision system consists of several phases:

- digitisation and signal processing
- edge and region detection
- object recognition
- image understanding

Not all will be present in any one system, as often acceptable results may be obtained with fewer levels of processing.

The raw image is usually digitised into pixels and may often be thresholded to give a simple black and white image. The image may be affected by noise. Digital filters can be used to smooth the image, which reduces noise, but can also blur edges. Different filters, including Robert's, Sobel's and the Laplacian, can be used to emphasise edges. Large digital filters can be expensive to apply, and so the simplest filter that gives acceptable results is used.

Having identified potential edge pixels, an edge-following algorithm must be used to collect them into lines. Some pixels may at this stage be discarded as noise if they fail to fit into any line. Alternatively, similar pixels can be collected together to form regions. Representing

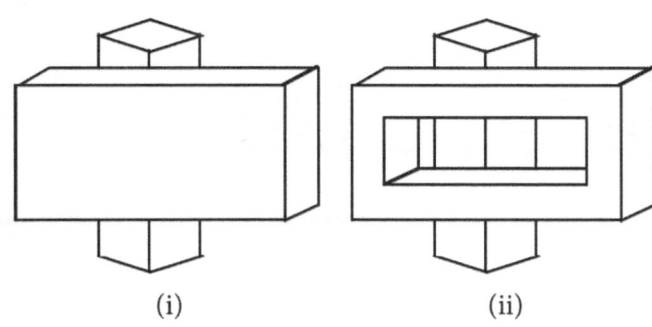

(i)   (ii)

FIGURE 12.36 For exercise 4.

regions can use a lot of space, and quadtrees are one way of efficiently storing regions.

Waltz's algorithm labels edges and vertices in a consistent manner, allowing lines to be built up into objects. However, some images are difficult to label unambiguously or may have one object split by an occluding object. Additional knowledge such as the use of shadows can help to resolve such ambiguity.

Identifying objects can be difficult because objects may be partly occluded, viewed at different angles or be in different positions from their templates. Techniques based on fuzzy matching of bitmaps, including the use of neural networks, can identify partially obscured objects, and the use of an object-centred co-ordinate system can help to reduce the effects of positioning. In some situations crude identification based on summary statistics may be sufficient. For more complex shapes matching of edges can be used to accommodate different viewing angles, but for paths without obvious vertices, such as handwriting, warping must be used to allow matching. For others, including face and body pose recognition, more model-based techniques can be used that use knowledge of human anatomy to guide algorithms.

Neural networks are also heavily used in vision-based systems. Sometimes vanilla networks are used, but several neural techniques are especially designed for or useful for images, notably autoencoders and convolutional neural networks. Multiple images from moving cameras or stereo vision can be used to obtain more information but can involve more processing time. Stereo vision can be used to calculate the relative distance of objects. Also, by tracking objects between frames of a moving image the object's speed can be calculated.

12.1   Take the digitised image in Figure 12.3(i). Threshold it at each of the following levels: 1, 5 and 8. Record your results on squared or graph paper, marking each square that exceeds the threshold. In fact, the threshold values above are not random. What does the picture look like thresholded at 6?

12.2   Again using Figure 12.3(i), apply Sobel's operator and then, choosing an appropriate threshold plot, draw the results.

12.3   Filters often lose information. To see this experiment with any popular image manipulation application such as Adobe Photoshop. These allow you to apply different kinds of smoothing and sharpening filters to captured images. Unfortunately you are not usually told the exact mathematical filter being applied, but they can give you a good feel for the possibilities of filtering. Compare the results of different filters. Try repeatedly applying smoothing and then sharpening filters to the same image.

12.4   Apply Waltz's algorithm to the image in Figure 12.36(i). Does it give the line labelling you would expect? What happens if you apply Waltz's algorithm to Figure 12.36(ii). Do you have any problems interpreting it?

FURTHER READING

R. Szeliski. *Computer vision: Algorithms and applications.* Springer, Cham, 2022.

A comprehensive textbook that includes state-of-the-art techniques in computer vision.

D. Andina, A. Voulodimos, N. Doulamis, A. Doulamis, and E. Protopapadakis. Deep learning for computer vision: A brief review. *Computational Intelligence and Neuroscience*, 2018:7068349, 2018. DOI: 10.1155/2018/7068349

A short review article that also acts an introductory primer for some of the major steps in the application of neural networks in computer vision.

D. Vernon. *Machine vision: Automated visual inspection and robot vision.* Prentice Hall, Hemel Hempstead, 1991. https://homepages.inf.ed.ac.uk/rbf/BOOKS/VERNON/vernon.htm

An early textbook that covers many of the more traditional algorithms of computer vision. Now available as an open access publication.

Natural Language Understanding

13.1 OVERVIEW

Natural language understanding is one of the most popular applications of artificial intelligence portrayed in fiction and the media. The idea of being able to control computers by talking to them in our own language is very attractive. Today this kind of speech control is no longer science fiction, but in everyday items from home automation controls to mobile phones; text-based understanding of language is common in chatbots as well as a key part of knowledge mining on the web. However, human language is ambiguous, which makes natural language

understanding particularly difficult. In this chapter we examine the major stages of natural language understanding – syntactic analysis, semantic analysis and pragmatic analysis – and some of the techniques that are used to make sense of this ambiguity. Big data and large-language models have had a major impact on many practical applications and appear to bypass some of this stepwise process, particularly at lower-levels.

13.2 WHAT IS NATURAL LANGUAGE UNDERSTANDING?

Whenever computers are represented in science fiction, futuristic literature or film, they invariably have the ability to communicate with their human users in natural language. By "natural language", we mean a language for human communication such as English,

French, Swahili or Urdu, as opposed to a formal "created" language (e.g. a programming language or Morse code). Unlike computers in films, which understand spoken language, we will concern ourselves primarily with understanding written language, rather than speech, and on analysis rather than language generation. As we shall see, this will present enough challenges for one chapter! Understanding speech shares the same difficulties but has additional problems with deciphering the sound signal and identifying word parts.

13.3 WHY DO WE NEED NATURAL LANGUAGE UNDERSTANDING?

Before we consider how natural language understanding can be achieved, we should be clear about the benefits that it can bring. There are a number of areas that can be helped by the use of natural language. The first is human–computer interaction, by the provision of interfaces for the user. This allows the user to communicate with computer applications in their own language, rather than in a command language or using menus. There are advantages and disadvantages to this: it is a natural form of communication that requires no specialised training, but it is inefficient for expert users and less precise than a command language. It may certainly be helpful in applications that are used by casual users (e.g. tourist information) or for novice users, and also for circumstances where hands free interactions are required, either through necessity (e.g. while driving a car) or preference (e.g. while sitting on a sofa).

A second area is information management, where natural language processing enables automatic management and processing of information, by interpreting its con-

DOI: 10.1201/9781003082880-16

tent. If the system understands the meaning of a document, it can, for example, store it with other similar documents.

A third possibility is to provide an intuitive means of database access. At present most databases can be accessed through a query language. Some of these are very complex, demanding considerable expertise to generate even relatively common queries. Others are based on forms and menus, providing a simpler access mechanism. However, these still require the user to have some understanding of the structure of the database. The user, on the other hand, is usually more familiar with the content of the database or at least its domain. By allowing the user to ask for information using natural language, queries can be framed in terms of the content and domain rather than the structure. We will look at a simple example of database query using natural language later in the chapter.

13.4 WHY IS NATURAL LANGUAGE UNDERSTANDING DIFFICULT?

The primary problem with natural language processing is the ambiguity of language. There are a number of levels at which ambiguity may occur in natural language (of course a single sentence may include several of these levels). First, a sentence or phrase may be ambiguous at a *syntactic* level. Syntax relates to the structure of the language, the way the words are put together. Some word sequences make valid sentences in a given language, some do not. However, some sentence structures have more than one correct interpretation. These are syntactically ambiguous. Secondly, a sentence may be ambiguous at a *lexical* level. The lexical level is the word level, and ambiguity here occurs when a word can have more than one meaning. Thirdly, a sentence may be ambiguous at a *referential* level. This is concerned with what the sentence (or a part of the sentence) refers to. Ambiguity occurs when it is not clear what the sentence is referring to or where it may legally refer to more than one thing. Fourthly, a sentence can be ambiguous at a *semantic level*, that is at the point of the meaning of the sentence. Sometimes a sentence is ambiguous at this level: it has two different meanings. Indeed this characteristic is exploited in humour, with the use of double entendre and innuendo. Finally, a sentence may be ambiguous at a *pragmatic* level, that is at the level of interpretation within its context. The same word or phrase may have different interpretations depending on the context in which it occurs. To make things even more complicated some sentences involve ambiguity at more than one of these levels. Consider the following sentences; how many of them are ambiguous and how?

1. I hit the man with the hammer.

2. I went to the bank.

3. He saw her duck.

4. Fred hit Joe because he liked Harry.

5. I went to the doctor yesterday.

6. I waited for a long time at the bank.

7. There is a drought because it hasn't rained for a long time.

8. Dinosaurs have been extinct for a long time.

How did you do? In fact all the sentences above have some form of ambiguity. Let's look at them more closely.

- *I hit the man with the hammer.*
  Was the hammer the weapon used or was it in the hand of the victim? This sentence contains syntactic ambiguity: there are two perfectly legitimate ways of interpreting the sentence structure.

- *I went to the bank.*
  Did I visit a financial institution or go to the river bank? This sentence is ambiguous at a lexical level: the word "bank" has two meanings, either of which fits in this sentence.

- *He saw her duck.*
  Did he see her dip down to avoid something or the web-footed bird owned by her? This one is ambiguous at a lexical and a semantic level. The word "duck" has two meanings and the sentence can be interpreted in two completely different ways.

- *Fred hit Joe because he liked Harry.*
  Who is it that likes Harry? This is an example of referential ambiguity. Who does the pronoun "he" refer to, Fred or Joe? It is not clear from this sentence structure.

- *I went to the doctor yesterday.*
When exactly was yesterday? This demonstrates pragmatic ambiguity. In some situations this may be clear but not in all. Does yesterday refer literally to the day preceding today or does it refer to another yesterday (imagine I am reading this sentence a week after it was written, for example). The meaning depends on the context.

- *I waited for a long time at the bank.*
There is a drought because it hasn't rained for a long time.
Dinosaurs have been extinct for a long time.
The last three sentences can be considered together. What does the phrase *for a long time* mean? In each sentence it clearly refers to a different amount of time. This again is pragmatic ambiguity. We can only interpret the phrase through our understanding of the sentence context.

In addition to these major sources of ambiguity, language is problematic because it is imprecise, incomplete, inaccurate and continually changing. Think about the conversations you have with your friends. The words you use may not always be quite right to express the meaning you intend, you may not always finish a sentence, you may use analogies and comparisons to express ideas. As humans we are adept at coping with these things, to the extent that we can usually understand each other if we speak the same language, even if words are missed out or misused. We usually have enough knowledge in common to disambiguate the words and interpret them correctly in context. We can also cope quickly with new words. This is borne out by the speed with which slang and street words can be incorporated into everyday usage. All of this presents an extremely difficult problem for the computer.

13.5 AN EARLY ATTEMPT AT NATURAL LANGUAGE UNDERSTANDING: SHRDLU

We met SHRDLU briefly in the Introduction. If you recall, SHRDLU is the natural language processing system developed by Winograd at MIT in the early 1970s [301]. It is used for controlling a robot in a restricted "blocks" domain. The robot's world consists of a number of blocks of various shapes, sizes and colours, which it can manipulate as instructed or answer questions about. All instructions and questions are given in natural language,

and even though the robot's domain is so limited, it still encounters the problems we have mentioned. Consider for example the following instructions:

> Find a block that is taller than the one you are holding and place it in the box
>
> How many blocks are on top of the green block?
>
> Put the red pyramid on the block in the box
>
> Does the shortest thing the tallest pyramid's support supports support anything green?

What problems did you spot? Again each instruction contains ambiguity of some kind. We'll leave it to you to figure them out! (The answers are given at the end of the chapter in case you get stuck.)

However, SHRDLU was successful because it could be given complete knowledge about its world and ambiguity could be reduced (it only recognises one meaning of "block" for instance and there is no need for contextual understanding since the context is given). It is therefore no use as a general natural language processor. However, it did provide insight into how syntactic and semantic processing can be achieved. We will look at techniques for this and the other stages of natural language understanding next.

13.6 HOW DOES NATURAL LANGUAGE UNDERSTANDING WORK?

So given that, unlike SHRDLU, we are not able to provide complete world knowledge to our natural language processor, how can we go about interpreting language? There are three primary stages in natural language processing: syntactic analysis, semantic analysis and pragmatic analysis. Sentences can be well-formed or ill-formed syntactically, semantically and pragmatically. Take the following responses to the question: *Do you know where the park is?*

- *The park is across the road.* This is syntactically, semantically and pragmatically well-formed, that is it is a correctly structured, meaningful sentence which is an appropriate response to the question.

- *The park is across the elephant.* This is syntactically well-formed but semantically ill-formed. The sentence is correctly structured, but our knowledge of parks and elephants and their characteristics shows it is meaningless.

- *The park across the road is.* This is syntactically ill-formed. It is not a legal sentence structure.

- *Yes.* This is pragmatically ill-formed: it misses the intention of the questioner.

At each stage in processing, the system will determine whether a sentence is well-formed. These three stages are not necessarily always separate or sequential. However, it is convenient to consider them as such.

Syntactic analysis determines whether the sentence is a legal sentence of the language, or generates legal sentences, using a grammar and lexicon, and, if so, returns a parse tree for the sentence (representing its structure). This is the process of *parsing*. Take a simple sentence, "The dog sat on the rug." It has a number of constituent parts: nouns ("dog" and "rug"), a verb ("sat"), determiners ("the") and a preposition ("on"). We can also see that it has a definite structure: noun followed by verb followed by preposition followed by noun (with a determiner associated with each noun). We could formalise this observation:

sentence =   determiner noun verb preposition
    determiner noun

Such a definition could then be tested on other sentences. What about "The man ran over the hill."? This too fits our definition of a sentence. Looking at these two sentences, we can see certain patterns emerging. For instance, the determiner "the" always seems to be attached to a noun. We could therefore simplify our definition of a sentence by defining a sentence component called noun_phrase.

noun_phrase = determiner noun

Our sentence definition would then become

sentence =   noun_phrase verb preposition
    noun_phrase

This is the principle of syntactic grammars. The grammar is built up by examining legal sentence structures and a lexicon is produced identifying the constituent type of each word. In our case our lexicon would include

dog : noun

the : determiner

rug : noun

sat : verb

and so on. If a legal sentence is not parsed by the grammar, then the grammar must be extended to include that sentence definition as well. Although our grammar looks much like a standard English grammar, it is not. Rather, we create a grammar that exactly specifies legal constructions of our language. In practice such grammars do bear some resemblance to conventional grammar, in that the symbols that are chosen to represent sentence constituents often reflect conventional word types but do not confuse this with any grammar you learned at school!

Semantic analysis takes the parse tree for the sentence and interprets it according to the possible meanings of its constituent parts. A representation of semantics may include information about different meanings of words and their characteristics. For example, take the sentence "The necklace has a diamond on it." Our syntactic analysis of this would require another definition of sentence than the one we gave above:

sentence =   noun_phrase verb noun_phrase
    prepositional_phrase
prepositional_phrase = preposition pronoun

This gives us the structure of the sentence, but the meaning is still unclear. This is because the word diamond has a number of meanings. It can refer to a precious stone, a geometric shape, even a baseball field. The semantic analysis would consider each meaning and match the most appropriate one according to its characteristics. A necklace is jewellery and the first meaning is the one most closely associated with jewellery, so it is the most likely interpretation.

Finally, in pragmatic analysis, the sentence is interpreted in terms of its context and intention.

For example, a sentence may have meanings provided by its context or social expectations that are over and above the semantic meaning. In order to understand the intention of sentences it is important to consider these. To illustrate, consider the sentence "He gave her a diamond ring." Semantically this means that a male person passed possession of a piece of hand jewellery made with precious stones over to a female person. However, there are additional likely implications of this sentence. Diamond rings are often (though of course not exclusively) given to indicate engagement, for example, so the sentence could mean the couple got engaged. Such additional, hidden meanings are the domain of pragmatic analysis.

13.7 SYNTACTIC ANALYSIS

Syntactic analysis is concerned with the structure of the sentence. Its role is to verify whether a given sentence is a valid construction within the language, and to provide a representation of its structure, or to generate legal sentences. There are a number of ways in which this can be done.

Perhaps the simplest option is to use some form of pattern matching. Templates of possible sentence patterns are stored, with variables to allow matching to specific sentences. For example, the template

< the ** rides ** >

(where ** matches anything) fits lots of different sentences, such as *the show-jumper rides a clear round* or *the girl rides her mountain bike*. These sentences have similar syntax (both are basically noun_phrase verb noun_phrase), so does this mean that template matching works? Not really. What about the sentence *the theme park rides are terrifying*? This also matches the template but is clearly a very different sentence structure to the first two. For a start, in the first two sentences "rides" is a verb, whereas here it is a noun. This highlights the fundamental flaw in template matching. It has no representation of word types, which essentially means it cannot ensure that words are correctly sequenced and put together.

Template matching is the method used in ELIZA [299], which, as we saw in the Introduction, fails to cope with ambiguity and so can accept (and generate) garbage. These are problems inherent in the approach: it is too simplistic to deal with a language of any complexity. However, it is a simple approach that has proved useful in constrained environments (whether such a restricted use of language could be called "natural" is another issue).

A more viable approach to syntactic analysis is sentence parsing. Here the input sentence is converted into a hierarchical structure indicating the sentence constituents. Parsing systems have two main components:

1. *a grammar*: a declarative representation of the syntactic facts about the language

2. *a parser*: a procedure to compare the input sentence with the grammar.

Parsing may be *top-down*, in which case it starts with the symbol for a sentence and tries to map possible rules to the input (or target) sentence, or *bottom-up*, where it starts with the input sentence and works towards the sentence symbol, considering all the possible representations of the input sentence. The choice of which type of parsing to use is similar to that for top-down or bottom-up reasoning; it depends on factors such as the amount of branching each will require and the availability of heuristics for evaluating progress. In practice, a combination is sometimes used. There are a number of parsing methods. These include grammars, transition networks, context-sensitive grammars and augmented transition networks. As we shall see, each has its benefits and drawbacks.

13.7.1 Grammars

We have already met grammar informally. A grammar is a specification of the legal structures of a language. It is essentially a set of rewrite rules that allow any element matching the left-hand side of the rule to be replaced by the right-hand side. So for example,

$$A \rightarrow B$$

allows the string XAX to be rewritten XBX. Unlike template matching, it explicitly shows how words of different types can be combined and defines the type of any given word. In this section we will examine grammars more closely and demonstrate how they work through an example.

A grammar has three basic components: *terminal symbols*, *non-terminal symbols* and *rules*. Terminal symbols are the actual words that make up the language (this part of the grammar is called the *lexicon*). So "cat", "dog" and "chase" are all terminal symbols. Non-terminal symbols are special symbols designating structures of the language. There are three types:

- *lexical categories*, which are the grammatical categories of words, such as noun or verb

- *syntactic categories*, which are the permissible combinations of lexical categories, for instance "noun_phrase", "verb_phrase"

- a special symbol representing a sentence (the *start_symbol*).

The third component of the grammar is the rules, which govern the valid combinations of the words in the language. Rules are sometimes called *phrase structure rules*. A rule is usually of the form

S → NP VP

where S represents the sentence, NP a noun_phrase and VP a verb_phrase. This rule states that a noun_phrase followed by a verb_phrase is a valid sentence.

The grammar can generate all syntactically valid sentences in the language and can be implemented in a number of ways, for example as a production system implemented in Prolog. We will look at how a grammar is generated and how it parses sentences by considering a detailed example.

13.7.2 An Example: Generating a Grammar Fragment

Imagine we want to produce a grammar for database queries on an employee database. We have examples of possible queries. We can generate a grammar fragment by analysing each query sentence. If the sentence can be parsed by the grammar we have, we do nothing. If it can't, we can add rules and words to the grammar to deal with the new sentence. For example, take the queries:

Who belongs to a union?

Does Sam Smith work in the IT Department?

In the case of the first sentence, *Who belongs to a union?*, we would start with the sentence symbol (S) and generate a rule to match the sentence in the example. To do this we need to identify the sentence constituents (the non-terminal symbols). Remember that the choice of these does not depend on any grammar of English we may have learned at school. We can choose any symbols, as long as they are used consistently. We designate the symbol RelP to indicate a relative pronoun, such as "who", "what" (a lexical category) and the symbol VP to designate a verb_phrase (a syntactic category). We then require rules to show how our lexical categories can be constructed. In this case VP has the structure V (verb) PP (prepositional phrase), which can be further decomposed as P, a preposition, followed by NP, a noun_phrase. Finally the NP category is defined as Det (determiner) followed by N (noun). The terminal symbols are associated with a lexical category to show how they can fit together in a sentence. We end up with the grammar fragment in Figure 13.1.

FIGURE 13.1   Initial grammar fragment.

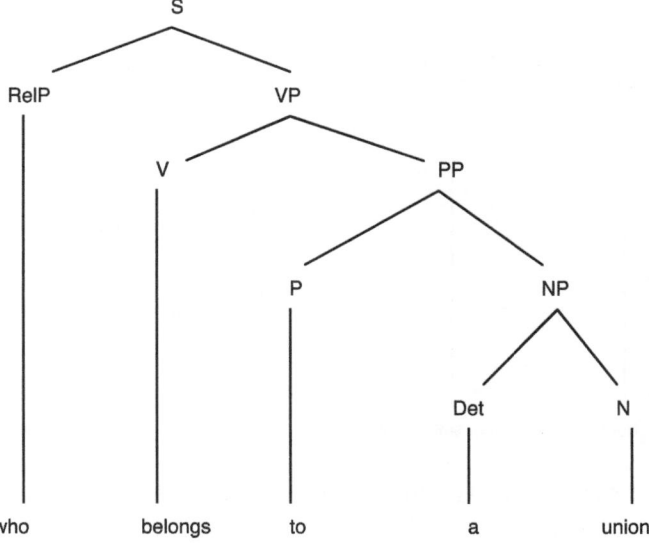

FIGURE 13.2   Parse tree for the first sentence.

This will successfully parse our sentence, as shown in the parse tree in Figure 13.2, which represents the hierarchical breakdown of the sentence. The root of the tree is the sentence symbol. Each branch of the tree represents a non-terminal symbol, either a syntactic category or a lexical category. The leaves of the tree are the terminal symbols.

However, our grammar is still very limited. To extend the grammar, we need to analyse many sentences in this way, until we end up with a very large grammar and lexicon. As we analyse more sentences, the grammar becomes more complete and, we hope, less work is involved in adding to it.

We will analyse just one more sentence. Our second query was *Does Sam Smith work in the IT Department?* First, we check whether our grammar can parse this sentence successfully. If you recall, our only definition of a sentence so far is

S → RelP VP

Taking the VP part first, *work in the IT Department* does meet our definition of a word phrase, if we interpret *IT*

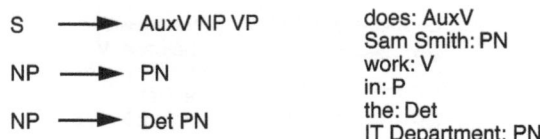

FIGURE 13.3   Further grammar rules.

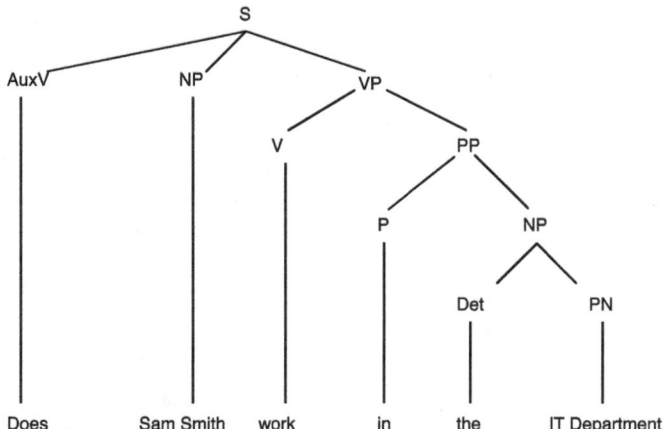

FIGURE 13.4   Parse tree for the second sentence.

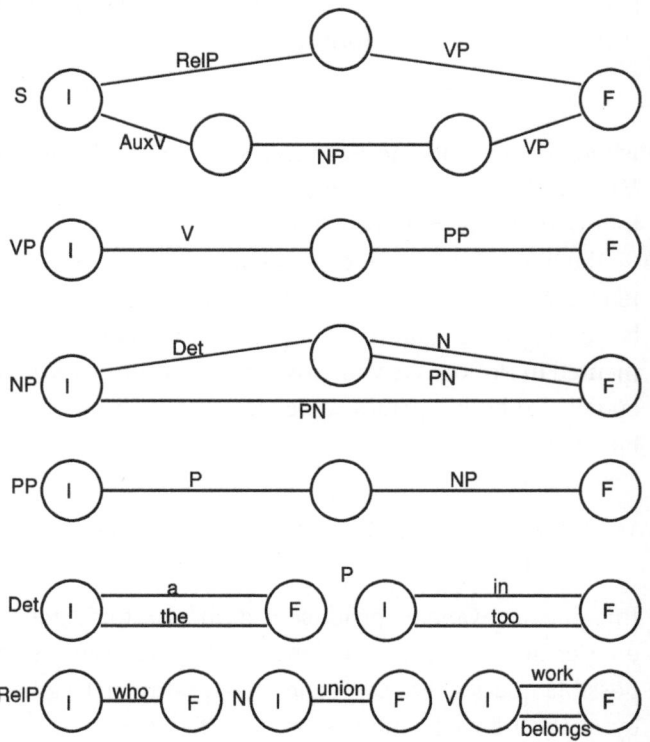

FIGURE 13.5   Transition network.

Department loosely as a noun. However, *Does Sam Smith* is certainly not a RelP. We therefore need another definition of a sentence. In this case a sentence is an auxiliary verb (AuxV) followed by an NP followed by a VP. Since Sam Smith is a proper noun we also need an additional definition of NP, and for good measure we will call *IT Department* a proper noun as well, giving us a third definition of NP. The additional grammar rules are shown in Figure 13.3.

Note that we do not need to add a rule to define VP since our previous rule fits the structure of this sentence as well. A parse tree for this sentence using this grammar is shown in Figure 13.4.

Grammars such as this are powerful tools for natural language understanding. They can also be used to generate legal sentences, constructing them from the sentence symbol down, using appropriate terminal symbols from the lexicon. Of course, sentence generation is not solely a matter of syntax; it is important that the sentence also makes sense. Therefore semantic analysis is also important. We shall consider this shortly. First we will look briefly at another method of parsing, the *transition network*.

13.7.3 Transition Networks

The *transition network* is a method of parsing that represents the grammar as a set of finite state machines. A finite state machine is a model of computational behaviour where each node represents an internal state of the system and the arcs are the means of moving between the states. In the case of parsing natural language, the arcs in the networks represent either a terminal or a non-terminal symbol. Rules in the grammar correspond to a path through a network. Each non-terminal is represented by a different network. To illustrate this we will represent the grammar fragment that we created earlier using transition network. All rules are represented but to save space only some lexical categories are included. Others would be represented in the same way.

In Figure 13.5 each network represents the rules for one non-terminal as paths from the initial state (I) to the final state (F). So, whereas we had three rules for NP in our grammar, here we have a single transition network, with three possible paths through it representing the three rules. To move from one state to the next through the network the parser tests the label on the arc. If it is a terminal symbol, the parser will check whether

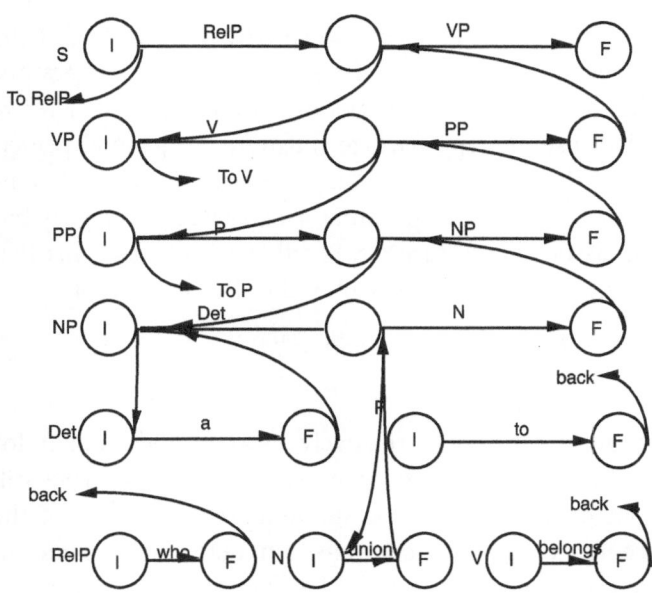

FIGURE 13.6   Navigation through transition network.

it matches the next word in the input sentence. If it is a non-terminal symbol, the parser moves to the network for that symbol and attempts to find a path through that. If it finds a path through that network, it returns to the higher-level network and continues. If the parser fails to find a path at any point, it backtracks and attempts another path. If it succeeds in finding a path, the sentence is a valid one. So to parse our sentence *Who belongs to a union?* the parser would start at the sentence network and find that the first part of a sentence is RelP. It would therefore go to the RelP network and test the first word in the input sentence "who" against the terminal symbol on the arc. These match, so that network has been traversed successfully and the parser returns to the sentence network able to cross the arc RelP. Parsing of the sentence continues in this fashion until the top-level sentence network is successfully traversed. The full navigation of the network for this sentence is shown in Figure 13.6.

The transition network allows each non-terminal to be represented in a single network rather than by numerous rules, making this approach more concise than grammars. However, as you can see from the network for just two sentences, the approach is not really tenable for large languages since the networks would become unworkable. Another disadvantage over grammars is that the transition network does not produce a parse tree for sentences and tracing the path through the network can be unclear for complex sentences. However, the transi-

tion network is an example of a simple parsing algorithm that forms the basis of more powerful tools, such as augmented transition networks, which we will consider in Section 13.7.6.

13.7.4 Context-sensitive Grammars

The grammars considered so far are context-free grammars. They allow a single non-terminal on the left-hand side of the rule. The rule may be applied to any instance of that symbol, regardless of context. So the rule

$$A \rightarrow B$$

will match an occurrence of A whether it occurs in the string ABC or in ZAB. The context-free grammar cannot restrict this to only instances where A occurs surrounded by Z and B. In order to interpret the symbol in context, a context-sensitive grammar is required. This allows more than one symbol on the left-hand side and insists that the right-hand side is at least as long as the left-hand side. So in a context-sensitive grammar, we can have rules of the form

$$ZAB \rightarrow ZBB$$

Context-free grammars are not sufficient to represent natural language syntax. For example, they cannot distinguish between plural and singular nouns or verbs. So in a context-free grammar, if we have a set of simple definitions

$$S \rightarrow NP\ VP$$
$$NP \rightarrow Det\ N$$
$$VP \rightarrow V$$

and the following lexicon

dog : N

guide : V

the : Det

dogs : N

guides : V

a : Det

we would be able to generate the sentences *the dog guides* and *the dogs guide*, both legal English sentences. However, we would also be able to generate sentences such as *a dogs guides*, which is clearly not an acceptable sentence.

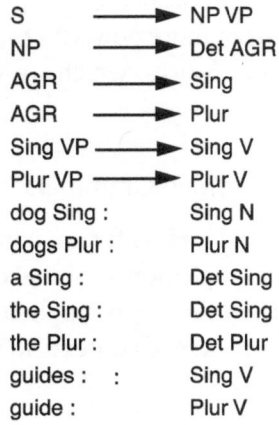

FIGURE 13.7 Grammar fragment for context-sensitive grammar.

By incorporating the context of agreement into the left-hand side of the rule we can provide a grammar which can resolve this kind of problem.

An example is shown in Figure 13.7.

The use of the symbols "Sing" and "Plur", to indicate agreement, does not allow generation of sentences that violate consistency rules. For example, using the grammar in Figure 13.7 we can derive the sentence "a dog guides" but not "a dogs guides". The derivation of the former is shown using the following substitutions:

S

NP VP

Det AGR N VP

Det Sing N VP

a Sing N VP

a dog Sing VP

a dog Sing V

a dog guides

Unfortunately context sensitivity increases the size of the grammar considerably, making it a complex method for a language of any size. Feature sets and augmented transition networks are alternative approaches to solving the context problem.

13.7.5 Feature Sets

Another approach to incorporating context in syntactic processing is the use of feature sets. Feature sets provide a mechanism for subclassifying syntactic categories (noun,

verb, etc.) in terms of contextual properties such as number agreement and verb tense. The descriptions of the syntactic categories are framed in terms of constraints. There are many variations of feature sets, but here we shall use one approach to illustrate the general principle – that of Pereira and Warren's Definite Clause Grammar [222]. In this grammar each syntactic category has an associated feature set, together with constraints that indicate what context is allowable. So, for example,

$$S \rightarrow NP \quad (agreement = ?a)$$
$$VP \ (agreement = ?b): a = b$$

Feature sets are a relatively efficient mechanism for representing syntactic context. However, we have still not progressed to understanding any semantics of the sentence. Augmented transition networks provide an approach that begins to bridge the gap between syntactic and semantic processing.

13.7.6 Augmented Transition Networks

The augmented transition network provides context without an unacceptable increase in complexity [302]. It is a transition network that allows procedures to be attached to arcs to test for matching context. All terminals and non-terminals have frame-like structures associated with them that contain their contextual information. To traverse an arc, the parser tests whatever contextual features are required against these stored attributes. For example, a test on the V arc may be to check number (i.e. plural or singular). The structure for the word *guides* would contain, among other things, an indication that the word is singular. The sentence is only parsed successfully if all the contextual checks are consistent. Augmented transition networks can be used to provide semantic information as well as syntactic, since information about meaning can also be stored in the structures. They are therefore a bridge between syntactic analysis and the next stage in the process, semantic analysis.

13.7.7 Taggers

We saw earlier that words can be ambiguous in terms of what they refer to "bank" as a financial institution or edge of a river. However, we also saw words that can be different parts of speech: "duck" can be the bird, a noun, or the act of dropping down, a verb. Indeed "bank" can also be used as a verb: "I'll bank the cheque".

This can get even more complicated, consider the sentence, "*I'll base my travel plans on the weather, either take a helicopter from the air base or climb from base camp.*" The first use of the word "base" is as a verb, the second as a noun and the third as an adjective.

When all the meanings are of the same syntactic category, disambiguation can be left to later stages of analysis, but if the same word has multiple possible syntactic categories, then ideally this level of disambiguation needs to happen at the syntactic stage.

Some forms of syntactic analysis can deal with this ambiguity as part of their normal functioning, but for others this is made easier or possible by prior use of a part-of-speech tagger, which allocates a POS (part-of-speech) category to each word. These often start with large digital dictionaries, such as WordNet [193, 194], which has multiple meanings of the word including the syntactic category of each. This provides a set of initial possible part-of-speech tags for each word. Figure 13.8 shows some of the meanings of the word "base" in WordNet; it can be a noun, verb or adjective, with sub-classes of each (e.g. 'noun.artifact').

These initial classifications may be augmented by word frequencies taken from large corpora to establish initial likelihoods for each meaning. This is followed by the use of techniques that can include partial semantic analysis, but also more sequence-of-words methods such as hidden Markov models, discussed in more detail in Chapter 14.

Figure 13.9 shows an example output of the CLAWS web tagger [232] on the (mistyped) sentence, "*I'll base by travel plans on the weather, either take a helicopter from the air base or climb from base camp*". Note how the first use of "base" is tagged "base_VVI" (infinitive of lexical verb) and the second "base_NN1" (singular noun). The third use of "base" is either tagged "base_SENT" or "base_NN1" depending on whether a full stop is added after the word "camp", but not "base_AJ0" (adjective); indeed CLAWS appears to struggle with most of the adjectival versions of "base" listed in WordNet. This is partly a matter of interpretation, note that "air" in "air base" is also tagged as a noun as it is effectively a noun used as adjective like "fish" in "fish market" ... even linguists disagree sometimes – the important thing is that the grammar and POS tagger use the same tagset.

Noun

- (4)02801449 <noun.artifact>[06]
S: (n) base#1 (base%1:06:04::), base of operations#1 (base_of_operations%1:06:00::) (installation from which a military force initiates operations) "*the attack wiped out our forward bases*"
... *plus 19 more entries*

Verb

- (75)00638550 <verb.cognition>[31]
S: (v) establish#8 (establish%2:31:03::), base#1 (base%2:31:00::), ground#12 (ground%2:31:00::), found#3 (found%2:31:00::) (use as a basis for; found on) "*base a claim on some observation*"
... *plus 2 more entries*

Adjective

- (6)01861961 <adj.all>[00] S: (adj) basal#3 (basal%3:00:00:basic:00), base#1 (base%3:00:00:basic:00) (serving as or forming a base) "*the painter applied a base coat followed by two finishing coats*"
... *plus 6 more entries*

FIGURE 13.8   Sample entries for the word "base" in WordNet.

I_PNP 'll_VM0 base_VVI by_PRP travel_NN1 plans_VVZ on_PRP the_AT0 weather_NN1 ,_PUN either_AV0 take_VVB a_AT0 helicopter_NN1 from_PRP the_AT0 air_NN1 base_NN1 or_CJC climb_VVB from_PRP base_SENT camp_NN1

FIGURE 13.9   Example output of CLAWS WWW tagger with three meanings for 'base'. Note that the tagger copes with the typing error 'by'.

13.8 SEMANTIC ANALYSIS

Syntactic analysis shows us that a sentence is correctly constructed according to the rules of the language. However, it does not check whether the sentence is meaningful, or give information about its meaning. For this we need to perform semantic analysis. Semantic analysis enables us to determine the meaning of the sentence, which may vary depending on context. So, for example, a system for understanding children's stories and a natural language interface may assign different meanings to the same words. Take the word "run", for example. In a children's story this is likely to refer to quick movement, while in a natural language interface it is more likely to be an instruction to execute a program. There are two levels at which semantic analysis can operate: the lexical level and the sentence level.

Lexical processing involves looking up the meaning of the word in the lexicon. However, many words have several meanings within the same lexical category (e.g. the noun "square" may refer to a geometrical shape or an area of a town). In addition, the same word may have further meanings under different lexical categories: "square" can also be an adjective meaning "not trendy", or a verb meaning "reconcile". The latter cases can be disambiguated syntactically but the former rely on reference to known properties of the different meanings. Ultimately, words are understood in the context of the sentences in which they occur. Therefore lexical processing alone is inadequate. Sentence-level processing on the other hand does take context into account. There are a number of approaches to sentence-level processing. We will look briefly at two: semantic grammars and case grammars.

13.8.1 Semantic Grammars

As we have seen, syntactic grammars enable us to parse sentences according to their structure and, in the case of context-sensitive grammar, such attributes as number and tense. However, syntactic grammars provide no representation of the meaning of the sentence, so it is still possible to parse nonsense if it is written in correctly constructed sentences. In a semantic grammar [37], the symbols and rules have semantic as well as syntactic significance. Semantic actions can also be associated with a rule, so that a grammar can be used to translate a natural language sentence into a command or query. Let us take another look at our database query system.

13.8.1.1 An Example: A Database Query Interpreter Revisited

Recall the problem we are trying to address. We want to produce a natural language database query system for an employee database that understands questions such as *Who belongs to a union?* and *Does Sam Smith work in the IT Department?* We have already seen how to generate a syntactic grammar to deal with these sentences, but we really need to derive a grammar that takes into account not only the syntax of the sentences but their meaning. In the context of a query interpreter, meaning is related to

the form of the query that we will make to the database in response to the question. So what we would like is a grammar that will not only parse our sentence but interpret its meaning and convert it into a database query. This is exactly what we can do with a semantic grammar.

In the following grammar, a query is built up as part of the semantic analysis of the sentence: when a rule is matched, the query template associated with it (shown in square brackets) is instantiated. The grammar is generated as follows. First, sentence structures are identified. Our sentences represent two types of question: the first is looking for information (names of union members), the second for a yes/no answer. So we define two legal sentence structures, the first seeking information and preceded by the word "who", the second seeking a yes/no response, preceded by the word "does". The action associated with these rules is to set up a query which will be whatever is the result of parsing the INFO or YN structures. Having done this we need to determine the structure of the main query parts. We will concentrate on the INFO category to simplify matters but the YN category is generated in the same way. Words are categorised in terms of their meaning to the query (rather than, for example, their syntactic category). Therefore, the words "belong to" and "work in" are semantically equivalent, because they require the same query (but with different information) to answer. Both are concerned with who is in what organisation. Similarly, "union" and "department" are also classed as semantically equivalent: they are both examples of a type of organisation. Obviously, such an interpretation is context dependent. If, instead of a query interpreter, we wanted our natural language processing system to understand a political manifesto, then the semantic categories would be very different. INFO is therefore a structure that consists of an AFFIL_VB (another category) followed by an ORG. Its associated action is to return the query that results from parsing AFFIL_VB. The rest of the grammar is built up in the same way down to the terminals, which return the values matched from the input sentence. The full grammar is shown in Figure 13.10.

Using this grammar we can get the parses

query: is_in(PERSON, org(NAME, union))

query: is_in(Sam Smith, org(IT, Department))

for the above sentences respectively. Parse trees for these sentences are shown in Figures 13.11 and 13.12. These show how the query is built up at every stage in the parse.

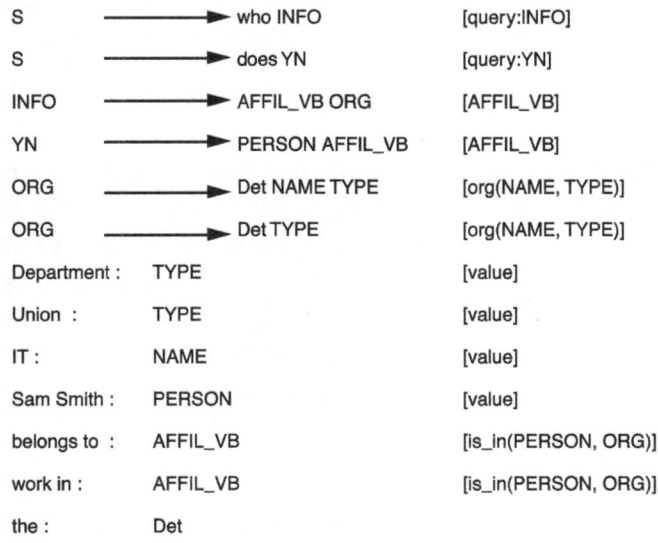

FIGURE 13.10   Semantic grammar fragment.

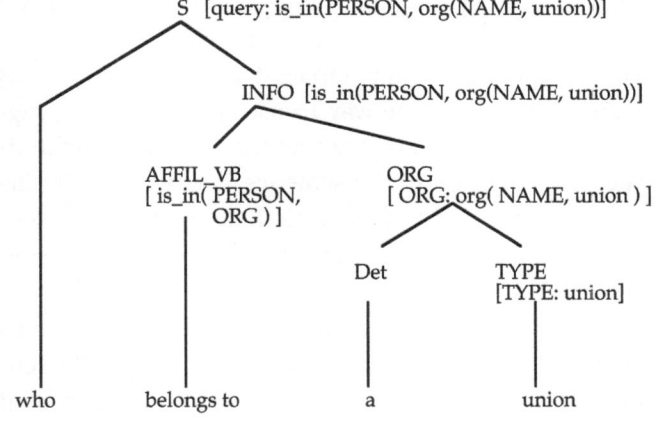

FIGURE 13.11   Parse tree for the first sentence.

Instantiation of the query components works from the bottom of the tree and moves up.

13.8.2 Case Grammars

Semantic grammars are designed to give a structural and semantic parse of the sentence. Grammars can get very big as a result. Case grammars represent the semantics in the first instance, ignoring the syntactic, so reducing the size of the grammar [102]. For example, a sentence such as *Joe wrote the letter* would be represented as

wrote (agent(Joe), object(letter))

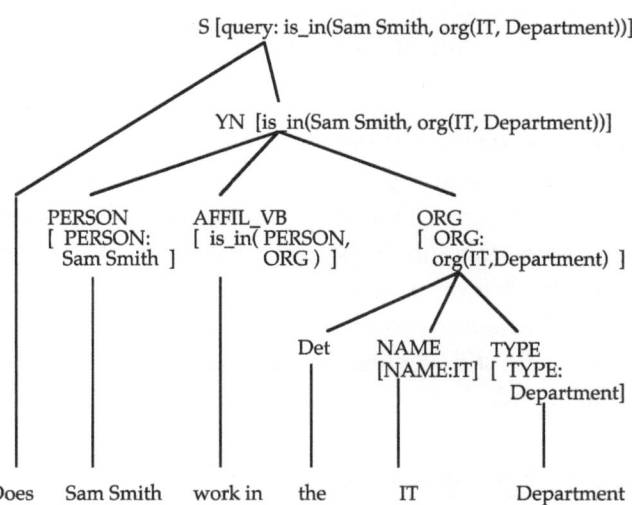

FIGURE 13.12   Parse tree for the second sentence.

This indicates that Joe was the active participant, the agent, who performed the action "wrote" on the object "letter". The passive version *The letter was written by Joe* would be represented in the same way, since the meaning of the sentences is identical.

Case grammars rely on *cases*, which describe relationships between verbs and their arguments. A number of cases are available to build case grammar representations. The following list is not exhaustive. Can you think of other cases?

- *Agent* – the person or thing performing the action.

- *Object* – the person or thing to which something is done.

- *Instrument* – the person or thing which allows an agent to perform an action.

- *Time* – the time at which an action occurs.

- *Beneficiary* – the person or thing benefiting from an action.

- *Goal* – the place reached by the action.

So, for example, the sentence *At 1 pm, Paul hit the gong with the hammer for lunch* would be parsed as

hit(time(1pm), agent(Paul), object(gong),
   instrument(hammer),
   goal(lunch))

If we changed the sentence to *At 1 pm, Paul hit the gong with the hammer for his father*, the case representation would be

```
hit( time(1pm), agent(Paul), object(gong),
     instrument(hammer),
     beneficiary(his father) )
```

The case structures can be used to derive syntactic structures, by using rules to map from the semantic components that are present to the syntactic structures that are expected to contain these components. However, case grammars do not provide a full semantic representation, since the resulting parse will still contain English words that must be understood.

13.9 PRAGMATIC ANALYSIS

The third stage in understanding natural language is pragmatic analysis. As we saw earlier, language can often only be interpreted in context. The context that must be taken into account may include both the surrounding sentences (to allow the correct understanding of ambiguous words and references) and the receiver's expectations, so that the sentence is appropriate for the situation in which it occurs. There are many relationships that can exist between sentences and phrases that have to be taken into account in pragmatic analysis. For example:

- A pronoun may refer back to a noun in a previous sentence that relates to the same object. *John had an ice cream. Joe wanted to share it.*

- A phrase may reference something that is a component of an object referred to previously. *She looked at the house. The front door was open.*

- A phrase may refer to something that is a component of an activity referred to previously. *Jo went on holiday. She took the early train.*

- A phrase may refer to agents who were involved in an action referred to previously. *My car was stolen yesterday. They abandoned it two miles away.*

- A phrase may refer to a result of an event referred to previously. *There have been serious floods. The army was called out today.*

- A phrase may refer to a subgoal of a plan referred to previously. *She wanted a new car. She decided to get a new job.*

- A phrase may implicitly intend some action. *This room is cold (expects an action to warm the room).*

One approach to performing this pragmatic analysis is the use of scripts [243]. We met in Chapter 2. In scripts, the expectations of a particular event or situation are recorded and can be used to fill in gaps and help to interpret stories. The main problem with scripts is that much of the information that we use in understanding the context of language is not specific to a particular situation but generally applicable. However, scripts have proved useful in interpreting simple stories.

13.9.1 Speech Acts

When we use language, our intention is often to achieve a specific goal that is reached by a set of actions. The acts that we perform with language are called *speech acts* [249]. Sentences can be classified by type. For example, the statement "I am cold" is a declarative sentence. It states a fact. On the other hand, the sentence "Are you cold?" is interrogative: it asks a question. A third sentence category is the imperative: "Shut the window". This makes a demand. One way to use speech acts in pragmatic analysis is to assume that the sentence type indicates the intention of the sentence. Therefore, a declarative sentence makes an assertion, an interrogative sentence asks a question and an imperative sentence issues a command.

This is a simplistic approach, which fails in situations where the desired action is implied. For example, the sentence "I am hungry" may be simply an assertion or it may be a request to hurry up with the dinner. Similarly, many commands are phrased as questions ("Can you tell me what time it is?"). However, most commercial natural language processing systems ignore such complexity and use speech acts in the manner described above.

Such an approach can be useful in natural language interfaces since assertions, questions and commands map clearly onto system actions. So if I am interacting with a database, an assertion results in the updating of the data held, a question results in a search and a command results in some operation being performed.

13.10 GRAMMAR-FREE APPROACHES

Sometimes it is possible to perform natural-language processing without the use of a formal grammar.

13.10.1 Template Matching

We noted earlier that the template matching used in ELIZA [299] has limitations in both accepted input language (it will attempt to interpret gibberish) and output. However, despite this minimal grammar these template matching algorithms have proved remarkably powerful, especially in chatbots which are widely used on the web.

In fact, both spoken language and instant-message-style conversations not only often include utterances which are formally 'ungrammatical', but this is the norm. Our own parsing of the things we hear and read is based partly on formal grammar but also many other mechanisms. Indeed, some orally based language teaching programmes, such as the "SaySomethingin" courses [263], make heavy use of small patterns that can be used with simple substitutions.

Various open source and commercial chatbots are available and allow you to both add to generic templates with domain-specific terms and phrases and also create structured conversations where needed.

Some systems force a restricted language, for example text-based chatbots may only allow you to create phrases using canned expressions, such as "I want to know about". Here the user is doing all of the work, and there is no real AI behind the language comprehension. However, even when the system has true AI behind it, for example in speech-based home automation systems such as Alexa, users soon learn the acceptable language and, often without realising, modify their speech patterns to suite the capabilities of the system ... just as we do when speaking with a small child.

13.10.2 Keyword Matching

At an even more basic level, simple keyword matching can be used to trigger actions or at least start more directed dialogues. One very early system for pre-consultation health discussions used an off-the-shelf chatbot but augmented it with a simple topic model. Keywords were used to determine what topics the patient had talked about and then, when there was a natural break in the conversation, the chatbot introduced topics that were still to be covered [218].

More complex versions of keyword matching can include LEGO-style language, where words have linked capabilities. For example if you hear "bone dog ate", you have no difficulty in recognising that it is the dog not the bone that is doing the eating. This is effectively a pragmatic level of understanding: a dog is a thing that does eating and a bone something that can be eaten. If you are a home automation system and hear someone say a series of words including "light" and "on", then (in the absence of obvious negatives) this would be interpreted as "turn the light on" where the light in question would be the one closest to the person unless there is an obvious room word such as "kitchen".

In some ways this is far more primitive language than full grammars – perhaps the way early humans first developed linguistic communication. However, in terms of the levels of language this is actually closer to pragmatics – the relatively simple keyword and association matching is understanding the user's intent, what they want to achieve. Ultimately language is about getting things done, not perfect grammar.

13.10.3 Predictive Methods

Other chapters deal with large-scale text processing and web search (Chaps. 8 and 17). Traditionally areas such as this were regarded as not real AI as they applied statistical techniques. However, these techniques have been found to be surprisingly powerful in some cases as alternatives to more semantically rich NLP or to supplement it. We have already seen this in the way that word frequency data can be used to help drive word disambiguation in POS taggers.

Many email systems and other forms of text entry offer suggestions of the next few words you might want to enter. These are often uncannily accurate and yet based on a very simple principle, the n-gram. A large corpus is analysed and frequencies calculated for every single word, every word pair (2-gram), every triple (3-gram) and maybe larger groups too. Then when you type "I want" the suggested ext word might be "a", "to" or "the" as the 3-grams "I want a", "I want to" and "I want the" are very frequent.

Large-language models such as GPT-3 [34] can be thought of as more sophisticated versions of the same thing. Rather than simply n-gram frequency a large sliding window (discussed in more detail in Chapter 14) is used to train a deep neural network that can then predict the next word. By iterating this the next 2, 3 or more words can be predicted or even running text.

These models are trained on vastly big corpora, but are also designed so that they are foundation models, that is the generic network can be used to generate more bespoke or domain specific models by a (relatively) small amount of additional training. This is possible because the set of network weights have in some way learnt general properties of language, not simply the specific word they have been exposed to. It appears that concepts such as parts of speech, and grammar, while not explicitly coded are, in some way, present as the text that is created is, on the whole, well-formed grammar.

There is ongoing debate as to whether larger and larger models with more and more parameters will be able to emulate all aspects of natural language understanding or whether there are fundamental limits at which more knowledge-rich methods will always be required. Possibly the answer is somewhere between with low-level network architectures that are in some way pre-programmed not with specific language rules, but with structures that are tuned to be particularly capable for different kinds of linguistic purpose.

These more statistical and data-driven machine learning techniques can certainly be used as part of more bespoke language systems, for example one could train a neural network to do POS tagging. In addition parts of neural-based language or text models can be used.

13.10.4 Statistical Methods

We have seen in other chapters the way principal components can be used to create reduced dimensional representations for recommender systems. A similar technique can be used for texts. A text is initially represented as a vector of word frequencies, possibly weighted by overall corpus frequency, as discussed in Chapter 10. Principal components or a similar method is used to create a lower-dimensional vector that acts a signature for the text, a process called latent semantic analysis [96]. The reduced dimensional space no longer represents words as such but in some way captures overall meanings, where documents that refer to similar topics lie closer in the latent space.

For document retrieval this is then used to help match search terms to documents based not on whether they directly contain the word, but more on the closeness of the overall gist of the query and document. More generally, this can be used for topic analysis, for example taking a 500-word sliding window over a document and creating a signature vector for each window in the latent space. Rapid movement in the latent space then corresponds to topic shifts in the text.

A similar technique, word2vec uses an internal layer of a neural network that has been trained to predict the close co-occurrence of words in a corpus [191]. This internal layer can be treated as a form of latent space (often called a word vector), that in a sense captures the meaning of a word. Crucially relationships are often maintained, so that the vector difference between Madrid and Spain is similar to that between Paris and France. These word vectors can be used instead of the original word as part of other language processing techniques.

13.11 SUMMARY

In this chapter we have looked at the issue of ambiguity, which makes natural language understanding so difficult. We have considered the key stages of natural language understanding: syntactic analysis, semantic analysis and pragmatic analysis. We have looked at grammars and transition networks as techniques for syntactic analysis; semantic and case grammars for semantic analysis; and scripts and speech acts for pragmatic analysis. We have also seen how levels of apparent language understanding can be achieved without the user of formal grammars using simple template and word matching or more complex large-language models and statistical methods.

13.1 For each of the sentences below generate the following:

- a syntactic grammar and parse tree
- a transition network
- a semantic grammar and parse tree
- a case grammar

What additional features would you represent if you were generating context-sensitive grammars for these sentences?

- My program was deleted by Brian
- I need a print-out of my program file
- The system administrator removed my files
- I want to create a new document file

13.2 Identify the ambiguity in each of the following sentences and indicate how it could be resolved.

- She was not sure if she had taken the drink

- Joe broke his glasses

- I saw the boy with the telescope

- They left to go on holiday this morning

13.3 Devise a script for visiting the doctor, and indicate how this would be used to interpret the statement: "Alison went to the surgery. After seeing the doctor she left."

FURTHER READING

N. Indurkhya and F. Damerau, editor. *Handbook of natural language processing*. CRC Press, Boca Raton, FL, 2010.

An edited handbook covering the complete process of NLP with experts in each area contributing the relevant chapters. It was published before the emergence of effective big data and deep learning techniques but is an in-depth and comprehensive view of traditional approaches to NLP.

D. Jurafsky and J. Martin. *Speech and language processing* (3rd edition, 2024 draft). (Update of 2nd edition, pub. Prentice Hall 2008). https://web.stanford.edu/ jurafsky/slp3/

This popular textbook has been updated to include latest developments including large-language models. The updated material is available through the authors' website.

T. Winograd and F. Flores. *Understanding computers and cognition*. Addison-Wesley, Ablex Corporation Norwood, NJ, 1987.

Includes a discussion of speech act theory and other aspects of natural language understanding from a more philosophical standpoint.

13.12 SOLUTION TO SHRDLU PROBLEM

1. Find a block that is taller than the one you are holding and place it in the box. This is referential ambiguity. What does the word "it" refer to?

2. How many blocks are on top of the green block? This is perhaps more tricky, but it involves semantic ambiguity. Does "on top of" mean directly on top of or above (i.e. it could be on top of a block that is on top of the green block)?

3. Put the red pyramid on the block in the box. This is syntactic ambiguity. Is it the block that is in the box or the red pyramid that is being put into the box?

4. Does the shortest thing the tallest pyramid's support supports support anything green? This is lexical: there are two uses of the word "support"!

Time Series and Sequential Data

14.1 OVERVIEW

The moves in a game of chess, annual Arctic ice extent since the beginning of the industrial era, words in a sentence, finger and hand positions during a mid-air gesture, hospital admissions reports for a patient – there are many types of data where it is not only the values of data items that matter but also the order in which they occur. Often, but not always, the sequence order reflects an underlying time order. Thinking of time order we'll often use the term 'event' for one of the sequential items, but this should be interpreted liberally to include, for example a word in written text.

There are some specific techniques used for particular domains, for example speech processing or games, but also some common features and techniques that can be applied across a number of domains.

In this chapter we look first at some of the general properties of temporal and sequential data and then at three main classes of algorithm: probability-based methods; grammar or pattern matching; neural networks and statistical methods. Grammars are, by their nature, specific to sequential data; however, we will see specialised techniques in all of these classes. Finally, we will see that data may often be viewed at multiple granularities (e.g. raw audio, phonemes, words), and different methods may be applied at each level.

14.2 GENERAL PROPERTIES

14.2.1 Kinds of Temporal and Sequential Data

Often more critical than the application domain are the characteristics of the data.

First we can look at the timing or sequence order of each data item, it may be:

Discrete events – for example hospital admissions, or words spoken. These may have times or periods in which they occurred, but there may also be gaps when nothing happens.

Samples of continuous time – for example hourly air temperature readings, where the air has a temperature between readings.

For discrete events, we may have a *complete record* of every event (e.g. each word in text) or may have an *incomplete record* with missing events, such as hospital visits that for some reason were not recorded. Note that in the latter case we may not even know that they are missing.

For sampled data we may have a uniform sampling rate, every second, hour or day, or the samples may be sporadic, perhaps only when some other event happens that triggers a reading. Missing vales are often more obvious but still cause problems with analysis.

If we look at the values associated with each event, these may be:

Homogeneous – Each event has the same type of data. For example, a word or a temperature reading.

Heterogeneous – Some events have different data associated with them than others. For example,

DOI: 10.1201/9781003082880-17

there may be different test results taken at different hospital visits.

In many scientific, engineering and economic applications one has homogeneous numeric data collected at a uniform sample rate, and there is a rich set of statistical and signal processing techniques targeted at these. In some cases, AI and ML algorithms offer alternatives for these, but in others the two work alongside each other, often with some form of numerical pre-processing of data.

Often data can be mixed, with different kinds of data in the same application, for example, in sales forecasting the daily or weekly sales of products form a uniform numerical time series, which is analysed using relatively basic statistical techniques. However, there are also unusual points, for example where there has been an advertising campaign. Here the analysts make adjustments based on their experience [9] – an obvious point also where a hybrid AI/statistical system could be useful.

14.2.2 Looking through Time

As with any form of data, one should always spend some time getting to know temporal data. Here are some behaviours you might notice:

Stationarity – Although the precise values change, the kinds of behaviour are relatively similar at any time.

Trends – Things get bigger or smaller over time, for example long-term inflation.

Periodicity – Some aspect of a process that repeats (possibly with small variation/noise) at fixed intervals, for example seasonal variations in ice cream sales.

Quasi-periodicity – Where changes have a nearly fixed timescale of change, but not tied to a precise 'tick' time. The 11-year sunspot 'cycle' is like this, and it is often the sign of a dynamic system with feedback.

Discontinuities – Points of sudden or unusual change, the cause of which may or may not be known. For example, a known or expected change would be the impact of an advertising campaign on sales, whereas an unexpected one would be the 1987 stock market crash (see Figure 14.1).

Phase changes – Points where the characteristics of the process change. For example, in finance it is relatively easy to predict future stock values during either bull or bear markets, but the transitions between bull and bear are where fortunes are made and lost.

Substructure – Portions of the signal may have characteristics of their own. For example, ECG data consists of a time series of electrical signals sampled at from 50Hz to 5kHz depending on the apparatus. However, within that it is the roughly once a second heart rate signal that is of most interest, and also the shape of this, as different shapes can reveal particular heart problems.

In addition, if you look at the same data at different timescales you might see different forms of activity. For example, if you look at stock values during a bull market they may appear to have an upwards trend, but if you look at a larger timeframe you will see flips between bull and bear markets. Similarly, if you look at ice cream sales from August through to October, you might detect a downward trend, but zooming out to view several years' sales you begin to see annual periodicity.

Although any actual dataset is finite, often what you have represents a portion of a longer, possibly indefinitely long series. For example, you might be processing a 20-minute ECG trace from a patient, who hopefully lives a lot longer. However, sometimes the underlying phenomenon has finite length, for example the stroke making up a character in handwriting recognition.

14.2.3 Processing Temporal Data

In the following sections we will look at algorithms of very different kinds but with underlying similarities. In the end all are trying to transform the indefinite sequence into a collection of finite data problems. The goal is usually to either predict the next data item from what came before or to classify the whole or part of the time series. Typically, they use one of two general techniques:

windowing – splitting the data into sub-sequences of a fixed length

hidden state – processing is more event-to-event, but assuming some underlying unobservable state

In addition, for certain sorts of data, particularly audio, the data may be transformed from time domain (event-by-event data points) into frequency domain (signal

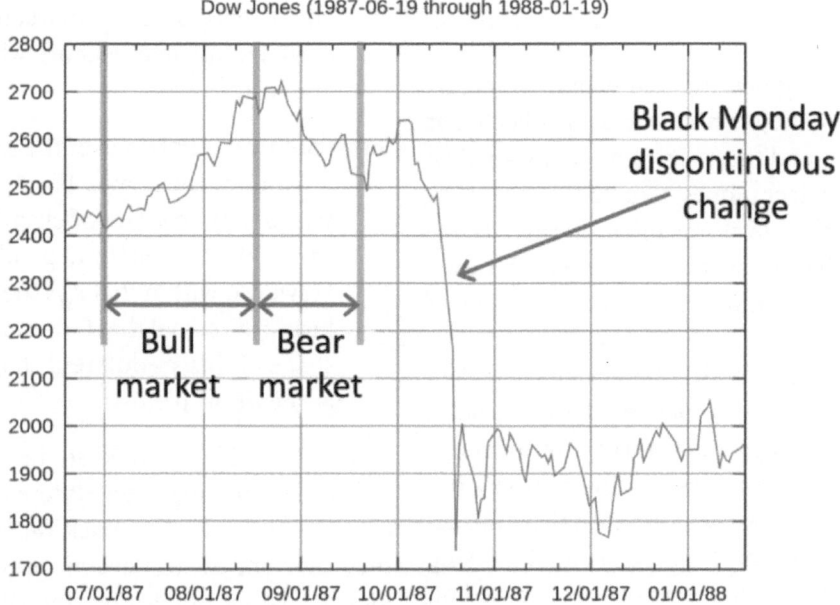

FIGURE 14.1   Stock market leading up to and after Black Monday, 19th October 1987 (adapted from https://commons.wikime dia.org/wiki/File:Black_Monday_Dow_Jones.svg).

strength at different frequencies). We'll look at each of these in more detail.

14.2.3.1 Windowing

The sequence is broken into a number of short pieces of fixed length. These can then, in principle, be fed into pretty much any algorithm.

Some algorithms will use non-overlapping windows, for example chopping the data sequence into one-hour segments. However, more common is to use moving windows, that is the last N items. For example, Covid case data was often presented as a 7-day moving average, which basically means take the last 7 days' figures and show the average of these.

Effectively algorithms using this technique are trying to work out:

last N steps ⇒ output
(classification / prediction of next step)

Occasionally the windows may be centred over regions of interest. For example, with ECG data, one might attempt to detect the peaks and centre the window over them. Also there may be start and end adjustments to prevent anomalies; for example a short linear ramp in/out.

Algorithms based on windowing have what is called finite impulse response (FIR), that is the effects of any sample point, no matter how extreme, only affect outputs for the next N steps. However, note that N can be extremely large; for example ChatGPT4 has a 128K token window and Google Gemini up to one million tokens – finite but for many practical purposes unbounded.

14.2.3.2 Hidden State

The algorithm assumes there is some additional state that is not immediately apparent in the observed data but which affects its behaviour. For example, the temperature and heart rate (observed data) of a patient will depend on the progress of the infection (hidden state).

Algorithms using hidden state are effectively trying to work out a function of the form:

current step × hidden state
→ output × new hidden state

In general, it is a lot harder for machine learning algorithms to learn this function as it depends on the hidden state, which, by definition, is not known!

In principle, it is sufficient to only use one step at a time as input – anything the algorithm 'wants' to remember can be stored in hidden state. However, as it is hard to learn the hidden state, it may sometimes be better to combine it with windowing, effectively finding a function of the form:

$$\text{last N steps} \times \text{hidden state}$$
$$\rightarrow \text{output} \times \text{new hidden state}$$

The hidden state can be simpler as it only has to 'remember' long-term things beyond the window's time frame.

Hidden-state techniques usually have an infinite impulse response (IIR), that is the effect of a single data item can last an indefinitely long time, albeit often typically becoming less significant the longer you wait. This is often what you want, as you are trying to account for long-lasting effects, but does mean that any sporadic mis-reading or extreme reading can affect performance for a long time to come.

14.2.3.3 Non-time Domain Transformations

Sequences of the form: data at $t = 0$, data at $t = 1$, ... where the index of the data values is a time value, is called time domain data and some algorithms work best on this form of data. However, sometimes the data is better transformed to replace or augment the raw data as an aid to subsequent processing.

There are many ways to do this, but the two most common are Fourier and wavelet transforms.

Fourier transforms are a more detailed version of what you see in the little bars on a high-quality sound system equaliser showing the bass and treble response in different ranges. The Fourier transform takes a signal and splits it into frequency components based on the power in particular frequency ranges. For example, a low-frequency Fourier transform might give 16 outputs: one average (0Hz); two at each of 1Hz power, 2Hz power, ... 7 Hz power; one at 8Hz. Note there are two components at each of the intermediate levels corresponding to *cos* and *sin* components, or, equivalently, the fact that a wave has a phase (where it starts) as well as a wavelength (how wide it is).

A Fourier transform can be calculated for a complete time series, but more commonly windows are used, usually having 2^M items for some choice of M as the fast Fourier transform (FFT) algorithm works best with a power of two length.

Figure 14.2 shows an example of a voice spectrogram, which is often used as the first stage of speech recognition systems. Here small windows of a fraction of a second are divided into 32 frequency bands from 0Hz to 10Hz. The gaps between words are very clear as are differences in the frequency characteristics of the various syllables.

Wavelet transforms also have a frequency element creating some outputs that are about the large-scale/slow structure. However, while the high-frequency components of a Fourier transform are about the whole of a time series, or window, the high-frequency components of Wavelet transforms are about short snippets of the signal. Basically, a signal is broken into a few slow-changing parts that capture the large-scale structure, plus a larger number of small short-timescale parts.

Figure 14.3 shows a family of wavelets, called the Haar wavelets, first proposed over a hundred years ago by Alfréd Haar [120]. It is easy to see that a discrete signal of length 8 can be broken into a constant part plus a sum of the wavelets. However, wavelets do not have to have sharp edges, and smoother wavelets are more often used as they have better mathematical properties but also more complex mathematical formulae.

14.3 PROBABILITY MODELS

One way to look at a time series is as a probabilistic process, where the probability of the next thing that happens depends on what has gone before. These are used particularly when the data values are discrete and finite, notably for text, and most common approaches are based around variants of Markov models.

14.3.1 Markov Model

The original Markov model, named after the Russian mathematician Andrey Andreyevich Markov, is specifically for systems with no memory, where the probability of the next item in a sequence is purely determined by the current item.

Here's a sequence of letters from the restricted alphabet A,B,C.

```
AAACCACCCCABBABBACAAAAACCACACACAACACAB
```

These were generated by the transition probabilities in Table 14.1. Reading through this table, row by row: if the current state is A, then there is a 1 in 2 chance it will stay as an A, and a 1 in 4 chance that the next will be a B and a 1 in 4 that it will be a C; if the current state is B, there is a 1 in 2 chance it will go back to an A and a 1 in 2 chance it will stay as B; finally if the current state is C, then there

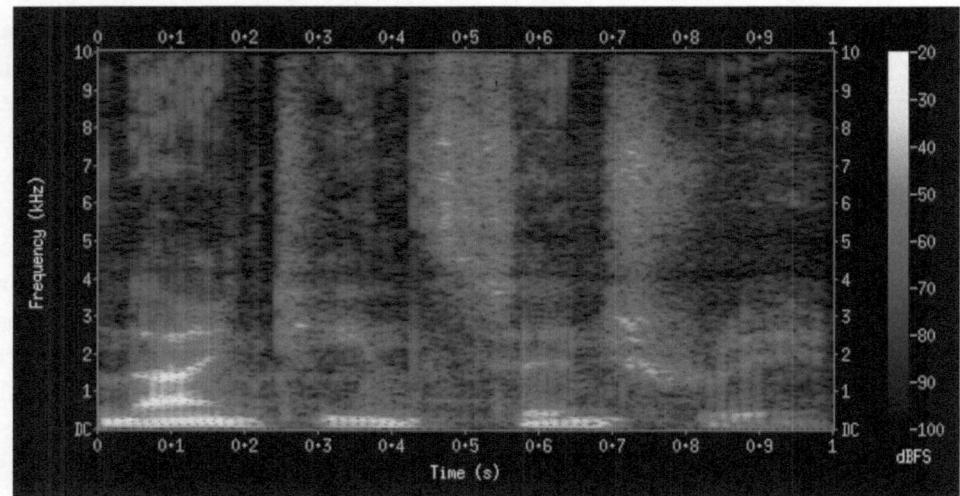

FIGURE 14.2   Digitally produced spectrogram of a male voice saying 'nineteenth century' (https://commons.wikimedia.org/wiki/File:Spectrogram-19thC.png).

is a 1 in 2 chance it will go to A and a 1 in 2 chance it will stay as C.

TABLE 14.1   Markov Model (upper) Transition Probability Table; (lower) Drawn as a Network.

	Next letter		
Current letter	A	B	C
A	0.5	0.25	0.25
B	0.5	0.5	0
C	0.5	0	0.5

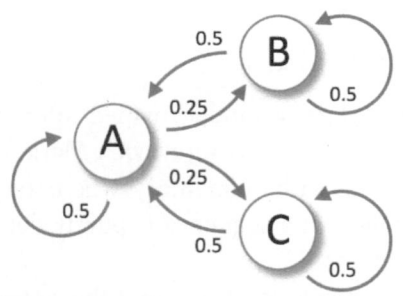

You can imagine building similar rules for the weather. Indeed a popular saying in Britain is "Rain before seven shine before eleven", which basically says that if you take four-hourly measurements the weather is likely to change!

Generating the sequence from the rules is easy (toss a coin!), but we need to first be able to learn the rules from the data. Happily this is simple for a Markov Model.

First you just count how often each pair appears in the sequence.

	Raw frequencies					Totals	
AA:	7	AB:	3	AC:	9	start with A:	19
BA:	2	BB:	2	BC:	0	start with B:	4
CA:	9	CB:	0	CC:	5	start with C:	14

These are then converted into estimated transition probabilities by dividing by the frequency it is in the relevant start state. For example, there were 19 times the sequence was an A in the current state and of these 9 times the next step was C (AC frequency), so the estimated transition probability is 9/19 (= 0.474). In full this gives us the estimated transition probabilities in Table 14.2.

TABLE 14.2   Estimated Transition Probabilities Based on Observed Transition Frequencies.

	Next letter		
Current letter	A	B	C
A	0.368	0.158	0.474
B	0.500	0.500	0.000
C	0.643	0.000	0.357

Note how this is not exactly the same as the actual transition probabilities that were used to generate the sequence. These are random rules, so short sequences can vary markedly. A far longer sequence is required to give robust probability estimates.

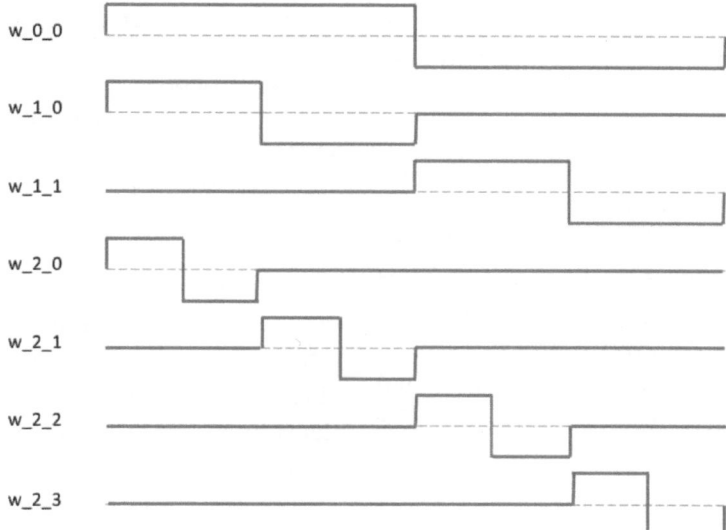

FIGURE 14.3   A simple family of wavelets – the Haar Wavelet.

Of course, the restriction of having *no memory at all* is a little unrealistic for many applications, indeed even the British weather is not that variable!

14.3.2 Higher-order Markov Model

Higher order Markov models allow one to peek slightly further back. Here's another sequence of letters, this time just As and Bs:

ABBAAAABBAAAAABBABBABAAABBAAA
BABAAAABBAAAAABBABABBABABABA

The transition probabilities that generated this are in Table 14.3. Note that this time the last two letters are used to generate the probability of the next letter. This is rather like using yesterday's weather as well as today's to predict the weather tomorrow, or if you are reading text, the words "it is" are likely to be followed by an adjective or gerund ('-ing' form of verb).

TABLE 14.3   Markov Model Transition Probability Based on Previous Two Letters.

| Last 2 letters | Next letter | |
	A	B
AA	0.75	0.25
AB	0.5	0.5
BA	0.25	0.75
BB	1	0

You can use the same methods to learn the two-step transitions, first of all counting the 3-grams:

Raw frequencies				Totals	
AAA:	11	AAB:	6	starting with AA:	17
ABA:	7	ABB:	8	starting with AB:	15
BAA:	6	BAB:	8	starting with BA:	14
BBA:	8	BBB:	0	starting with BB:	8

Then these are used to generate the estimate of the probabilities in Table 14.4. Note that again the estimates are in several cases well away from the actual values used to generate the sequence.

TABLE 14.4   Observed Transition Probabilities for Order 2 Markov Model.

| Last 2 letters | Next letter | |
	A	B
AA	0.647	0.353
AB	0.467	0.533
BA	0.429	0.571
BB	1.000	0.000

For language comprehension a two-step version like this is unlikely to be very useful, but you can grow the number of previous steps you take into account. Indeed, Google's text suggestions are based on models rather like this.

When you start growing the window, the number of entries in the transition table increases rapidly. In Table 14.3, we have an alphabet size (the number of values in each state) of two and a window size of two (in Markov model terms this is called the 'order' of the model). This led to eight (2^3) probabilities. In general, with an alphabet size of N and a window size of W, we have N^{W+1} probability entries in the table. Unless you are dealing with massive corpora, it becomes increasingly hard to have sufficient data to estimate the probabilities with any degree of accuracy.

Variable-order Markov models deal with this by having a window size that varies depending on the previous items. This is especially helpful if some items are a lot more frequent than others. In English text 'is' is a very common word, so one might have a simple table of one-step transitions for every word pair like in Table 14.1. However, in addition, there may additional probabilities for two-step sequences ending with 'is', for example.

$$\text{'cat is'} \rightarrow \begin{cases} \text{'on'} & p = 0.3 \\ \text{'purring'} & p = 0.2 \\ \text{'eating'} & p = 0.2 \\ \text{'sleeping'} & p = 0.3 \end{cases}$$

Sometimes there is some kind of classification of the alphabet; for example in language the part of speech of a word (verb, noun, etc.). In such cases it may be possible to build a Markov model at the abstract level of the classifications: for example how likely it is that a verb is followed by a noun, this can then be supplemented by more detailed rules for more common words. Note that these are all window-based variations of the Markov Model and so have finite impulse response; that is there is *no long-term memory*.

14.3.3 Hidden Markov Model

The hidden Markov model allows longer term memory by adding a hidden state. This is an additional state that is assumed to be present but not visible in the observed items.

As an example, Table 14.5 shows a simple model of weather in a British summer. The observable weather is either Sun or Rain, but in addition the weather has general tendency to be Changeable or Wet that affects the current weather. Try generating some weather sequences from this ... although do be warned they are likely to be depressingly wet!

Although it is just as easy to generate sequences from this kind of probability transition table, the fact that the Changeable/Wet state is hidden makes it harder to learn. As you don't know the state, you

TABLE 14.5 Hidden Markov Model for British Summer Weather – Hidden State in Brackets (Changeable/Wet).

Current weather	Tomorrows weather			
	Sun (Changeable)	Rain (Changeable)	Sun (Wet)	Rain (Wet)
Sun (Changeable)	0.3	0.4	0.1	0.2
Rain (Changeable)	0.3	0.3	0	0.4
Sun (Wet)	0.2	0.1	0.1	0.6
Rain (Wet)	0	0.1	0	0.9

cannot create transition frequencies for the observations. In general, you don't even know how many hidden states you need. There are various specialised algorithms for learning mostly based on iterative techniques.

Even when you know the model, applying it to observed data is also a little more complicated as you have to both work out the hidden state and use it to predict the future observations. However, it is often precisely this hidden state that is the most important thing. For example, from observations of a patient's day-to-day symptoms inferring whether or not their underlying ailment is improving.

14.4 GRAMMAR AND PATTERN-BASED APPROACHES

Instead of working out probabilities of various sequences, we may simply want to work out what is possible, the pattern or grammar underlying the sequence.

14.4.1 Regular Expressions

Look at the following (familiar) sequence:

```
AAACCACCCCABBABBACAAAAACCACACACAACACAB
```

Based on this alone, it looks like there are just As, Bs and Cs, and that while there can be various length runs of each, a C never follows a B, nor vice versa. Expressed as a regular expression (and assuming starting with an A), this is:

```
(A+(B+|C+))*A*
```

This says "one or more As, followed by either one or more Bs or one or more Cs; repeat that any number of times and then possibly some As at the end". Note, in a regular expression '+' after any item means 'one or more', '*' means zero or more, brackets group and vertical bar '|' designates alternatives.

In real applications, one would look at many such sequences, or for a single ongoing data stream a much longer sequence. It may be that in a longer sequence there are occasions with BC or CB, and also in the short sequence there are never more than 2 Bs in a row – is this part of the pattern or just accident?

Regular expressions are one of the simplest forms of grammar; yet it is still impossible to know for certain from a finite number of examples whether you have found exactly the regular expression that generated the examples. However, it is possible to infer relatively simple regular expressions that are consistent with available data.

The difficulty is finding a sensible point between a regular expression that is clearly too specific:

AAACCACCCCABBABBACAAAAACCACACACAACACAB

and one that is clearly too general:

(A|B|C)*

Every regular expression can be expressed as a finite state machine, rather like the hidden Markov model (HMM) but where there are no transition probabilities, merely possibilities: can it happen or not.

```
S0  =>  <A,S1>
            #  at least one A at start
S1  =>  <A,S1>  |  <B,S2>  |  <C,S3>
            #  anything
S2  =>  <A,S1>  |  <B,S2>
            #  B followed by B or A
S3  =>  <A,S1>  |  <C,S3>
            #  C followed by C or A
```

As with learning Markov models, one can start with the n-grams (Chap. 13), only here the question is about what is possible, so all one cares about is which sequences occur, and which never do. The aim is then to find the simplest possible state machine that can, from some state, generate every observed n-gram and never generate those that are absent. Although in many ways less complicated than a HMM, this is still not an easy task.

In some cases there are only positive examples. Non-presence is the only way to infer bad examples, but this becomes more difficult if the vocabulary grows as it becomes more likely that a particular sequence has never occurred by chance. In other applications there might be explicit negative examples, which can help a learning algorithm.

In text processing tasks there are often classes of symbols that are known to usually behave similarly such as letters and digits. Ideally the regular expression generated should use these when possible. For example, imagine you are given lots of names such as Jane, or Keith, but where none of them start with a 'Z'. Even though ([A-Y][a-z]+) might be a reasonable regular expression that matches every example, it would not match the name Zoe. The more general ([A-Z][a-z]+) should be preferred so long as there are no counter examples.

In some ways this bias towards 'sensible' regular expressions can make things more complicated, but it can easily be built into the fitness function of some ML algorithms. Also, it can help with the problems when there is insufficient learning data, as an example such as Jane is effectively treated as equivalent to all four-letter sequences of letters starting with a capital.

14.4.2 More Complex Grammars

Regular expressions are powerful and can match many kinds of data, such as email addresses or many kinds of identification codes. However, they cannot represent more complex linguistic structures such as nested classes, not even those found in programming languages or expression, such as matching brackets, which are often described with grammars such as this:

```
expr    ::=  number | expr '+' expr
             | expr '-' expr | '(' expr ')'
number ::=  digit | digit number
```

The extra expressive power of these hierarchical grammars makes them more powerful but correspondingly even harder to learn from examples.

One way is to operate bottom-up. A first pass looks for several relatively simple rules (such as regular expressions or simpler) where each rule matches multiple portions of the data sequence. A rule is deemed good if it is:

- reasonably simple

- matches reasonably long sequences

- matches many sub-sequences

- and doesn't overlap with other rules (or there is a precedence).

Once a good rule set has been developed, each rule is given a name, and portions of the data matching the rules are replaced with a single symbol based on the name. The process is then repeated on this abstracted sequence.

On its own this will generate multiple rules for each level:

```
expr4 ::=   expr3 | expr3 '+' expr3
            | expr3 '-' expr3
            | '(' expr3 ')'
expr3 ::=   expr2 | expr2 '+' expr2
            | expr2 '-' expr2
            | '(' expr2 ')'
expr2 ::=   expr1 | expr1 '+' expr1
            | expr1 '-' expr1
            | '(' expr1 ')'
expr1 ::=   numb | numb '+' numb
            | numb '-' numb
            | '(' numb ')'
numb  ::=   digit | digit number
```

So, some form of matching phase is then needed to generalise these rules. This would see that expr1, expr2, expr3 and expr4 all have similar expansions and merge them into a single non-terminal expr. Another approach is to use genetic programming. This is a form of genetic algorithm, but instead of simple 'gene' sequences, the individuals have program code, often represented as a tree. There are many variants of this. Typically, mutation may prune a whole subtree or take a node and expand it. Cross-over of two parent trees would take a subtree of one parent and use it to replace a subtree of the other. The fitness function will typically involve some form of simplicity as well as accuracy. One advantage of genetic approaches is that it is easy to incorporate known information such as the letter classes for regular expressions, or previous common patterns that can then be used as potential things to incorporate during mutation stages.

14.5 NEURAL NETWORKS

14.5.1 Window-based Methods

The simplest approach to using neural networks for temporal or sequential data is through windowing. The data is split into fixed or sliding windows and the resulting data used as input to a neural network.

If the aim is future predictions, then the output of the network needs to be the next item in the sequence. In other words, if the data sequence is $d_1, d_2, \ldots, d_t, \ldots, d_N$ and the window size is W, the sequence is treated as if it were the following N−W training examples:

input	output
$d_1, d_2, \ldots, d_W$	d_{W+1}
$d_2, d_3, \ldots, d_{W+1}$	d_{W+2}
...	
$d_{t-W}, d_{t-W+1}, \ldots, d_{t-1}$	d_t
...	
$d_{N-W}, d_{N-W+1}, \ldots, d_{N-1}$	d_N

For numeric data the input may be pre-processed using Fourier or wavelet transformation, but this may be omitted in deep neural networks depending on the structure of early layers. Classification for sequential data has some extra complications.

If the aim is to identify particular subsequences, for example instances of names in text, then we may use a binary classification labelling a window as "ends in a name" or not. Alternately we may simply have a label "contains a name". The former would be a more natural choice if you are using a sliding window as this will not only find parts of the sequence that contain names but pinpoint their exact location (see Figure 14.4).

If the classification is more diffuse, for example "poetic language" for text or "walking" for human activity data, there are likely to be portions of the data that are clearly labelled, and also portions that are in-between or perhaps bridging two kinds of classification (see Figure 14.5). Windows where the classification is less clear may either be omitted from the training set entirely or given a special 'transition' label. As we'll discuss later, these points of transition may be the most important aspect of a data stream.

These classifications can be given as part of supervised training, but classification algorithms may also learn in an unsupervised manner, looking for patterns in the data.

14.5.2 Recurrent Neural Networks

There are also forms of neural networks specifically designed for temporal data usually including some element of additional hidden state. These are collectively termed recurrent neural networks (RNN).

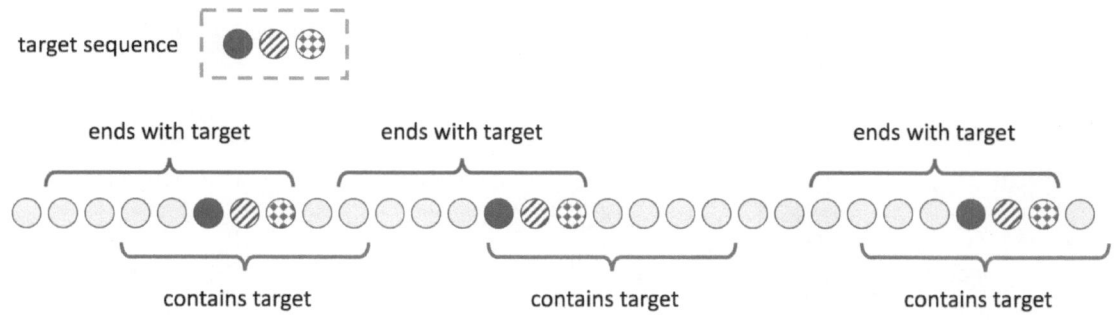

FIGURE 14.4   Locating sub-sequences in longer sequence.

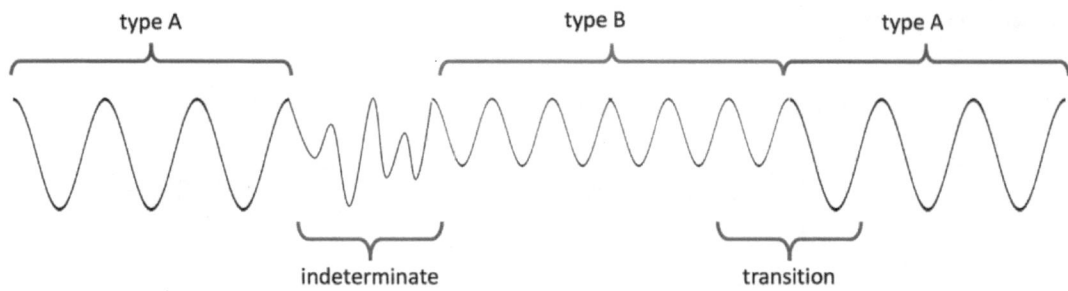

FIGURE 14.5   Diffuse classification of regions.

The simplest form of this simply takes the value of some output nodes at one time step and feeds these back into the input for the next time step. On the left of Figure 14.6 is a standard neural network, with inputs that feed into some sort of network of nodes and produce outputs. On the right is the recurrent neural network. The same number of extra input and output nodes are added and then the extra output nodes are fed back into the inputs. These extra nodes effectively form a state for the network.

Figure 14.7 illustrates this process once the RNN has been trained. We'll assume a starting state s_0. The data items for each time step are read from the input stream one at a time. The first item d_1 along with the initial state s_0 is fed as input to the network. This yields an output for this step o_1 and also a value for the 'state' output nodes s_1. The process is then repeated with d_2 and s_1 giving o_2 and s_2 and the process continues.

The state output at each time step becomes the state input for the next.

As with hidden Markov models, adding this state makes learning a lot more difficult as it is not known up front. The state can be initialised randomly or the net can be bootstrapped by starting with a windowed neural network and then gradually replacing the oldest data items with state nodes. Learning can be performed one step at a time, or by looking at several steps simultaneously, effectively pushing error terms backwards through time.

Note that as this is a state-based method it has infinite impulse response. In principle the effects of data items can be recalled indefinitely.

14.5.3 Long-term Short-term Memory Networks

Having some level of history in a neural network can significantly improve performance on time series. However, there are times when it is useful to forget and start again, for example, when there is a change in the stock market from bull to bear or some major event (see Figure 14.1).

Long-term short-term memory networks (LTSM) get round this problem by having special memory nodes that as well as an input and output also have a 'forget' input that resets the node. The neural analogy for this is long-term potentiation, chemical changes in cells that last from minutes to hours. They are longer lasting than the moment-to-moment flux of electrical activity but less

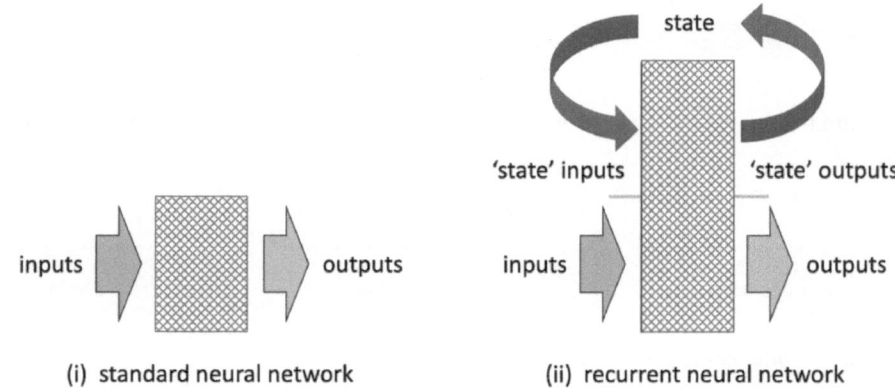

FIGURE 14.6 Recurrent neural network (RNN).

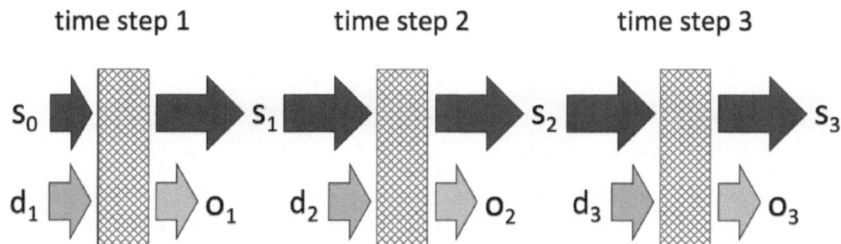

FIGURE 14.7 Running a recurrent neural network on data stream $d_1, d_2, d_3, \ldots$

so than true long-term memory encoded physically in synapse weights.

14.5.4 Transformer Models

Recall the LEGO-style matching for language described in Chapter 13; as a human reading text we are partly influenced by the last few words, but also parts of the text from much further back if they connect in meaning, like matching LEGO blocks, to the thing we are currently reading. For example, if we read the words "ate the bone", the immediately preceding text may be a very long subordinate clause, but our attention skips the clause and connects to the words "the dog" even though they are far back in the sentence.

Attention mechanisms try to emulate this by matching tokens to previous tokens, often at lower levels in a layer architecture. Each token has a 'key' and 'query' attached to it, where the key captures some aspect of the meaning of the token and the query the kind of things it wants to attach to. Both keys and queries are represented as vectors so the matching is imprecise.

Attention mechanisms can be combined with other sequence learning algorithms such as recurrent neural networks. However, transformer models, built almost solely using these attentional mechanisms, have proved both powerful and efficient [287] on language translation tasks. While originally developed for language processing, variants of transformer models have been used in other domains where there is sequential structure including predicting folding structures of proteins in AlphaFold2 [30].

14.6 STATISTICAL AND NUMERICAL TECHNIQUES

There are a wide range of highly successful statistical and numerical techniques used in time series analysis and signal processing. Often these are alternatives to AI and ML techniques and may be better for pure numerical data. However, in addition they may be used alongside AI-based techniques, for example as pre-processing stages as we've already discussed with Fourier and Wavelet transforms.

14.6.1 Simple Data Cleaning Techniques

Often raw time series data looks very noisy, sometimes due to random effects, sometimes more systematic ones. Figure 14.8 shows daily reported Covid-19 in the UK during 2020. Often weekend deaths did not get officially reported until sometime in the following week, leading to very noisy data. However, the solid line is a seven-day moving average. This is simply taking the last seven days' raw data and averaging that. The trends and patterns are immediately far more evident. It is the simplest example of a smoothing function.

Other examples of simple transformations are seasonal adjustment and trend removal.

Seasonal adjustments are applied when there is some form of fixed period often over a year, or week, that affects the data. For example, differing business district coffee sales between weekdays and weekends or woolly hat sales between summer and winter. We'll think of this in terms of monthly differences over the year, but a similar technique can be applied for any period.

To adjust for these one can take long-term averages over many years comparing each month with the overall year average. This month effect can then be subtracted from the observed data to remove the seasonal effect. It is then easier to see the impact of other changes, for example, if there is a September sales promotion on woolly hats, are increases in sales due to the promotion or just what you'd expect at that time of year?

Trend adjustments or de-trending are used when there is a long-term upwards or downward trend in the data. The aim is to in some way separate out the overall trend from finer-grain changes. One way to deal with this is to fit the line using linear regression and then remove this to leave the fluctuations from the trend. This works well in some circumstances but can suffer from sensitivity to outliers and effectively assumes the same trend lasts forever. Although, by definition, a trend is a long-term phenomenon, that may not mean forever. There are ways to update trend estimates incrementally and also to reduce the impact of outliers.

Often a simpler way to de-trend a time series is to focus instead on the difference between successive data items (first-order difference), how much it has increased or decreased. That is, you turn the data series $d_1, d_2, \ldots, d_t, \ldots$ into $d_2 - d_1, d_3 - d_2, \ldots, d_{t+1} - d_t, \ldots$. The linearly growing part is effectively cancelled out.

Sometimes this difference data still has a trend, for example when a car is accelerating, the velocity (change in position) is itself increasing. The differencing process can be repeated to yield second- or third-order differences, and many series that start off looking complex can be tamed this way. Indeed, this is precisely the principle behind Charles Babbage's Difference Engine!

14.6.2 Logarithmic Transformations and Exponential Growth

Many processes exhibit exponential growth or exponential decay, for example infections during the early stages of an epidemic, economic growth or feedback in a sound system. For these processes taking any order of differences still yields an exponential process. This is because they are fundamentally multiplicative rather than additive processes. The rate of growth is proportional to the level of disease, size of the economy or volume of noise.

It is possible to explicitly model the exponential part and remove it, similar to de-trending or seasonal adjustment. However, an easier approach and often more suited to pre-processing for other ML techniques is to perform a logarithmic transformation of the data. That is turn the series $d_1, d_2, \ldots, d_t, \ldots$ into $log(d_1), log(d_3), \ldots, log(d_t), \ldots$

The choice of base of the logarithm (common values $2, e, 10$) does not really matter as the logarithms are all just multiples of one another, and this scale factor is usually 'dealt with' by later stages of processing. In some cases a type of data suggests the most meaningful transformation: base 2 for computer-related data such as storage capacity, natural logarithm (base e) for biological processes and base 10 for acoustic data where the decibel (ten times log_{10}) is a common measure.

After a logarithmic transformation, exponential growth or decay becomes a simple linear upward or downward trend and so additional de-trending (and possibly seasonality adjustments) can be applied. The first-order difference of a logarithmic transformation is effectively proportional to a percentage increase figure, like a inflation figure.

Even where there is no clear exponential trend, it can often be the case that the variability in data is proportional to the size. For example, fluctuations in an adult's weight of a few pounds over holiday periods are quite common, but if a new-born baby were to vary by a similar

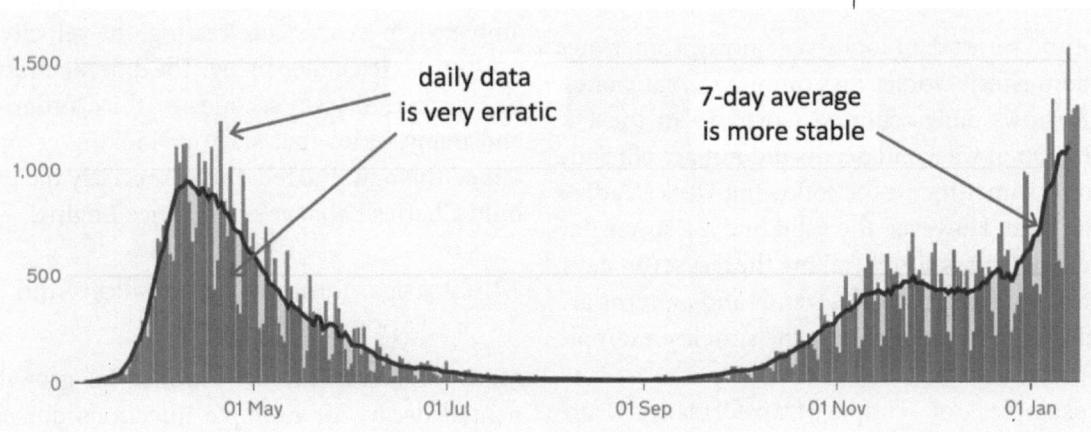

FIGURE 14.8   Covid-19 deaths for UK in 2020 – daily data is very noisy, but a 7-day moving average is far more stable (adapted from: https://coronavirus.data.gov.uk/details/deaths).

amount, one might be worried. This might also suggest a logarithmic transformation.

In both cases of exponential processes and scale-related variability, an alternative to a full logarithmic transformation is to actually work out a proportionate increase/decrease at each step, that is transform the data series $d_1, d_2, ..., d_t, ...$ into $d_2/d_1 - 1, d_3/d_2 - 1, ..., d_{t+1}/d_t - 1, ...$ If the changes are relatively small, this is effectively equivalent to the first-order differences of logarithmic data but may be easier to interpret.

14.6.3 ARMA Models

One of the most common types of standard statistical modelling is auto-regressive moving average (ARMA) models. There are lots of variations, but the basic assumption is that given an observed data series, the current value is determined by a combination of

- recent past states of the observed series. For example, your bank balance today is related to your bank balance yesterday.

- an unobserved series of random fluctuations. For example, all of those midnight internet purchases.

Following convention, we'll call the unobserved (hidden) series $\epsilon_1, \epsilon_2, ..., \epsilon_t, ...$ and these are assumed to be independent of each other and typically Normally distributed (Chap. 7).

This leads to two types of model, which can be combined to give the full ARMA model.

Moving average models assume that the data we see is a weighted average of the last M items of the unobserved series. That is:

$$d_t = w_1\epsilon_t + w_2\epsilon_{t-1} + ..., +w_M\epsilon_{t-(M-1)}$$

Because this is a finite window, the impact of a single unusually large or small ϵ_t only persists for M time steps, that is the process generated by this has finite impulse response. The types of data produced by this kind of model therefore have local structure but no longer term behaviour.

Auto-regressive models assume that the data we see is a linear combination of the previous N items plus a single error/noise term from the unobserved series. That is:

$$d_t = a_1d_{t-1} + a_2d_{t-2} + ..., +a_Nd_{t-N} + \epsilon_t$$

In this case although the term ϵ_t appears to only influence the current d_t, this in turn is fed back into d_t, which in turn influences d_{t+1}. The impact tends to decay exponentially, but has no fixed end, it is an infinite impulse response process. Auto-regressive models often create processes with larger scale pattern-like properties including quasi-periodic data.

Note that in both cases these are models of how the data is produced rather than how we process it to create predictions. In particular, the moving average model is a moving average of the unobserved series, so it is different from calculating a moving average of observed data in order to smooth it.

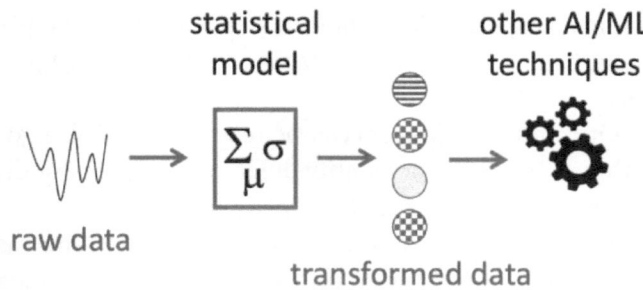

FIGURE 14.9 Using statistical processing as pre-processor for other AI techniques.

Auto-regressive models are easy to fit using a windowing technique, you simply take a training series, split it into windows of size N+1 and then apply standard linear modelling to find the coefficients a_i. Although they seem simpler, moving average models are a little more complicated as they involve the unseen series ϵ_t. However, there are ways to calculate these too and also mixed ARMA models.

Two final things to note. First the values for N and M in an ARMA model are often relatively small compared to the windows we are likely to use for many neural network or other machine-learning algorithms. Second these models are often fitted after seasonality adjustments and trend removal.

14.6.4 Mixed Statistics/ML Models

As noted a common use of statistical methods in AI is as pre-processing for other ML algorithms (Figure 14.9). There are three reasons for this, one, two or all of which may apply to a given application:

clearer – They are often easier to interpret than a series of weights in a neural network. By restricting AI-based ML techniques to aspects where they are most needed, we can end up with more explainable AI (see also Chapter 21).

better – If the transformations applied are based on real understanding of the data, they effectively feed knowledge into the ML process leading to more accurate and generalisable results.

faster – Sufficiently complex deep neural networks may well be able to learn for themselves the equivalent of the transformations produced by statistical techniques. However, appropriate pre-processing

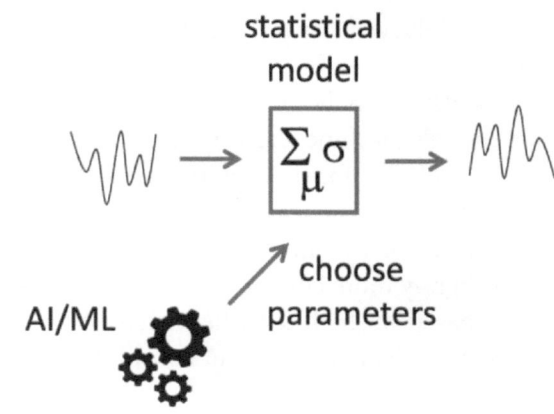

FIGURE 14.10 Using ML to choose parameters for statistical processing.

can reduce the complexity and hence learning time.

There are also downsides, they need more expert knowledge up front and also may run the risk of biasing the ML towards standard ways of viewing the data. The latter is mitigated if the transformations applied are information preserving; for example, if you sum first-order differences, you retrieve the original data series.

Another way to combine ML with statistical techniques is where the data is intrinsically well suited to traditional statistics, but where some form of choices are needed that usually require a statistical expert's judgement, or some form of trial and error. For example, in ARMA models the choice of N and M is critical in creating a good model but is a combination of experience, art and pure guesswork.

In these cases, an AI system can be used to make those choices, effectively functioning as a surrogate expert (Figure 14.10). This may be an expert system based on heuristics and rules of thumb derived from professional statisticians or can use machine learning tuned to a particular application area. This might include changing these parameters or switching models for different portions of the data (see below).

14.7 MULTI-STAGE/MULTI-SCALE

We've noted that data may have structure at different scales. This is true of numerical data, for example, the bull/bear market phases in Figure 14.1 and the heart beats in Figure 14.11, but also for textual data, for example the words in this sentence, the topic of the paragraph and the way these fit into the overall structure of the book.

It is possible to simply throw this kind of data into a complex enough machine learning algorithm. Some are designed to seek this form of multi-scale structure, but if not they may either struggle to work at all or be needlessly large with correspondingly long learning times.

Often a slightly more curated approach is adopted, or automated based on high-level heuristics. Let's look at the ECG data in Figure 14.11 and consider one way to process it.

1. We write custom code to identify the peaks in the signals

2. The time difference between this and the previous peak gives an inter-beat time for each

3. A moving average of the inter-beat is calculated to give a heart rate

4. Fixed-sized windows are extracted around each peak

5. These windows are treated as independent data items for training an unsupervised classifier

6. The classifier is then applied to every beat yielding a beat 'type'

7. The time difference, 'beat' and heart rate then form a new data sequence where each data item represents a single labelled heartbeat in order

8. This new time series is fed into one of the other techniques we've dealt with (e.g. HMM, RNN)

Note a few things about this process:

- It combines some expert knowledge such as the way the peak identifies a beat, the use of a moving average as peak-to-peak times can be quite variable, being affected by breathing or movement.

- It also allows the AI/ML to seek its own patterns at steps 5 and 8.

- It has turned fine-grained (kHz) samples of continuous time data into coarser (~1Hz) sequential discrete events which are more suitable input for many kinds of network.

Think of your own variations on this: for example, could an expert be brought in at step 5 to label the types identified or maybe some database of unusual heart patterns used?

Of course this process can be repeated, the ML algorithm at step 8 might identify different kinds of periods, some representing normal activity of different kinds (relaxed, strenuous exercise), but some abnormal, such as periods of arrhythmia. Applying this to time series data could reduce it further into even coarser sequential data such as in Table 14.6.

TABLE 14.6 High-level Classification of Periods of Data as a Coarse-grained Time Series.

D-723	high-rate, normal pattern (strenuous activity)	–	15 mins
D-724	low-heart rate (rest)	–	30 mins
D-725	tachycardia (problem)	–	2 mins
...			

Another layer of ML could then identify patterns in this, for example whether there are particular combinations of activity that are more or less likely to lead to arrhythmia.

This is an example of a bottom-up process with low-level data being used to create higher-level abstractions. There can also be top-down processes.

Imagine we have trained a stock market predictor A that works well during bull markets and a predictor B that works well on bear markets. We can use these predictors as (retrospective) classifiers by keeping track of how well each would predict the current market state based on previous days. If over a period A is better, we are probably in a bull market, if B is better, we are likely in a bear market. So far this is a bottom-up process, but it has given us a classification of the current market condition, which we can then use to choose which of the predictors A or B to use to estimate the next day's stock prices and so inform investment decisions.

Similar bottom-up and top-down approaches can be used in other sequential data such as text processing. Here's an example of such a process:

1. The origin of a document (e.g. web page domain) gives a default language and location

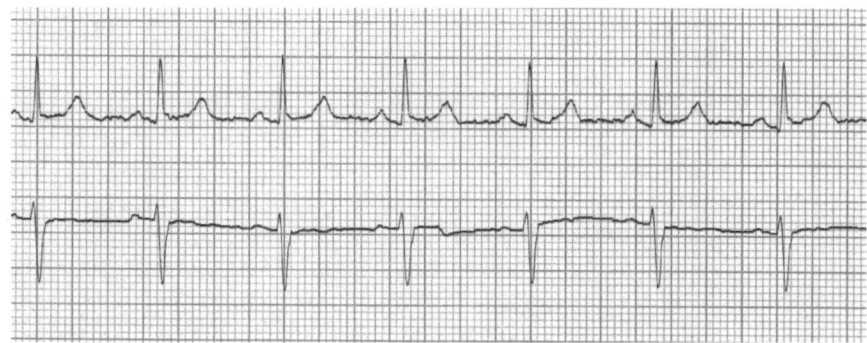

FIGURE 14.11   Extract of ECG trace (adapted from Ptrump16 – Own work, CC BY-SA 4.0, https://commons.wikimedia.org/w/index.php?curid=77817932).

2. This is used to select a dictionary, date and number formats, and maybe specialised lookups such as common names (top-down processing)

3. Text is split into 'word' units at spaces and punctuation

4. Units that are all alphabetic (probable words) are looked up in a dictionary

5. Units are matched against text patterns such as 'all letters', 'initial capitals' and also patterns such as dates or telephone numbers.

6. The resulting units are then coded: in/not in dictionary; all lower case/all caps/initial caps/mixed; etc. (coarser scale data series)

7. Higher level pattern recognisers (hand-coded or automatic) are applied, for example, initial caps words not in a dictionary suggest names, lists of names with dates and other words with initial capitals might be an article citation. (bottom-up processing)

Again, you might think of variants of this, for example using high-level items recognised at step 7 to feed back into revised context.

14.8 SUMMARY

We have looked at methods of analysing and predicting temporal and sequential data based on probability theory, in particular variants of Markov models; various forms of grammar; the use of generic and specialised neural networks, including recurrent and long-term short-term networks; and also statistical methods for more numeric data. Some of these are very specific, but there are also common aspects such as the use of windowing or hidden state. Crucially these methods are often combined, including different methods and different levels of abstraction in an event stream.

14.1 Generate weather sequences based on a hidden Markov model of the British weather as suggested in Section 14.3.3. If you are using a six-sided die, you can use the transition probabilities in Table 14.7, which has been modified from Table 14.5 so that the probabilities are multiples of 1/6 to make it easier.

TABLE 14.7   Weather Transition Probabilities for Use in Exercise 14.1

Current weather	Tomorrow's weather			
	Sun (Changeable)	Rain (Changeable)	Sun (Wet)	Rain (Wet)
Sun (Changeable)	1/3	1/3	1/6	1/6
Rain (Changeable)	1/3	1/3	0	1/3
Sun (Wet)	1/6	1/6	1/6	1/2
Rain (Wet)	0	1/6	0	5/6

14.2 Given the following sequences

(i)   AAACCA

(ii)   AAACCAB

(iii) AAACCBA

(iv)   AAACCACCCCABBABBBACAAAAA

        CCACACCACAAACCACCAB

(v) AAACCACCCCABBABBBACAAAAC

CACACCACAAACCACCAB

(vi) AAACCACCCCABBABBBACCAAAA

ACCACACCACAAACCACCAB

Which of the following regular expressions match them all?

a. (A+(B+|C+))*A*
b. (A+(C+(B+|A+))*B
c. (A+(B+|C+))*A*
d. (A+B+C+))*B
e. ((AA)*A+(B+|C+))*
f. ((AA)*A+(B+|(CC)+))*

Don't just use an online regular expression matcher! Try to understand *why*.

14.3 For this exercise use the sequence (iv) of As, Bs and Cs from Exercise 14.2.

a. Calculate the Markov model current-letter–next-letter probabilities similar to Table 14.1 using the technique described in Section 14.3.1.

b. Using an online random number app, spreadsheet or code, generate a sequence based on the measured probabilities.

c. Recalculate Markov model probabilities based on your generated sequence. How similar is this to step (a)?

14.4 As in the previous exercise, use the sequence (iv) of As, Bs and Cs from Exercise 14.2.

a. This time calculate the higher-order Markov model with a window of size two. That is previous two letters to next letter probabilities as in Table 14.3.2 in Section 14.4.

b. Were there any problems calculating this?

c. How confident would you be in the probabilities?

d. Repeat this using your generated sequence from Exercise 14.3, step (b).

e. Do things look different, and if so why?

FURTHER READING

C. Chatfield and H. Xing. *The analysis of time series: An introduction with R.* CRC Press, New York, NY, 2019.

In-depth coverage of statistical methods in time series analysis such as ARMA.

B. Lim and S. Zohren. Time-series forecasting with deep learning: A survey. *Philosophical Transactions of the Royal Society A*, 379(2194):20200209, 2021.

A short overview of state of the art of more neural network–related approaches to time series data.

A. Vaswani, N. Shazeer, et al. Attention is all you need. *arXiv 1706.03762*, 2017. DOI:10.48550/ARXIV.1706.03762

The paper that introduces the transformer model.

Planning and Robotics

15.1 OVERVIEW

In order to act in the world, we need to plan what to do. The same is true for computers and robots. Planning has long been an important part of artificial intelligence, and this chapter initially looks at two main aspects: planning actions and planning movements. Planning usually involves manipulating a model of the world in order to decide what actions will bring about the desired effects. However, in the real world we cannot model all the outcomes of our actions, either because the world is too complex, or because of external events over which we have no control, or both. We have already seen aspects of this when discussing games. Local planning deals with those situations where we can only plan so far ahead but must then respond to the circumstances we observe and the events that occur. In the next chapter we will look at software agents that act in the electronic world where planning is also necessary. However, designing robots that act in the physical world means we have to live within the limitations of reality. We will discuss some of the implications of this for cybernetics research and industrial robotics.

15.2 INTRODUCTION

15.2.1 Friend or Foe?

Robots have intrigued people since before the word existed. Plans were produced for clockwork and steam-powered humanoids, while moving manikins and automata adorn both fairgrounds and cathedral clock towers. The word "robot" means worker (or even 'serf' or forced labourer) and indeed they have become a major part of modern factory production.

However, popular images of robots are not so prosaic. Humanoid robots hold a particular fascination, with the promise of tireless service and even, like Data in *Star Trek* or R2D2 in *Star Wars*, friendship. However, there is a dark side as well, and in science fiction robots are often the mortal enemy of humankind (with the added *frisson* of not being mortal!). It is interesting to note that the most dreaded enemies have been those that are only partly robot: Frankenstein's creation in Mary Shelley's novel was constructed from dead flesh and the Daleks have something slimy within. In Karel Čapek's play "R.U.R.", in which the term 'robot' was coined, the robots were android slaves, and ended up rebelling against their human overlords [40]. Strangely enough, artificial life (albeit mostly virtual) has become a respectable area of AI!

For the foreseeable future there is little danger from independently malevolent robots; although much research in robotics has military funding, and semi-autonomous drones and battlefield robots are increasingly common. Outside a human war zone, for most of us, accidents and misadventure are a far more likely danger. Isaac Asimov foresaw this with his Laws of Robotics, setting limits on robots' freedom to act [10]. Although real-life robots do not wield ray guns, they often have lasers, and an encounter with a ton of industrial robot, whether "armed" or not, could be unfortunate. In fact, it is likely that the less intelligent the robot, the greater the danger – it will not be able to tell the difference between drilling a hole in an engine block or in your head!

DOI: 10.1201/9781003082880-18

15.2.2 Different Kinds of Robots

The simplest industrial robots obey a pre-programmed sequence of commands. They have no intelligence whatsoever – although AI might be used in planning their movements. An example of this is spray painting of cars. An operator initially trains the robot by moving an instrumented robot arm to perform the task. The movements are recorded and then the production line robot repeats the movements indefinitely – rote learning. The lack of intelligence becomes obvious when there is any change in the circumstances. If there is a gap in the production line, the robot will happily spray thin air! Although such robots do not include any intelligence, they are very important in industrial applications.

A slightly more complex example would be a drilling machine. The machine needs to detect when a drill bit breaks in order to report the damage to a human operator (as there may be a part-drilled hole or a piece of drill bit left on the work piece) and load a new bit. This behaviour is pre-programmed but may involve some planning – perhaps using a different drilling machine when one goes offline.

Finally, we get to robots where the need for AI is obvious. These may be stationary: for example, on a production line where parts come in different orientations (vision needed), perhaps piled on top of one another, and the robot needs to select parts to assemble. Alternatively, they may need to move around in their environment: for example, an automated forklift moving things around a factory, a smart vacuum cleaner in the home, or an autonomous car on the road.

In the first edition, the reviewers thought the authors at very best quaint for suggesting that readers might be inside a robot (referring to a lift) before they met one. However, now while lifts still have the autonomous quality of a moving robot (albeit restricted), examples of robots that surround us are far more common: some moving, autonomous road vehicles as well as lifts and auto-piloted planes; or stationary, smart buildings and cities. On the latter HAL, the AI in *2001 Space Odyssey*, was controlling the whole space ship or perhaps was the whole space ship. At present while you talk to a home automation system such as Alexa, and ask it to do things for you, it is normally seen as an actor external to the devices themselves, so more a software agent as will be discussed in Chapter 16. However, as more parts of your home are automated and sensed, and the relationships

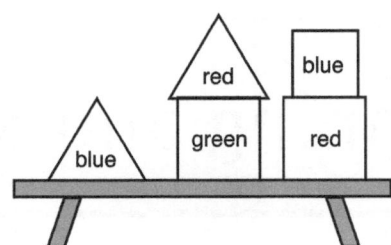

FIGURE 15.1   Blocks world.

between them become more complicated, at what point does it feel more as if the whole house is an intelligent entity?

15.3 GLOBAL PLANNING

15.3.1 Planning Actions – Means–Ends Analysis

When we have considered state space search with moves between states, we have simply assumed that there is some oracle that gives us the set of possible moves from a given state. In fact, many problems are far more structured than that.

One general class of problems can be attacked by a technique called means–ends analysis. This is based on operators that transform the state of the world. Given a description of the desired state of the world (the end) it works backwards working out operators that will achieve it (the means). This is not done as a single step, but instead works incrementally: in order to apply operators which would achieve the goal state, conditions must apply to the previous state and so the algorithm is applied recursively. Note that this is a special sort of knowledge-rich search as discussed in Chapter 4.

States are described in some structured way (e.g. by predicates), and moves are performed by the operators. Each operator has a precondition, which constrains the states it can be used in, and a postcondition, which says what will be true when it has finished. In a state described by predicates the postcondition must say both what is additionally true and what ceases to be true in the new state.

As an example, consider a blocks world similar to that used as the domain of the historic natural language AI program SHRDLU [301]. This world consists of blocks of different shapes and colours, which can be piled on top of one another or placed on a table top. An imaginary robot inhabits this world and can pick up and move blocks to try and get to any desired state. The states can be displayed graphically (Figure 15.1) or described using predicates. Our world has two kinds of shapes, pyramids and boxes, in various colours.

on_table(blue_pyramid)
on_top(red_pyramid,green_box)
on_top(blue_box,red_box)
on_table(green_box)
on_table(red_box)

The predicate "on_top(A,B)" says that block A is on top of block B and "on_table(A)" is self-explanatory. We also require another predicate "in_hand(A)" which says that A is in the robot's (single) hand. There are four operators with the pre- and postconditions shown in Table 15.1.

TABLE 15.1   Blocks World Operations.

operation	precondition	postcondition
pick_up(A)	on_table(A)	in_hand(A)
	$\land \neg$ on_top(C,A)	$\land \neg$ on_table(A)
	$\land \neg$ in_hand(X)	
put_down(A)	in_hand(A)	on_table(A)
		$\land \neg$ in_hand(A)
pick_off(A,B)	on_top(A,B)	in_hand(A)
	$\land \neg$ on_top(C,A)	$\land \neg$ on_top(A,B)
	$\land \neg$ in_hand(X)	
put_on(A,B)	in_hand(A)	on_top(A,B)
	$\land \neg$ on_top(C,B)	$\land \neg$ in_hand(A)

As an example, we can read the first rule as saying:

> In order to pick up the block A, it must be on the table, must have nothing on top of it and there must be nothing in the robot's hand.

> When it has been picked up, the block A is in the robot's hand and no longer on the table.

Notice how the first operator (pick_up(A)) makes some things true that weren't before (in_hand(A)), and some things false that were previously true (on_table(A)). These are called the add list and the delete list, respectively.

In order to simplify the rules, there are two operators to pick things up, one to pick up things from the table "pick_up" and one from other blocks "pick_off".

Now imagine that our goal is to have a pile on the table consisting of the blue triangle on top of the red box. We don't care about any of the other blocks:

on_top(blue_triangle,red_box) $\land$ on_table(red_box)

Now the operator information could be used to do a simple depth or breadth first search for this goal state. From the start state we could generate all operators whose precondition was true of the current state, and then search the children in the manner determined by the search. However, this does not effectively use the structural knowledge in this representation. For example, we would examine useless moves like moving the red triangle off the green box.

Means–ends analysis does use this knowledge. It looks at the current state and the goal state and works out the difference between them – not just a numeric measure of distance as used in heuristic search, but an analysis of which things need to be changed. Consider the current state. We see that "on_table(red_box)" is already true, so the difference from the goal state is

on_top(blue_triangle,red_box)

We can then match this difference against the postconditions and look for an operator that reduces the difference. In the example this can be achieved by "put_on(blue_triangle,red_box)". We check its preconditions against the current state. Unfortunately, they are not met. So we make these preconditions a new goal state, calculate the difference and look for a new operator.

This movement from the goal state towards the current state is called backward chaining. In this example, it is more efficient than moving from the current state to the goal (forward chaining), as the forward branching factor is much larger than the backward branching factor.

If we look at the next stage in this means–ends analysis, we find there are two terms in the new goal state:

in_hand(blue_triangle) $\land \neg$ on_top(C,red_box)

We can either work on both simultaneously or instead work out a way to get to each part separately. For example, we could first work out a way to achieve "in_hand(blue_triangle)":

pick_up(blue_triangle)

and then seek to achieve "$\neg$ on_top(C,red_box)" by

pick_off(blue_box,red_box)

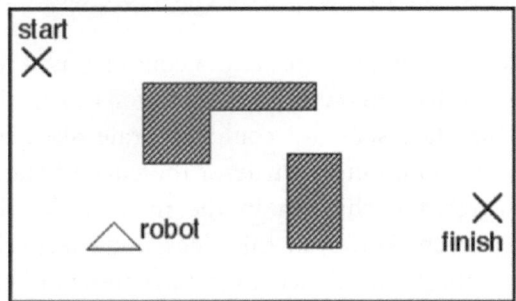

FIGURE 15.2 Navigation challenge for a robot.

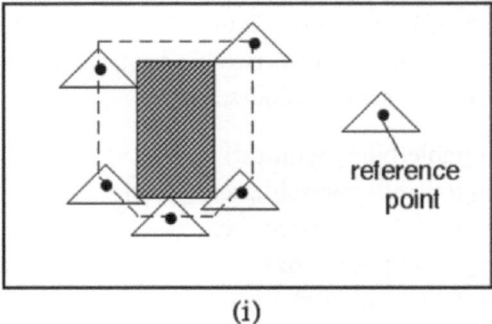

(i)

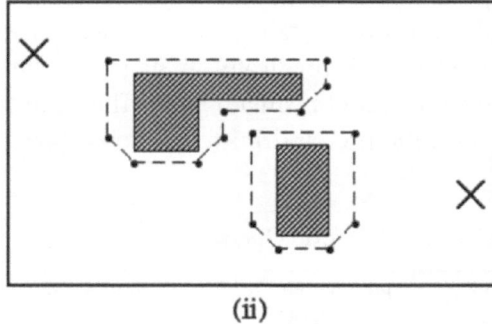

(ii)

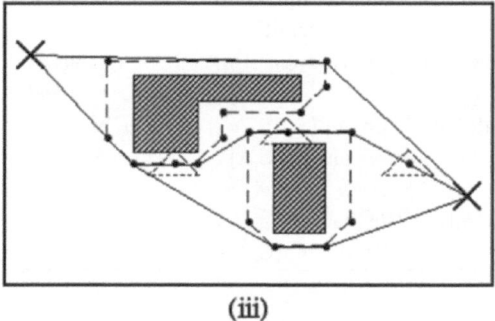

(iii)

FIGURE 15.3 Using a configuration space to plan a route.

Unfortunately in this case we cannot simply combine these plans, as they interfere with one another. This will not always be the case, and splitting up a problem into subproblems (divide and conquer) is a powerful solution technique. Even where interference is found, it is often more efficient to produce two interfering subplans and then modify them than to work on the whole problem at once.

The process of finding a single sequence of operators that follow one after the other is called linear planning. In contrast, non-linear planning builds a partially ordered collection of actions. The actions are each application of operators, and dependencies are recorded between actions. This reduces the amount of backtracking required while searching for a plan. However, in the end even non-linear plans must be reduced to a linear sequence of operators to be performed. This is done by finding a linear ordering that is consistent with the dependencies in the plan.

15.3.2 Planning Routes – Configuration Spaces

Suppose the little triangular robot in Figure 15.2 wants to get across the room from the place marked start to the one marked finish. In the room are two obstacles. A straight-line path between the two points will not work – the robot will collide with the obstacles. One cannot simply find a line that avoids the obstacles because the robot is wide and may not be able to squeeze through every gap. When we plan a path through the obstacles, we must take into account the size and shape of the robot at each point. This makes the planning task quite difficult.

One way to tackle this is using a configuration space. Recall how in Chapter 2 we saw how a change in representation can make a hard problem easier. The configuration space is just such a change of representation. Each object is expanded so that we can regard the robot as a single point and then we can find a simple path across the room.

Figure 15.3 shows the stages of route planning using a configuration space. First, a reference point is chosen on the robot (i). We then imagine moving the robot around each object tracing the path of the reference point. This is shown for a single object in (i). We then regard these paths as being the boundaries of the expanded objects (ii) and plan a path past these. The shortest path must graze past some of the objects and hence must pass through a series of vertices of the expanded objects. Three such routes are shown in (iii), and some search algorithm, such as A* (Chap. 4), can be used to select the shortest route.

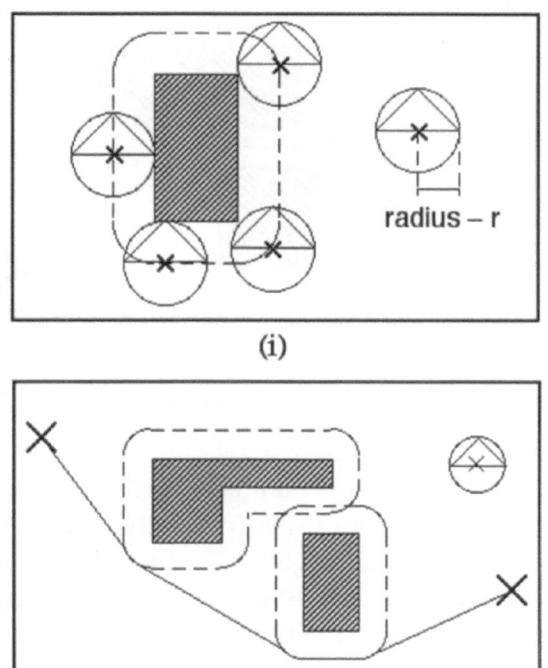

FIGURE 15.4   Circle-based configuration space.

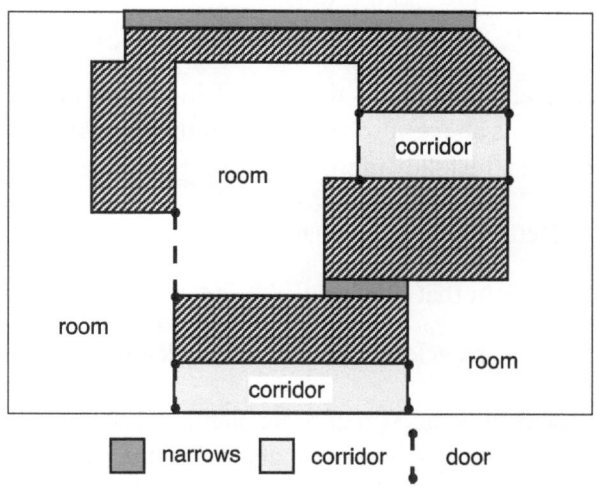

FIGURE 15.5   Corridors and rooms between obstacles.

The above algorithm depends on the robot maintaining its orientation. This can lead to it being both optimistic and pessimistic in its chosen paths. We all know that twisting an object round can make it easier to get through an opening. However, some robots may only be able to move forward. The configuration space solution might include the robot moving sideways, crab-like, and so be impossible. The first problem, finding out whether some combination of movements and rotations can get an object past obstacles, is very difficult. However, we can tackle the second by modifying the configuration space.

In Figure 15.4, we see a configuration space based on a circumscribed circle drawn around the robot. The centre of the circle has been chosen so as to make it as small as possible. If the robot can only turn about a particular point, then this should be chosen instead. The circle is then used to generate the expanded obstacles and new paths can be chosen round these. One of these paths is drawn in (ii). Notice how the path between the two objects is not possible in this configuration space. This is because the gap was wide enough for the triangle to navigate sideways, but not point first. The circle-based space is conservative. If you find a path through it, you can def-

initely do it, but it may disallow some paths that are possible. For example, if the robot were long and narrow, like a truck, it would say that corridors (roads) need to be as wide as the length of the truck!

One way to get round this is to examine the paths based on circles and those based on unexpanded objects. The former is very pessimistic, the latter optimistic. If a promising path exists in the latter, but not the former, one can use more sophisticated methods to check whether the route is possible or not given the particular movements available to the robot.

A similar approach is to generate possible paths based on the narrowest points between obstacles; that is, to concentrate on the gaps rather than the obstructions. Some of these gaps will be so small that they can't possibly be navigated: they can effectively be "filled in" and ignored. Other gaps will be narrow enough that care is needed (say, narrower than the diameter of the circumscribed circle). Finally, some gaps will be so large that they can be considered as rooms – large enough that free movement is possible. The meetings between these gaps can be thought of as doors between the corridors and rooms. Possible routes criss-cross the rooms going from door to door. One can work out which are navigable and which turns are possible and then use a search algorithm to choose a route. This has the advantage that paths can be constructed to run as far as possible from obstacles and so avoid near misses. It also means that we can use different heuristics for navigating across rooms and down corridors (Figure 15.5).

15.4 LOCAL PLANNING

15.4.1 Local Planning and Obstacle Avoidance

Get up and walk across the room. Did you use one of the algorithms above? Probably not! Your behaviour will be more like the following:

- Determine approximate direction to take.

- Walk in that direction.

- If an obstacle is encountered, walk around it.

Notice how this sort of route planning has two phases: global planning – when you determine the approximate route – and local planning – overcoming obstacles as they occur. The same sort of thing happens at other levels. If you are planning a mountain walk, you may plan the route using maps and guides, but you still have to watch where you are going.

In real life, one does not preplan all one's activities. Instead there is a hierarchy of plans at different levels ranging from overall goals of life to automatic reactions. The global level of planning has to know what is reasonable to expect at the local level but does not have to plan the low-level details. However, there has to be some sort of monitoring to revise global plans should problems occur at the lower level. For example, when hill walking you might find that a path has been washed away and have to replan your route to avoid the unforeseen obstacle.

Planning at multiple levels has computational advantages (several small problems rather than one large one) and is also far more flexible, especially if the environment changes. It is not only useful in route finding but also in other problems such as assembly tasks.

One way to handle local planning is to give a robot a desired direction of travel and a set of avoidance rules for obstacles. For example, a rule could be:

1. where possible move towards target
2. if you encounter an obstacle:
   2.1. move back 1 unit
   2.2. move sideways for 5 units
   2.3. resume preferred direction

By "back" and "sideways" we mean that the robot determines (with sensors) in which direction the obstructing object lies and moves first directly away from it (back) and then at 90° to it (sideways). Such a path is illustrated in Figure 15.6.

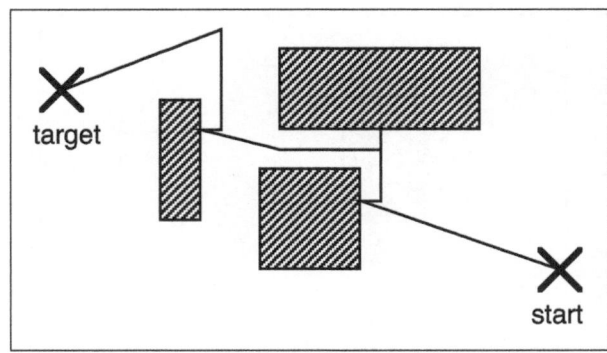

FIGURE 15.6   Local planning to avoid obstacles.

Notice how this algorithm could get stuck in a deep alcove. If it entered an alcove (concavity), the backward movement at step 2.1 might not be enough to get it out, and so the robot would never escape. It is precisely this behaviour that would have to be detected in order to restart global planning.

In the above example, the robot only noticed the obstacle when it hit it. This may be sufficient in some applications, but more generally some remote sensing would be employed, perhaps vision or sonar. This is particularly important if the obstacles are not stationary. It is no good avoiding a bus after it has hit you! In fact, local avoidance algorithms can be adapted quite easily to handle moving objects. The exact form of the algorithm will depend on the sensors available. Let's assume that the robot can detect objects that are within a certain distance of it and can determine their speed and direction. An avoidance algorithm could be:

1. where possible move towards target
2. when an object is detected and a collision is imminent
   2.1 either (i)   move directly away from it (escaping)
       or   (ii) move normal to it (dodging)
   2.2 when the collision has been avoided resume preferred direction

First of all, when an object is detected, the robot must determine whether a collision is likely. This depends on the velocity of the object and the current velocity of the robot. It may be that the object will not cross the robot's path or that the robot can move in front of the object before the object arrives. One way to perform this calcu-

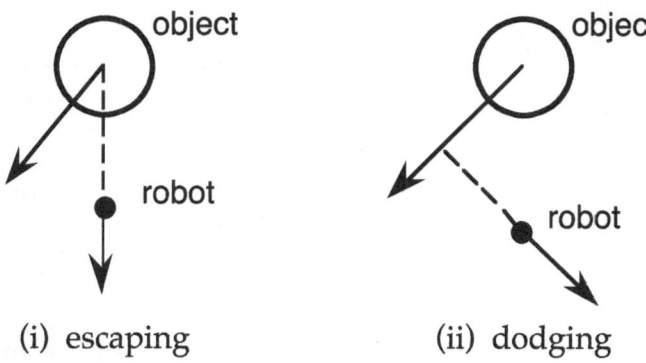

(i) escaping   (ii) dodging

FIGURE 15.7 Avoiding moving objects.

lation is to use configuration space techniques to expand the object and then to subtract the two velocities to give the velocity of the robot relative to the object. If the line in the direction of the relative velocity meets the expanded object, then a collision would occur. The distance to the meeting and the magnitude of the velocity (speed) allow one to work out how soon the collision will occur. There may be several potential collisions, so rule 2.1 is applied to the most imminent.

At step 2.1, two alternative avoidance mechanisms are suggested. These are illustrated in Figure 15.7. The first tries to get away from the oncoming object as fast as possible. It may not be optimal, but it is generally a good approximation to the fastest escape. The second is less drastic: avoiding the path of the oncoming object rather than running away. The second is more like stepping back onto the pavement when you see a bus coming rather than running down the road in front of it. It is suggested [8] that (i) is better when a collision is imminent whereas (ii) is better when the collision is some way off (and hence one has more time to avoid it).

Notice how these local algorithms are all more approximate than the global ones (which themselves were inexact). They are reactive and rely on heuristics rather than using prepared plans based on models of the world. The local algorithms must typically execute in real time; hence the need for simplicity. Also these vague algorithms are more likely to be robust when the assumptions they are based on are violated.

15.4.2 Finding Out about the World

Global planning algorithms depend on a model of the world. Local algorithms do not build a global model, but they react to local information. However, consider what happens when the local algorithm reaches an impasse,

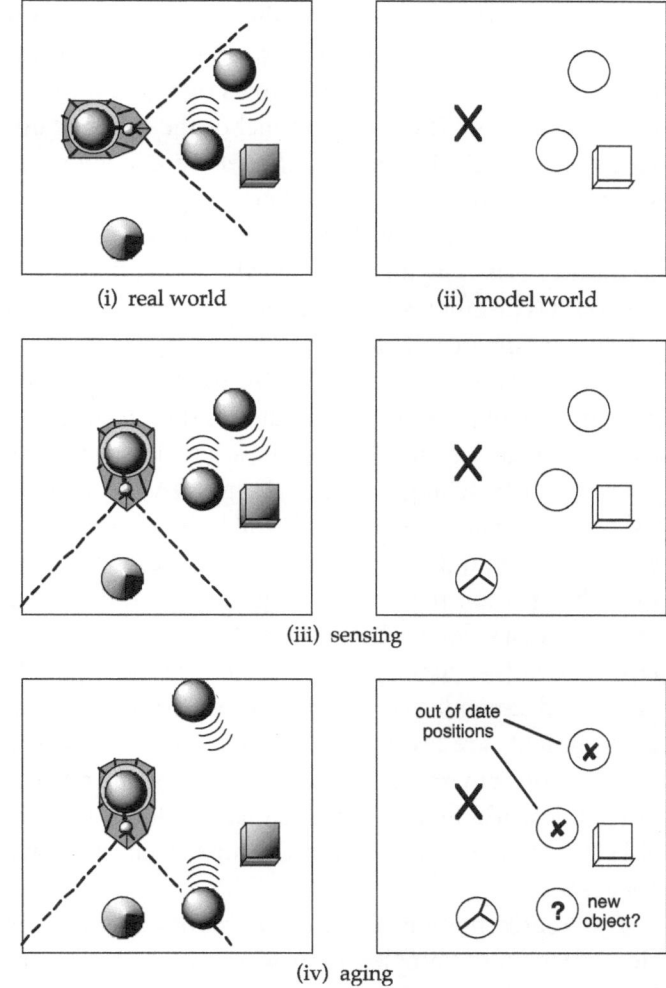

(i) real world   (ii) model world

(iii) sensing

(iv) aging

FIGURE 15.8 The real world and the model world.

say if the robot enters an alcove. At this stage some more model-based planning is again required. However, it must make use of the additional information gained during the robot's movements. Thus we see that the robot's model of the world is not static but changes as it encounters and senses the environment. This sensing could be deliberate (looking around) or a side effect of locally planned movement.

The robot's knowledge of the world grows as it senses but is constantly getting out of date. Things move or change, and so objects sensed some time ago may not be where they were or even exist at all. The model is therefore uncertain as well as incomplete. We can see these two processes are constantly working against one another: knowledge increases through sensing and decays through ageing.

In Figure 15.8 we see these processes in action. The robot in (i) initially can only see the two moving spheres and the cubes. Its model of the world in (ii) is thus incomplete. The robot then rotates to its right and the cone comes into view. The robot's model of the world is updated accordingly. However, the spheres are moving and so a few moments later the situation is as depicted in (iv). Both spheres have moved and hence the robot's model of the world is now incorrect. Furthermore, one of the spheres has just moved into the robot's range of vision. Is it the same sphere as before?

Obviously the robot's model of the world must include not only the objects' positions but also their speeds and some estimate as to whether these are likely to stay constant. If watching a game of snooker, the table is likely to stay where it is, the balls will keep moving in the same direction until they hit something, but the players may change their positions and speeds erratically. For a mobile factory floor robot, information about the floor layout and other fixtures (shelves, etc.) can be explicitly given and amended when necessary. However, other environments are less predictable. One use of robots is in hazardous environments, perhaps after some nuclear or chemical accident. In such a situation, floor plans are at best tentative: walls or even the floor itself may have collapsed!

The representation within the robot's memory is clearly far more complex than when the environment is fixed and known. The exact choice of representation will depend on a variety of factors, not only the internal AI-related ones (reasoning style, search algorithms, etc.) but also external factors (the types of objects in the environment, the nature and accuracy of sensors). However, we can consider two broad classes of representation: historical and current state.

A historical representation will keep track of what has been observed and when, together with the accuracy of sensor. At any moment, the robot can estimate the current positions of any objects based on their known past locations. For example, the model of the world at step (iv) in Figure 15.8 could be represented as a collection of location and velocity data for each object, as shown in Table 15.2.

Notice how the position and velocity of each object have accuracy measures. In this case, the error in velocity is greater for most objects, presumably because the sensor is less accurate at measuring velocity. Both the box and the cone have a measured velocity of zero, but the

TABLE 15.2 Object Location and Velocity Data. (Note error values also recorded for each.)

time	object id	type	position		velocity	
			(x,y)	error	(x,y)	error
1	#317	ball	(3.3,3.2)	0.1	(−0.5,1.0)	0.2
1	#318	ball	(2.8,2.0)	0.1	(0,−0.5)	0.2
1	#319	box	(3.7,1.7)	0.1	—	0.0
2	#320	cone	(1.5,0.5)	0.1	(0,0)	0.2
3	#321	ball	(2.7,0.7)	0.1	(0.1,−0.8)	0.2

box's error figure for velocity is zero because (in this environment) it is known that boxes never move, whereas the cone is a potentially mobile object that just happens to be (sensed as) stationary.

In the alternative, current state representation, the robot would keep similar information, but at each time step, it would update the current position of each object based on its last known velocity.

For example, at time 2, the state of ball #318 would be recorded as:

id	type	position		velocity	
		(x,y)	error	(x,y)	error
#318	ball	(2.8,1.5)	0.3	(0,−0.5)	0.2

See how the position of the object has been updated by adding the velocity. However, its error figure has also been increased as the velocity itself is uncertain. Notice that the representation does not keep track of the time the object was observed. This is unnecessary as the passage of time is recorded implicitly in the updated position. The updates in estimated position and velocity might be more complicated. For example, some objects (such as people or other robots) are likely to change velocity spontaneously and for such objects the uncertainty in velocity would increase accordingly.

With the historical representation, one is never committed permanently to an interpretation of the evidence. If at time 3 the robot decided that balls #318 and #321 were the same object, it could still change its mind when at time 4 another ball (even more similar to #318) appeared in view. In contrast, a single model of the world constantly commits the robot to particular interpretations. Once it had decided that the new ball was the same as #318, it would simply update ball #318's position and velocity to reflect the new observation. All memory of the original two observations would be lost.

However, the historical representation is very inefficient. It requires the robot constantly to recalculate the same projections from past data. Also, at some point it will need to forget past observations. The current state representation has some pruning problems in that it can't track every object it has ever seen but it is clearly far easier and more efficient to manage.

In practice, a system might involve a combination of a model of the current state for rapid real-time response together with some limited historical information in case it needs to reconsider past judgements. What about you? Which representation do you use as you walk in a busy street or drive a car?

15.5 LIMBS, LEGS AND EYES

We have discussed how robots plan how to perform tasks involving picking up and moving objects, how they plan how to move about in the world and how they look at different parts of the world in order to build up their knowledge. Each of these activities involves the control of physical transponders and sensors. The construction of these is a significant engineering problem, especially where the robot is expected to operate in a hostile environment. However, this is general robotics and beyond AI, so we will just consider the issues of control.

15.5.1 Limb Control

Consider a simple robot arm as illustrated in Figure 15.9. This has three main degrees of freedom: the arm is mounted on a rotating section of the robot which can be set to any angle θ, the arm can move up and down by an angle ϕ and can extend by moving the smaller cylinder in and out of the larger by a distance b. In addition, the "hand" at the end can open and close. Other important dimensions are marked on the diagram: the radius of the centre section r, the length of the unextended arm a and the height h of the centre section from the floor.

In order to pick up an object we need to move it to a particular position in space. We can calculate the co-ordinates of the end of the arm using trigonometry:

$$
\begin{aligned}
x &= [r + (a + b)\cos\phi]\sin\theta \\
y &= [r + (a + b)\cos\phi]\cos\theta \\
z &= h + (a + b)\sin\phi
\end{aligned}
$$

Thus, if we have a particular position we want to move to, we can solve these equations to find the right values of θ, ϕ and b. This is messy but not too difficult:

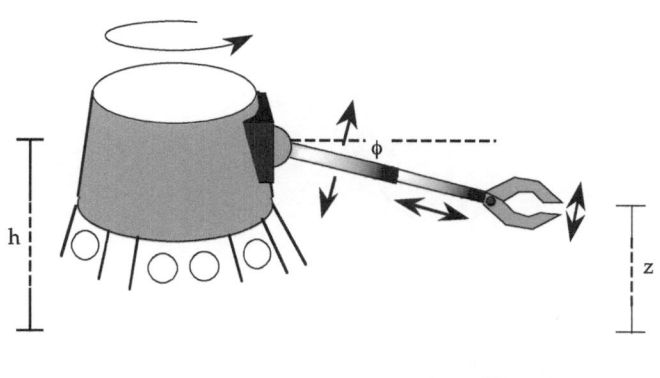

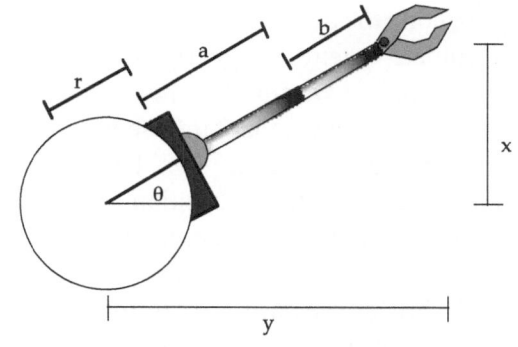

FIGURE 15.9   Calculating limb positions.

$$
\begin{aligned}
\theta &= \arctan(x/y) \\[4pt]
\phi &= \arctan[(z - h)/(\sqrt{x^2 + y^2} - r)] \\[4pt]
b &= \sqrt{(\sqrt{x^2 + y^2} - r)^2 + (z - h)^2} - a
\end{aligned}
$$

Unfortunately, it gets worse. This is the very simplest an arm can be; you need at least three degrees of freedom to have any chance of reaching within a three-dimensional world. In addition, one often needs some control of the orientation of the hand at the end – some kind of wrist. Let's assume we add some more movement at the wrist. Say we want to pick up a suitcase. The hand must be pointing directly down and opened so that the "fingers" close on the handle. However, it is no good simply turning the wrist at 90° to the arm. We need to take into account the angle ϕ of the arm. Also, the robot on which this diagram is based has two arms, offset from one another (see Figure 15.10), each with a full ball joint instead of just up and down movement. To make matters worse, the robot will move about and change its orientation.

Working out the final position and orientation of the hand, given all these movements, is a nightmare. Reversing the process to work out the desired movements to get

FIGURE 15.10   Robot with two limbs.

to any position is even worse! The problem can be simplified by breaking the process down into steps using different co-ordinate systems. You start with a position in the world's co-ordinate system. You then take into account the position and orientation of the robot to work out the desired position relative to the robot. If the robot has several joints, you translate the position into co-ordinates relative to each joint in turn, eventually getting the position relative to the hand. Similarly, you can reverse the process to work out the position of the hand in the world. There are special languages for programming industrial robots that include particular constructs for moving between different co-ordinate systems.

Translation between co-ordinate systems does not solve the problem entirely. The equations for calculating the relevant joint angles and extensions for any desired position are still complicated. However, even if one solves the equations exactly the results may not be perfect. Joints have play in them, limbs may flex under strain. When people pick things up, they rely not so much on accurate calculation but on feedback. We are constantly monitoring and correcting our behaviour. Preplanning everything is called open-loop control, whereas relying on feedback is called closed-loop control. If the environment is very controlled and predictable, say on some production line robots, open-loop control can be effective. However, in general, closed-loop control is far more robust.

For a robot, there are two kinds of feedback: local feedback on a particular joint, which can be used to ensure that the joint is positioned as you want; and global feedback, perhaps through visual sensors, of the relative position of the hand and the target. Local feedback is effectively giving you a more accurate and reliable motor, so does not dramatically affect the style of planning. Global feedback, however, allows more goal-directed behaviour. It is often easier to solve the reverse equations for small movements, and so one

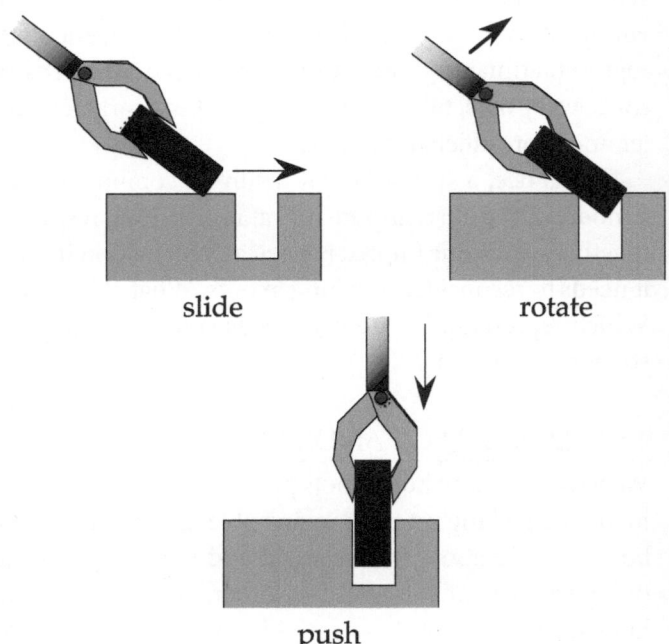

FIGURE 15.11   Compliant motion.

can incrementally move the hand towards the desired position.

Of special importance is the pressure feedback when the hand grasps an object or when a held object is being placed. Imagine picking up an egg without such feedback! If such sensors are too expensive, or impractical because of the environment, the robot's hands must be padded or sprung to avoid damage to itself or the work piece. Effective use of this feedback can make positioning of objects far easier. For example, to place a peg in a hole, one can push the peg along the surface until it catches in the hole (see Figure 15.11), and then rotate it until it slides in. Without pressure feedback the robot would gouge the peg into the surface! This use of feedback to allow things to be naturally slid into place is called compliant motion.

15.5.2 Walking – On One, Two or More Legs

The robot featured in Figures 15.8–15.10 moves around skimming the ground using electrostatic levitation. Other robots use wheels or tracks. However, all have a distinct problem with stairs, and there has long been an interest in various designs of robots with legs. Pragmatically, this allows the robot to manage in very rough terrain, but also the study of robots with large numbers of legs has given some insight into the way lower animals function. One can identify four styles of robot by counting legs:

one leg – good for trying out ideas but not very practical

two legs – for humanoid robots and animations

four to eight – practical robots for difficult terrain

lots of legs – study of lower animals and distributed control

The earliest attempts at walking robots used detailed physical models of the robots' dynamics. If one knows the masses of all the components that make up the robot and can control the forces that it exerts on the floor, then it is possible to predict how a particular movement of the legs will move the robot as a whole and hence work out which forces on which joints will make the robot walk in a particular direction. Of course, as in the case of limbs, this involves co-ordinate translations based on all the joint angles and lots of trigonometry.

The robots usually fell over.

In fact humans fall over too, but we are expecting it and can catch ourselves before we go too far. Indeed, one way of thinking about walking is that you are constantly standing on one leg and falling over in the right direction, and then moving the other leg forward to catch yourself before you fall too far. This form of reactive movement is based again on feedback and closed-loop control and is thus far more robust than the use of detailed dynamic models.

The trouble is that even if a model is entirely accurate, it is expensive to calculate in real time (if you don't do the calculations fast enough, you fall over!) and is difficult to adapt to changing circumstances. Consider what happens when you pick up a heavy object, move your arms so that your centre of gravity moves or walk out of a building into a high wind. In each case, the dynamics of your body have changed which, for a model, would require extensive recalculations.

In contrast, reactive motion is based on fuzzy rules. For example, some of the rules for standing still might be:

- if you are falling forwards, slowly push down on the front of your feet

- if you are falling sideways, take a step in the relevant direction.

The rules are not designed to stop you from moving at all (equilibrium) but to keep you constantly moving back towards the desired position (homeostasis).

Detailed physical models still have a part to play in robotic movement. Building robots is expensive and time consuming. Furthermore, trying out new control algorithms on possibly fragile experimental robots is not to be recommended. A detailed model of a robot's dynamics can be used to simulate different designs of robot and the effects of different control algorithms. Indeed, sometimes the simulation is all that is required, for example early work in this area was developed to address the need for lifelike animations in CGI movies [135]. However, simulations can also be used to train algorithms which will eventually be used in physical robots. For example, the AIs in the robots that compete in the annual RoboCup football tournament are usually initially trained on simulated football matches [109].

This approach is also used for training autonomous vehicles. The vision systems for these are trained using many thousands of hours of actual in-vehicle video, but infrequent yet critical incidents, notably accidents, are by their nature, and desirably, rare. Simulations based on games and VR engines can be used to augment the training, especially for these critical incidents. This example brings up several issues for any such simulation-based training of robots or autonomous vehicles:

ground truth – The world in the simulation is generated and thus known perfectly. This is in some ways better than real-world training, where a large volume of material is unlabelled.

sensing – While the simulator knows everything, it is important that the robot in the simulation does not. The simulation must therefore emulate the sensing capabilities of the robot, including inaccuracy and noise. For the case of autonomous cars this is usually the video view, which VR and games engines are designed to do anyway for the human player.

other agents – The other agents in the environment must not only look real but behave in realistic ways so that the training is appropriate. For example, someone crossing the road and then noticing a car approaching may freeze or take a step back. If the

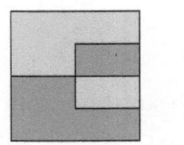 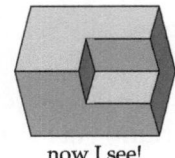

what is it?   now I see!

FIGURE 15.12   Resolving ambiguity.

simulated person simply continued at a fixed pace, the in-vehicle system could learn behaviours (such as cutting close behind the pedestrian) that would be dangerous in real life. We will return to this issue in Chapter 16.

15.5.3 Active Vision

Computer vision is discussed in detail in Chapter 12. However, the discussion is focused on algorithms to interpret an image or set of images that were presented to them – a *fait accompli*. However, where the camera or cameras are fitted on a moving robot, the movements can be planned deliberately to aid the vision process, for example to peek around corners. Similarly, even when cameras are fixed, there may be some control of their direction, zoom and focus.

In Chapter 12 we saw how some scenes were difficult to interpret. In Figure 15.12 we see one such scene. The original image on the left is ambiguous. There is no obvious three-dimensional interpretation of the scene. However, if the camera is moved slightly to the left, the resulting image is far more easily understood. The confusing cross-junction in the middle has been resolved into two separate fork junctions.

This disambiguation may occur naturally as the result of stereo vision or pre-programmed camera movement, but if the vision system can control the camera, it can deliberately seek the necessary evidence. In fact, Figure 15.12 is rather like those intriguing photographs of everyday objects taken from strange angles. In ordinary life one rarely encounters such effects, as when they occur one automatically moves slightly to obtain a better perspective.

Recall that in Figure 15.8(ii), the robot moved its perspective in the world and was thus able to build more complete model of the world. This change in perspective might be an accident of actions that the robot is doing anyway, but might be a deliberate attempt to find out more about the world, just like the human moving their

head. In ecological psychology this is called epistemic action, actions in order to gain information. This might be as simple as moving your head but might include moving around the environment to find out more. Often in algorithms, both physical and virtual, one makes choices between exploration and exploitation.

The scenario above used horizontal movement of the camera. Camera heads may allow control over several other degrees of freedom:

- *fixation* – the point at which the camera is "looking"

- *vergence* – the horizontal angle between two cameras in a stereo head, which allows both cameras to fixate on one object

- *cyclotorsion* – the ability to rotate the camera and thus the horizon

- *zoom* – increasing the size of distant images

- *focus* – the distance at which objects are sharp images

- *aperture* – controls the amount of light entering the camera and also the depth of field (what is in focus simultaneously).

All of these can be used to give additional information for image processing, or to make the raw image more easily processed. Controlling the point of fixation allows one to track a moving object, which might otherwise move out of view. This is especially important when it is not easily matched, say a human figure that changes shape as it walks. The angle of vergence when two cameras are fixated on the same object allows easy calculation of distance by triangulation. The matching of objects in the two stereo images is important for this and other stereoscopic effects. However, if the cameras are not perfectly horizontally aligned, this can be very difficult. Cyclotorsion allows the cameras to compensate for inaccuracies and any flexing in their supports in order to align the horizon in the two images.

The remaining three effects allow one to examine particular objects in detail. Use of zoom can allow one either to scan a large area at low resolution or to examine a particular object in detail, as a small part of the image is spread over all the pixels in the image. Controlling the focus especially allows one to sharpen up the edges in an object of interest and even obtain an estimate of depth from monocular vision. By adjusting the thresholds for edge detection, the blurred edges will not be registered, hence aiding the separation of an object from its background (called figure–ground separation). This is enhanced if the aperture can be adjusted. Once the object is in focus, the aperture can be opened up to make all other objects more blurred. However, the aperture is probably more important for level control, ensuring that neither too little nor too much light gets to the camera. In many cameras this is automatic, but if the aperture can be controlled by the vision system, then it can be adjusted to favour interesting parts of the scene. (NB: Our eyes do not allow this degree of high-level control.)

Most camera heads do not have all of these degrees of freedom, but many mobile phones have multiple cameras, game controllers often have stereo cameras for depth perception and security cameras are often motorised to allow panning.

15.6 PRACTICAL ROBOTICS

The leading edge of robotics research is designing vehicles that guide themselves over the surface of Mars or micro-robots to travel through human arteries scraping and cleaning, but the majority of robots are far more prosaic. Rather than designing general purpose robots that can operate in unforeseen environments, it is usually better to aim for specific jobs and to control the environment. Indeed, as we have seen, real robots may not look like robots at all. Autonomous cars are designed to work in the (relatively) constrained environment of a road systems, including using lane markings and other features designed for human-controlled cars. They are pushing the limits of AI, but you would not expect a car to wash the dishes.

15.6.1 Controlling the Environment

We have already discussed in Chapter 12 how control of light levels and object positioning can make industrial vision easier and more cost effective. The same applies to other areas of robotics. Suppose you want a robot to move materials around in a warehouse. First of all, you are unlikely to choose a robot with two arms and two legs. A wheeled robot with a forklift is a much more practical arrangement.

What about navigation? A general purpose route planner with sophisticated visual input seems like a neat idea, but why not simply paint white lines on the floor that a

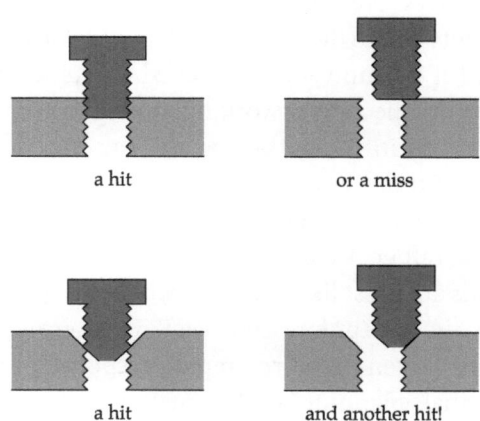

FIGURE 15.13 Designing for easy assembly.

trivial image processing system can follow. The only real disadvantage of such a system is that the lines get dirty, but there are various electronic alternatives.

There is a similar tale for manipulation tasks. Just as with manual assembly, a simple redesign of a component may make assembly tasks far easier. For example, consider screwing a bolt into a threaded hole. If the hole is simply drilled into the metal, the screw has to be positioned very accurately in order to ensure it fits properly; a slight inaccuracy to either side will mean that the screw simply spins against the surface of the metal. However, taper the end of the screw slightly and countersink the hole and suddenly the accuracy required reduces dramatically (see Figure 15.13). Basically, one is designing the system so that compliant motion is successful. The greater margin for error means higher reliability and cheaper robots.

There is of course a trade-off between flexibility and economy. It is usually the case that a specialised tool costs less than a general purpose one. But whereas at one stage production lines involved many highly specialised tools, now the move is towards more flexible manufacture. Tools are still specialised, but far more flexible and easily reprogrammed. Successful industrial robotics requires robots that are just general enough to do the range of tasks that they are likely to encounter. However, as more general purpose robots become cheaper and more reliable, the balance of economics may swing even further along the path of generality.

15.6.2 Safety and Hierarchical Control

Industrial robots, like all industrial equipment, are dangerous. They can hurt people, damage their workplace or themselves. However, they act within controlled environments where the issues of safety are at least well established if not universally adhered to. As we begin to consider mobile robots working among other workers, perhaps even in the outside world, the issues of safety become central.

First of all, this re-emphasises the importance of feedback rather than open-loop control. Dangerous situations are most likely to arise when the environment changes in an unforeseen way. Furthermore, such situations often require rapid responses; possibly the normal planning cycles may be too slow. We may not even trust the planning totally. As we have seen, the best algorithms usually involve a mixture of heuristics and uncertain reasoning. They do not always guarantee correct behaviour even when programmed correctly, and who really trusts several thousand lines of LISP code, let alone a billion weights in a neural network.

One solution is to establish a software ring-fence around the normal planning activities. When a dangerous situation is detected, a high-level control process takes over and performs some special action. This may be some form of avoidance behaviour or most likely stopping the machine dead. The safety sensors may be based on proximity sensors or based on unexpected resistance to movement. You will almost certainly have encountered such sensors built into lift doors (another robot to get inside of!). The important thing is that the higher levels of control are simple and reliable. We can afford to use clever algorithms at the lower levels so long as we know that we are protected from their malfunction.

15.7 SUMMARY

Real robots do not usually walk about on two legs and fire rayguns. Most are in fixed positions on assembly lines, or moving along marked tracks in warehouses. Many have little "intelligence", but obey pre-programmed actions.

Global planning operates by having a complete model of the world, planning what to do, and only then doing it. Means–ends analysis can be used to plan sequences of actions to achieve a desired end state. This can include knowledge about the positions of objects and physical constraints. Configuration spaces can be used to plan routes where obstacles block the way.

Local planning is more opportunistic. The robot has a general goal and tries to move towards it reacting to

problems as they arise. Routes can be found past obstacles by having a desired direction and then simply changing direction when an obstacle is encountered. Avoidance rules can be added to allow for moving obstacles. While a robot moves about, it can find out more about the world (sensing), but also its model of the world may become inaccurate as objects move about (ageing).

Controlling a robot's limbs is not intrinsically difficult but typically involves a complicated series of translations between co-ordinate systems. Feedback can be used to compensate for slackness and inaccuracy and also facilitate local planning. It allows closed-loop control, which is more robust than preplanned open-loop control. Pressure feedback is especially useful, as it allows compliant motion to be used to position objects. Many mobile robots use wheels or tracks, but some walk on one, two or more legs. Again, it is usually best not to preplan movements but instead constantly start to fall over and recover. Active vision uses the movement of the robot or camera adjustments to give more information about a scene and resolve ambiguities.

In practical situations it is often better to design a suitable environment for a simple robot than to use a more complicated one. Simpler robots are usually cheaper but will be less flexible. Industrial robots can be dangerous, and several levels of control may be necessary.

15.1 Produce an operator table for the Towers of Hanoi problem similar to the blocks world one in Table 15.1. To make it similar to the blocks world think of it as the Tables of Hanoi problem with three tables rather than three towers. Use the same two operations as in Table 15.1, but the "on_table" predicate will have an extra parameter: "on_table(T,R)" where "T" will be a particular table and "R" a ring. You will also need a predicate "bigger_than(R1,R2)" to record which rings are bigger and will have to ensure that you do not put more than one object on top of another.

15.2 Use your operator table and means–ends analysis to solve the three ring problem given the starting state

on_table(1,big) ∧ on_top(small,middle) ∧ on_top(middle,big)
∧ bigger_than(big,middle) ∧ bigger_than(middle,small)
∧ bigger_than(big,small)

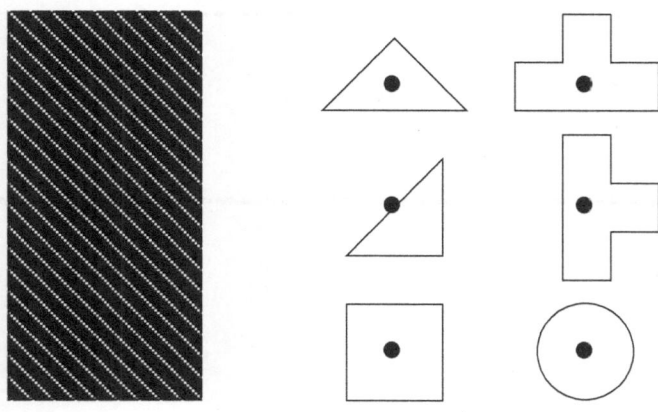

FIGURE 15.14   For exercise 3.

and goal state

on_table(2,big) ∧ on_top(small,middle) ∧
on_top(middle,big)

15.3 Consider the shaded rectangle in Figure 15.14. Draw configuration spaces for each of the robot shapes to the right of the rectangle.

15.4 Collect different everyday items, screws, bolts, plugs, lids. Do they exhibit good design for compliant motion – like the screw in the lower illustration of Figure 9.12 or are they more like in the top pictures! (Good class exercise)

FURTHER READING

B. Siciliano O. Khatib and T. Kröger, editor. *Springer handbook of robotics.* Springer SHB:200, Berlin, 2008.

Definitive edited collection in the area, the first part 'Robotics Foundations' offers greater detail on the topics covered here while succeeding parts look at more advanced topics such as soft-robotics, specialised application areas and human–robot interaction.

Newell and H. A. Simon. GPS: A program that simulates human thought. In E. A. Feigenbaum & J. Fieldman, editors, *Computers and thought*, pages 279–293. McGraw-Hill, New York, 1963.

A. Newell and H. A. Simon. *Human problem solving.* Prentice-Hall, Englewood Cliffs, NJ, 1972.

GPS, the General Problem Solver, was an early model that used means–ends analysis to emulate aspects of human planning.

R. E. Fikes and N. J. Nilsson. STRIPS: A new approach to the application of theorem proving to problem solving. *Artificial Intelligence*, 2: 189–208, 1971.

STRIPS applied techniques developed in GPS to planning in the blocks world as described in this chapter.

T. Lozano-Pérez. Spatial planning: A configuration-space approach. *IEEE Transactions on Computers*, 32(2), 108–120, 1983.

Original work on configuration space.

H. I. Christensen, K. W. Bowyer, and H. Bunke, editors. *Active robot vision: Camera heads, model based navigation and reactive control.* World Scientific Singapore, 1993.

An early collection of articles that deal with both movement (local planning and obstacle avoidance) and vision (the problems and also the leverage that can be obtained by using active vision).

Rodney A. Brooks. *Cambrian intelligence: The early history of the new AI.* MIT Press, 1990. DOI:10.7551/mitpress/1716.001.0001

An alternative view of AI from a robotic perspective, in particular considering machines that behave intelligently without explicit internal representations.

Agents

16.1 OVERVIEW

In the previous chapter, we considered individual robots planning and acting in the real world. In this chapter, we look at three related areas.

Software agents are autonomous entities that inhabit and act on the electronic world on our behalf. Obviously route planning is not usually necessary for such agents, although they do need some means to choose their actions. The examples we shall consider use a mixture of *ad hoc* rules and machine learning techniques.

Agents often need to be able to function in new environments, physical and digital, by experimentation. We shall see how reinforcement learning intersperses action and learning.

In the last part of the chapter, we shall look at what happens when agents (whether electronic or physical) act together. We shall see that they are more than the sum of their parts and can work together to achieve co-operative purposes.

16.2 SOFTWARE AGENTS

The word robot means worker. As well as working for us in the real world, AI can be used to develop independent entities that work for us in the virtual world of information spaces. These are often called software agents, or simply agents (although the word agent is rather overloaded in AI). The term robotic process automation is also used in commercial systems. These agents can be used to sort your (electronic) mail, perform repetitive tasks, search databases for interesting information or manage your diary. These are all applications where the agent is "visible" to the user of a system, a sort of helper. The word agent is also often used where a system is split into several co-operating subprograms or agents. We will discuss this case, where the agents are co-operating with one another, later in this chapter. In this section we'll confine ourselves to agents that interact with and work on behalf of a computer user.

The simplest agents are not really intelligent at all. Imagine you perform the same routine every week to back up your computer files to an optical disk. So, instead of performing the same actions again and again, you write a script that is automatically invoked at the same time each week. This is the simplest kind of software agent. The ability to write such scripts has been around since the earliest operating-system macro languages and is available in various scripting systems such as automator for MacOS. More end-user web automation is also possible through tools such as IFTTT (If This Then That) [141, 285], and the proliferation of internet-connected devices (IoT – the internet of things) will drive the need for the equivalent for home-automation.

In the rest of this section we will consider the reasons for the interest in agents and the different sorts of events that trigger agents to act. We will then look at email filtering agents as an example of learning, and agents for searching large information spaces.

16.2.1 The Rise of the Agent

Is the use of the word "agent" just a buzzword or sales gimmick? In fact, there are some recent developments

DOI: 10.1201/9781003082880-19

that make the connotations of the word agent appropriate:

- *End-user orientation* – Whereas the writing of scripts in traditional operating systems was confined to the system gurus, the emphasis is now on the ordinary user. The user's control may be exercised using simple scripting languages or by direct demonstration.

- *Embodiment* – Most graphical interfaces project a very passive model. The user acts upon objects in the interface. When the application does anything, it is as a "tool". However, complex and repetitive tasks do not really fit into this model of the world. It is thus natural to inhabit this virtual world with agents to perform these tasks. In principle, to fit within the model-world paradigm, these agents should be visible. Indeed, over the years there have been attempts to make this explicit including Hewlett-Packard's NewWave interface where agents were presented as icons designed to look like a secret agent; Microsoft's Clippy [16] and the helpful cat in the early programming-by-demonstration system Eager [64]. However, none of these are still used and successful systems seem to have textual or graphical manifestations.

- *Lostness* – Most people only use and understand a fraction of the functionality of a modern application, and even where they know about features it is often far from clear how to combine them to obtain a particular effect. Hence the use of Wizards in many products to guide and help the user, especially for installation and set-up tasks. Similar trends can be seen in information spaces. Even early hypertext systems, with relatively small numbers of documents, had problems with users getting lost, and so some incorporated various forms of "guide" to show you round. The problem is far worse on the web and other internet information sources where there are billions of web pages – where do you go to find the information you want? Ask an agent.

- *Intelligence* – The scripting languages used for programming agents are sometimes more like code, sometimes more template/form-like and sometimes more like natural language (although the same was said of COBOL in the 1960s).

However, agents are also learning what to do in more intelligent ways. The user may explicitly demonstrate the required behaviour which the agent later copies. Alternatively, the agent may watch the user and learn the user's habits and preferences. It may then use this knowledge when asked to perform a task, or even volunteer help.

Agents that address one or more of these issues are seen by the users of a system to have some level of independence. They are not just part of the system but act in and on the system on behalf of the user. Where the sense of embodiment is low but aspects of independent activity and/or intelligence are apparent, it is perhaps better to regard the system as exhibiting agency. For example, the word processor being used for this chapter periodically suggests saving the work so far. Another example is in range selection in the Microsoft Excel spreadsheet. When the user invokes the sum operator, the system suggests (in the form of a highlighted selection) the range it thinks the user wants. This is based on simple heuristics but has the appearance of intelligent behaviour.

16.2.2 Triggering Actions

One key difference between a program or macro and an agent is that the former only acts when told to, whereas the latter acts independently. Another way of looking at this is that a program is characterised by *what* it does, whereas an agent is characterised by both *what* it does and *when* it does it.

In addition, because agents have some form of continued existence over time, they usually have a persistent state. Because of the similarities with object-oriented programming, the scripts for actions that an agent performs are often called methods, and the communications between agents are called messages.

A typical life-cycle for an agent will be as follows. It remains in a quiescent state until some event triggers it into action. Depending on the nature of the event it then performs one of its methods. This method will update the internal state of the agent, and also possibly change the state of other things, send messages to other agents or interact with the user.

The event that triggers the method may be caused by various things:

- *User events* – The user may explicitly ask the agent to perform some task. This may result in some instant action on the part of the agent (e.g. searching a database) or may simply change the internal state of the agent (e.g. setting criteria for sorting incoming mail). Alternatively, the user may engage in some action that the agent is monitoring. For example, an intelligent tutor may notice that a pupil has performed a task in an inefficient or incorrect fashion and suggest alternatives.

- *System events* – Other events may occur that are not directly caused by the user. For example, email may arrive that the agent sorts into folders. Another example would be where the user has initiated a long-running computation (perhaps running a large AI program!). When the execution is finished, the agent examines its output and informs the user if it is not as expected.

- *Changes in status* – The agent may constantly monitor parts of the rest of the system and act when certain changes occur. For example, if free disk space falls below a certain level, an agent may compress infrequently used files. In a factory setting, an agent may monitor various processes and warn the operator if the values fall outside acceptable limits.

- *Timed events* – The agent may perform repetitive actions at regular intervals or at particular times. For example, an agent may monitor your diary and download necessary files to your laptop computer for the next day's meetings. Timed events may also be used to trigger monitoring activities, such as those discussed under the previous heading. This polling activity should be distinguished from true timed activity.

In addition to triggered actions, an agent may act continuously to gather information. This information may come from the user or from the rest of the system. For example, an agent may monitor the user's interaction with the system and notice frequently repeated actions. Later, when the agent detects the user beginning a complex action sequence, it can offer to complete the task for the user.

16.2.3 Watching and Learning

We'll look now at email filtering agents as an example that you are likely to have encountered yourself. Studies of the early introduction of email into institutions found that there is little reduction in other forms of communication but a continued growth in email messages. This is partly because of the ease of replication. The photocopier and the word processor each made their contribution to junk mail, but neither so effectively as email! It is simply too easy to include a large number of names when sending an email or to mail to a distribution list of hundreds or thousands of individuals. Sifting through the email each morning is a major task, and that is even before looking at numerous forms of messaging applications and social media.

Just the job for an agent!

Most email systems allow the user to set up filters. The email message has specific fields ("To:", "From:", "Subject:", etc.) and the user fills in a template, which is then matched against incoming messages. If the template matches, then whatever action the user has specified (say filing the mail in a particular folder) is carried out. For example, a colleague of the authors got fed up with receiving seminar announcements and so set up a filter to delete all incoming messages that contained the word "seminar". Unfortunately, the agenda of an important meeting included "seminars" as one of its items. The announcement was discarded and the meeting missed!

Filtering may also be carried out by fixed rules. An early research email system at Stirling University organised all mail messages into conversations – linked, often branching, streams of messages [53]. If enough people had had compatible mailers, these could have been used to ensure that when a message was sent, it was added to the appropriate conversation at the recipient's end. However, email comes from so many disparate sources that this was considered an over-restrictive method. Instead, the system used simple rules to sort incoming messages into conversations. An example rule was

> **if** the new message (N) is '**From:**' person A
> **and** the last message (M) sent to person A is in conversation C
> **then** add N to conversation C linked after message M

There were also rules concerning multiple recipients so that messages from the same person to different distribution lists could be filed successfully. Although the rules were simple, they worked most of the time, and the cost of a misclassified message was low (it was easy to track all recent messages). Note that this is a general rule – intelligent agents will not always be right, so make sure it doesn't matter when they are wrong (see also Chap. 19).

Simpler versions of this kind of rule are used in Gmail to group messages into threads, although looking at the gap between the early work in this area and when it was adopted in Gmail is an interesting story of the rate at which new ideas are adopted, even in the digital world.

So, we've seen examples of agents that are told what to do by the user and by the system's designer. A more ambitious kind are those that attempt to work out what to do themselves. Many email systems do this to help sort mail into urgency categories. These are often seeded by fixed rules, for example favouring mails with fewer recipients, but over time they learn the user's own preferences.

The user interacts with the mail system as normal. Each mail message (or at least its header!) is read, and the user performs some action to it, files it in a particular folder, deletes it or possibly marks it as urgent (if the system supports marking of messages). The agent watches. After a while the system has a collection of examples of the form: message⇒action. This is ideal input for machine learning algorithm. The agent can learn patterns in the user's actions and then automatically sort the mail – intelligent filtering

Of course, it's not quite that simple! First, it is very important that the user retains a sense of control, especially when the action is to delete a message! There are various ways to achieve this. One way is for the agent to construct filter templates and present these to the user for approval. That is, the job of the learning agent is to simplify the task of creating templates for the agent that does the actual filtering. Another option is for the agent not actually to perform the actions on the messages as they arrive but simply to add a classification and offer a simple means for the user to accept or reject the agent's offered choice. These issues of control and grace of interaction between agent and user are common to any system that involves learning user actions.

Another problem with learning filtering rules is that the data within email fields are quite complicated. The algorithm needs to be quite knowledgeable about email addresses, since two different addresses may refer to the same person (e.g. alan@hcibook.com and alanjohndix@gmail.com). Also some of the fields may contain lists of addresses or email distribution lists. A simple application of machine learning would give poor results without some of this information being

taken into account. Furthermore, the most important information is all in free text fields requiring complex text matching algorithms. These are discussed in the next section.

In practice these limitations mean that intelligent filtering is still fairly limited, for example Gmail has four fixed categories and Outlook only two (in addition to spam filtering, which is managed differently). There is a cost–benefit calculation here, missing important mails is often worse than scanning a small number of less relevant ones, so if you occasionally scan the less urgent categories, the cost of categorisation mistakes are low. Social media filtering and ordering is much more aggressive as the cost of failure, missing one among hundreds of posts by contacts, is low ... until the day you miss a critical post by a close friend.

16.2.4 Searching for Information

The amount of information available online is enormous. The problem is finding what you want without wasting time on the even more enormous amount of dross. Agents have been posed as a solution to this problem, and you may use these yourself, for example Google alerts for interesting web content, job notifications from employment sites or research articles recommended by academia.edu. The remit of such agents is simple – find interesting information and tell me about it. Satisfying this is less straightforward. Agents can help in three ways:

1. They can **find** where suitable documents are stored.

2. They can **mediate** between the user and different information sources.

3. They can **choose** appropriate documents from a large document set.

The first step is necessary as there are too many information sites to search them all in detail. An agent may find sites by consulting a simple preferences file, perhaps created by hand or built up as a record of sites that the user has visited. A more sophisticated agent may consult a directory of information sites. It will need to match the description of what the sites contain with the interests of the user. This process is similar to the document matching in step 3. Finally, the agent may find sites by following a trail.

Consider a bibliographic search for articles on intelligent agents. The agent looks first at the *Journal of Artificial Intelligence*. It finds some articles of interest (using step 3) and then looks at the articles cited in their reference lists. This will yield potentially interesting articles but also the journals and conferences where those articles are found are good candidates for searching. A similar process can be followed on the World Wide Web. If you know that a particular document is of interest, the agent can look at the links from that document and search the sites where those documents are found. Indeed, this is precisely what Google effectively pre-processes in the PageRank algorithm to build its web indices.

Another sort of trail is one based on usage. Suppose that each document server keeps track of who looks at what. Your information agent notices that many of the documents that you read have also been read by another user, so your agent asks the other user's agent about other sites it visits. These are then candidate sites for interesting documents. You are effectively following the other user's path through information space. Although the details are proprietary, it is likely that many search engines use this form of algorithm, and certainly it is common for advertising trackers and recommendations in internet shopping.

There are of course privacy problems here. In the first edition of this book we wrote, "*it is rather like browsing someone else's bookshelf*", an interesting analogy before this became common in Amazon and then later Spotify and other media sites. We'll see later that there are methods to try to gain the advantages of collaborative learning without the privacy concerns. However, you could turn the personal element to your advantage: your agent could negotiate with advertising sites to get good deals in exchange for your information, or 'talk' with other users' agents and introduce you – computer dating?

Note that following a trail may lead to both interesting sites and also specific documents. However, once you find an information source you are faced with understanding and navigating a new interface. One of the reasons for the success of the World Wide Web is the common interface to all information. However, this is not shared with other services, and online bibliographic databases are particularly renowned for their obtuse user interfaces. Agents can help you here too. You ask the agent for what you want, and it converts this into the required commands for different information

services. This is often viewed in terms of multiple agents, one for each type of information service, and these communicate with a single user interface agent. For example, IFTTT has many hundreds of plugins to integrate with different web services but offers a single interface to the user.

There are problems of control, for example if an online service has charging, you need to decide whether you want to incur the costs. Also an automated agent may not be able to understand all of your requirements, or be able to assess the quality of a source, for example it can be very difficult to find a hotel's own website on search engines among the hundreds of booking sites.

Finally, we have found an information source and can communicate with it, but it contains thousands of documents. How can an agent work out which ones will be of interest? Assume that the agent has access to some collection of documents which you have previously found interesting. One way to use these is to use some form of concept learning and generate a rule for interesting files. This would typically be based on key words or other summary information. For example, the agent might decide that you are interested in all documents with "agent" and "intelligent" in the key word list. However, this sort of precise rule is often not suitable for handling imprecise ideas such as "interesting". For these more fuzzy forms of matching are often preferred based on statistical methods, such as those discussed in Chapters 7 and 8. These are often based on the complete set of words in the abstract of an article or on the whole article itself. The aim is to have some measure of closeness between documents. Then a document is deemed interesting if it is close to one or more of the documents you have previously found interesting. Other measures of similarity may use semantic features of the documents; for example, citations in common for articles, or links between objects in a hypertext.

Let's look at one measure of closeness in detail. Take two documents d_1 and d_2 and generate the complete list of words in each, w_1 and w_2. Let the number of words in each document be n_1 and n_2 and the number that are in both n_{12}. That is, n_1 is the size of w_1 and n_{12} is the size of $w_1 \cap w_2$. Then a measure of similarity is:

$$similarity(d_1, d_2) = \frac{n_{12}}{n_1 + n_2 - n_{12}}$$

This formula has a value of 1 when the documents have exactly the same words and 0 when they have none in common. Similar, but more complicated, measures

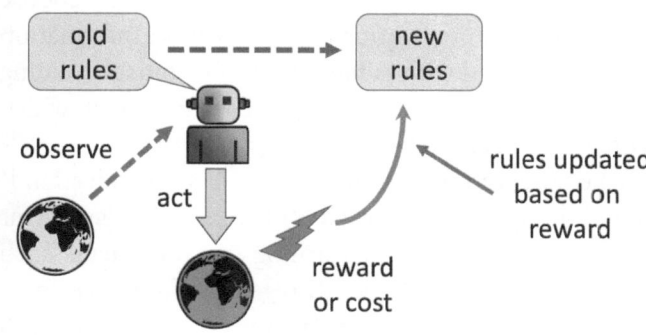

FIGURE 16.1 Simple reinforcement learning – single action-learning step, repeated many times.

take into account various factors. Common words such as "the" and "it" may be ignored or given a low weight in the match; on the other hand, words that occur frequently in both documents may count more highly. Also the word lists may be processed to reduce words to their simplest forms (e.g. simplest → simple) – called stemming – or to equate different words for the same thing using a thesaurus.

16.3 REINFORCEMENT LEARNING

When faced with a new situation, we sometimes have to act without knowing in advance the best thing to do. As we try things we gradually learn what does or does not work well and hence learn. We can do the same with an AI system. This may occur in a purely simulated environment, such as playing a computer game, or in the real world, such as a robot in a factory. The AI agent needs to act based on the state of the world and its current rules and then learn from the consequences of its actions; this is called reinforcement learning, and we've seen an example already with the pole balancing task in Chapter 5.

16.3.1 Single Step Learning

Figure 16.1 illustrates a single step of reinforcement learning. The AI agent has a current rule set and observes the state of its environment. This observation may be perfect and complete or may in some way be partial or noisy. Based on the observation and the current rules, it acts. The action affects the environment and creates some sort of reward or incurs a cost. The agent uses this reward or cost to update its rules, reinforcing rules that led to positive consequences and inhibiting those that led to negative consequences. This single step is repeated many times, with the rule set improving with each cycle.

Reinforcement learning is often seen as a special case of unsupervised learning as there is no prior set of classified examples on which to base the learning. However, in pure unsupervised learning, such as the self-organising map in Chapter 6, there is no feedback at all apart from an internal measure of a good model. In contrast, reinforcement learning has an external reward (continual or intermittent), so could be considered a form of supervised learning. In practice, reinforcement learning systems may use a combination of supervised and unsupervised learning algorithms internally, maybe using unsupervised learning methods to simplify observations, and supervised algorithms to update the action rule set.

Note too that this is different from the email learning discussed in Section 16.2.3, which is normally based on observing user actions and learning rules to emulate those, so is closer to pure supervised learning. Of course, in practice there may be a mixture of learning by observation and more experimental actions.

A very simple reinforcement learner could keep a list of observations and actions with a weight that is incremented when the response is positive and decremented when it is negative. Here is such a table for behaviour at traffic lights – in a virtual environment; experimentation at real traffic lights is not recommended. We assume it has been learning for some time and that getting across faster is rewarded, but accidents or near-misses are punished.

observation	action	weight
green	wait	-1
green	drive	10
amber	wait	-1
amber	drive	-4
red	wait	-1
red	drive	-23

Note that driving on green gets a positive reward for speed, while waiting slows the journey, so is negative. Of course we know that waiting on green is at best unnecessary and at worst annoys the car behind you, but the AI system doesn't know this until it has experimented. During training, waiting on either amber or red lights would be given a negative weight as this slows the journey; however, the negative weight for driving on amber and red would be greater due to the increased risk of accidents.

Note that in deterministic environments with a fixed consequence, the table above simply needs to be filled in with known consequences after each trial. However, more often the consequences themselves are stochastic or depend on hidden aspects of the environment, so the observation–action mapping is about probabilities and likely effects. Because of this, sometimes Bayesian methods (Chap. 3) are used to work out the probable responses based on the evidence at each stage.

16.3.2 Choices during Learning

When the environment is fully learnt, the AI agent simply needs to choose the action that has the best expected outcome. However, earlier in learning, or when learning in a changing environment, the agent may be faced with more difficult decisions:

novel situations – Sometimes the situation in the environment is completely or partially new, there is little or no past experience on which rules are based. Gaps may be filled at this point in different ways. Unknown observation–action slots can be filled with random or default values, or some form of similarity to known results might be used to generalise from previous situations. In some forms of learning system, such as neural networks (Chap. 6), the representation is more diffuse and the network may be initialised with random values before learning cycles start.

exploration vs. exploitation – When the agent has encountered a situation previously with a positive consequence, why would it ever try anything else? Or if it has once had a negative consequence for an action, it might avoid it ever after – a form of local maxima. Because of this, reinforcement learning systems may sometimes *exploit* past experience by taking the best choice action, but sometimes might instead *explore*, try something new or retry something that did not previously give a good result.

Managing the exploration–exploitation trade-off is one of the key problems in reinforcement learning and depends a lot on the criticality of the choice – simulated environments offer opportunities to try and fail, but physical environments, such as driving on real roads, typically require a more risk-averse strategy. In fact there is always a cost of action, this may be in terms of fuel and risk on the road, or computational time in a simulation. Choosing the best opportunities for experimentation is

thus critical, for example one might deliberately choose to experiment in situations that maximise information learnt compared with potential cost. Of course, deciding this itself involves knowledge of the environment, which potentially needs to be learnt.

The observation–action mapping has no understanding of the system; it is based on the very simplest forms of animal reinforcement learning in experiments such as Pavlov's dogs or Skinner's pigeons. If one has some knowledge, richer learning methods may be used, such as those described in Chapter 5. In Chapter 22, we will see how emulating human regret can harness complex counterfactual reasoning to guide lower-level reinforcement learning and in so doing guide the exploration–exploitation trade-off and avoid local maxima.

16.3.3 Intermittent Rewards and Credit Assignment

The reinforcement learning process illustrated in Figure 16.1 assumes that the reward or cost happens immediately and is directly related to the last action taken. In animal reinforcement learning, delays between action and reward severely reduce the ability to learn, and this is even true of humans, hence a dashboard fuel consumption monitor is more effective at encouraging eco-friendly driving than the weekly shock of the bill when filling the tank.

Figure 16.2 illustrates the more realistic situation where successive actions have an impact on the world, but only intermittently create rewards or costs. The reward at any point may not only be stochastic but is determined by any one or a combination of past actions. As noted in the pole balancing task in Chapter 5, credit assignment is difficult – which actions should be reinforced or inhibited. Typically AI agents in such environments need some internal memory of past observations and actions, and ideally a richer model of the environment to help connect rewards with the possible cause.

16.4 CO-OPERATING AGENTS AND DISTRIBUTED AI

In the previous section we talked about agents communicating with one another. Also, when we discussed route planning in Chapter 15, we noted that some of the objects one robot might encounter may be other robots. The communication and interaction between agents is an exciting area offering several interrelated benefits:

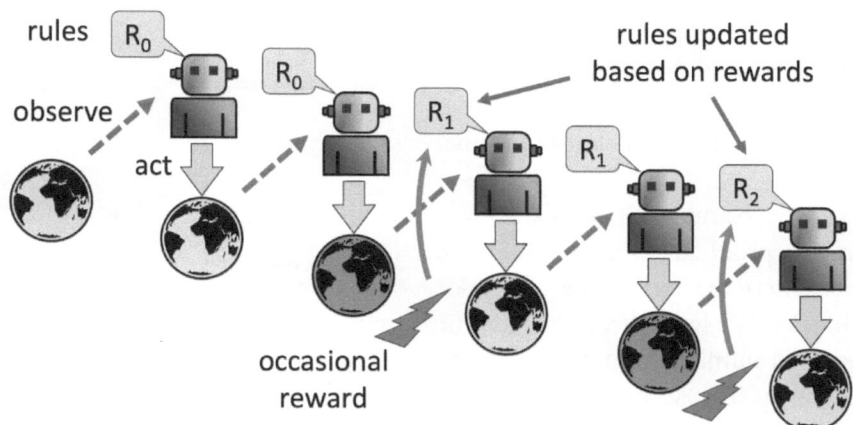

FIGURE 16.2   Reinforcement learning with intermittent rewards.

- Structuring an intelligent system into several communicating but largely independent parts can reduce development costs, increase run-time efficiency and ease maintenance.

- Different parts of the system may reside in physically distinct places. It may be impractical or impossible to perform totally central planning.

- Maintaining separate agents may be important to preserve personal privacy or protect commercial data.

- The interactions between agents can give insight into the social interactions between people or between animals. There are also theories of individual cognition which stress the co-operation between semi-independent "agents" within our own minds [197].

The study of interacting intelligent components is called distributed artificial intelligence. When the aim is understanding living creatures, it is even called artificial life, although this term also includes other aspects of computer-generated life forms.

16.4.1 Blackboard Architectures

The use of multiple semi-independent knowledge bases is not new and predates the now ubiquitous use of the word "agent". As we noted, this has obvious software engineering benefits. Each knowledge base can be built, tested and updated individually. Furthermore, when tackling a problem in a particular area, only the relevant knowledge for that area is used. Each knowledge base contains only the knowledge needed for its purpose and may employ representations and reasoning methods appropriate for its particular domain.

However, to solve a common problem, the knowledge bases have to communicate. In an object-oriented architecture this is likely to be via message passing. When a knowledge base/object/agent needs information it sends a message to another to ask for it. When the other one has found the answer, it sends a message in reply. This approach can be very powerful, as can be seen in the information-seeking agents described in the last section. However, it has the disadvantage that each object needs to know which other one has the required information or knowledge.

A traditional form of co-operation that avoids this problem is the blackboard architecture. The object-oriented architecture is similar to lots of people working in separate rooms occasionally sending memos to one another. In contrast, the blackboard architecture is rather like a group of people in the same room who are jotting down ideas on a blackboard. As one person writes something down another sees it, perhaps in conjunction with other items on the blackboard, thinks about it and then writes a new idea based on it.

As well as ideas (or solutions), the computer blackboard will also contain unsolved problems. When an agent sees a problem that it can tackle, it solves it and then removes the original problem, posting up the solution. If in trying to solve the problem the agent hits an impasse, it can post up a subproblem on the blackboard in the hope that another agent will see the problem and be able to tackle it. Unlike the object-oriented architecture, it does not have to know which agent can solve the subproblem, merely post it to the board. When an agent sees that the subproblem has been solved, it can continue to tackle the original problem.

Figure 16.3 shows an example of a blackboard architecture in action. The problem concerns adding up using counting blocks. There are two kinds of blocks, ten-blocks and one-blocks. We have three agents. One agent, the reader, can read numbers and convert them into blocks and vice versa. The second, the grouper, knows how to add up blocks by simply pushing the piles of blocks together. The third, the swapper, knows how to swap a ten-block for ten one-blocks and back again. The initial problem is posed in terms of numbers, "add 13 to 8", and the answer is also required as a written number. The initial representation is shown in the first frame of the figure. The three agents then solve the problem in the following steps:

1. The reader converts the numbers 13 and 8 into the equivalent blocks.

2. The pusher clumps the two together to give an answer in blocks – one ten-block and 11 one-blocks. At this stage the reader might try to convert the answer back into digits but would fail as there are more than ten one-blocks.

3. However, the swapper can work on the blocks and change ten of the one-blocks into one ten-block, giving two ten-blocks and one one-block.

4. Finally, the reader has a pile of blocks that can be converted into a number, 21, which is the final answer.

Notice how no agent needs to know what the other agents can or can't do. However, it is important that there is a common representation on the blackboard so that the outputs of one agent can be recognised by another.

One disadvantage of a pure blackboard architecture is that there is no central control whatsoever. This can lead to problems. For example, after stage 1, the reader might have looked at the blackboard and thought "Ah, piles of blocks, I'll change them into digits". The system could easily have thrashed about indefinitely changing things back and forth. As a model of cognition this is not

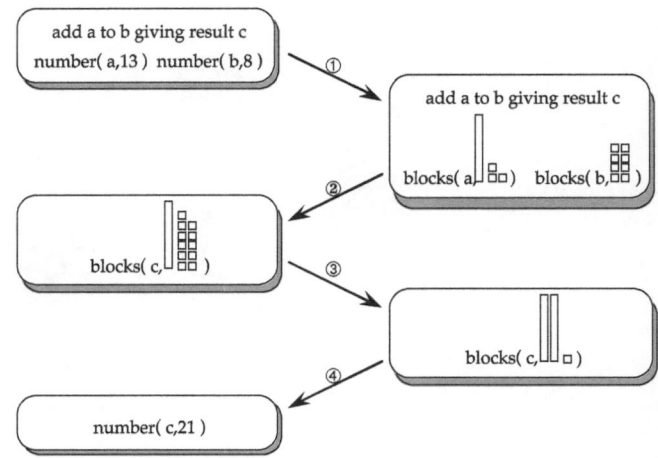

FIGURE 16.3 Blackboard architecture.

too far from the truth: a frequent error in mathematical proofs is to use a series of equalities, but end up where you began. However, when one becomes practised in a domain, one uses higher-level heuristics. So, for adding with blocks one would learn always to apply the agents in the following order: reader, pusher, swapper and then reader again. A half-way approach between pure blackboard and totally centralised control is to have some sort of co-ordinator agent that activates different agents at different times. The co-ordinator does not need to understand everything on the blackboard but simply a high-level plan of when to do what.

16.4.2 Distributed Control

A production line is producing cream cakes. As the cakes go past, a machine squirts a dollop of cream onto each one. For a few minutes there is a problem on the line and the supply of cakes stops. However, the machine goes on placing a dollop of cream on the line where each cake should have been. This is perfectly understandable, the way machines work. However, if the cream was being put on by a human, the supervisor would be very annoyed to see a hundred dollops of cream on the conveyor belt. People are supposed to use some common sense even in the most repetitive jobs. In a more enlightened factory, the employees may be given more autonomy. Perhaps a group of workers is given targets for the productivity of its particular subprocess and is free to organise its work in whatever fashion it chooses so long as its goals are achieved. Similarly, in an army the commanders make strategic decisions about the deployment of troops and the lines of attack; lower levels of command make tactical decisions; but it is ultimately each soldier who decides precisely when to pull the trigger.

The models of planning we presented in Chapter 15 were largely monolithic. Ultimately the planner knew everything. However, in a large factory such control becomes impractical. Attempting to preplan each machine tool and robot will lead to problems like the cream on the belt. Instead, the central planner must give orders to each tool or robot that it will obey using its own planning systems. This is similar to the issues of hierarchical control we discussed in the context of a single robot, but here we are thinking of many robots co-operating together under some central co-ordinator. Also, unlike the agents co-operating under the blackboard architecture, the major interactions here are physical rather than electronic.

The problem with this form of decentralised control is that the central planner needs to be able to predict global properties from the local properties of agents. Imagine a factory has just two processes: baking and dolloping. Both baker and dolloper have an average throughput of 1800 cakes per hour. So, if both are placed on the same production line, can one assume that the line can produce 1800 cakes per hour? Only if both baker and dolloper can promise a continuous and reliable rate of one cake every two seconds. If the machines sometimes work faster, but sometimes have to pause to refill, then we end up with either the line stopped for a proportion of the time or more dollops on the conveyor belt. Clearly, the central planner would need to have more information about the individual machines. However, it would not need to know about the precise details of each machine. For example, the dolloper might use sophisticated image processing to determine the positions of the cakes on the line, followed by a planning algorithm to decide the order in which to place a dollop onto the cakes. The planner need know none of this, simply that the machine is capable of putting a dollop on each cake.

16.5 LARGER COLLECTIVES

Most of the agents we have discussed so far have been relatively sophisticated. In addition there are methods which depend on the collective behaviour of large numbers of quite simple agents.

16.5.1 Emergent Behaviour

In the example of the blackboard architecture, where was the knowledge of adding up? Similarly, in the factory, where was the knowledge of how to make a cream cake? In neither case can we point to any particular agent and say "that one knows". The knowledge and ability are distributed between the agents. So, no one of the adding agents can add up, but together they can. This is called emergent behaviour.

Emergent behaviour is not just a feature of the electronic world but is present in nature at many levels of life. Consider a swarm of bees building a hive. There is no architect, no plan to follow, but the individual actions of each bee together create a purposeful activity. Similarly, when disease infects your body, there is no central control which says "fight that organism"; instead the various cells and chemical messengers within your body each perform individually in a way that fulfils a common purpose.

The same sort of emergent behaviour is found in humans. This can be seen at a gross level in crowd movements, or in the flow of traffic along a road – lots of individual decisions together giving rise to a global behaviour. On a smaller scale, there is a growing acceptance that the thinking of individuals and groups cannot be isolated in their heads but is instead distributed between the people and even their environment. This approach is called distributed cognition [139]. It is similar to the adding up example. In the building of a skyscraper, where are the thought processes that lead to its construction? In the architect, the engineer, the financier? The answer is in none individually, but in them as a group, and not solely in the people but also in the representations they use, plans, models, even the building itself. In the adding example, the blackboard itself is crucial in the adding task.

16.5.2 Cellular Automata

Some computational models are built purely to study these emergent behaviours. Groups of agents each act out their own individual, and often very simple, behaviours, but together give rise to complex patterns in the large. Possibly the simplest example is Conway's Game of Life. In Life, the world consists of a rectangular matrix of cells (Figure 16.4). We consider each cell to be either populated or not. At each step we consider the neighbours of a cell. If an empty cell has three or four

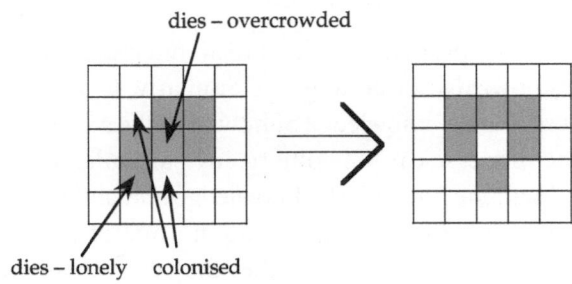

FIGURE 16.4 Game of Life.

of its neighbours populated, then the cell is colonised. However, if a populated cell has more than five of its neighbours populated, it "dies" through competition. It also dies if fewer than three of its neighbours are populated – loneliness! Most readers will have seen this game animated. Some initial configurations die out completely, others seem to go on indefinitely changing. Some become stable, and others, the gliders, swoop across the screen. These patterns are not coded explicitly into the rules but emerge from the conjoint behaviour of all the individual cells.

Life is a simple example of a general class of models called cellular automata. In general, the state of each cell can be more complicated, not just populated or not, as can the rules. Also the cells need not be in a rectangular mesh but may have some other topology.

16.5.3 Artificial Life

The rules used in cellular automata are usually quite simple and not very intelligent. However, there are other models that give each agent more complex rules, often based on social phenomena. For example, one model has agents wandering over a rectangular playing field, meeting other agents. When they meet, the agents engage in a "prisoner's dilemma"-style interaction. Each agent has a different disposition to "trust" other agents and may reinforce that trust or reduce it depending on the result of the interaction. If the agents are able to choose where to go, there is a tendency for trusting agents to group together, building up mutual trust – societies in microcosm.

Another system models robots in a physical environment with obstructions and also simulated locomotion (they are pogoing robots). The individual robots all have a desired direction to travel but also have rules to avoid bumping into each other or getting too far apart. The robots exhibit flocking behaviour rather like birds. The flock moves in the general direction of travel but occasionally sweeps one way or another, or a small group may temporarily break away from the main flock. When an obstacle is encountered, the flock divides around it and then, when past the obstacle, the two streams criss-cross and intermingle before merging fully.

These models are not just of theoretical interest. Models of traffic flow on motorways can improve the safety and efficiency of roads and cars. Also, models can tell us about phenomena that cannot be observed directly, for example models of crowd movements in burning buildings. Models have even been formulated of the movements and social structures of hunter-gatherers in prehistoric France, hoping to explain some of the phenomena during the change from nomadic to settled existence. This form of simulation has been called Artificial Society.

16.5.4 Swarm Computing

We have already mentioned swarms of bees as an example of emergent behaviour in the natural world. In fact bees, ants and other simple animals that exhibit apparently intelligent behaviour have proven a fertile ground for many artificial algorithms. These have several defining features:

1. Large numbers of relatively simple and similar agents.

2. Semi-independent and (initially) highly random individual behaviour.

3. Some ability to communicate weakly with others as to the quality of places/states they have visited.

4. Modification of individual behaviour to favour paths/places that others have found fruitful.

Think of ants initially foraging widely and largely randomly. One ant finds a boiled sweet, gathers a tiny piece itself and then heads back to its nest, leaving a this-was-good pheromone trail on the way back. Other ants pursue their own semi-random paths, but if they encounter the pheromone trail, may choose to follow it. If they in turn encounter the boiled sweet, they add to the pheromone trail on their own way back to the nest.

For bees the communication step 3 is via waggle dances as they fly rather than crawl, but a similar principle applies. In both cases the initial explorations (step 2) may be guided by scents of vegetation in the air or other factors, so are not entirely random.

It is important that step 4 is not about an ant always following others to the most favoured spots, but sometimes going to and fro between good but less optimal ones. This continued exploration means that if the environment changes, or the best source of food is exhausted, there are secondary sites waiting to be exploited.

Artificial versions of this, effectively following steps 1–4 as pseudocode, have been used in applications such as network routing. Note in this case the optimal route is the critical feature – finding the path through the network with least latency and best quality. The network is flooded with small packets that follow random paths. Those that get to their target node retrace their steps leaving a 'trail' on the way back.

As well as methods based on animals, lower-level biological processes have learning-like behaviours. In particular the immune system learns in order to distinguish the many types of cells that are part of your own body from those that shouldn't be there – that is infections. There are multiple mechanisms, some of which work by building up responses after infection has started, these are the kind that vaccines target to give them a head start and they effectively react to examples of bad organisms in the body and learn to attack them. However, we also have a base immune response that works the other way round.

The full immune system is complex, but a simplified version is as follows. When you are born your bone marrow starts off in a learning mode and generates a form of exploratory immune cells with each programmed to detect a wide variety of different surface proteins. As they circulate in the blood stream, some encounter cells that match their target protein in your own body; those that match then replicate and find their way back to the bone marrow. Here their presence suppresses further generation of that kind of cell. After a period the only immune cells produced are those that do not match any of your own cells, and these become locked in as a form of memory and the body changes its mode of response to a defence state. Now if one of the cells sent into the blood stream encounters a matching organism, it must be foreign to the body and an attack is triggered.

It is evident how this kind of model can be emulated as a form of artificial immune response. A network security AI monitors network traffic initially generating large numbers of random rules for matching individual packets of sequences of packets. Assuming that in this initial learning state, the network is benign, only those rules that match nothing are retained. Later, during the defence stage, if any of these rules match passing packets, they are flagged as potentially malicious.

16.5.5 Ensemble Methods

In Chapter 5 we saw how decision trees could be constructed as a form of machine learning. For large datasets with large numbers of features there is a danger of overfitting, that is the trees match the particularities of the sample that comprises the dataset rather than the more generalisable aspects. There are various ways to counter this, for example limiting the depth or complexity of the trees. However, one simple solution that has proved surprisingly successful is random forests.

In random forests one randomly splits the dataset into smaller subsets and possibly also chooses a different random subset of features for each (Figure 16.5). A simple decision tree algorithm is then applied to each subset of the data items using its respective subset of features. This yields a large number of trees, the forest, each trained on the different fragments of the initial dataset. The individual trees are not necessarily very good when applied over the whole dataset, that is they are individually weak learners. However, one can combine their individual outputs to create a single strong learner that is better than any individually. This is the automated equivalent of the wisdom of the crowds.

The methods of combination can vary. For binary classification often a simple majority is used, for more numerical outputs simple weighted sums can be used. More complex methods can be used for this combination of many weak learners into a single strong learner, often called boosting methods. These may use probabilistic decision algorithms, neural networks or other forms of machine learning.

Another way to think about this is that the original feature set has been replaced with the vector of outputs of the individual trees. Seen like this it has similarities with the various non-linear transformation techniques we saw in Chapter 7 such as support vector machines and reservoir computing.

The individual elements need not be decision trees, and other ensemble methods use different forms of algorithms for the lower-level learners. For example, one

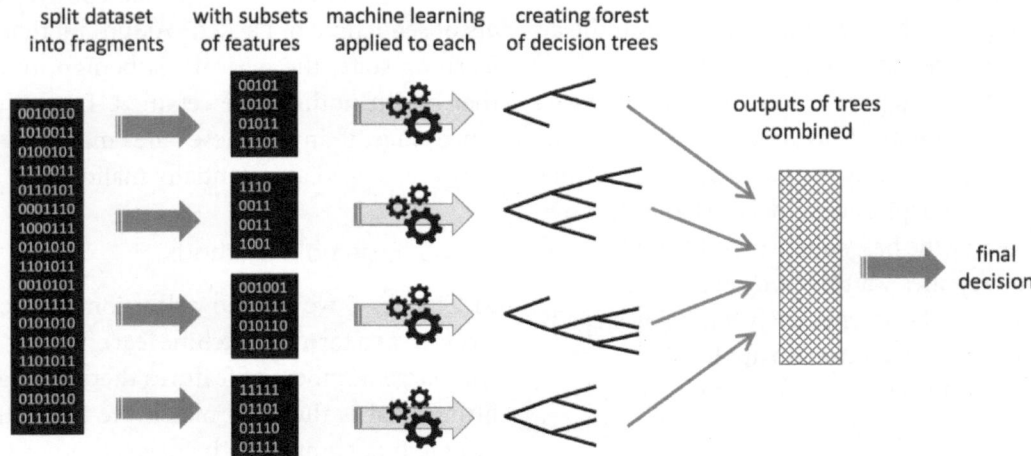

FIGURE 16.5   Random forest.

might use simple regression or linear classifiers or concept learning alongside or instead of decision trees. If all of the lower-level learners are of the same kind, the system is homogeneous, but if a mix is used, it is heterogeneous. A random forest is a particular example of a homogeneous ensemble.

Neural-network-based models are often monolithic, but they can be structured in similar ways. In particular, the DeepSeek LLM uses a mixture-of-experts (MoE) approach [173]. There is a central general-purpose portion of the network that is always active, and other parts that are specialised for different topic areas. This means that only a relatively small portion of the network needs to be activated for any particular query, enabling performance comparable to other LLMs while using far lower computing power.

16.6 SUMMARY

Software agents are at the heart of several recent and current developments to make software easier to use and to help people find their way through complex information spaces. They act in response to different kinds of events and can gather information to use later. Agents can watch a user's actions, use machine learning algorithms to identify common actions and then offer to do them for the user. Agents can use knowledge about a user's interests to search for information and may interact with other user's agents in the process. Similarity measures are one technique used for identifying potentially interesting documents.

Reinforcement learning can be used when agents encounter new situations where they need to act in order to obtain the information they need to learn. Critically they need to balance exploitation (using the knowledge they have already learnt to maximise benefits) with exploration (seeking out new knowledge).

Co-operating agents can be used to structure an intelligent system, to divide a system between different places or to simulate human co-operation. Blackboard architectures can be used to allow simple agents to solve problems co-operatively. In an industrial setting distributed control can allow groups of simple robots and machinery to work together without a central planner, but some overall direction is usually necessary. The joint actions of many simple agents often give rise to more complex effects, called emergent behaviour. This is observed in humans: a group of people have a shared knowledge and ability that no individual possesses (distributed cognition). Also cellular automata, very simple computational agents, can produce complex, lifelike phenomena. Applications include the simulation of social behaviour and crowd movements.

EXERCISES

This is another chapter where exercises are more discursive, suitable for individual projects or group discussion.

16.1 Find out about web "crawlers" and "spiders", which rove the internet looking for useful resources. Some are related to the major search engines but you should try to find examples of more specialised crawlers and how they work.

16.2 Similarly, collect examples of "intelligence" in popular applications such as word processors, spreadsheets and drawing packages. Classify the examples you find into those where the intelligence is hidden or working behind the scenes, and those where it is explicitly embodied in some form of agent. Compare your list with other students. Do you agree on what constitutes intelligence and agency?

16.3 Experiment with different rules similar to the Game of Life. You can start using paper and pencil or draughts on a chess board but may find it faster to write a program to do it. (It is said that Conway used plates on a tiled kitchen floor!)

Writing on software agents is quite widely dispersed. Papers can be found in conferences and journals on human–computer interaction, the internet and even sociology, as well as traditional AI sources and proceedings of specialised conferences and workshops on distributed AI such as the European Conference on Multi-Agent Systems and International Conference on Autonomous Agents and Multiagent Systems.

FURTHER READING

M. Wooldridge. *An introduction to multiagent systems.* John Wiley & Sons, 2nd edition, 2009.

Influential textbook with a formal flavour.

R. Sutton and A. Barto. *Reinforcement learning: An introduction.* MIT Press, 2018.

Comprehensive treatment of reinforcement learning.

M. Minsky. *The society of mind.* Simon and Schuster, New York, 1985.

Classic work viewing cognition as interaction.

I. Hitoshi. *AI and SWARM evolutionary approach to emergent intelligence.* CRC Press, Boca Raton, FL, 2020, ISBN: 9780367136314

An accessible and beautifully illustrated overview of current swarm intelligence.

P. Maes. Agents that reduce work and information overload. *Communications of the ACM*, 37(7): 30–40, 1994.

Classic article on the promise of software agents in the user interfaces.

Web-scale Reasoning

17.1 OVERVIEW

We use AI and big-data enabled algorithms every time we do a web search, use social media or go to an internet shopping site. However, the vast volume of material available on the web can also be used as a resource for AI. We have seen some of this, especially the machine learning potential in Chapter 8. In this chapter we will see how the idea of the Semantic Web allows the web to be a locus for machine reasoning, both from special machine-readable data and from human-readable web pages with additional markup. We will also see how various forms of external semantics and text mining can be used to extract information from web pages and social media to allow them to better support our own day-to-day interactions and also to make them available as a source of large-scale information.

17.2 THE SEMANTIC WEB

The web now is all about video streaming, social networking, shopping and instant information. However, Berners-Lee's original design for the web was focused primarily on the sharing of scientific data at CERN [24]. This vision of the web as a place for computer readable data is still very much alive including in large-scale open data initiatives by governments and others. Some of this is simply about using the web to share data, but there is also a more profound mission to provide a 'web of data' through a range of technologies that are collectively known as the semantic web [25].

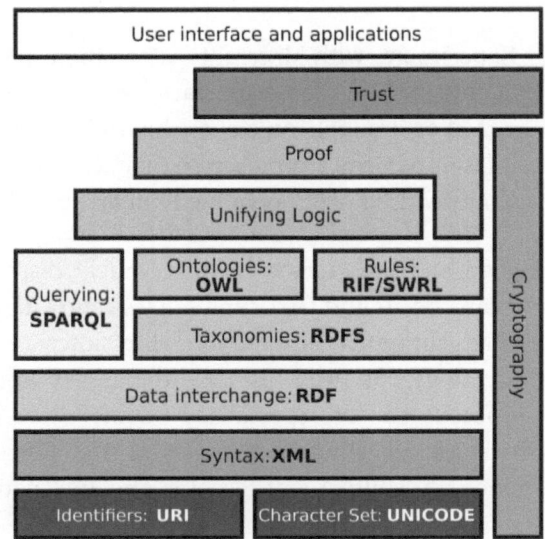

FIGURE 17.1 The Semantic Web Stack (user: Marobi1, CC0, via Wikimedia Commons, https://commons.wikimedia.org/wiki/File:Semantic_web_stack.svg).

These technologies build upon one another, from ways to represent global character sets to complex reasoning (see Figure 17.1). However, the foundations, the most mature, are RDF and ontologies, which are forms of network representation as discussed in Chapter 2.

17.2.1 Representing Knowledge – RDF and Triples

RDF, the Resource Description Framework, was originally developed to talk about properties of web pages, for example that the author of `https://alandix.com/` is Alan Dix, but has been extended to become a general knowledge representation tool.

Core to RDF is the notion of subject–predicate–object triples, for example:

DOI: 10.1201/9781003082880-20

id	name	balance	limit
1637	Jo Doe	3500	1000
2578	Alex Smith	2300	2000
...	...	...	...

(i) Relational database table

subject		predicate		object
record_1637	–	has_name	–	Jo Doe
record_1637	–	has_balance	–	3500
record_1637	–	has_limit	–	1000
record_2578	–	has_name	–	Alex Smith
record_2578	–	has_balance	–	2300
record_2578	–	has_limit	–	2000
...		...		...

(ii) The same data as triples

FIGURE 17.2   Encoding arbitrary length records as triples.

subject		predicate		object
https://alandix.com/	–	has author	–	Alan Dix

Look back to Figure 2.6 and see how some of the properties in that could be represented in this way:

subject		predicate		object
Scooby Doo	–	instance_of	–	Great Dane
Dog	–	number_of_legs	–	4

Notice that there are two kinds of triple, ones that relate two kinds of entity (objects or concepts) and those that express some sort of value of a property.

Triples are very simple and yet also very expressive. All the standard data structures can be represented in triples. For example, where a traditional database might have a table with many fields/columns for each row, with triples you would give the row a unique id and then have lots of triples, one for each field (see Figure 17.2). However, while you can represent pretty much everything with triples, they are not always the most efficient or easy to understand representations; sequences and arrays are particularly difficult, with multiple representations, each having problems.

Because RDF is a web notation, all the objects, concepts and predicates are expressed using URIs (Uniform Resource Identifiers). You will be familiar with web addresses given as URLs (Uniform Resource Locators). URIs have a similar style, but as well as referring to web pages, images, etc., they can include

namespaces that refer to things outside the web itself. The last three lines in Figure 17.3 show some of the examples we've already discussed, expressed as RDF triples. The document is written in Turtle, one of several concrete syntaxes for expressing abstract RDF.

See how these examples include:

1. A URI that is an actual URL of a web page: `<https://alandix.com/>`

2. Literal string value: `"Alan Dix"`

3. A typed string value representing an integer: `"4"^^xsd:integer`

4. A 'built-in' RDF URI `<rdf:type>` (defined in `http://www.w3.org/1999/02/22-rdf-syntax-ns#`)

5. A URI `<dc:creator>` from a standard ontology, the Dublin Core [297], which is used to represent 'meta data' about information resources, such as authorship.

6. URIs from a specialised vocabulary: e.g. `<aibk:Scooby_Doo>`

The namespace prefixes 'dc:', 'rdf:', 'aibk:' and 'xsd:' are shorthand for longer URI prefixes, which would normally be defined at the start of an RDF document. These are normally declared in the document that contains the RDF, as can be seen in Figure 17.3.

Because URIs build on the naming system of the web, they can be globally unique. This is very different from notations and systems that use local names. This is partly an 'accident' of the development of RDF for describing web resources but is crucial in allowing RDF to be very extensible, in several ways.

1. If an entity is defined in one RDF document, say if an RDF example in this book references `<aibk:alanjohndix>`, anyone can refer to this from their own RDF document adding their own statements.

2. In general, RDF documents can be merged as they refer to globally defined entities.

3. As the vocabularies are also defined by URIs, the things that can be talked about (predicates,

```
@prefix rdf: <http://www.w3.org/1999/02/22-rdf-syntax-ns#>,
@prefix aibk: <https://alandix.com/aibook/uri/ex/>.

...


<https://alandix.com/>   <dc:creator>          "Alan Dix" .
<aibk:Scooby_Doo>        <rdf:type>            <aibk:Great_Dane> .
<aibk:Dog>               <aibk:number_of_legs> "4"^^xsd:integer .
```

FIGURE 17.3   Fragment of RDF document in Turtle syntax showing namespaces and triples.

classes) are also both open (new ones can be added) and common (if two documents refer to xsd:integer, they both mean the same thing).

In general, RDF effectively adopts an open world assumption. This is the opposite of the closed world assumption adopted by Prolog in Chapter 2. In Prolog, the list of facts was assumed to be precisely and only what was true. If something is not stated, it is assumed to be false. In contrast an RDF document containing lots of RDF statements is understood to mean that these things are asserted to be true, but not that they are the only things that are true. Other documents may be added at a later time that talk about new entities or say new things about existing entities.

17.2.2 Ontologies

In philosophy 'ontology' is the study of being: what makes a cat a cat, and indeed what does it mean to even be an 'it', to be talked about as a thing? Within computing 'ontologies' have a more mundane meaning as ways to express relationships between concepts, ideas and properties. How we might express formally the *kinds of relationship* we saw in Chapter 2.

As well as the basic RDF predicates defined in 'rdf:', there are a larger number of predicates in RDFS (Resource Description Framework Schema), which include class–subclass relationships between concepts (a dog is a kind of animal) and the kinds of values that are acceptable for different predicates (the number of legs is an integer).

```
<aibk:Dog>         <rdfs:subClassOf>  <aibk:Animal>
<aibk:number_legs> <rdfs:range>       <xsd:integer>
```

However, this is still fairly limited, sufficient to say that statements in RDF are well-formed, but not for more complex reasoning. For this a variety of more complex ontologies have been defined, some for specific domains, such as legal concepts, some to simply say more expressive things about general statements.

The nature of RDF is to be extensible, but the more general vocabulary is represented in the Web Ontology Language, OWL. For example, this allows you to say that 'is owned by' is the opposite of 'owns', so that one can infer from 'Shaggy owns Scooby Doo' that equivalently 'Scooby Doo is owned by Shaggy'. In RDF triple notation:

```
<aibk:owns> <owl:inverseOf> <aibk:is_owned_by>
```

OWL comes in several variants that vary in:

expressiveness – how complex are the things you can specify

tractability – how hard or even possible is it to verify statements

In general the more expressive a notation, the harder it is to automatically check things. For example, the simplest variants of OWL only allow you to specify that the cardinality (how many values something can have) is zero, one or unlimited. This is effectively making checking more about binary decisions. If one is allowed to express a constraint such as "number of legs <= 4", then any reasoning system checking the statement needs to be able to do arithmetic proofs.

OWL inherits the open world assumption of RDF, but some of the variants do allow you to make strong statements that effectively include some level of closedness. For example, if there is a cardinality statement that says a dog has only one owner, we know it is not possible to consistently have a statement in another document that adds a second owner.

17.2.3 Asking Questions – SPARQL

RDF can be put in a simple text file in one of the standard formats. However, there are also a number of research and commercial triplestores. These are like databases for RDF.

SPARQL is the query language for RDF. It holds a similar place to RDF and triple stores as SQL does to relational databases. It was crucial in allowing Semantic Web technology to be used in end-user applications. One of the early examples of this was the BBC's website for the 2010 World Cup, which used an RDF triplestore and SPARQL to allow different facets of the tournament data to be easily explored [231].

SPARQL allows queries to be expressed using variables to denote unknown entities. For example, the following is a query to find the email addresses of all people with the nickname "Shaggy".

```
SELECT ?e
WHERE {
    ?p  vcard:nickname  "Shaggy"  ,
    ?p  vcard:hasEmail  ?e  .
}
```

Note how this query uses the vcard ontology, developed for the transfer of information between address books.

The query above is a simple conjunction; it effectively matches all pairs of triples that match the two pattern statements. The pattern variables ?p and ?e are instantiated during the matching.

More complex queries are possible, including disjunctions (either match this or that) and crucially negation. Figure 17.4 shows an example of SPARQL with a negation, 'NOT EXISTS'. This query looks for all people who have no name specified. It is clear from this example that this could be useful.

However, the 'NOT' here is adopting a negation as failure model, like Prolog; that is, SPARQL is using a closed world assumption. This makes sense as SPARQL is effectively used as a query language over triplestores, which contain a fixed set of statements. However, this causes some conflict with the general open world nature of RDF, a situation that can lead to endless academic discussion.

17.2.4 Talking about RDF – Reification, Named Graphs and Provenance

Sometimes we just want to say something "Scooby Doo is a Great Dane". However, we often also want to talk about who said something, whether something is true, how you know it is true. For example, I might want to say "It is true in the film 'Ghostbusters' that Scooby Doo is a Great Dane".

There are two ways this is dealt with in RDF, the first is reification, which basically takes a statement and treats it as an entity in its own right, with the subject, predicate and object of the statement defined as triples. This is rather like putting "quote marks" round a statement in text to turn it into a 'thing' you can talk about (as in the paragraph above). We can then make statements about the statement!

```
aibk:_stmt_42  rdf:type       rdf:Statement.
aibk:_stmt_42  rdf:subject    aibk:Scooby_Doo.
aibk:_stmt_42  rdf:predicate  rdf:type.
aibk:_stmt_42  rdf:object     aibk:Great_Dane.

aibk:_stmt_42  aibk:true_in_film
                        aibk:Ghostbusters.
```

Notice how the statement itself has been given a sort of anonymous id, this is rather like numbering equations in a mathematical proof. The URI is only there to enable us to talk about the statement, so would probably not be given a meaningful name unless it were a special statement such as Descartes' [72] "*I think therefore I am*".

When RDF descriptions are created in a triplestore or through some form of automatic tools, there will be many of these generated URIs, even for meaningful things. This is rather like record ids in a traditional database. For example, Alan has a Facebook handle alanjohndix, but internally within Facebook's database and API, this has an id 635054223.

This is powerful, but cumbersome – OK to talk about one or two things, but looks like a pain if you want to talk about *everything* that happens in *Ghostbusters*. Similarly, if you want to import RDF statements from several documents into a triplestore, you may want to keep track of which statements came from which document, especially if you have different levels of confidence in the reliability of the different sources.

```
PREFIX  rdf:     <http://www.w3.org/1999/02/22-rdf-syntax-ns#>
PREFIX  foaf:    <http://xmlns.com/foaf/0.1/>

SELECT ?person
WHERE
{
    ?person  rdf:type  foaf:Person .
    FILTER NOT EXISTS  { ?person foaf:name ?name }
}
```

FIGURE 17.4   SPARQL example with negation (from https://www.w3.org/TR/sparql11-query/#negation).

To support this, many triplestores are actually quadstores, labelling each triple with a graph identifier ... which in true RDF fashion is itself, of course, a URI.

```
<graph_uri> <subject_uri> <predicate_uri>
                    <object_uri_or_literal>
```

These are called 'named graphs', although the name may simply be a URI. They can be used in various ways:

1. To make it easier to update the store if an RDF document has changed and needs to be reloaded

2. To allow queries against particular graphs as well as the whole store

3. To allow statements to be made about the named graph as a whole

The first is just bookkeeping (albeit important). The second can be used if, for example, you were not entirely sure about the reliability of one of the sources, so some queries might want to be only against the trusted source. However, the last is the most interesting.

One can use this to talk about all the statements in the graph, in a similar way to reified statements above:

```
aibk:graph_37 aibk:true_in_film
                    aibk:Ghostbusters.
```

Perhaps more important, it can be used to talk about provenance, where the data came from, its reliability, etc.

```
aibk:graph_37 ex:derived_from
                    https:/ex.com/a_doc.rdf.
aibk:graph_37 ex:reliability "medium"
```

17.2.5 Linked Data – Connecting the Semantic Web

Having URIs that represent things not actually on the web is really powerful, they allow RDF to talk about physical things such as Mount Everest, ideas such as 'truth' and fictional things such as Harry Potter. However, this does not help discoverability. We might know that the URI aibk:alanjohndix is an unambiguous way to talk about the particular Alan Dix who is the author of this book, but how do we find out more?

Linked data achieves this by making URIs link to actual web accessible documents that then contain RDF that describes the thing. That is, Linked Data is RDF where the URIs are URLs pointing to machine readable data. Two of the central examples of this are GeoNames and DBpedia.

The GeoNames website and database collects open geographic data from a number of sources and has a web interface and a number of APIs. Each country, city, mountain, village and stream has a unique GeoNames id. For example, 6077243 is the id for Montreal, and geonames:6077243 can be used in RDF documents to refer to Montreal. Figure 17.5 shows a portion of the RDF available at GeoNames about Montreal.

Note that 'geonames:' is a common prefix, which is short for https://sws.geonames.org/; so geonames:6077243 expands to https://sws.geonames.org/6077243/.

One of the entries in Figure 17.5 is a 'see also' link to https://dbpedia.org/resource/Montreal, the entry for Montreal in DBpedia. DBpedia is an extract of the parts of Wikipedia that can be easily codified as data. If you look at this page in an ordinary web browser, it will render it in a readable manner (not raw RDF!), and you can see, among

```
<gn:Feature rdf:about="https://sws.geonames.org/6077243/">
  <rdfs:isDefinedBy rdf:resource="https://sws.geonames.org/6077243/about.rdf"/>
  <gn:name>Montreal</gn:name>
  <gn:alternateName xml:lang="ht">Monreyal</gn:alternateName>
  <gn:alternateName xml:lang="la">Mons Regius</gn:alternateName>
  <gn:featureCode rdf:resource="https://www.geonames.org/ontology#P.PPLA2"/>
  <gn:countryCode>CA</gn:countryCode>
  <gn:population>1600000</gn:population>
  <wgs84_pos:lat>45.50884</wgs84_pos:lat>
  <wgs84_pos:long>-73.58781</wgs84_pos:long>
  <gn:parentCountry rdf:resource="https://sws.geonames.org/6251999/"/>
  <gn:nearbyFeatures rdf:resource="https://sws.geonames.org/6077243/nearby.rdf"/>
  <gn:locationMap rdf:resource="https://www.geonames.org/6077243/montreal.html"/>
  <gn:wikipediaArticle rdf:resource="https://en.wikipedia.org/wiki/Montreal"/>
  <rdfs:seeAlso rdf:resource="https://dbpedia.org/resource/Montreal"/>
</gn:Feature>
```

FIGURE 17.5   Portion of RDF about Montreal (from https://sws.geonames.org/6077243/about.rdf).

other things, that there are typically 11.5 rainy days in November (predicate dbp:novRainDays) and that the album Queen Rock Montreal was recorded there (dbo:recordedIn dbr:Queen_Rock_Montreal).

So, if you use the GeoNames URI as an identifier for Montreal in your own RDF, it is possible for code to follow this through to the GeoNames RDF, and from that to the DBpedia RDF. A reasoning engine can then, if it wants to, find that Queen Rock Montreal was recorded there, that it was recorded in November 1981 (from its DBpedia RDF), and that means there was approximately a 1 in 3 chance that it was raining the day it was recorded.

GeoNames and DBpedia were two of the first datasets in the 'Linked Data Cloud' in 2007, but by November 2020 there were more than 1200 listed (see Figure 17.6) as well as many individual RDF documents that use URIs from Linked Data without being publicly listed themselves. In addition to more generic resources, such as DBpedia, there are also densely interconnected sub-clouds, for example around medical and biological data.

The promise of this as a data source for knowledge-rich reasoning is clear, but in practice there are additional steps needed.

First, if you just type a Linked Data URI into a browser you will normally get a human readable HTML page. Code accessing the page needs to say explicitly that it wants RDF as a response. In

the case of GeoNames, if your code asks for RDF from https://sws.geonames.org/6077243/, GeoNames will give an HTML 303 redirect to https://sws.geonames.org/6077243/about.rdf where the actual RDF about Montreal can be found. This is so that there is a difference between the URI that represents Montreal the city and the URL of the file containing RDF about Montreal. These distinctions sound subtle but make it possible, for example, to talk about provenance.

Second, although this use of the HTTP request–redirect is now fairly standard, there are variations between datasets, for example some just deliver RDF directly, some have a SPARQL endpoint. Often you need to know a little more about the particular dataset.

Third, the RDF at the DBpedia URI tells you what DBpedia knows about Montreal but does not know about every dataset that has a link to it. Even within a dataset often the RDF at the URI only tells you the RDF statements where the thing you are referring to is the subject, not those where it is the object. If the dataset has a SPARQL endpoint, then it is easy to ask for everything that refers to a specific URI, but not everything has this. In general, full discovery needs an additional Semantic Web search/index rather like Google does for the human readable web.

Finally, GeoNames uses a 'see also' link to DBpedia but doesn't explicitly say they are the same thing. There is a

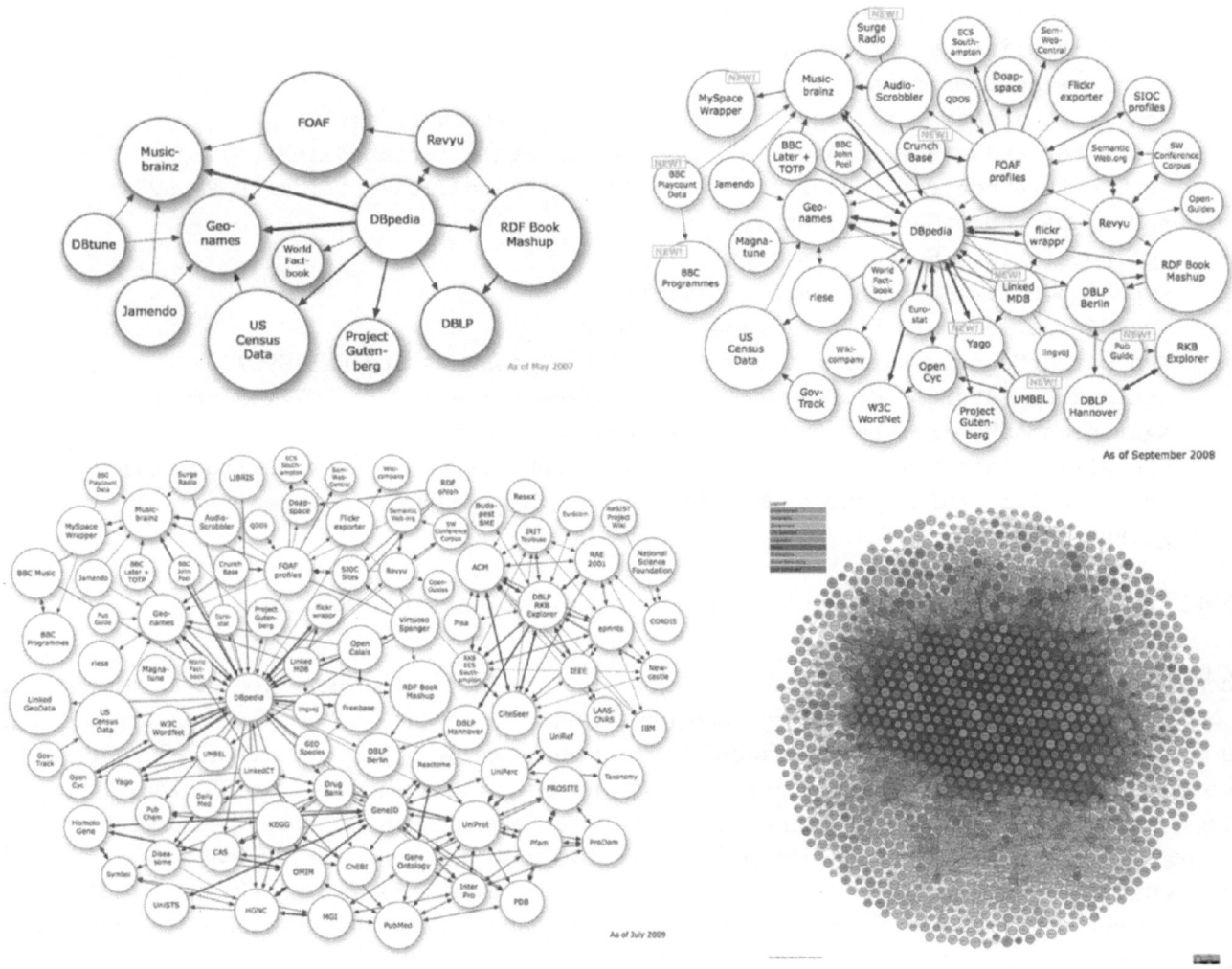

FIGURE 17.6   Growth of the Linked Data Cloud 2007–2020 (source: https://lod-cloud.net/, CC-BY).

standard RDF predicate owl:sameAs, which means that two URIs refer to the same thing. However, as is evident, it is often easier to 'mint' your own URIs rather than refer to one from another dataset. In general, there is a lot of effort involved in verifying that two things are actually identical, and datasets may differ in this judgement. For example, GeoNames makes a distinction between Montreal the city and Montreal the larger administrative region. DBpedia includes RDF owl:sameAs statements for its Montreal URI but declares it to be the same as two GeoNames URIs, which GeoNames would regard as distinct!

On the last point, while this is a specific issue for Linked Data, it is also a common problem when attempting to reason using multiple datasets. As humans we manage to deal quite easily with concepts that are similar but not identical. For example, when we refer to 'Portugal', we could be referring to the administrative country, including Madeira and the Azores, we may mean only the geographic region overlooking the Bay of Biscay or we may even be referring to the Portuguese football team. Often the distinction doesn't matter, and we make it unambiguous only when necessary. For computer knowledge bases, we are often forced to be completely unambiguous from the beginning and not all human knowledge easily fits into such hard and fast categories.

17.3   MINING THE WEB: SEARCH AND SEMANTICS

Only a fraction of the information in Wikipedia is captured in DBpedia, and in general much of the web is in

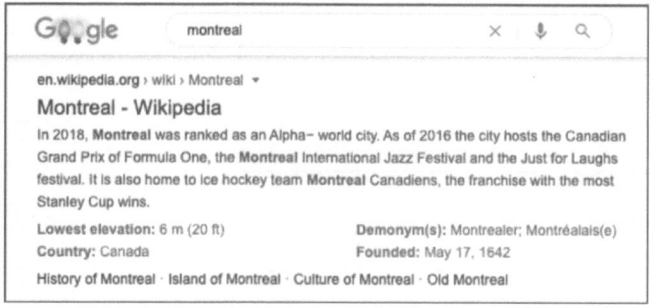

FIGURE 17.7 Half a billion search results for 'Montreal', how does Google rank more useful ones first.

human readable text that is not immediately available for automated reasoning. Applications that want to use the data on the web need to mine this material in some way.

17.3.1 Search Words and Links

Search engines are the most obvious way in which web information is mined to make it more useful. The normal purpose is to give web pages back to the user, although similar techniques are used to help deliver effective adverts.

The earliest search engines were simply indexes, web pages are scanned, the text extracted, broken into words and then the words used to create an index that links a single word, say 'Montreal', to all web pages that contain the word. If several words are used, then the results returned are instead the set of web pages that contain some of the words, typically sorted so that the pages that contain more of the terms appear first.

For a small collection of documents, finding those that refer to terms of interest is sufficient, but once we get to the web with 10 billion or so pages, things get more difficult. If you had to look through nearly 500 million result pages to find the one you were interested in (see Figure 17.7), the web would not be very useful! How does Google search manage to find the most relevant ones to put first?

One of the techniques is to use *properties of the text itself*. If you are searching for the term "Montreal bus station", then 'bus' and 'station' are more common words than 'Montreal'. The pages that mention 'Montreal' and 'bus', but not 'station', will get ranked more highly than those that just mention 'bus' and 'station'; even though both have the same number of search terms, the pages with the more rare terms get ranked more highly. Also

pages that include the words close to one another, say the sentence starting "This bus station is not to be confused ..." will count more highly than if 'bus' and 'station' appear anywhere on the page.

These techniques work for any collection of documents. However, the web is not simply a collection of text documents, but it is a hypertext, the documents are linked to one another. This can be used in two ways.

First, this can be used as extra relevance information for specific words and phrases. For example, if web page A references web page B using a link in the text "Montreal", then this will add to the "Montreal" index for page B as well as page A; possibly even if page B does not explicitly mention Montreal itself. The fact that for the author of page A, B is the page to go as a reference for Montreal counts in page B's favour.

Second, the link structure can be used to generate a measure of importance of a page in general. If lots of pages link to page A, then it suggests that page A is an important page. In the web the link counts in and out of pages tend to follow a power law distribution; there is a small number of pages with lots of link connections and a large number with far fewer, so even this simple measure can be used to help order pages.

Google's PageRank algorithm takes this a step further, by ranking the importance of incoming links by the importance of the pages they come from. This is a circular definition of importance, but the algorithm deals with this by starting with a uniform measure of importance and then iterating, this is effectively a variant of spreading activation (Chap. 3). There are various equivalent ways of thinking about this. One is to imagine an army of people randomly clicking links on pages and then keeping track of how often they end up on different pages. Another, more mathematical, analogy is to think about a massive matrix with one row and one column for each page and the number in the matrices representing the links between the pages. The PageRank is then the principal eigenvector of the matrix.

The actual algorithms used by Google and other search engines add many other factors, for example, there are some well trusted sites (such as Wikipedia), so these can be used to boost the initial weights of PageRank.

```
<p>
A new book is being published An Introduction to Artificial Intelligence,
second edition, by Alan Dix (born Cardiff, 1960)
</p>
```

(i) Original HTML for human readable content

```
<p itemscope itemtype="http://schema.org/Book">
A new book is being published <span itemprop="title">An Introduction to
Artificial Intelligence, second edition</span>,
by <span itemprop="author" itemscope  itemtype="http://schema.org/Person">
  <span itemprop="givenName">Alan</span>
  <span itemprop="familyName">Dix</span>
  (born <span itemprop="birthPlace">Cardiff</span>,
  <span itemprop="birthDate"  datetime="1960-07-28">1960</span>)
</span>
</p>
```

(ii) Same web page with microdata markup

```
{
  type: "http://schema.org/Book",
  title: "An Introduction to Artificial Intelligence, second edition",
  author: {
    type: "http://schema.org/Person",
    givenName: "Alan",
    familyName: "Dix",
    birthPlace: "Cardiff",
    birthDate: "1960-07-28"
  }
}
```

(iii) The extracted information (as JSON)

FIGURE 17.8   Using schema.org markup to make text meaningful for automated tools.

17.3.2 Explicit Markup

HTML pages are not just text but structured documents. This can be used to boost the text techniques above, for example mentions of words in headers will count more highly than in ordinary paragraphs. However, the structure can be used in other ways.

Some of this is a side effect of human readable HTML. Tables are explicitly marked using the `<table>` tag, so they can be extracted as data. Similarly form fields have hidden names for each field, which are used by web browsers to suggest pre-filled values. A special case of this is password input fields which have a special tag `<input type="password">` and often the previous input field is a user name – this is used by the browser to save your passwords.

However, these are fairly limited, for example a company web page may well include a phone number or address, but these are likely to be organised using the same layout tag as ordinary tags.

In order to help search engines and other web tools, some web pages include additional information that is invisible to the human reader but enables automated tools to extract structured information. A number of specific emergent standards (called microformats) arose to tackle specific kinds of content, such as contact details (address, email, etc.) and events (time, place, etc.), but these have now mostly been merged into a single industry standard called schema.org.

Schema.org allows a web page to include markup that is rather like (and in some cases exactly like) a form of lightweight RDF. Existing tags in the HTML have properties added to say what they enclose and where additional markup is needed HTML <span> tags are used, which have no impact on human-readable layout, but allows extra information to be added. Figure 17.8 shows an example of an initial paragraph describing the publication of this book (i), as might appear in, say, a blog or press release. The same information is then shown (ii) with the microdata added that says that the paragraph is talking about a book, that the author is Alan Dix and that he was born in Cardiff. Finally, the extracted information is shown as JSON data in (iii).

This information can be used to help augment searching but also to create structured representations of arbitrary web pages in a similar way to what DBpedia does for Wikipedia pages. It would of course be a lot of work to add this information by hand to the HTML, but authoring tools can help for one off pages. Many pages are themselves generated from databases, and in this case it is a relatively simple matter to add this kind of markup automatically and also to provide parallel RDF versions.

17.3.3 External Semantics

Not every web page has explicit semantic markup, so often structure has to be inferred from the outside. This can use the structural tags of the HTML to help, for example, a part that is set in italic, bold or given a class with a <span> will often represent some sort of name or title.

There are two main ways of looking for this structure, but they often interact.

entity identification – finding representations of things such as people's names, book titles or dates

structured data identification – looking for structure such as tables or relationships between entities

Named entity recognition may use lookup tables, for example lists of people's names or places. This may be based on large public datasets, or local ones, such as your address book. This can be used to tag a word or phrase as a possible location or film title. For places or names there are often several possible matches for the same name, for example there is a region called Montreal in France as well as the city in Canada, not to mention numer-

ous Montreal Hotels across the world. For some purposes having a list of possible matches, or just knowing it is a place name is sufficient, for example if the reader can make the selection. However, often we at least want an order of likelihood. Context can often be used in these cases, one starts with a weighting proportional to common uses, so that Montreal the city in Canada is most highly weighted, but if other entities or the page origin suggest it is about places in France, then Montreal the French region would gain extra weight. Given other named places on the same page might also be ambiguous this might be an iterative process.

Other forms of entity recognition are based on patterns. This can be used to suggest possible named entities, for example a series of words with initial capitals within a body of text might suggest that it is a book or film title. Pattern-based recognition can also be used for various forms of non-textual codes, dates, etc. These are often called data detectors and are found in various applications and built into the Apple operating systems since the late 1990s.

Figure 17.9 shows an example of a pattern to match potential ISBNs in text. It mainly consists of a regular expression that looks for an optional prefix such as 'ISBN10:', followed by a suitable number of digits and dashes and possibly a final 'x' in the case of older length 10 ISBNs. Note that the presence of dashes means that the regular expression is only a possible match, and a further verification stage is needed to count the digits and calculate the checksum is correct; in the Snip!t system that this example is drawn from, this additional verification is performed by a JavaScript constructor for the 'isbn' datatype.

Extensions of this technique can be used to detect more complex structures such as tables, or lists, even when this is not immediately apparent in the HTML markup, or if the document is text or a PDF. For example, if multiple lines contain a high proportion of numbers and additional spaces or tabs, this might trigger a table detector. Similarly if something that looks like a personal name occurs in a paragraph with an email address and telephone number, then this might trigger a 'contact' detector, binding the various elements together and creating an 'add to address book' suggested action. This kind of detector may also use cue phrases in the text, for example, 'let's meet' close to a date or time would cue a meeting recogniser that would look for a place, etc. and trigger an 'add to calendar' suggestion.

```
<regexp>
  <name>RE_ISBN</name>
  <regexp>(ISBN(10|-10|13|-13){0,1}[:.]{0,1}(\s)*){0,1}([0-9]([0-9]|-){8,15}[0-9xX])</regexp>
</regexp>
    <!-- 13 digit ISBN up to 4 dashes as well as digits, so min 8 and max 15 digits
         or dashes between initial digit and final checkdigit
    -->


<simpleregexprecogniser>
  <name>isbn_recogniser</name>
  <title>ISBN recogniser</title>
  <pattern>
    <pre_context>\W</pre_context>  <match>$RE_ISBN</match>  <post_context>\W</post_context>
  </pattern>
  <fields>
    <field name="isbn" value="ALL" />
    <field name="prefix" value="FIELD 1" />
    <field name="rawnumber" value="FIELD 4" />
  </fields>
  <match>
    <type>isbn</type>
    <description>ISBN $$</description>
    <match_name>isbn</match_name>
  </match>
</simpleregexprecogniser>
```

FIGURE 17.9   Data detector for ISBNs in text (from https://snipit.org/tellmeabout/).

As well as this form of generic pattern recogniser, hand-crafted tools are often used for scraping data from specific websites. For example, the reference manager Zotero has more than 500 crafted JavaScript 'translators' for extracting information from websites that might mention referenceable items including those of journals, news sites and booksellers. Figure 17.10 shows a portion of the Zotero translator for Amazon; it is scanning the site for tags with particular names or classes.

These various forms of pattern matching, data detectors, recognisers and translators can be seen as a form of AI in their own right. Some are simple, but others involve complex matching algorithms or machine learning. In addition, it is quite common for various AI or machine learning projects to start with web scraping in order to gather data.

17.4 USING WEB DATA

We have seen how structured data can be made available on the web, how web pages can have additional markup to make them more machine readable and how even plain web pages can be analysed to extract meaning. However, we want to use this information.

17.4.1 Knowledge-rich Applications

Both structured and unstructured data can be used as part of human-in-the-loop applications. For example, if you want to know the height of the Eiffel Tower, you could be presented with several snippets of the text from the web that appear to show the height of the Eiffel Tower. If one of the snippets starts "*This model of the Eiffel Tower stands 3.2 metres tall ...*", the human reader can easily skip this and move on to more relevant results. Similarly, if a web assistant extracts fields from a web form and presents them as a suggested calendar entry, the user can choose to ignore or amend the fields. It is also clear how the structured data in RDF and in semantic markup can be used as part of a knowledge base for automated reasoning.

Part of the power of human reasoning is our ability to draw on a wide range of knowledge accumulated dur-

```
var productClass = attr(doc, 'div[id="dp"]', 'class');
if (!productClass) {
    productClass = attr(doc, 'input[name="storeID"]', 'value');
}
if (!productClass) {
    if (doc.getElementById('dmusic_buybox_container')) {
        productClass = 'music';
    }
}
```

FIGURE 17.10 Fragment of Zotero translator for scraping an Amazon product page (from https://github.com/zotero/translators/blob/master/Amazon.js).

ing our lives, some very specific and precise (*"The Eiffel Tower is 324m high."*), some more qualitative (*"The Eiffel Tower is a lot taller than most buildings."*) and some more analytic (*"If building A is taller than building B and building B is taller than building C, then building A is taller than building C."*). In early AI systems the latter was easier to create than the former. There have been a number of projects over the years to gather declarative knowledge of the first kind in volume (see Chap. 2), but many have faltered due to the scale of the task. The web has changed that.

Crucially there is a trade-off in human and AI reasoning between using declarative knowledge and analytic rules. For example, you might remember your multiplication tables or just work things out quickly on your fingers. The web allows automated reasoning to have far more knowledge and therefore rely less on analytic methods. This may make it possible for a system to produce answers that appear to be very intelligent, without deep understanding.

In 2011 the IBM AI program Watson beat human contestants in Jeopardy!, a US general knowledge quiz show. This involved techniques to understand what was being asked [101] but also many terabytes of data gathered from digital sources including traditional dictionaries and encyclopaedias as well as selected high-quality web resources, in particular Wikipedia. It used a number of specialised natural language processing techniques to extract useful knowledge from free text and turn it into a form that could be interrogated to answer questions faster than the human contestants.

Watson was very careful in its selection of sources, and thus could have high confidence in the answers. However, many of the kinds of automated extraction described in Section 17.3.3 are uncertain, offering possible data but with less confidence. This is problematic for totally automated reasoning systems. However, the sheer volume of web information comes to our rescue.

Imagine you have an automated reasoner that needs to know the height of the Eiffel Tower but doesn't have it in high-quality sources. The system can gather web pages that mention Eiffel Tower and scan these for the word height near a length number. This is then used to create a collection of 'heights' for the Eiffel Tower. This would include 3.2m from the web page with the sentence *"This model of the Eiffel Tower stands 3.2 metres tall …"*, but there would be many more that say around 324–330m. Choosing a mid-point of the most common cluster of values would almost certainly give a good estimate.

We can see that even very uncertain semantics can be boosted by volume.

Note too that between the first draft of this chapter and publication, the true figure changed from 324m to 330m as the radio antenna at the top of the tower was replaced with a longer one [49]. At the time of writing there were still ten times as many Google search results for *'height of the Eiffel Tower 324m'* compared with *'height of the Eiffel Tower 330m'*; however, the top results for *'height of the Eiffel Tower'* all show 330m as Google's PageRank algorithm is favouring the more up-to-date and reliable sources. That is while volume is important, some knowledge of authority can also help.

17.4.2 The Surprising Power of Big Data

This effect of volume can be used to create data-oriented solutions to problems which would appear to need deep understanding.

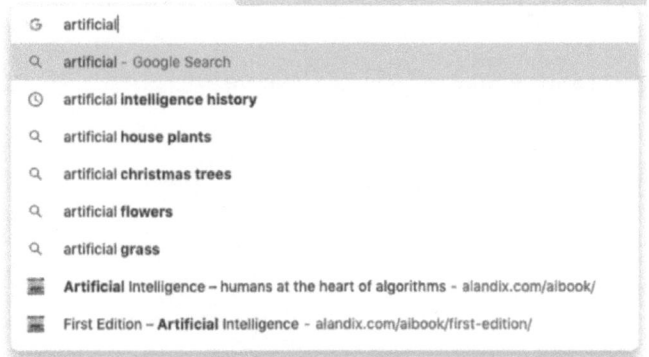

FIGURE 17.11 Search suggestions after typing 'artificial' into Google. The first five suggestions will be based on popular search terms, the last two are probably using the author's previous browsing history.

A simple example of this is search suggestions. As you begin to type into a search box, your typing is compared with millions of previous search terms and the most common are reflected back as suggestions (see Figure 17.11). This appears 'intelligent', guessing what you want before you type it but in fact can be built using a simple index (albeit very large) and counting. Similar techniques are used to offer auto-completion when typing an email message, although this is also often personalised.

This type of technique has proved very powerful in tackling natural language processing tasks without needing to create the rich grammars that are needed in the techniques described in Chapter 13. Statistical techniques have a long history in NLP, for example Hidden Markov Models (see Chapter 14) use a probabilistic state machine to predict the next word from the current state and word. HMMs are trained using large amounts of continuous text. Other methods are based on adjacency statistics, most commonly n-gram frequencies (see also Chap. 13).

An n-gram is simply a collection of n successive words in the text. The simplest 1-gram is simply the frequency of each word. For example, in the last paragraph the top word is 'the' with frequency 3 followed by six words that occur twice (a, are, in, see, state, word) and then another 50 or so that occur just once. Two-grams are word pairs; there are no pairs that appear more than once in the previous paragraph, but at the point of writing the 2-gram "are used" occurs five times in the chapter as a whole, so it would be in the chapter 2-gram list with a frequency of 5. Similarly at the same moment of time the chapter

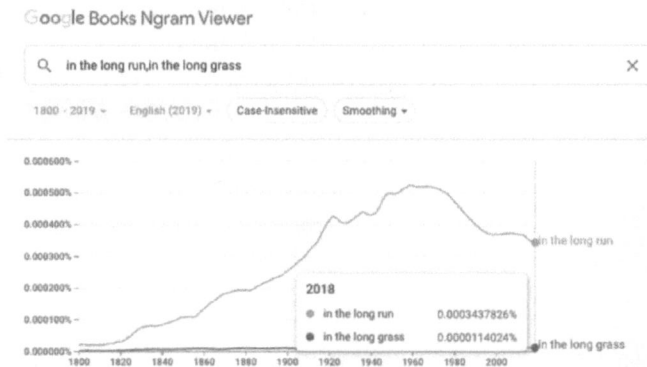

FIGURE 17.12 Google Books Ngram Viewer (captured 28/12/2020).

contains the 3-gram "a web page" three times, including this one.

As is evident even 2-grams (bigrams) become quite rare unless the document or document collection is large. Of course the web is very large, and so it is possible to obtain frequency statistics for quite long n-grams. Google also does this for n-grams in the scanned text in Google Books. With large n-grams it is possible to perform apparently intelligent tasks relatively easily. For example, to obtain predictive text in an email, the last n words can be looked up as the first words in the n+1-gram statistics. If there is a frequent enough n+1-gram, then the last word (or top few) is used as the next prediction. If there are no matching n+1 grams, the last n-1 words are considered.

Figure 17.12 shows statistics from Google Books n-grams. You can see how there are a substantial number of 4-grams for "in the long run" (0.000344% of all 4-grams), although far less than in the 1960s. In contrast "in the long grass" only occurs about 1/30th as often (0.000011% of all 4-grams). If you typed "In the long", then "run" would be an obvious next suggestion.

As well as text prediction n-grams can be used to build sophisticated language models. For example imagine clustering words based on whether they are preceded or followed by the same words. So that "in" and "on" might get clustered together as they are often followed by "the" or "that". If each cluster is given an id and then the words are replaced by the cluster ids, we can do the same thing to build higher level clusters. This can be used to create emergent grammatical categories for any language.

Similar techniques are used for translation services based on corpora where there are existing translations (say English and Welsh versions). Large numbers of examples are used to match phrases in the two languages, so that "bore 'ma" often occurs in the Welsh corresponding to sentences with "this morning". This will not be one-to-one correspondence as the same phrase may translate in different ways, but with a sufficiently large number of examples, translators can be built on a phrase-by-phrase basis. By using overlapping phrases grammatical changes, such as different sentence orderings, can be managed.

17.5 THE HUMAN WEB

While some of the content we see on the web is generated directly from data, the majority of our day-to-day use will be with material that has been crafted by human hands: news items, Wikipedia pages and social media. However, as social media makes clear, the web is not merely a static output of human authoring, but a place where there is a constant stream of human activity. Indeed the web science community refer to the web as a social machine [129, 252].

This human activity, the things we do on the web, is both an opportunity for AI, and especially ML, to help us, but also the acts themselves are a source of input into many AI algorithms. Some of these are built using data that is passively gathered from web users, such as usage patterns, some is explicitly gathered, for example, when you give a star rating. Sometimes the results of this feed back into the web itself, for example in targeted advertising or prioritising of social media posts, but the data can also be used for other purposes.

17.5.1 Recommender Systems

When you look at a product on Amazon or a film on Netflix, you are presented with suggestions of other products and films you might also like. Some of this is based purely on raw popularity, the top ten films or music in the genre. However, we know that much of this is far more targeted, sometimes implicitly (those suggestions that just seem prescient) and sometimes explicitly "people who watched X also watched Y". Collectively, and for obvious reasons, these are called recommender systems, and we've already mentioned them in Chapters 7 and 8.

There are very many algorithms used in recommender systems and also different kinds of data, but largely they are based around an event stream of the form:

```
At time <t> user <u>
    interacted with product <p>
    in an event <e>
```

The events may be of different kinds, for example viewing a product and buying it, or checking the information about a film vs. watching it. In addition, there may be some extra information attached to the event, for example if you are asked to star rate a film, or how long you viewed the information about a product.

When users are asked to rate a product or film, this is called relevance feedback. You will also see this used in other kinds of interactions, such as customer support sites which ask you to rate the answer to a query. We look in detail at intelligent interactive systems in Chapter 19.

This stream of $<t,u,p,e>$ events is complex to handle in a raw state, so is often reduced to relatively simple matrices. Let's look at just one way to do this.

We'll turn the event stream into a matrix, let's call it M, where the rows are people and the columns products. Each cell is some sort of score measuring the level of positive engagement with the product. The score will often be a weighted combination of factors, with some factors, such as whether the person bought/watched it or a star rating, weighted more highly than others, such as viewing the information. Indeed, viewing the information on a film and then never watching it might be included as a negative score.

Note a few things:

1. We have completely ignored some information, such as the order we viewed two different items.

2. We have reduced the stream of events for each product to a single score.

3. The resulting matrix is sparse, that is most of the entries are zero. On a large site each person will only engage with a relatively small number of products and vice versa.

The first two are a form of initial data reduction, which is a common theme in all big data analysis. The volume of data is often too big even with massive computational resources, so we have to find ways to progressively simplify it (see also Chapter 8).

```
score(product A, product B)
  sum = 0
  foreach user u
    sum = sum + M(u,A) * M(u,B)
  return sum
```

FIGURE 17.13 Pseudocode for product–product score based on multiplying user scores. Note this is schematic, not optimised for sparse matrix M.

The last is more about data representation but also influences the kinds of algorithms we apply. Let's look at the scale of this matrix:

N – total number of users (many millions in a typical e-commerce site, billions for internet giants)

P – products (many thousands on movie sites, millions on large eCommerce sites such as Amazon)

E – typical number of products engaged with by a single user over the time frame of the modelling, say a few hundred per year

The total size of the matrix M is $N \times P$, but the number of non-zero entries is only N×E, that is only E/P of the cells have numbers in. If we take a million users, with 10,000 products, the sparse matrix has perhaps 100 million non-zero entries (big enough), but the complete matrix has 10 billion cells. It is clear we want to avoid any sort of algorithm whose time taken, or memory used is based on the complete matrix.

As there are typically more users than products, so often it is the users that are 'simplified' further using some form of clustering into classes of users or representation as a vector of characteristics. Both of these can be obtained through automated processes, for example, the self-organising maps in Chapter 5, although at each stage algorithms often have to be adapted to deal with scale.

Another approach is to create product–product scores by summing over users. Figure 17.13 shows a simple example of this, each user's scores for a pair of products are multiplied together and add to the relevant product–product score. If you know your matrix algebra, you might recognise this as transpose(M)×M.

This code is rather oversimplified in two senses. First, it is not optimised for the sparseness of the matrix M, but there are ways of doing this. Second, one might want to combine the users' scores in other non-linear ways, not just multiply them. For example, if the same user has a positive score for two products, we might count this more strongly than a positive and a negative, or two negative scores.

Even this score matrix is quite large (P×P), so further data reduction might be necessary. However, let's assume t is manageable. We can then see how this can be used in practice.

When the user is looking at product A, and finds other products B_1, B_2, ... where $score(A, B_i)$ is large, this is the simple "customers who liked/bought A also liked B". Adding a little more sophistication, one could rank the B_i by also looking at other products P_1, P_2, ... the user had recently engaged with and using the values of $score(B_i, P_j)$ to enable a more personalised ranking of suggestions.

17.5.2 Crowdsourcing and Human Computation

One of the defining features of the web has been its use of crowdsourcing, engaging very large numbers of people in the creation of knowledge as opposed to small numbers of professionals. Sometimes this is voluntary as in the case of Wikipedia with over a quarter of a million active contributors. However, it may also be paid, as in the case of Amazon Mechanical Turk and many other sites offering 'human intelligence tasks' to the lowest bidder. Finally, there are tasks that happen, possibly without you being aware that you are doing them as they are part of a game or login process.

These human tasks vary in their complexity. Some involve expertise or creativity including design and writing work. Some are more low-level such as checking/correcting text generated by OCR (optical character recognition). The latter are sometimes also called human computation, a phrase that conjures up images of the worker as a cog in the machine; in fact not an unfair portrayal of much of the work.

One of the earliest examples of human computation was reCAPTCHA codes, used as part of an authentication process to ensure that the user was human and not an automated bot. These asked the user to type the text in several slightly broken or blurry images. The images were derived from scans of books and news articles where OCR had either failed or had low confidence. Two images were shown, one had known text and was used to verify the user was human, the other had unknown text

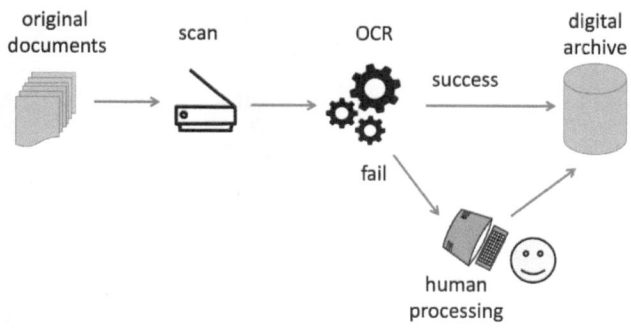

FIGURE 17.14 Early reCAPTCHA – human as gap filling for automated systems.

and the user was effectively providing a transcription. If enough users typed the same text, it became 'known' and the relevant OCR updated.

This early example was effectively filling in gaps in the automated system (see Figure 17.14). OCR has become increasingly sophisticated, often including rich language models to help choose where the text is ambiguous, even for poor quality or hand-written text. This is true of many kinds of automated systems. If the gaps are occasional and can be detected, then it may be more cost effective to have the task completed by people rather than trying to create ever more complex automated systems, even where this is possible.

You may also have used reCAPTCHA codes that show you a number of images and ask you whether these include cyclists or road crossings. This is also being used as part of an intelligent system, but rather than filling in gaps in the processing you are preparing training materials.

Deep learning and other machine learning algorithms have allowed the development of highly accurate image recognition systems. However, to do this they require large quantities of training data. It is easy to gather vast quantities of image data, from dashboard cameras in cars to satellite imagery. If these are tagged to describe the features seen 'person', 'cloud', 'mountain', then the machine learning system can work on them. The difficult thing is adding tags as this is a huge quantity of images. When you complete the image reCAPTCHA codes, you are doing precisely that, adding tags so that Google can use these to train the image processing systems for autonomous cars (Figure 17.15).

Both of these can happen in the same system. For example, many apparently AI-driven chatbots have human backup. First they attempt automatic recognition of your

question. If this succeeds with high enough confidence, then the system responds. However, if the confidence is low, there is a pool of back-up humans who attempt to interpret the user's request. The human-generated response is then used as further training data so that if the system encounters a similar request in future it may be able to respond automatically.

17.5.3 Social Media as Data

We have already seen how the large corpora of text in the web and social media can be used to derive language models and structured knowledge simply by its volume. In addition, social media can be used as a source of raw data for sophisticated (and simple) real-time analysis.

The national health services and the WHO have extensive monitoring processes for doctors and health workers to report potential outbreaks of endemic diseases from the common cold to Ebola. This can then be used to direct resources or generate emergency responses as necessary. During the Covid-19 pandemic, the same authorities were trying to gauge public sentiment as to whether they would respond well to various public health measures and to predict uptake of vaccination.

However, often changes in search behaviour, Twitter/X and other social media content precede official channels. When seasonal flu begins, there will be more web searches for symptoms and more phrases in tweets and posts suggesting poor health. Similarly language analysis of tweets from early in the Covid-19 outbreak included levels of fear or surprise [Medford, 2020]. This data cannot always be taken at face value as small numbers of people account for large volumes of activity, and these are often, depending on the platform, slanted towards younger and more affluent users. However, with sufficient processing these data sources can be used as indicators of change and trends without relying on them uncritically.

Similar techniques have been used by national power regulators. It may take several minutes for local power generation and distribution companies to inform their national grid about outages. This time can be critical to prevent catastrophic knock-on failures when problems in one area cause power surges in neighbouring regions. However, spikes in '#powercut' tweets can start within seconds allowing early warning of impending problems.

A slightly different version of similar techniques has been used to monitor public opinion shifts during political campaigns at a far finer granularity than is

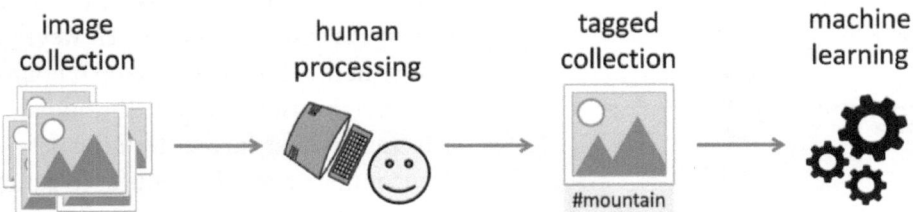

FIGURE 17.15   More recent reCAPTCHA – human creating training data for machine learning.

possible with traditional opinion polls. Many online betting sites allow users to place bets on political campaigns such as the US presidential election, and the sites change the odds based on betting patterns. So, the odds during a live debate can give a near instant reading of whether watchers think this has shifted the overall sentiment [293]. Of course, the raw result is focused on those who are both watching the debate and also online gamblers, but again this can be raw data for more sophisticated machine learning algorithms.

17.6 SUMMARY

We've seen that the Semantic Web and in particular RDF and linked data can facilitate web-scale reasoning. Triplestores can be used to store RDF including meta-information and provenance, while SPARQL allows queries. Various forms of markup can also make human-readable web pages in HTML available for automated reasoning, but text-mining techniques can also be used. The vast quantity of information allows new forms of knowledge-rich reasoning, but also simple statistical techniques can be surprisingly powerful, allowing apparently intelligent understanding of text based on vast quantities of essentially frequency data. The web is ultimately a social machine where people and machines interact, this leads to opportunities to learn automatically from similarities and communications between people and to explicitly include people in data production and management through crowdsourcing.

17.1 Following the pattern in Section 17.2.1 represent all of the *entity–entity relationships* in Figure 2.6 as triples. Note, in the semantic network diagram in Figure 2.6, the entities are the objects or concepts enclosed in boxes. For this exercise, ignore the properties that are not relationships between entities, that is: 'is carnivorous', 'is wild', 'has/has no tail', 'barks/doesn't bark', 'has 4 legs', 'is tall', 'is brown' and 'is drawn'.

17.2 Now represent the *properties* in Figure 2.6 as triples. Think carefully about whether the property is boolean (true/false) or has some numeric or other type of value.

17.3 Imagine you are trying to create a name recogniser, rather like the ISBN recogniser in Figure 17.9. You want it to match simple names such as 'Alan Dix', 'J Finlay', 'J.E. Finlay' and 'Alan John Dix'. That is any number of forenames including optional initials followed by a surname. Initials may or may not be followed by a full stop. You can assume that names always start with a capital letter but do not match fully capitalised names such as "Alan DIX".

a. Create a small number of regular expression that together will match names. That is any valid name should match at least one of the regular expressions, but no non-name should match any of them. Try to do this using as few regular expressions as possible.

b. Can you think of names of people that this would not match?

Note: If you are not sure of how to write regular expressions, look up the entry for regular expression in the book's online glossary.

FURTHER READING

S. Brin and L. Page. The anatomy of a large-scale hypertextual web search engine. *Computer networks and ISDN Systems*, 30(1–7):107–117, 1998. Elsevier.

This article, by the founders of Google, describes the rationale, development and architecture of the first Google prototype including the first version of the PageRank algorithm. As well as the historical interest, the article describes how the authors dealt with issues of scale as well as pragmatic issues such as size-speed trade-off for compression and emails from site owners due to misunderstandings about web-crawlers.

T. Berners-Lee, J. Hendler, and O. Lassila. The semantic web. *Scientific American*, 284(5):34–43, 2001.

The article that lays out the vision for the semantic web.

G. Antoniou and F. Van Harmelen. *A semantic web primer.* MIT Press, 2004.

The classic book about semantic web technology.

C. Aggarwal, editor. *An introduction to social network data analytics.* Springer, Cham, 2011.

Edited collection covering many aspects of the analysis of network data as found in the web and social media.

R. Zafarani, M. Abbasi and H. Liu. *Social media mining: An introduction.* Cambridge University Press Cambridge, 2014.

More specialised volume on the specific issues of social media data. This includes some general network-analysis techniques but also more pragmatic issues related to social media data gathering.

IV

Humans at the Heart

Expert and Decision Support Systems

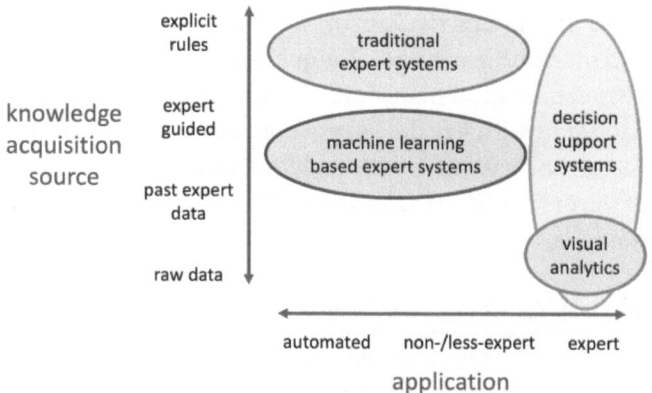

FIGURE 18.1 Different forms of expert involvement.

18.1 OVERVIEW

Expert systems were one of the early success stories of AI, with medical advice systems dating back to the 1970s. In this chapter we will look at systems where AI in various ways seeks to aid, learn from and in some cases replace experts. As well as expert systems, which attempt to capture knowledge explicitly, we will consider decision support systems, which use a variety of statistical and AI techniques to offer advice or supporting information for experts making their own decisions. In particular, visual analytics combines interactive machine learning and advanced visualisation. We will see that knowledge acquisition may be explicit, seeking to draw out the expert's understanding of a domain, or implicit based on the expert's past behaviour and data. Explanation is critical at various stages of this process, an issue we will return to in Chapter 21.

18.2 INTRODUCTION – EXPERTS IN THE LOOP

The core feature of the systems and techniques we will cover in this chapter is that experts are in some way involved. The term 'expert' is itself broad as we are all expert at something. We will principally be dealing with uncommon expertise, such as medical knowledge, but many of the techniques would also apply for more day-to-day but still individual or unique perspectives, such as the way your grandma makes pasta. Experts can be in-

volved in two different ways: they can be used as sources during knowledge acquisition in the construction of an AI system; or they can be users applying the outputs of the system (Figure 18.1).

Traditional expert systems (Section 18.3) ask users to explicitly encode their knowledge in rules or one of the other forms of knowledge representation dealt with in Chapter 2. This may then be used entirely automatically or by people with less expertise. For example, you might follow the instructions from a cookery application based on your grandma's expertise. This process of knowledge elicitation can be difficult, and we will look at this in more detail in Section 18.4.1. Machine learning can be used either alongside this or to implicitly capture expert knowledge from past data. For example, a system might use data about past patient symptoms combined with (human) treatment decisions and use this to build automated recommendations.

Decision support systems (Section 18.6) provide information and visualisations to help experts make deci-

DOI: 10.1201/9781003082880-22

sions or predictions. This may be based on the expert's own knowledge or the knowledge of peers, for example natural language processing techniques to retrieve past cases similar to a current patient's symptoms. More often they are based more on pure data analysis, for example the (highly controversial) systems used by some courts that offer predictions of the likelihood of a felon re-offending [7].

A special case of decision support is visual analytics (Section 18.6.3). Many expert systems and decision support systems pre-process past data and knowledge and then use this as a relatively static resource. In contrast, visual analytics systems offer the expert the ability to explore data interactively using a combination of statistical and machine learning techniques and rich visualisations. This can be used historically to gain understanding of a dataset or with live data as part of decision making. Note that in the case of visual analytics, one of the outcomes of the process is to increase expert understanding of phenomena.

18.3 EXPERT SYSTEMS

An expert system is an AI program that uses knowledge to solve problems that would normally require a human expert. The knowledge is collected from human experts and secondary knowledge sources, such as books. The knowledge is represented in some form, often using logic or production rules, although forms of neural networks are increasingly common. The system includes a reasoning mechanism as well as heuristics for making choices and navigating around the search space of possible solutions. It also includes a mechanism for passing information to and from the user. Even from this brief overview you can probably see how the techniques that we have already considered might be used in expert system development.

We can think of an expert system as operating in two main phases (see also Figure 18.2):

knowledge acquisition – A knowledge engineer works with experts in order to elicit and then represent their knowledge.

application – The running system based on that knowledge is used by non-experts to help them make judgements or may be applied automatically.

In this section we will look at the latter, assuming that expert knowledge has been captured and represented, and we will return to the question about how this is acquired in Section 18.4.

18.3.1 Uses of Expert Systems

If an expert system is a program that performs the work of human experts, what type of work are we talking about? This is not an easy question to answer since the possibilities, if not endless, are extensive. Commercial expert systems have been developed to provide financial, tax-related and legal advice; to plan journeys; to check customer orders; to perform medical diagnosis and chemical analysis; to solve mathematical equations; to design everything from kitchens to computer networks; and to debug and diagnose faults. And this is not a comprehensive list. Such tasks fall into two main categories:

diagnosis and advice – Those that use evidence to select one of a number of hypotheses; and

design and planning – Those that work from requirements and constraints to produce a solution which meets these.

So why are expert systems used in such areas? Why not use human experts instead? And what problems are candidates for an expert system? To take the last question first, expert systems are generally developed for domains that share certain characteristics.

rare expertise – First, human expertise about the subject in question is not always available when it is needed. This may be because the necessary knowledge is held by a small group of experts who may not be in the right place at the right time. Alternatively it may be because the knowledge is distributed through a variety of sources and is therefore difficult to assimilate.

problem clarity – Secondly, the domain is well defined and the problem clearly specified. At present, as we discovered in Chapter 2, AI technology still struggles to handle common sense or general knowledge very well, but expert systems can be very successful for well-bounded problems.

willing experts – Thirdly, there are suitable and willing domain experts to provide the necessary knowledge to populate the expert system. It is unfeasible

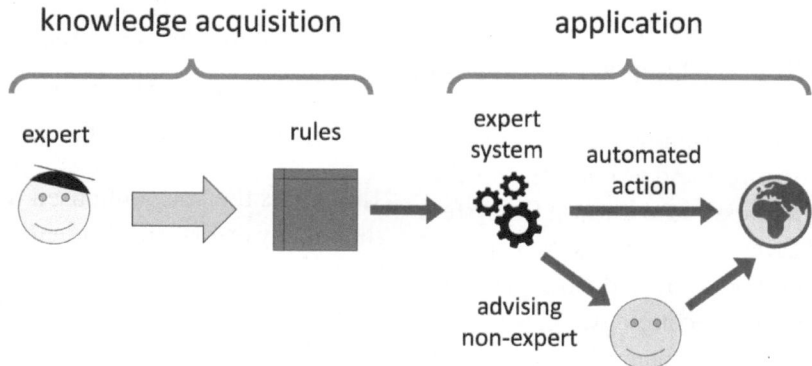

FIGURE 18.2   Expert system capturing and applying knowledge.

to contemplate an expert system when the relevant experts are either unwilling to co-operate or are not available.

limited scope – Finally, the problem is of reasonable scope, covering diagnosis of a particular class of disease, for example, rather than of disease in general.

If the problem fits this profile, it is likely to benefit from the use of expert system technology. In many cases the benefits are in real commercial terms such as cost reduction, which may go some way to explaining their commercial success. For example, expert systems allow the dissemination of information held by one or a small number of experts. This makes the knowledge available to a larger number of people, and less skilled (so less expensive) people, reducing the cost of accessing information. Expert systems also allow knowledge to be formalised. It can then be tested and potentially validated, reducing the costs incurred through error. They also allow integration of knowledge from different sources, again reducing the cost of searching for knowledge. Finally, expert systems can provide consistent, unbiased responses. This can be a blessing or a curse depending on which way you look at it. On the positive side, the system is not plagued by human error or prejudice (unless this is built into the knowledge and reasoning), resulting in more consistent, correct solutions. On the other hand, the system is unable to make value judgements, which makes it more inflexible than the human (e.g. a human assessing a loan application can take into account mitigating circumstances when assessing previous bad debts, but an expert system is limited in what it can do).

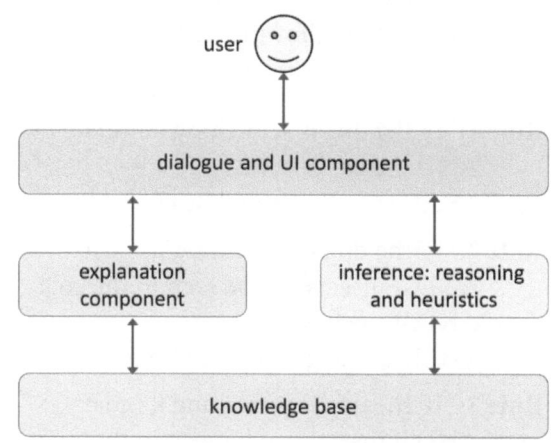

FIGURE 18.3   Typical expert system architecture.

18.3.2 Architecture of an Expert System

An expert system comprises a number of components, several of which utilise the techniques we have considered so far (see Figure 18.3).

Working from the bottom-up, we require: (i) knowledge; (ii) a reasoning mechanism and heuristics for problem solving (e.g. search or constraint satisfaction); (iii) an explanation component; and (iv) a dialogue component or user interface. We have considered the first two of these in previous chapters and will come back to them when we consider particular expert systems. Before that, let us look in a little more detail at the last two.

18.3.3 Explanation Facility

It is not acceptable for an expert system to make decisions without being able to provide an explanation for the basis of those decisions. Clients using an expert system need to be convinced of the validity of the conclusion drawn before applying it to their domain. They also need to be convinced that the solution is appropriate and applicable in their circumstances. Engineers building the expert system also need to be able to examine the reasoning behind decisions in order to assess and evaluate the mechanisms being used. It is not possible to know if the system is working as intended (even if it produces the expected answer) if an explanation is not provided. So explanation is a vital part of expert system technology.

There are a number of ways of generating an explanation, the most common being to derive it from the goal tree that has been traversed. Here the explanation facility keeps track of the subgoals solved by the system and reports the rules that were used to reach that point. For example, imagine the following very simple system for diagnosing skin problems in dogs.

Rule 1: IF the dog is scratching its ears
AND the ears are waxy
THEN the ears should be cleaned

Rule 2: IF the dog is scratching its coat
AND if insects can be seen in the coat
AND if the insects are grey
THEN the dog should be treated for lice

Rule 3: IF the dog is scratching its coat
AND if insects can be seen in the coat
AND if the insects are black
THEN the dog should be treated for fleas

Rule 4: IF the dog is scratching its coat
AND there is hair loss
AND there is inflammation
THEN the dog should be treated for eczema

Imagine we have a dog that is scratching and has insects in its coat. A typical consultation would begin with a request for information, in an attempt to match the conditions of the first rule "is the dog scratching its ears?", to which the response would be no. The system would then attempt to match the conditions of rule 2, asking "is the dog scratching its coat?" (yes), "can you see insects in the coat?" (yes), "are the insects grey?". If we respond yes to this question, the system will inform us that our dog needs delousing. At this point if we asked for an explanation, the following style of response would be given:

It follows from rule 2 that

If the dog is scratching

And if insects can be seen

And if the insects are grey

Then the dog should be treated for lice.

This traces the reasoning used through the consultation so that any errors can be identified and justification can be given to the client if required. However, as you can see, the explanation given is simply a restatement of the rules used, and as such is limited.

In addition to questions such as "how did you reach that conclusion?" the user may require explanatory feedback during a consultation, particularly to clarify what information the system requires. A common request is "why do you want to know that?" when the system asks for a piece of information. In this case the usual response is to provide a trace up to the rule currently being considered and a restatement of that rule. Imagine that in our horror at discovering crawling insects on our dog we hadn't noted the colour – we might ask to know why the system needs this information. The response would be of the form

You said the dog is scratching

and that there are insects.

If the insects are grey

then the dog should be treated for lice.

Notice that it does not present the alternative rule, rule 3, which deals with black insects. This would be useful but assumes look-ahead to other rules in the system to see which other rules may be matched.

This form of explanation facility is far from ideal, both in terms of the way that it provides the explanation and the information to which it has access. In particular it tends to regurgitate the reasoning in terms of rules and goals, which may be appropriate to the knowledge engineer but is less suitable for the user. Ideally, an explanation facility should be able to direct the explanation towards the skill level or understanding of the user. In addition, it should be able to differentiate between the domain knowledge that it uses and control knowledge, such as that used to control the search. Explanations for users are best described in terms of the domain; those for engineers in terms of control mechanisms.

In addition, rule tracing only makes sense for backward reasoning systems, since in forward reasoning it is not known, at a particular point, where the line of reasoning is going.

For these reasons researchers have looked for alternative mechanisms for providing explanations. One approach is to maintain a representation of the problem-solving process used in reaching the solution as well as the domain knowledge. This provides a context for the explanation: the user knows not only which rules have been fired but what hypothesis was being considered. More complex explanations may include a domain model [269] (rather like the blocks world model of SHRDLU, but more complex for the real world) or meta-knowledge. In order to do this successfully, expert systems must be designed for explanation.

18.3.4 Dialogue and UI Component

The dialogue component is closely linked to the explanation component, as one side of the dialogue involves the user questioning the system at any point in the consultation in the ways we have considered. However, the system must also be able to question the user in order to establish the existence of evidence. The dialogue component has two functions. First, it determines which question to ask next (using meta-rules and the reasoning mechanism to establish what information is required to fire particular rules). Secondly, it ensures that unnecessary questions are not asked, by keeping a record of previous questions. For example, it is not helpful to request the model of a car when the user has already said that they don't know its make.

The dialogue could be one of three styles:

- system controlled, where the system drives the dialogue through questioning the user

- mixed control, where both user and system can direct the consultation

- user controlled, where the user drives the consultation by providing information to the system.

Most expert systems use the first of these, the rest the second. This is because the system needs to be able to elicit information from the user when it is needed to advance the consultation. If the user controlled the dialogue, the system might not get all the information required. Ideally a mixed dialogue should be provided, allowing the system to request further information and the user to ask for "why?" and "how?" explanations at any point.

Expert systems for large-scale use deployed over the web or in phone apps may use other forms of interaction, including menus or forms for initial input possibly combined with chatbots.

18.3.5 Examples of Four Expert Systems

To illustrate how the components that we have looked at fit together we will consider four early expert systems. Although these systems are far from up-to-date, they were systems that were groundbreaking when they were built, and they have all been successful in their domains. As such they rank among the "classics" of expert systems and therefore merit a closer look. In each case we will summarise the features of the expert system in terms of the key components we have identified. This will help you to see how different expert systems can be constructed for different problems. In each case, consider the problem that the expert system was designed to solve, and why the particular components chosen are suited to that task.

18.3.5.1 Example 1: MYCIN

MYCIN was an expert system for diagnosing and recommending treatment of bacterial infections of the blood (such as meningitis and bacteremia) [258]. It was developed at Stanford University in California in the 1970s and became a template for many similar rule-based systems. It was intended to support clinicians in the early diagnosis and treatment of meningitis, which can be fatal if not treated in time. However, the laboratory tests for these conditions take several days to complete, so doctors (and therefore MYCIN) have to make decisions with incomplete information. A consultation with MYCIN begins with requests for routine information such as age, medical history and so on, progressing to more specific questions as required.

- *Knowledge representation.* Production rules (implemented in LISP).

- *Reasoning.* Backward chaining, goal-driven reasoning. MYCIN uses certainty factors to reason with uncertain information.

- *Heuristics.* When the general category of infection has been established, MYCIN examines each candidate diagnosis in a depth first manner. Heuristics are used to limit the search, including checking all premises of a possible rule to see if any are known to be false.

- *Dialogue/explanation.* The dialogue is computer (or system) controlled, with MYCIN driving the consultation through asking questions. Explanations are generated through tracing back through the rules that have been fired. Both "how?" and "why?" explanations are supported.

18.3.5.2 Example 2: PROSPECTOR

PROSPECTOR was an expert system to evaluate geological sites for potential mineral deposits, again developed at Stanford in the late 1970s [95]. Given a set of observations on the site's attributes (provided by the user), PROSPECTOR provides a list of minerals, along with probabilities of them being present. In 1984 it was instrumental in discovering a molybdenum deposit worth 100 million dollars!

- *Knowledge representation.* Rules, semantic network.

- *Reasoning.* Predominantly forward chaining (data-driven), with some backward chaining. Bayesian reasoning is used to deal with uncertainty.

- *Heuristics.* Depth first search is focused using the probabilities of each hypothesis.

- *Dialogue/explanation.* The dialogue uses mixed control. The user volunteers information at the start of the consultation, and PROSPECTOR can request additional information when required. Explanations are generated by tracing back through the rules that have been fired.

18.3.5.3 Example 3: DENDRAL

DENDRAL was one of the earliest expert systems, developed at Stanford during the late 1960s [170]. It infers the molecular structure of organic compounds from chemical formulae and mass spectrography data. It is not a "stand-alone" expert, more an expert's assistant, since it relies on the input of the human expert to guide its decision making. However, it was successful enough in this capacity to discover results that were published as original research.

- *Knowledge representation.* Production rules and algorithms for generating graph structures, supplemented by expert user's knowledge.

- *Reasoning.* Forward chaining (data-driven).

- *Heuristics.* DENDRAL uses a variation on depth first search called generate and test, where all hypotheses are generated and then tested against the available evidence. Heuristic knowledge from the users (chemists) is also used to constrain the search.

- *Dialogue/explanation.* The dialogue uses mixed control. The user can supply information and the system can request information as required.

18.3.5.4 Example 4: XCON

XCON was a commercial expert system developed by Digital Electronics Corporation to configure VAX computer systems to comply with customer orders [14]. The problem is one of planning and design: there could be up to 100 components in any system and XCON had to decide how they could best be spatially arranged to meet the specification. The design also had to meet constraints placed by the functionality of the system and physical constraints.

- *Knowledge representation.* Production rules.

- *Reasoning.* Forward chaining (data-driven). Since it is possible to specify rules exactly no uncertainty is present.

- *Heuristics.* The main configuration task is split into subtasks which are always examined in a predetermined order. Constraint satisfaction is used to inform the search for a solution to a subtask.

- *Dialogue/explanation.* The dialogue is less important than in the previous situations since the customer's requirements can be specified at the beginning and the system contains all the information it needs regarding other constraints.

These examples illustrate how the different techniques we have considered in previous chapters can be combined to produce a useful solution, and how different problems require different solutions.

18.3.6 Building an Expert System

We have looked at some of the applications for which expert systems have proved successful, and what components an expert system will have. But how would we go about building one? First, we need to be certain that expert system technology is appropriate to solve the problem that we have in mind. If the problem falls into one of the categories we have already mentioned, such as diagnosis, planning, design or advice giving, then it has passed the first test. The second consideration is whether the problem can be adequately solved using conventional technology. For example, can it be solved statistically or algorithmically? If the answer to this is no, we need to ask whether the problem justifies the expense and effort required to produce an expert system solution. This usually means that the expert system is expected to save costs in the long term, perhaps by making an operation more efficient or making knowledge more widely available. The problem should also be clearly defined and of reasonable size, since expert system technology cannot handle general or common-sense knowledge.

18.3.7 Limitations of Expert Systems

We have looked at expert systems, what they are used for and whether to build one. But what are the current limitations of expert system technology that might affect our exploitation of them? We have already come across a number of limitations in our discussion, but we will reconsider them here.

First, there is the problem of knowledge acquisition: it is not an easy task to develop complete, consistent and correct knowledge bases. Experts are generally poor at expressing their knowledge, and non-expert (in the domain) knowledge engineers may not know what they are looking for. Some tool support is available, and using a structured approach can alleviate the problem, but it remains a bottleneck in expert system design.

A second problem is the verification of the knowledge stored. The knowledge may be internally consistent but inaccurate, due to either expert error or misunderstanding at the acquisition stage. Validation of data is usually done informally, on the basis of performance of the system, but this makes it more difficult to isolate the cause of an observed error. Knowledge elicitation techniques such as critiquing, where the domain expert assesses the knowledge base in stages as it is developed, help to alleviate this problem, although the verification is still subjective.

Thirdly, expert systems are highly domain dependent and are therefore brittle. They cannot fall back on general or common-sense knowledge or generalise their knowledge to unexpected cases. A new expert system is therefore required for each problem (although expert system shells can be re-used) and the solution is limited in scope.

An additional problem with brittleness is that the user may not know the limitations of the system. For example, in a Prolog-based system a goal may be proved false if the system has knowledge that it is false or if the system does not have knowledge that it is true. So the user may not know whether the goal is in fact false or whether the knowledge base is incomplete.

Finally, expert systems often lack meta-knowledge, that is knowledge about their own operations, so they cannot reason about their limitations or the effect of these on the decisions that are made. They cannot decide to use a different reasoning or search strategy if it is more appropriate or provide more informative explanations.

18.4 KNOWLEDGE ACQUISITION

So we have examined our candidate problem and decided that an expert system would be an appropriate solution; what next? Assuming that we have considered our domain of interest carefully and defined the boundaries of the expert system, our first and most crucial stage is *knowledge acquisition*. Knowledge acquisition is the process of getting information out of the head of the expert or from the chosen source and into the form required by the expert system. We can identify two phases of this process:

knowledge elicitation – where the knowledge is extracted from the expert; and

knowledge representation – where the knowledge is put into the expert system.

In this section we will focus primarily on knowledge acquisition and representation for symbolic/rule-based expert systems. Many aspects do not change when dealing with hybrid systems that also incorporate machine learning, but there are some differences which we will discuss in Section 18.5.

18.4.1 Knowledge Elicitation

The knowledge engineer (the title often given to the person developing the expert system) is probably not an expert in the domain of interest. The engineer's first task is therefore to become familiar with the domain through talking to domain experts and reading relevant background material. Once the engineer has a basic level of understanding of the domain he or she can begin knowledge elicitation. There are a number of techniques used to facilitate this. It is the job of the knowledge engineer to spot gaps in the knowledge that is being offered and fill them.

The problem of knowledge elicitation is not a trivial one. To help you to understand the magnitude of the problem, think of a subject on which you would consider yourself expert. Imagine having to formalise all this information without error or omission. Think about some behaviour in which you are skilled (a good example is driving a car): can you formalise all the actions and knowledge required to perform the necessary actions? Alternatively, imagine questioning someone on a topic on which they are expert and you are not. How do you know when information is missing? This is where concrete examples can be useful since it is easier to spot a conceptual leap in an explanation of a specific example than it is in more general explanations.

The interview can capture qualitative information, which is the crux of knowledge elicitation, and therefore provides the key mechanism for acquiring knowledge. There are a number of different types of interview, each of which can be useful for eliciting different types of information. We will consider a number of variants on the interview: the unstructured interview; the structured interview; focused discussion; role reversal; and think aloud.

18.4.1.1 Unstructured Interviews

The unstructured interview is open and exploratory: no fixed questions are prepared and the interviewee is allowed to cover topics as he or she sees fit. It can be used to set the scene and gather contextual information at the start of the knowledge elicitation process. Probes, prompts and seed questions can be used to encourage the interviewee to provide relevant information. A probe encourages the expert to provide further information without indicating what that information should be. Examples of such questions are "tell me more about that", "and then?" and "yes?". Prompts are more directed and can help return the interview to a relevant topic that is incomplete. Seed questions are helpful in starting an unstructured interview. A general seed question might be: "Imagine you went into a bookshop and saw the book you wished you'd had when you first started working in the field. What would it have in it?" [149].

18.4.1.2 Structured Interviews

In structured interviews a framework for the interview is determined in advance. They can involve the use of check-lists or questionnaires to ensure focus is maintained. Strictly structured interviews allow the elicitor to compare answers between experts whereas less strict, perhaps more accurately termed semi-structured interviews combine a focus on detail with some freedom to explore areas of interest.

Appropriate questions can be difficult to devise without some understanding of the domain. Unstructured interviews are often used initially, followed by structured interviews and more focused techniques.

18.4.1.3 Focused Discussions

A focused discussion is centred around a particular problem or scenario. This may be a case study, a critical incident or a specific solution. Case analysis considers a case study that might occur in the domain or one that has occurred. The expert explains how it would be or was solved, either verbally or by demonstration. Critical incident analysis is a variant of this that looks at unusual and probably serious incidents, such as error situations.

In critiquing, the expert is asked to comment on someone else's solution to a problem or design. The expert is asked to review the design or problem solution and identify omissions or errors. This can be helpful as a way of cross-referencing the information provided by different experts and also provides validation checks, since each solution or piece of information is reviewed by another expert.

18.4.1.4 Role Reversal

Role reversal techniques place the elicitor in the expert's role and vice versa. There are two main types: teach-back interviews and Twenty Questions. In teach-back interviews the elicitor "teaches" the expert on a subject that has already been discussed. This checks the elicitor's understanding and allows the expert to amend the knowledge if necessary. In Twenty Questions, the elicitor chooses a topic from a predetermined set and the expert asks questions about the topic in order to determine which one has been selected. The elicitor can answer yes or no. The questions asked reflect the expert's knowledge of the topic and therefore provide information about the domain.

18.4.1.5 Think-aloud

Think-aloud is used to elicit information about specific tasks. The expert is asked to think aloud while carrying out the task. Similarly, the post-task walk-through involves debriefing the expert after the task has been completed. Both techniques are better than simple observation, as they provide information on expert strategy as well as behaviour.

18.4.2 Knowledge Representation

When the knowledge engineer has become familiar with the domain and elicited some knowledge, it is necessary to decide on an appropriate representation for the knowledge, choosing, for example, to use a frame-based or network-based scheme. The engineer also needs to decide on appropriate reasoning and search strategies. At this point the engineer is able to begin prototyping the expert system, normally using an expert system shell or a high-level AI language.

18.4.2.1 Expert System Shells

An expert system shell abstracts features of one or more expert systems. The shell comprises the inference and explanation facilities of an existing expert system without the domain-specific knowledge. This allows non-programmers to add their own knowledge on a problem of similar structure but to re-use the reasoning mechanisms. A different shell is required for each type of problem, for example to support data-driven or goal-driven reasoning, but one shell can be used for many different domains.

Expert system shells are useful if the match between the problem and the shell is good, but they are inflexible. They work best in diagnostic and advice-style problems rather than design or constraint satisfaction and are readily available for most computer platforms. This makes building an expert system using a shell relatively cheap.

18.4.2.2 High-level Programming Languages

High-level programming languages, designed for AI, provide a fast, flexible mechanism for developing expert systems. They conceal their implementation details, allowing the developer to concentrate on the application. They also provide inbuilt mechanisms for representation and control. Different languages support different paradigms, for example Prolog supports logic, LISP is a functional programming language and OPS5 was a production system language designed specifically for expert systems. Python has become popular for AI particularly for machine learning. As well as handling data, there are implementations of various reasoning mechanisms in Python, including Prolog-like logic rules.

However, high-level languages do demand certain programming skills in the user, particularly to develop more complex systems, so they are less suitable for the "do-it-yourself" expert system developer. Some environments have been developed that support more than one AI programming language, such as POPLOG which incorporates LISP and Prolog, and there are configurable expert system shells available in many languages. These provide a blend of flexibility and some programming support but still require programming skills.

18.4.2.3 Ontologies

Expert systems often include specific reasoning rules such as: "if the patient has a headache and loss of smell suspect Covid-19". However, this is often backed up by large amounts of declarative knowledge such as "Covid-19 is a kind of coronavirus" or "headaches are a symptom of flu".

This declarative knowledge is most often encoded in some form of ontology as we first saw in Chapter 2. This can be encoded in a bespoke fashion within tools or hand-edited using a standard such as OWL/RDF (Chap. 17). However, large ontologies are best created using a purpose-built tool as these include ways to edit, visualise and often verify properties of the ontology. Ontology editors will typically be able to import/export in standard formats that make them easy to share with other projects and use by different reasoning engines.

Expert system shells may include some form of ontology, but there are also many commercial and open standalone ontology editors. Probably the most well known and used is Protégé [204], not least because it has a highly extensible architecture, so its large developer community can create their own plugins which expand its capabilities. Figure 18.4 shows a screenshot of

WebProtégé, which allows both web-based editing and also cloud-hosted sharing of ontologies.

18.4.2.4 Selecting a Tool

There are a number of things to bear in mind when choosing a tool to build an expert system. First, select a tool which has only the features you need. This will reduce the cost and increase the speed (both in terms of performance and development). Secondly, let the problem dictate the tool to use, where possible, not the available software. This is particularly important with expert system shells, where choosing a shell with the wrong reasoning strategy for the problem will create more difficulties than it solves. Think about the problem in the abstract first and plan your design. Consider your problem against the following abstract problem types:

- problems with a small solution space and reliable data

- problems with unreliable or uncertain data

- problems where the solution space is large but where you can reduce it, say using heuristics

- problems where the solution space is large but not reducible.

Each of these would need a different approach. Look also at successful systems, try to find one that is solving a similar problem to yours and look at its structure. Only when you have decided on the structure and techniques that are best for your problem should you look for an appropriate tool. Finally, choose a tool with built-in explanation facilities and debugging if possible. These are easier to use and test and will save time in implementation.

18.5 EXPERTS AND MACHINE LEARNING

One possible solution to some of the limitations of expert systems is to combine the knowledge-based technology of expert systems with technologies that learn from examples, such as neural networks and inductive learning. These classify instances of an object or event according to their closeness to previously trained examples and therefore do not require explicit knowledge representation (see Chap. 22 and Chap. 5 for more details).

Some machine learning systems learn purely from real-world data, using ground truth. For example, looking at large-scale data on the way risk factors such as smoking relate to reported lung-cancer deaths. Even then there is a level of expertise in that the choice of what data to collect and what outcomes to address comes from somewhere. However, here we'll look at ways in which the expert involvement is richer.

There are several ways in which experts can be involved in the creation of a hybrid expert system incorporating machine learning (Figure 18.5):

implicit capture of expertise – Here the data collected incorporates some form of expert assessment, behaviour or knowledge. For example, we might look at data on initial patient symptoms and tests ordered by physicians and then use this to streamline hospital admissions by automatically ordering the most common tests (so long as they are not too costly). Similarly, we might train a system using the eventual diagnosis by senior clinicians and then use this to create an expert system to guide less-experienced practitioners.

labelling – Experts may explicitly label data items. This can be with a final outcome measure such as a medical diagnosis, "has influenza". However, the labelling could also be of intermediate features such as "tachycardia" for an ECG trace.

feature selection – The choice of features is often critical, both in initial data collection and those used as part of machine learning. If important features are omitted, then the machine learning will not be accurate, and if too many features are present, there may be overlearning for smaller datasets.

knowledge and rules – The experts may still encode knowledge and/or rules to be combined with more automated techniques. For example a taxonomy of disease types may make it easier to automatically train a system as it can diagnose to higher level disease types when data is sparse for more precise diagnoses.

synthetic data models – We saw in Chapter 8 that synthetic data used in training requires domain knowledge. At its simplest it may be about saying what kinds of distortion are realistic (e.g. blurring, rotating). However, it may be more complex, for example using images of a tumour from one X-ray to be artificially added to others in anatomically correct places, or creating models of human anatomy and

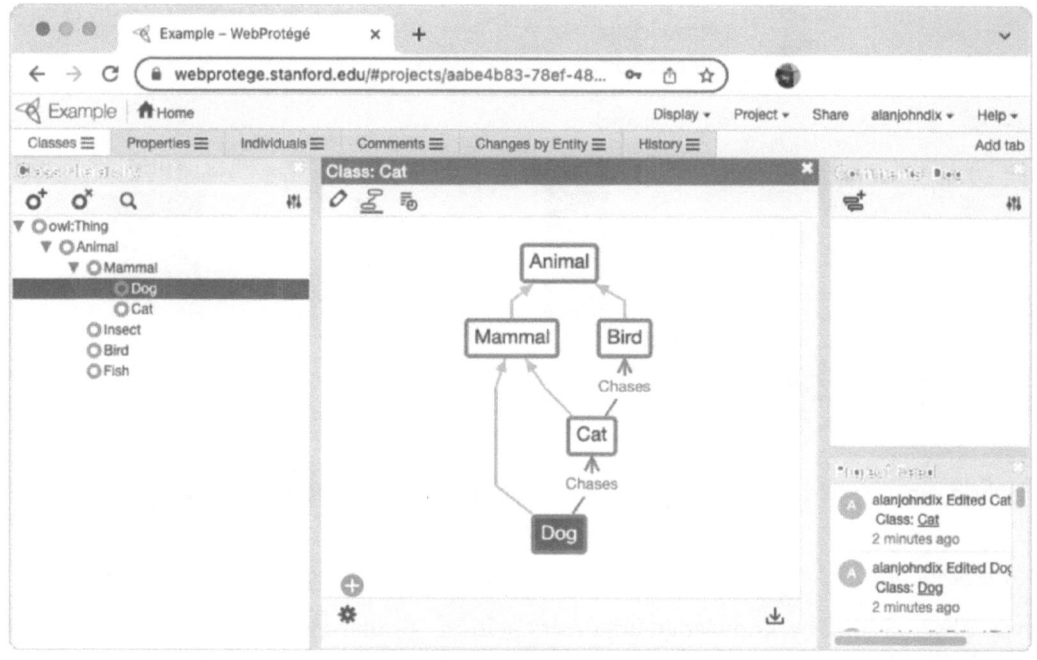

FIGURE 18.4   WebProtégé – a web-based version of the Protégé ontology editor [204].

cancer growth to generate completely artificial images.

Note that this expert-based data and knowledge may be combined with ground truth data such as eventual clinical outcome, especially in the cases of feature labelling.

Some aspects of expert systems are the same no matter whether machine learning is involved, but there are some crucial differences, in knowledge elicitation and validation, and perhaps most crucial is the way algorithmic choice impacts explanation.

18.5.1 Knowledge Elicitation for ML

18.5.1.1 Acquiring Tacit Knowledge

Some expert knowledge is explicit, the expert knows it, and they know that they know it. There may be problems in dealing with the volume of information or in encoding it in ways that a machine can understand, but still this is the easiest kind of expert knowledge to acquire. However, it is usually only a fraction of the experts' full knowledge, much of which is implicit. There are many things they 'know' in the sense that they are implicitly used in their decision making, but which are tacit knowledge, that they cannot easily tell you that they know.

Tacit knowledge may be physical, such as the way we move our arms and legs while walking. Typically only elite sports players or those undergoing some sort of rehabilitation have a deep knowledge of their movements, and then often only through external movement experts recording them and discussing their gait or technique. Tacit knowledge may also be cognitive, ways of addressing problems or the way one just gets an impression that someone is unwell.

An expert knowledge engineer can use the techniques in Section 18.4.1 to gain some insight into this tacit knowledge and then externalise it, for example by asking for explanations of decisions. However, it is hard and not always successful. It is precisely in these circumstances that machine learning techniques can be most valuable.

Deriving the training set purely from implicit behavioural data entirely bypasses the need for the expert to articulate their knowledge; it is captured entirely from the outcomes of their previous conscious and unconscious decision making. There are disadvantages of this, not least that our behaviours may be influenced by unconscious bias (see also Chapter 20).

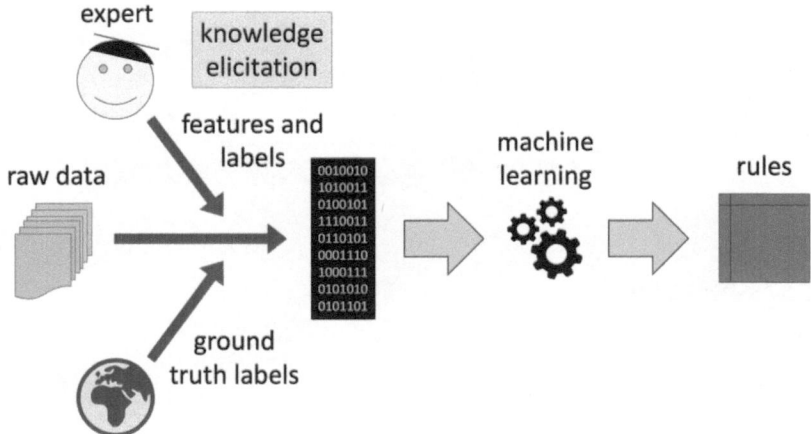

FIGURE 18.5   Hybrid expert system incorporating machine learning guided by human expertise.

18.5.1.2 Feature Selection

One of the early observations by knowledge engineers engaged in knowledge elicitation was that experts may struggle to tell you the precise rules they employ; it is often tacit knowledge, not explicitly available to the experts, but guiding their decision making. However, they are often far better at telling you which *features* they consider. This is fortunate; as the appropriate choice of features is critical for effective machine learning.

18.5.1.3 Expert Labelling

Asking experts to label data sits somewhere between purely behavioural data and more explicit knowledge elicitation. However, an issue with expert labelling is that many machine learning techniques require large volumes of data. This is not a problem when the data is implicitly captured or even explicitly captured as part of normal practice, for example diagnoses in medical records. However, experts' time is, by definition, limited and costly, so it is not usually possible to ask them to label vast quantities of data simply to create a training set.

Sometimes it is sufficient to use semi-supervised learning over partially labelled data, as discussed in Chapter 9. Alternatively it may be possible for some aspects of labelling to be carried out by those with more limited expertise, for example one might delegate the task to junior doctors rather than senior physicians. However, it is then critical that those doing the task recognise when they don't know enough and so are able to flag the more problematic cases for expert review.

18.5.1.4 Iteration and Interaction

In a similar way, the machine learning algorithm itself may be able to identify difficult, low certainty or boundary cases. This then enables an iterative process where a relatively small initial labelled dataset is combined with partially unsupervised learning and then, in a series of cycles, the difficult cases are presented to the expert for verification or labelling. In a similar way, the user can provide an initial, but partial, collection of rules, which can then be used to bootstrap inductive learning.

In all cases it is crucial to use efficient and easy user interfaces for this process, to allow rapid scanning and labelling. In the cases when there is less expert human labelling or semi-automated labelling, it can often be faster to have an interface for the experts that is verification-based, '*this X-ray appears to have a tumour here – Y/N*', rather than open-ended entry.

18.5.2 Algorithmic Choice, Validation and Explanation

Because of the advantages in simplifying knowledge acquisition, hybrid expert systems have been used since the early days of AI and with a wide variety of algorithmic approaches.

Symbolic machine learning techniques such as concept learning, decision trees and rule induction lead to knowledge representations that are similar in form to those generated by purely human-based knowledge elicitation. They can therefore use the same forms of dialogue management and explanation.

In addition, the scrutability of this kind of rule makes it amenable to various forms of iterative expert validation where the rules are presented back to the expert who can verify whether they make sense or not. However, they should be presented in language or visual form that is familiar in the expert's own domain, rather than AI-related terms. For example, note the way that the Query-by-Browsing interface described in Chapter 5 shows an SQL representation of the decision tree generated by machine learning (Figure 18.6).

Concrete examples are also important and often easier to understand than generic rules. Examples can be presented to the expert with the classification or advice that would be generated by the expert system for the expert to verify. We have already discussed how this can be used during knowledge elicitation in order to ask for labels of boundary cases or difficult to classify data items. Exactly the same techniques can be used during validation as it is precisely the boundary cases or those with little training data and large uncertainty where errors are likely to occur.

Using Query-by-Browsing again as an example, note the way that the user can see both the SQL query and the list of selected and unselected records (Figure 18.6). The combination of intentional (rules) and extensional (examples) representations makes it easier to verify whether the generated decision rule is what the user wants. This is particularly important for logical connectives such as AND/OR, as the formal meaning of these can differ from day-to-day use.

We have discussed the phenomenal success of deep learning and other large-data-based techniques in many areas that were previously thought to require far richer knowledge. While neural networks and other sub-symbolic approaches have been used for many years, their popularity has increased. However, they tend to be "black-box" techniques, which are poor at providing explanations for their decisions even to data scientists let alone domain experts or end users. This is also true to a large extent with other techniques that create very large or complex rule sets, including random forests.

Example-based methods of validation and explanation can also be used for the final system, and we will discuss other ways to address these issues in Chapter 21. In fact part of the expert's role may be to help the knowledge engineer to craft explanations for the end-user, for example creating meaningful labels for classes generated by unsupervised learning. These explanations are not only important when the expert or end-user asks for them but also to avoid automation bias [62, 63], the tendency to accept blindly the outputs of computers. The more 'intelligent' the algorithm, the greater the tendency to assume the computer knows best. Explanations, especially when combined with some form of confidence rating, can help to encourage a more sceptical use of automated evidence.

Hybrid architectures have a role here; that is systems that encompass both symbolic and sub-symbolic techniques. For example, a machine learning system may use labelled ECG data to classify different forms of arrhythmia. This classification could then form the input to a more rule-based diagnosis. An explanation might then say "drug X is being proposed as the patient is an ex-smoker and has tachycardia" – the smoking history may come from a form input or questions generated by the dialogue component, and the rules that generate the advice *given* the tachycardia will be explainable in the ways discussed in Section 18.3.3. The categorisation of tachycardia itself may not be explainable except in the form of a fixed patient-oriented description, "rapid heart rate". However, this would be similar to the explanation a doctor would have given; it is sufficient reason.

18.6 DECISION SUPPORT SYSTEMS

The term decision support system includes many forms of data dashboard or visualisations; however, here we are only considering those that include some form of AI component.

Sometimes the AI is based solely on machine learning from data with no expert input. This is the case with the controversial COMPAS system that has been used in US parole decisions [7]. However, often the line between expert systems and decision support is blurred as systems implicitly or explicitly include elements of expert knowledge combined with historic and current data.

As an example, consider the online tool that medical practitioners in the UK use when assessing risk associated with high blood pressure. They enter a number of factors into a web form including blood pressure, cholesterol level, smoking and drinking; this then returns a risk factor for the person. The tool embodies the best expert knowledge, itself drawing on statistical data and published medical results. This is a good example of evidence-based medicine. In many cases the practitioner may directly follow this as an

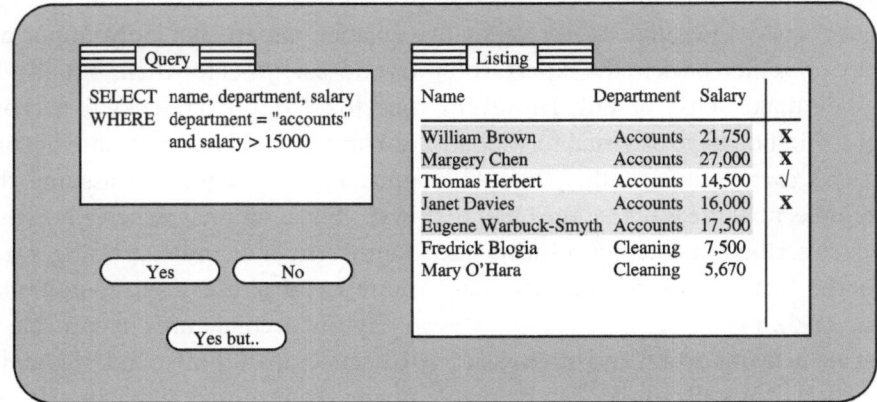

FIGURE 18.6   Query-by-Browsing – shows inferred rule both as SQL query and highlighted listing.

indicator combined with prescribing norms from NICE (National Institute for Health and Care Excellence) to, say, prescribe statins. However, especially on boundary cases, a physician may also take into account additional risk or mitigating factors that are not part of the tool, perhaps a recent change in lifestyle.

We can see from this example that the decision support tool is aiding the medical practitioner to make a decision. However, the medical practitioner is still using their own discretion and judgement based on their own expert knowledge, both general and situational.

In 1960 Licklider wrote about a 'symbiosis' between humans and computers harnessing the complementary abilities of each [168] (also known as synergistic interaction, see Chapter 19). Decision support systems can be thought of in these terms. The list of different abilities in Table 18.1 was produced in 2004 in an influential paper on decision support [62]. Twenty years later, some of the abilities on the left, particularly pattern recognition and the ability to recall pertinent information are certainly within the realms of AI, but even then in *different ways* to humans.

The challenge of decision support is to harness the abilities of the machine (in the right-hand column) but then to present these in ways that maximise the abilities of the human expert (on the left).

In some cases this is managed using textual or form-based interactions, as in the blood pressure example. In other cases there may be very domain-specific methods such as a kitchen-planning aid that knows about the kitchen triangle (sink, cooker, refrigerator) and warns the planner if they are not suitably placed. However, often decisions may be based on larger datasets, for

TABLE 18.1   Strengths of Humans and Computers in Decision Making, from [62].

Humans are better at:	Computers are better at:
Perceiving patterns	Responding quickly to control tasks
Improvising and using flexible procedures	Repetitive and routine tasks
Recalling relevant facts at the appropriate time	Reasoning deductively
Reasoning inductively	Handling many complex tasks simultaneously
Exercising judgment	

example a government planner using past and projected population and traffic trends to help determine transport policy.

We will look at visualisation and associated management issues, before looking at visual analytics, when visualisation and machine learning are interactively linked.

18.6.1 Visualisation

You will undoubtedly have seen many forms of static visualisation or infographic, from simple histograms or pie charts, to geographic images that colour areas depending on some factor such as population or carbon footprint. In addition, visualisation research has created many more, some used only fleetingly by the researchers who developed them and then forgotten, others that have become part of the toolkit of visualisation, for example parallel coordinates [127]. In addition to general purpose visualisations, others are more bespoke such as the ROC curves we saw in Chapter 9 and Nyquist diagrams, used in control engineering. Images and text documents can be particularly hard to visualise in bulk as there is no obvious 'average' value, but central examples can be used for clusters, or rapid serial visualisation shows images or words in flip-book fashion.

Some visualisations, such as the last above, already include some form of animation. However, the power of visualisation is often increased dramatically by interaction. At its simplest interactivity can be used to make otherwise static decisions dynamic; for example, when drawing a scatter graph of multi-dimensional data, one has to choose which numeric fields to plot and this can be made selectable.

Figure 18.7 shows dancing histograms [87], simple stacked histograms, but where the user can select which attribute to align to the x-axis. In other systems interaction is used to zoom into areas of a data plot, to show the details of any specific data item or to interactively filter results based on sliders [3, 277].

Ben Shneiderman's 'visual information seeking mantra' [255] summarises many of the common forms of interactive visualisation:

Overview First – Show the whole dataset even if this means in some way reducing detail such as amalgamating close elements.

Zoom and Filter – Allow the user to zoom spatially into areas of interest and also to apply filtering criteria.

Details on Demand – Open up individual data items for inspection when needed.

There are additional interaction possibilities when there are several visualisations of the same data (see Figure 18.8). One of the simplest is to select a data item on one visualisation and then see the corresponding item highlighted on another visualisation. An extension of this is to sweep a range of values that are close on one visualisation and see the whole set highlighted on the other. If they are using different ways to visualise, this can often reveal rich patterns.

18.6.2 Data Management and Analysis

This form of interactive visualisation is easily possible on desktop workstations or even personal devices when the dataset is relatively small. However, to feel interactive the response to continuous actions such as rotating, zooming or moving a slider needs to be in the order of 100s of milliseconds, with more discrete actions such as a major change in view, within a couple of seconds. For more complex visualisations even medium-sized data (millions of items) may make such interactive visualisation impossible. In addition, in an AI-powered decision support system, these visualisations may themselves require data items to be run through a pre-trained network or similar algorithm, further exacerbating the problems in the data-pipeline.

This can partially be tackled by some of the data reduction techniques we saw Chapter 8. For example, if we have geographical data at one metre resolution, we might down-sample to a kilometre grid, or show averages over geographical areas such as postal districts. Note that the former reduces the dataset size by a factor of one million.

However, data reduction techniques may need to be recomputed as the user interacts with the visualisations. In the case of the geographic data if the user zooms in one might want to have a higher resolution sample of the smaller area. Similarly in the earlier example, if one had randomly sampled documents to visualise and then made a sub-selection based on chosen words, more documents might need to be sampled from the smaller set.

These changes need to be managed in ways that are comprehensible to the user of the system, especially when the analyst is a domain expert, but not a data scientist. In the case of sampled data it may be important to keep track of which samples have previously been shown so that if the user zooms out and then zooms back again into the same part of the data space, the same items are shown.

Sometimes it is possible to perform some kind of precomputation, either generic or bespoke. For example, full ECG data is often recorded at 100–500 Hz, that is hundreds of samples a second, but often one is more interested in heart rate. This can be pre-calculated and a custom lower-resolution data stream created with, say, average heart rate and heart rate variability every minute. Just as with the geographic down-sampling, it would be possible to zoom into a region and retrieve the raw ECG data.

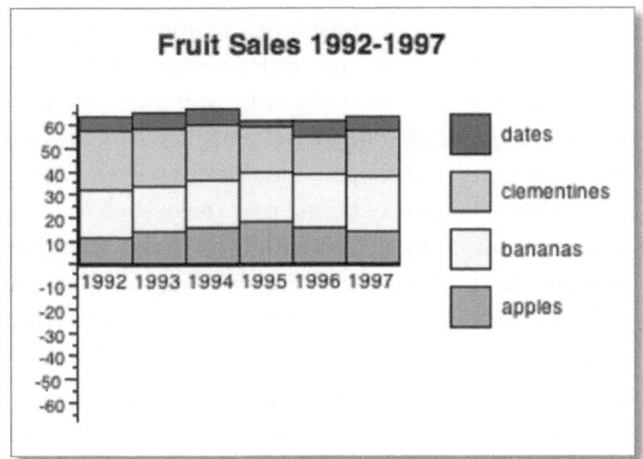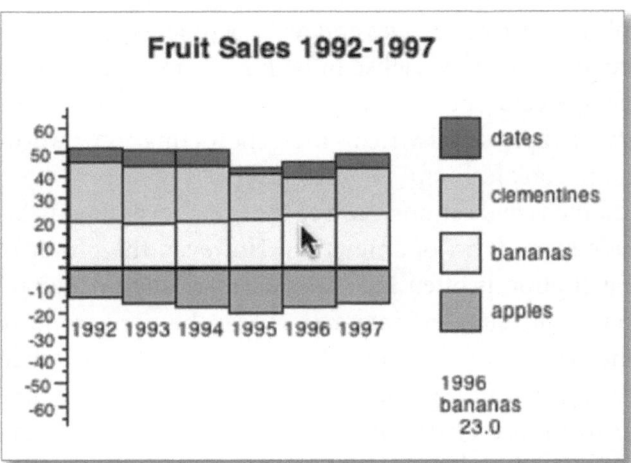

FIGURE 18.7   Dancing histograms: (left) plain stacked histogram – easy to discern overall trends and trends in the baseline category (apples), but other categories less clear; (right) add interaction – click on a category (bananas) to alter the baseline and make trends in that category easier to see. Available at https://www.meandeviation.com/dancing-histograms/.

18.6.3 Visual Analytics

Visual analytics has been defined as, "the science of analytical reasoning facilitated by interactive visual interfaces" [271]. The term was initially coined in the aftermath of the 9/11 terrorist attacks on the United States. Research funded in part by US Homeland Security sought to understand how they might be better able to deal with the vast volumes of information available to security services and hence prevent future attacks [54]. However, it was recognised at the time that this was giving a name to a style of interactive visualisation of data analysis that was present in earlier systems and that the research agenda was applicable across a wide range of areas including medical and environmental data [152]. The volume of available data has of course grown even more since that time, and visual analytics can be seen as an aspect of the broader area of data science.

The core of the idea of visual analytics is to create ways in which human experts can better analyse and understand large volumes of data and/or complex simulations and models. Central is a tight interactive loop where the analyst can select and modify parameters of analysis tools, focus in on specific parts of a dataset and in real time see the results in multiple visualisations. The aim is to explore the data and in so doing gain insights, and ideally *actionable* insights, that can lead to better decisions.

For example, suppose you are studying a large document set. One view of the data might represent the documents interlinked in a graph based on text-based similarity measures. Selecting a document might recentre the network visualisation at the focus document, but also in a second window show the word cloud associated with the document (see Figure 18.8). In the word cloud you might choose a subset of the terms and then tell the system to restrict further analysis to documents containing the chosen terms, and then get it to do an unsupervised clustering on the selected documents.

Notice that even in a relatively simple domain there is the need to move back and forth between different visualisations and to combine human analytic choices with algorithmic analysis. The range and type of such visualisations and algorithmic data analysis varies between application domains, and also on the level of technical expertise of the analyst. Some applications are tailored so that they are more usable by a domain expert, but consequently may need to have a more restricted set of capabilities than an application designed for more open-ended exploration.

This creates additional challenges for both visualisation and data analysis.

18.6.3.1 Visualisation in VA

All of the techniques in Section 18.6.1 can be applied here, with the main difference being that the algorithms

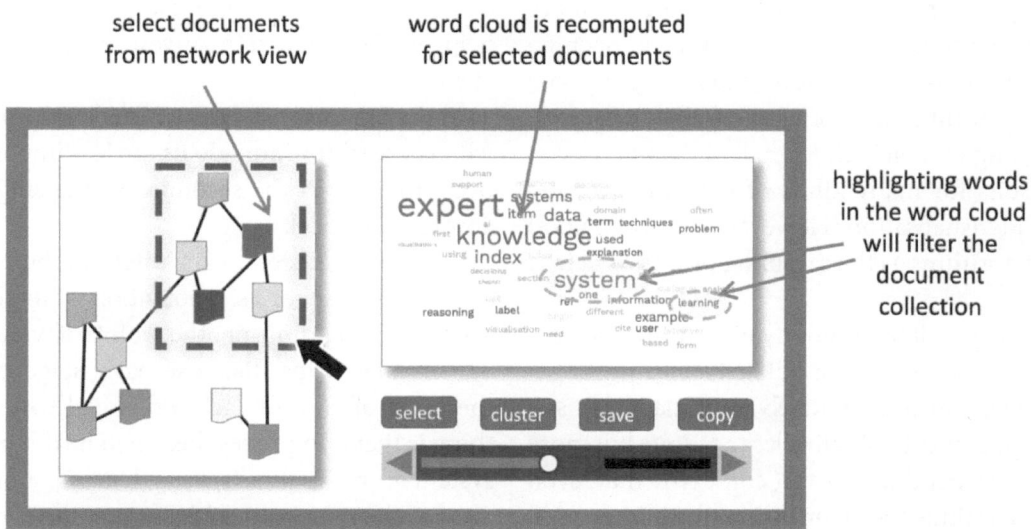

select documents from network view

word cloud is recomputed for selected documents

highlighting words in the word cloud will filter the document collection

select · cluster · save · copy

FIGURE 18.8   Connected visualisations.

generating the data being presented may need to be recomputed as the user interacts with the visualisation. This creates a two-way path:

From algorithm to visualisation – where the computation changes the processed data presented.

From visualisation to algorithm – where the user interactions change the parameters of the algorithm or the filtered data being processed.

In Section 18.6.1 we saw ways in which interaction between multiple visualisations can be used to cast light on both. In visual analytics, multiple visualisations are the norm; however, here the visualisations are typically the result of some form of AI or statistical algorithm. The two paths above then work back-and-forth between the visualisations.

This was clear in the example above. There are two visualisations, one shows the documents spatially arranged based, say, on Jaccard similarity and the other a word cloud. Lassoing a subset of documents in the spatial layout cloud selects these and then the word cloud is *recomputed* based on the selected documents. Alternatively, as in the initial example for this section, selecting words in the word clouds could filter the documents which would then need a fresh 2D layout based on an unsupervised algorithm.

This creates computational issues, discussed below, but also interaction ones. It can be easy to lose track of the path of interaction. Imagine the user in the example looking at the word cloud for a particular set of filtered data and algorithm settings, thinking "I recall these terms in an earlier cloud", but being unable to recall which particular set of choices had led there. Some visual analytics systems include the ability to see multiple copies of the same visualisation side-by-side with different parameter choices or filters for each. In addition, history mechanisms become important, recording past settings and allowing the user to review these, mark interesting ones, and then, using the side-by-side techniques above, compare, contrast and above all gain insight.

18.6.3.2 Data Management and Analysis for VA

The problems we discussed in Section 18.6.2 are further exacerbated in visual analytics. As well as the visualisation being interactive, we are dealing with AI and statistical components that are not fully determined and pre-trained but typically require the user to change parameters or re-run algorithms based on interactively filtered data. As we saw in Chapter 8, many successful machine learning and data analysis techniques obtain their power through applying massive computation based on large-scale data. These datasets may not fit within the memory of an ordinary computer, and computation that even takes a few seconds can feel interminable when working interactively.

The techniques described in Section 18.6.2 are more complicated when they also need to work with machine learning or other data analysis tools interactively. Some combinations of techniques work quite well, for example random sampling of largely independent data items can give good results for many kinds of machine learning as well as rapid visualisation. However, this is not good for network algorithms as the sampling process will lose connections.

It may also be possible to modify algorithms. For example, some methods, such as simulated annealing or genetic algorithms, improve iteratively, so it may be possible to stop them earlier to obtain a less accurate but more timely answer. Depending on the context it may even be possible to continue the algorithm in the background and have the outputs gradually update. Some algorithms or variants of them can operate incrementally adding or removing data as the user updates a selection. For example, a neural network can simply be given additional items and weight them more highly during subsequent cycles of learning, and even negatively weight items to be removed.

18.7 STEPPING BACK

It is easy in AI to focus on the techniques and algorithms, even when dealing with human expertise. However, it is important to step back and look at how the system fits into the world. A medical diagnosis support system affects the lives of patients as well as the functioning of the hospital as a whole. We need to ask ourselves who the system affects, what it is intended to achieve and how this fits into the larger organisational or social context. All of these influence the cost–benefit balance when we trade-off precision and recall.

18.7.1 Who Is It About?

When considering what a system is for, one question to ask is whether the impact is *individual* or about a *population* as a whole.

Imagine designing a vision-based system to sort fruit in a packing factory. The system might be based on a combination of expert rules and lower-level vision based on labelled data for categories such as bruising, ripeness and shape. You do not expect the system to be 100% accurate but do want a high level of accuracy *on average*. In contrast, imagine you are the owner of the factory and using an AI planning tool to help decide whether to invest in the new technology at all.

In the first case we are interested in the population as a whole, errors for individuals are less important than the overall levels. In the second case the single decision is absolutely critical.

The same system can often embody both these aspects, individual vs. population – single decision vs average, depending on the point of view. Consider a system that helps diagnose early-stage cancers. From the point of view of each individual patient the crucial thing is that it improves their own health outcome; if the system improves this, it is worth using. From the point of view of the national health system, the improved health outcomes are important, but they also have to consider the costs of the system and perhaps additional tests and procedures for any false positive diagnosis. Additional costs for cancer care mean less money elsewhere.

18.7.2 Why Are We Doing It?

Another question is what we intend to learn from using the system. Is it to make an assessment of the current state, to give a prediction about the future or to generate insight about the phenomenon as a whole?

In the case of a medical system the question is often "does the patient have disease X?" that is knowledge about the (hidden) current state of an individual or the world. In other cases, for example climate modelling, we are interested in predictions, "what happens next?"

Of course both assessments of current state and predictions are estimates and both typically have some level of uncertainty or, in some cases, probabilities attached to them. Indeed, for algorithmic purposes the two do not differ that much. In fact, for certain diseases an absolute definitive diagnosis can only be made post mortem, so any diagnosis on the living patient could be argued to be a prediction about the findings of that future post-mortem. Even when definitive tests or investigations are possible, the system is operating before that point on what is, at that point, hidden and unknown.

However, while being technically similar and indeed in many ways epistemologically similar (that is in terms of knowledge), there is a fundamental difference between prediction and assessment of current or past state. The future is mutable, the past is not. One can take actions based on past knowledge and the current state, indeed this might even change the state, perhaps curing an infection, but the state as it is now cannot be changed. In contrast, acknowledging another AI story, "*the future is not set*" [38]. This means the way we *use* the system's outputs is often different.

Sometimes either instead of or alongside some form of decision making, the outcome of the use of an AI system is to increase human knowledge or give insight. This is particularly obvious for visual analytics systems, which can help the user to make decisions, but are principally about exploring the data, working out the best way to understand it, and then based on that knowledge maybe making some sort of decision.

This can also be the case during knowledge elicitation. The very act of externalising tacit expert knowledge changes the expert's own explicit knowledge and maybe their future behaviour. This may be amplified when the results of AI are fed back to the expert, possibly showing logical consequences of rules, or inferred rules or cases based on data.

Some years ago, data was collected in the heart unit of a US hospital. When a heart attack patient is admitted and stabilised, the doctors need to make an assessment of future risk and hence treatment. To do this they used to record around 30 factors including ordering multiple tests. The historic data included this information and the final patient outcomes (ground truth). This was used to perform data analysis using some form of machine learning or statistical analysis. As with many expert systems in medicine, this was never deployed – the legal and medical barriers are often too high. However, the physicians learnt from the exercise. The analysis showed that the influence of the majority of the factors was extremely low, only four of the factors had any predictive power. The physicians changed their practice and only focused on the relevant factors but otherwise used their own human judgement as before. With less information, but the best information, their clinical outcomes improved.

In the natural sciences symbolic regression has been used to re-learn fundamental physical equations based purely on data [247, 279]. This is a form of genetic programming that learns formulae such as $F = q(E_f + B v \sin \theta)$ by building, mutating and combining trees of basic operators. More recently this has been combined with deep learning to create

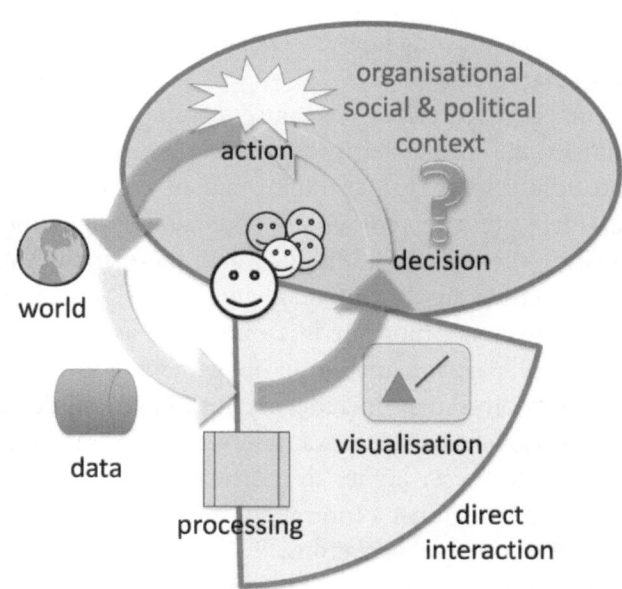

FIGURE 18.9 Visual analysis – the big picture: data coming from the world is used to generate insights, which influence decisions, which then change the world.

new hypotheses, for example concerning galactic evolution in the early universe [58]. It may be that in time these systems will create truly new physics, but at present the new formulae that are generated are not the final outcome; instead they are used to prompt more theoretical analysis [166].

18.7.3 Wider Context

Expert systems and decision support, indeed any application of AI, does not exist in a vacuum; the insights or predictions gained from the process will be used to inform decisions in a wider organisational or social context (Figure 18.9). For example, any large supermarket will have a team of sales forecasters whose job is to analyse past sales data in order to make forecasts on the future sales of different products [9]. However, this is not the end of the story; they will communicate their predictions to the sales team or company board who will then make marketing and stocking decisions.

This may mean that additional forms of visualisation may be needed in order to inform others of the insights gained. These may be simple tables or static graphs in a PowerPoint presentation or may include animated or interactive visualisations. Where more complex analysis techniques have been used, for example black-box machine learning, the analyst may need to be able to explain the reasons for the choice of the technique to an appropriate level, showing how it works in general and for the particular forecast.

For sales forecasts and any complex prediction domain, the forecasting process includes many tweaks and adjustments, for example manually increasing a past data point to remove the anomalous impact of a stocking shortage. In addition, where there are a range of predictions (as in climate change models), one can choose to make more or less optimistic assumptions. These are partly a matter of judgement and so the precise choices may include organisational politics as well as data analysis. For example, a manager may want forecasts to be set higher to motivate sales staff to try harder or low so that a future report can show the team is outperforming expectations.

Often forecasts are measured based on accuracy. How close is the predicted value to what actually happens? For example, you might tell your friends that you think your football team will win this week, and after the match you will know how accurate you were. This is unproblematic as a metric because you are not part of the team and have not had influence on the team's behaviour. However, if you are a player and think the match will be an easy win, you may relax and then perform less well. In the sales forecasting example, if the forecaster predicts an increase in the sales of speciality cheeses, the store may display the cheeses more prominently and the marketing team have a special advertising campaign. If the cheese sells better, is this because the forecast was right or simply because of the better marketing?

18.7.4 Cost–Benefit Balance

Think back to the fruit factory example. There is a cost to rejecting items that are saleable, but also we don't want to accept too many items that are blemished or misshapen. The vision system can be tuned to make a trade-off between false negatives and false positives (recall the ROC curve from Chapter 9). This tradeoff depends on costs. If the system occasionally misclassifies an item of fruit then the impact is perhaps a disgruntled shopkeeper or customer who has a poor quality banana. However, if this were instead analysing wheat for ergot, then the result of a failure might be a dead customer.

These trade-offs are also different if we are looking at a population as a whole or at an individual. For the population, we might consider some sort of overall measure of costs and benefits (sum or average), which in public policy include putting a value on human lives. For example, in the UK NICE (the National Institute for Health and Care Excellence) uses a figure of £20–30,000 per QALY [211]. A QALY is a Quality Adjusted Life Year, which measures both quality of life (e.g. coma vs full health) and also extension of life.

In contrast for an individual entity we may be more risk averse. For example, imagine we have used an expert system to help guide the investment decision for the factory. It may suggest there is a 80% chance that using new technology will increase net profits by 50% even taking into account investment costs, but a 20% probability that increased productivity will be insufficient to service the interest repayments on the costs of refitting leading to bankruptcy. On a simple average returns basis (maybe a venture capital firm deciding on the investment), this looks like an easy decision, on average we would have a 40% increase in profitability, but the factory owner and the workforce may not agree.

We also have to consider how this fits into the wider organisational and social processes, often in timescales beyond the simple decision point. A clinician looking at a cervical smear may notice an inconclusive mark that could be cancerous but ignore it as they know that there will be another smear test in two years' time and that if it is cancerous it will be slow developing and will be caught at a subsequent point. Here the cost–benefit trade-off is not just about the immediate severity of a false negative but that the test now is just one of a series.

Sometimes, these cost–benefit calculations are built into the expert rules or training data. For example, if an image system is based on smears together with the human labelling, the judgements about re-testing intervals will already be part of the data. In contrast, if a machine learning system is built using historic smears and post-hoc knowledge of cancer outcomes, then it may generate too many false negatives leading to unnecessary stress for the patients and wasted hospital procedures.

In the latter cases we can sometimes use wider knowledge of the costs of different kinds of misdiagnosis together with models of the process to build more appropriate advice. Alternatively, we might design an overall decision support system that takes the raw learnt outcomes and presents them in a form that allows the user of the system to make expert judgements. For example, if a system labels a test "80% chance of cancer, time to develop 6 years", this may be judged less critical than "20% chance of cancer, time to develop 6 months".

Note also that where systems are based on machine learning or complex statistics, the algorithms themselves will have uncertainty within their own training and operation which may need to be factored into decisions. This can be hard to assess, again pointing to the need for more explainable AI ... see Chapter 21.

18.8 SUMMARY

In this chapter we looked at the main applications of expert systems and the components that we would expect to see in an expert system. We considered the stages in building an expert system, concentrating on knowledge acquisition and choosing appropriate tools. Machine learning can be used alongside expert knowledge or make use of data that embodies prior expert decisions or behaviour. This reduces some of the difficulties of expert knowledge elicitation but may lead to less comprehensible rules making explanation difficult.

Decision support systems use AI to provide guidance and data that helps experts employ their own human judgement. This often includes elements of visualisation as well as data analysis in order to make the outcomes of algorithms more comprehensible to the expert. This can be made more interactive in visual analytics where the choices and parameters of machine learning and statistical transformations are both displayed and manipulated through multiple visualisations.

In all cases we need to keep in mind a wider picture including the nature of the decisions which will be made and the organisational and social processes within which it will be used. Both will affect cost–benefit choices which may be embedded in various ways within algorithms.

18.1 You are asked to advise on the use of expert systems for the following tasks. Outline appropriate reasoning methods and other key expert system features for each application.

   a. a system to advise on financial investment (to reduce enquiries to a bank's human advisor)

   b. a medical diagnosis system to help doctors

   c. a kitchen design system to be used by sales personnel

18.2 Working in small groups and using the textual information below about causes for a car overheating (extending it where necessary)

   a. Formalise the knowledge as a set of rules (of the form IF evidence THEN hypothesis)

   b. Calculate certainty factors (see Chap. 3) for each hypothesis given the evidence (estimate measures of belief and disbelief from the statements made)

   c. Use an expert system shell or bespoke code to implement this knowledge.

Car overheating: There are a number of reasons why a car might overheat. If the radiator is empty, it will certainly overheat. If it is half full, this may cause overheating but is quite likely not to. If the fan belt is broken or missing, the car will again certainly overheat. If it is too tight, it may cause this problem but not always. Another possible cause is a broken or jammed thermostat, or too much or too little oil. If the engine is not tuned properly, it may also overheat, but this is less likely. Finally, the water pump may be broken. If none of these things is the cause, the temperature gauge might be faulty (the car is not overheating at all). Also the weather and the age of the car should be considered (older cars are more likely to overheat). A combination of any of the above factors would increase the likelihood of overheating.

18.3 Individually or in a group, find examples of graphs, diagrams and infographics used in magazines, newspapers or academic papers.

   a. Look for potential ways in which each could be made interactive as described in Section 18.6.1.

   b. Consider ways in which the underlying data might make use of AI, perhaps machine learning, or clustering.

   c. Now think of how the AI could interact with the visualisation (or an alternative visualisation), to show its outcomes and to be controlled.

If you found a lot of examples, you can focus on the most promising as you work through the steps.

FURTHER READING

P. Jackson. *An introduction to expert systems.* Addison Wesley, Workingham, 2nd edition, 1990.

Detailed coverage of many of the topics introduced here as well as other aspects of expert systems. An excellent next step for anyone wanting to know more about the subject.

L. Medsker and J. Liebowitz. *Design and development of expert systems and neural networks.* Macmillan, New York, 1994.

A book that attempts to provide a balanced view of the role of traditional and connectionist techniques in the practical development of expert systems.

S. Goonatilake and Khebbal, editors. *Intelligent hybrid systems.* John Wiley, Chichester, 1995.

A collection of papers detailing some of the research in using hybrid techniques in expert systems and knowledge acquisition.

A. Kidd, editor. *Knowledge acquisition for expert systems: A practical handbook.* Plenum Press, New York, NY, 1987.

A collection of papers discussing a range of knowledge elicitation techniques. A worthwhile read for anyone wanting to gather information to build an expert system.

T. Segaran and J. Hammerbacher, editors. *Beautiful data: The stories behind elegant data solutions.* O'Reilly Media Sebastopol, CA, 2009.

This book includes many case studies of visualisations used in real-world projects, written by the visualisation experts who created them. The book draws out general lessons and principles from the specific examples.

D. Keim, D. J. Kohlhammer, G. Ellis, and F. Mansmann, editors. *Mastering the information age: solving problems with visual analytics.* Eurographics Association, 2010. https://www.vismaster.eu/book/

Arising from the European VisMaster programme, this book both summarises the key issues in visual analytics and lays out a roadmap for future research. Many of the case studies, although now some years old, still look futuristic, emphasising the gap between state of the art and day-to-day practice.

A. Dix. Information visualization. In *Information retrieval meets information visualization*, pages 1–27. Springer LNCS 7757, 2013. https://alandix.com/academic/teaching/Promise2012/

A short introduction to information visualisation principles and techniques. A video course based on this is at: https://hcibook.com/hcicourse/2013/unit/08-infovis

AI Working with and for Humans

19.1 OVERVIEW

This chapter looks at the issues that arise when people interact closely with AI. It begins by looking at some broad dimensions on which this can vary including who is in control and the level of automation. We will then look in more detail first at intelligent user interfaces, where AI is very explicitly part of an application, followed by smart environments, where AI is less obvious, but more intimately embedded in sensors and devices around us. Just like people, the results of AI are not always 100% correct, so we look at interaction techniques which are robust to individual errors. Finally we will see how synergistic interactions between humans and AI may require changes in the design of both AI algorithms and user interfaces.

19.2 INTRODUCTION

As we noted at the beginning of this book, every AI-based system will in the end need to work with people. In this chapter, we will look at cases where this is very direct such as an intelligent website, a semi-autonomous car or a smart home. In the last chapter we were dealing with expert use, whereas here it is more everyday applications for anyone.

In some ways AI systems are yet another kind of technology and so to understand the relations between humans and AI it is sufficient to look at the history of new technology in general or consult human–computer interaction texts [88].

However, there are key differences, which we can summarise in terms of three 'C's (we'll see a different three 'C's later).

complexity – When we pick up a rock, we can see the rock and know what will happen. When we turn a door handle or flick a light switch, the mechanism is hidden but there is a straightforward cause-and-effect that we can learn. Standard (non-AI) computer programs have higher complexity, every developer has encountered unexpected behaviour of the code that they wrote, but the intention when it is written is that the coder knows what they want to happen! Even traditional rule-based AI systems have added complexity, as the order and interactions between rules is hard to track, and when we have neural networks with billions of parameters, it is impossible to know unambiguously what is going on.

(un)certainty – With the exception of mechanical breakdown or electrical failure, turning the door handle *always* opens the door, flicking the switch *always* turns the light on. However, the complexity of many machine learning systems means we are often uncertain as to precisely what the outcome will be. Furthermore, many AI systems themselves take input from sensors that either have uncertainty in themselves or where the interpretation of the sensor is open to doubt. If a heart sensor shows a high rate, is that because the person is excited or has just walked up three flights of stairs?

co-adaptation – Humans always adapt to the technology we use, learning to work with its benefits and work around its limitations. Often we are creative

DOI: 10.1201/9781003082880-23

appropriating technology in ways that designers never considered [81], such as a shop-keeper using a mobile phone contact list as a place to remember who owes money. However, with the exception of wear patterns, traditional devices do not themselves change. In contrast, many AI systems constantly adapt to their users, for example the way autocompletion gets to know your common phrases. This co-adaptation can be powerful: voice-based systems learn your voice, but also you unconsciously adapt the way you speak to be more understandable. However, when two systems adapt to each other, there can be new problems, think about the little dance when you try to pass someone on a narrow sidewalk.

In each case there are non-AI systems that also share one or more of these characteristics, for example the operation of a large chemical plant is very complex, and even simple, sensor-based systems have considerable uncertainty. This is helpful as we can think of related but simpler systems that share some of these properties as we think about creating AI systems to work with people.

In the rest of the chapter we'll see many examples of systems working with people. We'll consider how to both design human interactions that work with AI taking into account the three 'C's and also how AI algorithms can be modified to work better with people.

In the next section we'll look at some of the different ways in which people work with AI before looking at more specific areas in more detail.

19.3 LEVELS AND TYPES OF HUMAN CONTACT

There are a wide range of ways in which AI-based systems can interact with people. We'll explore four dimensions of this:

Social scale – From single users to the whole of society.

Visibility and embodiment – From screens and robots to smart environments.

Intentionality – From pressing a button to unobtrusive support.

Who is in control – From saying what you want to being told what to do.

We will describe each in a little more detail, but we will also see aspects of them emerge in the various examples in the rest of the chapter.

19.3.1 Social Scale

Some systems are focused on a single individual, for example biometric authentication on a phone or recommender systems in a website. This is the most obvious form of human interaction with AI, and Section 19.4 will consider several examples.

At the other extreme is the use of AI at a social scale, for example, when AI is used in mass surveillance, or to help governments plan healthcare. Chapter 23 will look at these issues in more detail.

Between the two are systems that operate where a small group of people are involved, for example, the lift that needs to decide between all the humans' floor requests, or the digital party-hats or other filters that can be applied in video chat.

This may be partially hidden, for example when you chat with a salesperson on a website, your interactions may sometimes be handled by AI and sometimes by a human, and you may even be unaware which it is. More often when AI mediates human–human communication, it is clear which is which. For example, in the pre-consultation system described in Chapter 13, the patient was interacting with a chatbot and knew it was not human, but the patient's responses were designed to be available to help a later face-to-face consultation with a human clinician.

The individual and group concerns can interact, particularly if the level of personalisation of the system is high. In many case studies of control rooms, such as in the London Underground, it has been found that shared displays and also casual overhearing or seeing activity in peripheral vision are key to effective collaboration [126]. Furthermore in collaboration between teams of differing kinds of expertise, boundary objects have been found to be crucial; that is physical or information artefacts that connect in different ways to each person's specialised domain [265]. Even informal helping assumes that the same application behaves the same for everyone. There are many benefits in the use of machine learning to adapt systems to each individual user, but also care needs to be taken to ensure opportunities for human–human collaboration are not hindered.

19.3.2 Visibility and Embodiment

The device embodying AI can also differ. Sometimes it will be very explicit: visualised on a screen, or a physical device in the home to which you speak. However, sometimes it may be invisible, more part of the environment. We will look at the latter in more detail in Section 19.5.

Often smart systems, such as lights or home music, are controlled by a separate device, most commonly through a phone app. In these cases we need to think both about the design of the app and the way in which the user makes sense of its, sometimes invisible, effects on the environment.

In some cases the AI system may be embodied in ways that have considerable autonomy. At its simplest a lift can be thought about as a robot with a single degree of freedom that we happen to ride in ... and we can sometimes find lifts hard to control. Industrial robots, autonomous cars and drones all create situations where it is not just about telling the AI device what to do, but fluidly working alongside it.

19.3.3 Intentionality

AI based systems also differ in their level of intentionality, that is whether the user explicitly instructs the system to do something, or whether it chooses to do so. This can be thought of as a continuum (note the use of the term 'intentionality' here is different from philosophical notions of intentionality discussed in Chapter 23):

explicit – Here the user issues some form of explicit command, for example telling a home automation system, "turn off the lights". Note that here the user consciously and explicitly plans what will happen.

implicit – Here the user performs some form of action that triggers the system to respond, but it is more a natural action such as tipping an e-book to turn a page [246]. The action is still in a sense planned or triggered by the user, but more unconsciously.

expected – Here the user doesn't do anything specific to trigger the action, but they know it will happen and would be surprised if it didn't, for example automatic doors opening as you approach them or lights going on when you enter a room.

incidental – Here the user is doing some action as part of other activities which the system uses to perform

some other action [79]. This is often at a completely different time, for example learning email habits; or it may be to help a different person such as the use of one customer's book buying to make suggestions to another. The action is planned entirely by the AI, the user may not even be aware that it has happened.

accidental – Finally there are cases when neither human nor AI plan that things happen, but they arise as an unexpected side effect or emergent behaviour. In the telecoms industry feature interaction has been a recognised issue for many years, where several features each of which seem reasonable have an unexpected, and possibly damaging, effect together. AI and ML often intensify this due to the complexity of the algorithms; for example the personalisation of news and social media seemed overwhelmingly positive, but has given rise to filter bubbles.

Note that the implicit, expected and incidental levels all make heavy use of context in order to make sense of the users' activity. This can involve the user of physical sensors in the environment or monitoring of digital interactions. This is because the AI system has to interpret the users' actions in order either to understand the users' intentions or at least make sense of their actions.

19.3.4 Who Is in Control

Closely related to levels of intentionality is the question of how the level of control between human and AI can vary:

Human as cog (in the machine) – In the web chapter (Chap. 17) we have seen how reCaptcha codes simply regard the end-user as a 'recogniser' to be used as part of a larger machine learning system. Similarly in many gig-economy applications, such as ride or delivery services, the driver is told by the machine who or what to pick up when.

Human as controller – In other applications, the human is definitely in control, telling the AI what to do. For example, you tell an autonomous car where you want to go and then let it drive you there.

Human as partner – At other times the relationship is more collaborative, for example a system might suggest options to you, or semi-automate processes, but let you fine tune or guide. This is sometimes called hybrid AI, as we saw in Chapter 18; however, the term 'hybrid' is also used for discrete/continuous problems (Chap. 4) or combining different kinds of AI algorithm (Chap. 6).

In these we can see different options as to who has the **initiative**, does something that starts things off; who makes the **decision** as to what should be done; and who actually takes **action** to do it (see Table 19.1 for some examples). When the human acts as a cog (in the machine), the initiative and decisions are made by the system and the human merely performs the action that has been selected. Whereas in the human as controller situation, the initiative and high-level decisions are taken by the human, with the system dealing with details.

Another term you may hear is human in the loop. This can refer to situations in all three of the above classes, wherever the results of the AI are not applied without additional human interaction. This is particularly clear where the human is used as part of a broader algorithm (human as cog) or working alongside (human as partner).

TABLE 19.1 Who Has Control and Who Does the Work, Some Examples.

initiative	decision	action	example
machine	machine	human	delivery service
human	machine	machine	autonomous car
human	joint	human	decision support
machine	joint	joint	recommender system
human	joint	machine	engine management system

19.3.5 Levels of Automation

When considering automobile automation, the Society of Automotive Engineers have defined five levels [238, 239], which have been adopted or adapted by many national and international standards:

0. No Driving Automation

1. Driver Assistance

2. Partial Driving Automation

3. Conditional Driving Automation

4. High Driving Automation

5. Full Driving Automation

In levels 0–2 the driver is still 'driving' the car. Level 0 includes basic features such as ABS (anti-lock braking system) and various forms of warning such as blind spot or lane changing alarms. Levels 1 and 2 include more advanced features such as automated steering within lanes, or maintaining distance from the vehicle in front, but the driver is still expected to maintain attention and oversight even when the AI system is doing much of the fine work of driving.

Levels 3–5 include times when the automation is actually driving the car. In Level 3 the driver has to take over when requested by the system (which creates hand-over challenges for the driver-car interface). In levels 4 and 5 the car may have no human operated controls at all; the difference is that level 4 is only within constrained environments, such as an airport shuttle service.

Although the levels were developed for cars, we can see other kinds of human–AI system in a similar vein. For example in an AI-assisted programming environment, level 0 would include things such as syntax highlighting or auto-completion of variable names and function templates, whereas level 2 and 3 would include auto-completion of more substantial chunks of code, as with GitHub Copilot [48]. In level 3 we can imagine a system that writes full programs but maybe stops and requests help from the programmer when it gets stuck; level 4 would be something that codes entirely autonomously, but within a limited context, such as configuring IoT rules in a domestic setting; while level 5 would be a fully autonomous general-purpose artificial coder.

Shneiderman argues that this one-dimensional view of automation is too simplistic and instead suggests considering a two-dimensional framework with higher and lower levels of human-control compatible with higher and lower levels of automation [257].

Figure 19.1 illustrates this with camera design. Early mass-market film cameras (bottom right: low human-control, low automation) had no automation beyond springs in the shutter and little human control beyond pushing the button to take the picture and winding on the film. In contrast older (pre-digital) SLR cameras (top left: high human-control, low automation) allowed the photographer to manipulate many settings such as the focus, exposure time, aperture size and level of zoom

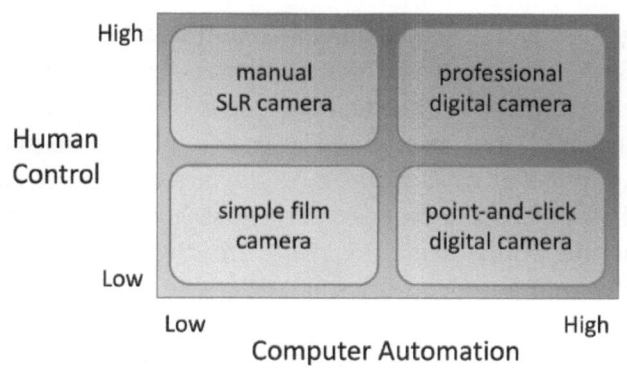

FIGURE 19.1 Shneiderman's two-dimensional human–centred AI framework.

as well as switch lenses, or attach different forms of flash devices or lens filters for different conditions. Early mass-market digital cameras and the basic mode of most phone-based cameras are essentially point-and-click (bottom right: low human-control, high automation), where the embedded computer automatically focuses and determines the exposure time and other settings available to the camera. Finally high-end professional cameras allow photographers to manage settings when they want to but can also include rich automation to enhance the photographer's experience and capabilities, especially when in challenging circumstances such as needing to take many rapid shots at a wedding or during wildlife photography.

The meaning of 'control' in this framework is slightly different from the kinds of control we were considering in Section 19.3.4, but the more synergistic human–AI interaction, which we will return to in Section 19.7, tends to fall into the upper-right quadrant of Shneiderman's framework which he considers the sweet spot for "reliable, safe and trustworthy" AI systems.

19.4 ON A DEVICE – INTELLIGENT USER INTERFACES

Throughout this book, we've already seen many examples of user interfaces that make use of some form of AI including recommender systems and data detectors in Chapter 17. User interfaces where AI forms a significant element are often called intelligent user interfaces, and there is a long-running ACM conference dedicated to the area.

19.4.1 Low-level Input

Many low-level input methods make use of some form of AI or ML. Some are obvious such as the speech recognition used in voice assistants, but others less so.

In Chapter 12 we discussed gesture recognition that is often used in pen-based systems. This includes recognising letters and numbers to enable handwriting recognition to avoid the use of a keyboard. In addition, there are usually action gestures, such as crossing out for delete or circling text to select it. These systems come pre-trained but will also usually adapt themselves to the user's own styles of writing.

In many situations the goal of the recognition engine is to translate the free-flowing gestures into a series of predefined codes or tokens such as letters or editing commands. However, for some strokes such as lassoing text to edit, the precise path is important. In games often the speed of the movement is also critical.

Even smartphone virtual keyboards embody quite complex algorithms to minimise (inevitable) mistyping. Often this is not in the public domain as it is either commercially sensitive or just buried in the code. However, some years ago, Microsoft engineers gave a rare glimpse into the detailed engineering of the Windows Phone 7 virtual keyboard [119]. The design combined fixed rules to ensure that the centre portion of every key is consistent with more predictive techniques that expand the effective size of each key depending on the likelihood it will be pressed next.

Even less visible are the algorithms used for picture stabilisation and auto-focus in a smartphone camera. These are only noticed when they go wrong.

19.4.2 Conversational User Interfaces

The origins of conversing with a computer date back many years, not least ELIZA in the mid-1960s [299] that we discussed in Chapter 1. By the mid-1980s the MIT Speech Interface Group used various state-of-the-art techniques to create an envisionment 'Phone Slave', which still looks remarkable given the 30 years it took to become commonplace [245]. It is interesting that many of the issues that can still cause problems today were foreseen at that point, for example how to distinguish deliberate commands vs accidentally overheard speech.

A home automation device has a relatively high-quality microphone and also gradually learns your speech patterns. Voice-based systems over telephones

FIGURE 19.2   Talking to Siri.

that have to work with anyone can still be difficult and can be particularly problematic for certain accents, or older people. For this reason, interactive voice response (IVR) systems are still often built around a computer-driven dialogue with quite complex parts spoken by the system and key presses or very simple spoken responses by the user.

IVR systems and indeed many chatbots typically operate using a simple flow-chart-based dialogue, often called a 'decision tree', but in a different sense to the decision trees in machine learning. The decisions here refer to the choices made by the user, for example what kind of query they have, or choosing between different kinds of problem.

Text-based chatbots can also use more sophisticated NLP using the kinds of techniques discussed in Chapter 13. This is in part because the actual words are well defined even though the extraction of meaning can be more difficult.

In all cases there will be errors and uncertainty and so it is critical that any conversation-based interactions have plenty of provision for error detection and repair, for example the system repeating back what it thinks the user has said. We'll return to these issues of repair in Section 19.6.2.

Home automation systems or voice assistants on smartphones or in a car have to work by voice alone but typically do not operate in a fully conversational mode, instead adopting a question-and-answer or command-and-action pattern. Figure 19.2 shows an attempt to get MacOS's Siri to engage in conversation.

Sometimes conversational interfaces merge with gesture-based interaction in multimodal interfaces. Both the speech and gesture aspects have their own algorithms and uncertainties, but it is also important that they are synchronised to enable interactions such as "*put that circle there*" in a graphics editing system. The different locations of the user's pointed finger at the words 'that' and 'there' are crucial. Often the recognition process has small delays, so it is important that the speech-recognition part is able to link the recognised words with the time they were uttered not the time the recognition is complete.

19.4.3 Predicting What Next

The key-level algorithm used in the Windows Phone keyboard and the word-level prediction that offers suggested completions are performing a simple form of lookahead prediction: the next token in a sequence. Older systems often used Markov models or similar statistical techniques, and simple frequency-based methods are often sufficient for more discrete short text such as search auto-complete. However, for sequence-based prediction, such as sentence completion in an email client, transformer models (Chap. 14) are increasingly used.

In some ways these are simply more sophisticated versions of older statistical or probability-based methods, taking a window of past text and predicting the next few words. However, there is a point at which the simple change in quantity of text feels like a qualitative difference. GPT-3 was trained on half a trillion words, many thousands of times more than a child would encounter as they grow up, and this huge figure is only increasing with subsequent versions. Our own language is often a mix of stock phrases and adaptations to the precise situation. To some extent the transformer models are doing this. Crucially they maintain a substantial amount of context and so are able to stay 'on topic' for extended periods, to the extent that whole essays are being produced with publicly available tools such as ChatGPT.

Often GPT-based applications are tuned to specific areas, and there is a considerable art in priming the models, that is giving them some starting text that establishes the right topic area. For example the Codex model is a version of GPT-3 trained on 54 million public repositories at GitHub [48]. This has been used to offer code auto-completion in GitHub Copilot [108], which often offers amazing coding suggestions, but also can create reasonable but wrong code, or possibly worse, return code that duplicates whole chunks of specific code from GitHub with implications for IP.

The same underlying Codex has also been to drive a chatbot in 'The Programmer's Assistant' offering conversations that are remarkably like one might expect with a fellow programmer [236]. Crucially however, the developers of the Programmer's Assistant prime Codex with a prompt that shifts it towards effective interactions. This prompt contains some fixed text that has been tuned to nudge Codex in a helpful direction, which is combined with phrases drawn from the specific current coding context.

19.4.4 Finding and Managing Information

Researchers in personal information management (PIM) often focus on four activities; the user needs to acquire, organise, maintain and retrieve information [23, 150]. Organisational information management has similar activities. Various machine learning techniques are used in at least three of these – relatively little for the 'maintain' side.

Thinking first on *acquiring* information, we have already discussed forms of recommender system in previous chapters. These use your own and other people's past selection of material (books, music, news items) in order to offer suitable items to suggest to you. Similar techniques are used for both targeted advertising and dating apps. Note these systems often depend on relevance feedback, some indication of whether the recommendations are indeed what you want to see or hear. This can be explicit, as in star ratings for films you have watched or favouriting an item to return to; or it can be implicit, for example that you chose to listen to a recommended piece of music to the end.

The *organising* side has also been a major success, at least in email management, where the volume of incoming information is greatest. As well as spam filters, the main commercial systems exploit machine learning for classification to pre-sort mail into major categories, trying to identify the most important emails for immediate attention. These systems are so good that people rely on them and will often miss emails that have not been correctly categorised. Outside of email, the picture is more patchy; hierarchical filing systems on personal computers and the cloud have changed little since the 1970s. Of-

ten the assumption is that if retrieval is good enough, careful filing is unnecessary.

As noted there seems to be less use of AI for *maintaining* information, that is ensuring it is kept up-to-date and filing systems are efficient. As with organisation, if retrieval is good enough, having things poorly sorted is not a problem. One of the reasons for deleting old information is to reduce the clutter that gets in the way of effective human retrieval, but if AI can identify candidates for deletion, it may as well not bother and simply ignore or low-rank these items during retrieval.

There are two times when deletion is important. First is for resource management. Phones may identify apps to archive, as running apps reduce battery life and cloud services may automatically migrate items to and from your personal computer. However, full deletion is rare except for disk clean-up applications, which use relatively simple rules, not least because mistakes are hard to fix. The other reason is privacy and security, both personal and corporate. Indeed, many organisations still find it hard to comply with data protection legislation that requires personal information not to be kept longer than necessary.

In some ways *retrieving* information is similar to *acquiring* it, except that the sources are internal rather than external. Indeed at an organisational level very similar techniques are used. At a personal level things are a little more complicated as by definition the information is not personal and therefore simple popularity-style metrics cannot be used to learn from other people's behaviour. The search systems in use for personal information do indeed seem to be relatively simple.

Crucially personal information search requires transfer learning where things learnt about one person's behaviour in an abstract sense can be transferred to another person's documents or emails, even though the actual items are completely different. The main example of this at present is at the language level where synonyms for terms learnt from document corpora or other people's (public) searches can be used to tune searching of personal or corporate repositories, but as work on transfer learning develops this may soon change.

Search and retrieval systems differ in terms of accuracy. Many of the highly successful systems developed for acquiring information, particularly recommender systems, work on the principle that there is too much information available and so giving good enough information is sufficient, you do not need to have every single piece of information that is relevant, and a few irrelevant items don't matter as you can skip them.

However, sometimes it does matter that you have precisely the items you want. This can be the case for public information, for example it is important that a lawyer retrieves 'all' of the relevant case law. It is even more common for retrieval of personal or organisational information – you want precisely the document you were working on three weeks ago, or exactly the records of employees that are nearing retirement. This was one of the drivers for Query-by-Browsing (QbB) described in Chapter 5; if you are going to perform an action, such as giving a pay rise to employees, or sending a loan reminder to clients, it is important that you select exactly the right ones. Note that QbB was designed to create an understandable query which could be verified by the user – we will return to these issues in Chapter 21.

While non-AI based systems tend to merely retrieve information when asked, more intelligent systems can proactively suggest potentially useful information. Recommender systems do this for acquiring information, while some personal knowledge management may suggest relationships between items as a form of creativity or inspiration aid. Note that personal knowledge management is a term often used in commercial systems rather than PIM, which is used more in the research domain.

19.4.5 Helping with Tasks

Various forms of more (or less) intelligent algorithms can be used to help make day-to-day digital tasks easier.

At the simpler end is automated form-filling in email browsers. Research systems have created sophisticated versions that use contextual information to tune suggestions for each field, but most systems used in practice rely on simple rules such as the names of the web form fields and match these to previous entries in fields of the same name, especially on the same web form.

One step up are data detectors and named entity recognition introduced in Chapter 17. For the former, simple patterns are used to match structured content such as a date. The latter match specific words or sequences of words to known entities such as places, people or titles. These are sometimes combined in more complex recognisers such as the way some email systems will notice that if a time, a date and place are mentioned, then this may be an appointment. If the source of an email is recognised, then more specific rules may be

applied, for example parsing the automated emails from flight or hotel booking sites. Most of these systems are based on pre-specified rules rather than learning from user interactions.

There is also a long history of systems that observe and then automate user tasks. Some of the earliest examples were related to programming user interfaces [169, 205], and hence the term programming-by-demonstration is often used even when the applications involve no code.

Systems such as Apple Automator simply record the user's actions on an interface and then allow the user to replay them, but more sophisticated systems attempt to match potential variable elements, so that they can repeat the same task for a different object or look for repetitions (or loops in a coding analogy). An early example of this was EAGER [64], which was able to automate iterative tasks in HyperCard, a common hypertext and coding system on Apple computers at the time. As well as being relatively sophisticated EAGER was embodied in the interface in the form of a small cat.

Despite these early research systems being relatively successful, the level of sophistication of actual deployed systems is very basic. This is partly due to the complexity of correctly inferring both parameterised and repeated tasks, not least the completion criteria for the latter. The success of code-completion systems such as Copilot in inferring complex coding structures is because they are based on large bodies of code containing loops as text, not because they understand coding structures themselves.

Another barrier to effective task automation is access to the underlying data. First the units of activity on a graphical system are not obvious. If you simply record user actions you end up with a series of mouse-click events, whereas for effective task learning this needs to happen at a more semantic level such as "MOVE FILE august.xls TO FOLDER monthly_reports". Apple attempted to encourage developers to code applications using this form of intermediate representation so that they were recordable for automation, but few followed this advice including many of Apple's own applications.

Furthermore both task automation and simpler aids such as form filling ideally would have access to rich data about the user and current context such as contacts and different messaging systems. There have been attempts to capture this in what has been called a semantic desktop applying semantic web technologies to personal computers [242]. However, these efforts have never found their way into commercial systems, for both practical reasons and also issues of privacy.

19.4.6 Adaptation and Personalisation

Most computer applications have settings that can be customised by the user, but few ever change the default settings. This is partly because customisation requires upfront effort in order to improve future experience. Furthermore the precise implications of customisation decisions can be hard to predict, further reducing our willingness to put in effort for uncertain outcomes.

Various forms of intelligent adaptation can address these problems. For example, a word processor might highlight 'colour' as a potential spelling mistake but, when you open the pop-up menu, will offer both changing it to 'color' and changing the default language setting to 'English (UK)'. Note that this:

- reduces the knowledge required – how to change language settings,

- reduces the effort required – agreeing to the suggestion rather than navigating to the settings,

- is timely in terms of understanding – the user can see how the change would have helped the interaction,

- is timely in terms of value – because of this the user can appreciate the value of the (small) effort to agree the settings change.

While there are many advantages to this form of timely suggestion, some adaptations may be performed entirely automatically. For example, if you adjust your screen backlight, the system could use this to improve its algorithm that sets the backlight depending on ambient light. Similarly, if a user frequently opens then closes files, it may be a sign that double-click delay needs to be slightly adjusted. Note that these are both examples where it is very hard to say precisely what is a good setting except in the precise situation and where the user may be unaware of the adaptation except in so far as it subtly improves their experience.

In between are adaptations that are automated but visible. One of the oldest examples of this is long menus. An obvious way to adapt a long menu is to order it by frequency of use so that the most commonly used items are at the top. Indeed, this is the approach taken in some file browsers where the default view is of recently accessed files. In the case of the file system you still can access them in an alphabetic or hierarchical view. However, if the menu is entirely sorted by recency, then it becomes very hard to find the less frequently accessed items as their location in the menu keeps changing.

This illustrates a general issue for adaptation, the need for deterministic ground, having things that do not change as well as things that do and knowing which is which. In the filing system the hierarchical folder view is the deterministic ground, meaning you can rely on it being stable and hence learn to access things in it without them being confused by adaptations. In fact, most adaptive menus, such as font selection, do have a deterministic ground as they only place a small number of most likely items at the top (adaptive area) and then have a fully alphabetic menu below (deterministic ground).

Note too that an effective deterministic ground can also help manage the conflict between personalisation and cooperation, which we highlighted earlier.

19.4.7 Going Small

Many of the machine-learning approaches we have seen in this book require large datasets and extensive computational power for initial training. We've already discussed how transfer learning can help deal with the limited data available for a single user. However, in addition the user's device cannot perform the massive computation that is available for the most sophisticated algorithms. Execution is usually less computationally expensive than learning, but still the many billions of parameters of some deep learning models would challenge the memory and processing capacity of smaller devices.

This is one of the reasons that many of the systems we have described use relatively simple rules rather than more complex machine learning. One solution is to use large computational resources to train models and then use some form of simplification to create smaller models that perform nearly as well. This simplification is itself a computationally expensive task but only needs to be done once and is performed centrally with its results then downloaded into individual devices. Of course, if the device is internet connected, then some computation can be off-loaded to the cloud, and this is precisely the approach taken in some speech-based systems.

19.5 IN THE WORLD – SMART ENVIRONMENTS

Most of the examples so far have been where the intelligence is in some form of information system and where there is a single user largely in control. However, we are increasingly living in environments where many aspects are digitally controlled and monitored. This is true in the home, in hospitals, on the road and at work. These devices are often network connected creating what is known as the internet of things (IOT) and furthermore these devices often either include local intelligent algorithms or connect to cloud-based intelligent services.

The nature of human interactions with these diffuse smart environments (smart home, smart city) is fundamentally different from information systems.

We can think of this from both the computer and the human side. First of all the designer of smart environment algorithms needs to consider:

initiative – These systems often involve the computer taking the initiative, that is in the incidental end of the intentionality spectrum.

interpretation – Because of this the computer algorithm also needs to interpret the environment which may include one or many people.

sensing – Often the inputs for the system come from sensors in the environment, not user commands. However, these sensors may offer incomplete or uncertain data due to placement or inherent accuracy of the electronics.

physical – The output or impact of computer actions may be more informational (such as the colour of lights) but also may include actual physical actions such as the raising of a barrier or an autonomous car driving on the road.

human action – Some outputs such as traffic lights only have an effect indirectly through human actions, which may not be reliably predictable.

distribution – The computation for AI may be distributed in sensors or actuators as well as in centralised systems, which can then be seen as interacting with one another digitally and physically.

In addition, these systems feel very different for the human:

uncertainty – The uncertainty of sensors is compounded with the uncertain behaviour of complex AI systems.

hiddenness – Some system actions are immediately obvious, for example changing the music, but others are not immediately obvious, such as turning on an outside security light.

time lags – Furthermore there are often delays between an action being performed by the system and its effect on the environment, for example between turning on heating and feeling the air warm up the room.

conjoint action – Changes in a smart environment are often related to broad human activity, such as traffic volume, rather than individual user actions.

We'll look at a selection of issues in more detail.

19.5.1 Configuration

If IOT devices are being installed in an industrial environment, then expert installation engineers can set the system up, configuring the locations and settings of the individual sensors and connecting them into the overall system. However, in a domestic environment this has to be managed by an ordinary person. In both cases mechanisms to ease the installation process are important.

The smart environment typically needs to take into account the relative locations of devices, which may be used by specialised algorithms or fed as additional input alongside the sensor values as part of machine learning. This is relatively easy to configure in an industrial plant where there are plans and schematics, but harder in a home or for devices installed outside. However, algorithms can be designed to help this process. This can use time of flight for wireless signals or sound, or triangulation using cameras. For example, the Firefly system in Figure 19.3 allows lights to be positioned randomly but then works out their locations after installation by twinkling each light in a unique code [44].

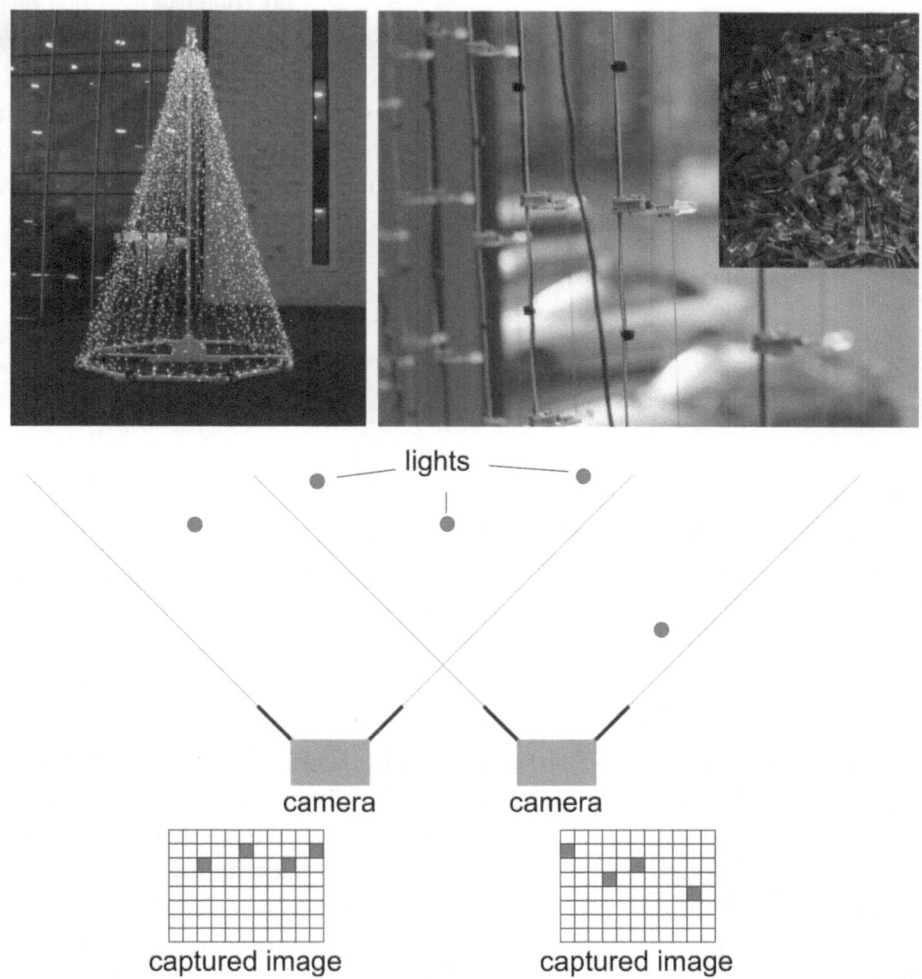

FIGURE 19.3   Firefly – individual lights flash a unique code and are then positioned using triangulation from multiple camera angles (see https://lucidina.com/).

After initial installation, sensors may fail or new sensors are deployed. Ideally this should not require configuring from scratch. Where there is redundancy in the sensor network this can be used to fill in the gaps for broken sensors and to train on the fly for new sensors.

19.5.2 Sensor Fusion

There are often many sensors in an area. These may be homogeneous, for example many air-quality sensors in a city street, or heterogeneous, for example temperature sensors in a conjunction with cameras. Some sensors may return very different data about the same thing, for example a camera and air-quality sensor in the same street; some similar data about different things, for example temperature sensors in a distant room.

However, sensors can also offer redundant data, that is they are measuring the same or nearly the same thing. This sounds wasteful, but allows for sensors to fail, or also to potentially deploy large numbers of low cost and low quality sensors that together are accurate, sometimes called smart dust.

The data from the sensors needs to be brought together to generate a coherent view of the environment. This is called sensor fusion. At its simplest all the sensor input could be used as inputs to a single large neural network; however, this approach is rarely best as datasources are so different both in terms of sample rate (e.g. sound sampled as 44kHz vs video at 60 frames per second) and size (a single volume level or two stereo channels for sound, vs millions of pixels for video). Because of this, sensors will typically have some initial media-specific processing, for example a convolutional neural network for im-

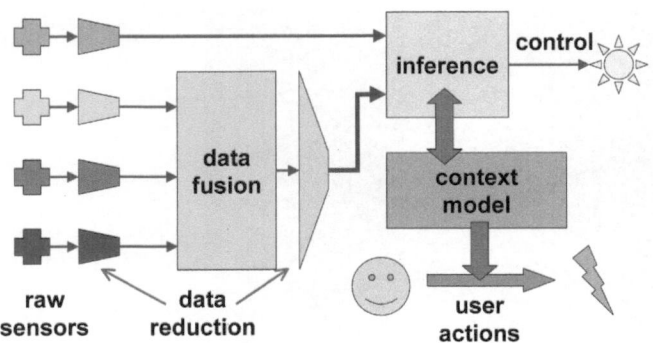

FIGURE 19.4 Sensor fusion: from multiple raw sensors to rich context model.

ages or Fourier analysis for sound sources, that reduces the heterogeneous signals to forms that can be brought together either through more rule-based approaches or machine learning.

19.5.3 Context and Activity

The final aim of processing sensor data will be to do something. This might be to autonomously make decisions, for example to adjust the pattern of heating and ventilation in the house as a particular area is too hot. It may also be used to modify or augment explicit user interactions, for example taking into account the packet of food you have taken out of the freezer when you say "put the oven on" to the home computer. This latter is called context-aware interaction.

In the case of the freezer pack the system might be using a vision system to read the packet label or bar code, or scanning an RFID tag. This is relatively straightforward. However, more complex automatic decisions or context aware interactions may require the system to recognise what you are doing. This activity recognition typically involves observing over a time period, for example the difference between stirring and beating an egg, or running vs walking.

The final output of such a system may be a single activity, say 'running', possibly with a confidence value, say 53%; or it may be a vector of activities with associated confidence levels: say [running:53%, hopping:32%, walking:15%]. Of course the latter can be reduced to a single activity by selecting the activity with highest confidence as often a single choice may be needed.

When we have several possible activity types (or indeed several kinds of classification in general for ma-

chine learning), then there is not a simple accuracy value, nor even simple false positive/negative, as a classification of 'running' could be wrong in several ways. A confusion matrix captures this uncertainty giving the probability, say, that a person who is classified as running is actually walking; in the example confusion matrix in Table 19.2 this is 12%.

TABLE 19.2   Confusion Matrix for Activity Recognition.

		actual activity		
		running	walking	hopping
classified activity	running	67%	12%	21%
	walking	82%	15%	3%
	hopping	57%	7%	36%

Note that you can draw this the other way round, asking that if someone is actually running, how likely they are to be classified as walking, but this is *not* simply the transpose of the matrix in Table 19.2 as it needs to take into account the probability that a person is running in the first place, that is the base rate.

Note finally that this uncertainty in the final classification in the output of sensor processing (whether an activity label or some other kind of measure) includes both uncertainty due to raw sensor measurement and due to the recognition/processing process. Sensor redundancy and sensor fusion may reduce some of the raw sensor uncertainty, but there is almost always significant uncertainty remaining.

19.5.4 Designing for Uncertainty in Sensor-rich Smart Environments

Designing in smart environments is almost always about designing within uncertainty, creating a system that works as a whole even if elements cannot be 100% accurate. One way to do this is to try to match the accuracy of the sensor and AI. We'll see how to do this using a simple example of car courtesy lights.

Figure 19.5 shows a scenario of getting into a car ready to drive off. Each step is marked with how important it is that the lights are on ('+'s) or off ('−'s), the more plusses or minuses there are, the more important it is that the lights are on or off in the state. So for example it is more important that the lights are on when looking up the route than when adjusting the seat.

This kind of representation can be useful in itself as part of a discussion with potential users. Note especially the little bomb symbol for walking up to the car. This

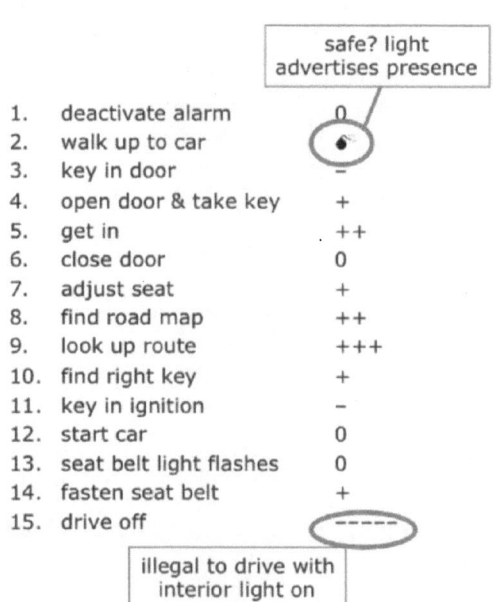

safe? light advertises presence

1.	deactivate alarm	0
2.	walk up to car	●
3.	key in door	–
4.	open door & take key	+
5.	get in	++
6.	close door	0
7.	adjust seat	+
8.	find road map	++
9.	look up route	+++
10.	find right key	+
11.	key in ignition	–
12.	start car	0
13.	seat belt light flashes	0
14.	fasten seat belt	+
15.	drive off	– – – – –

illegal to drive with interior light on

FIGURE 19.5 Courtesy lights – when do you want them on?

was initially marked with plusses as it seemed good to have a little light as one was opening the door, but in a discussion about this one person pointed out that when alone at night, it was dangerous to have the light turn on as one approached the car as it let muggers know where you were going.

Having got this it can be used to match with available smart technology. This might include simple door sensors, weight sensors in the seats (available for seat belt warnings) and data from the internal infra-red security sensors that can help differentiate the driver and passenger's activity.

Let's imagine that we have trained a neural network to identify the various activities in Figure 19.5 and produced the confusion matrix. If there were only two activities, then we would need to trade-off false positives and false negatives using a ROC curve as we saw in Chapter 9. However, here we have a multi-way classification and in particular some of these are less critical than others. These less critical cases give us leeway to improve the behaviour in the most critical situations.

Suppose the classification system is 60% certain you are trying to put the key in the ignition (step 11), when it is slightly easier with no light as it is in shadow anyway; but with a 30% certainty that you are finding the right key on your key ring (step 10), when it is a little better to have light, but possible by feel; and only 15% certainty

that you are still looking up your route (step 9). A 'highest certainty' rule would turn the light off, as you are most likely to be trying to get the key in the ignition, but a more importance weighted rule would keep the light on as the state of the light in step 11 is of low importance whereas step 9 is critical.

19.5.5 Dealing with Hiddenness – A Central Heating Controller

With the car courtesy light the problem is that the system doesn't know precisely what you are doing. However, if there is a time lag, or the state of the system is hidden, the user may be in a position of uncertainty about the system. This can sometimes be used constructively by taking into account the context and the user's inferred meaning because of that.

One of the problems with heating controllers is that people think the level on the controller is about the quantity of heat that is being produced. Imagine you have only recently arrived home and the heating has only just started to warm the room, that is the current temperature of the room is below the thermostat temperature and the heating is already working at full capacity warming it up. However, you feel cold, so you turn the thermostat right up. This has no immediate effect as the heating is already working as hard as it can, but some while later the room ends up far too hot. You then turn the thermostat right down ... and the cycle continues!

We are going to replace the system with a smart system that tries to adjust the temperature to your preferences but still allows you to override the system's chosen temperature – indeed it is precisely these times you override it that allow it to learn. If the new system simply copies the old model of a thermostat temperature, it will inherit the same problems.

One solution is to make the state less hidden, perhaps showing the target temperature the system is trying to reach and the current temperature, with an icon such as a fire to show the heating is working hard (Figure 19.6, left).

Another option is to have no display at all and simply a control with plus/minus or up/down arrows meaning "I'd like it warmer" or "I'd like it colder" (Figure 19.6, left). Maybe pressing the plus button twice might mean "I'd like it a lot warmer". If we return to the scenario of the heating being below the current system target temperature and you press the plus, the system will sim-

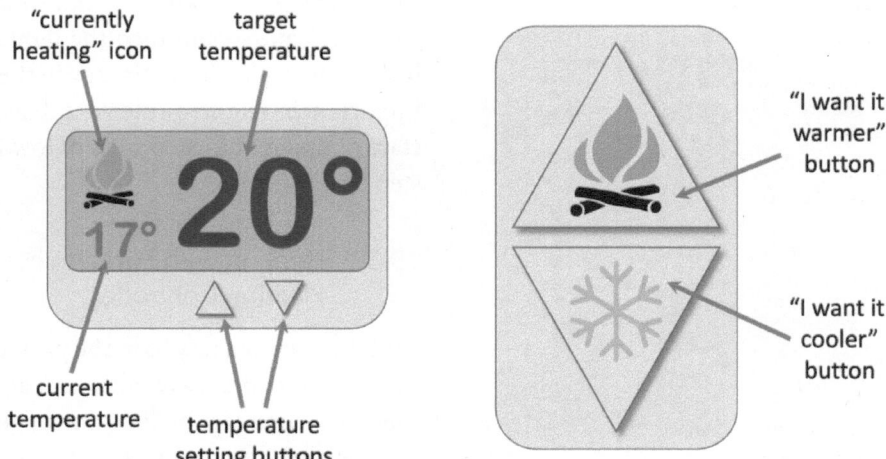

FIGURE 19.6 Heating controller (left) first design option more information; (right) second design option less information, more intelligence.

ply ignore your press – it is doing it already. However, if the temperature is at the system setting, it will both turn up the target temperature and use this to improve its learning.

Thinking further imagine you have installed a carbon-conscious AI system. Over time the system can gradually lower the room temperature from its best estimate of your desired temperature. It may be that you don't notice or if asked would ideally like it a little warmer, but not enough to be bothered to do anything. However, if it gets too cold you press the plus button. This means that the system can not only learn over time your ideal temperature but also your tolerance either side and deliberately save energy by keeping the room only just above the point where you care. Maybe over time you even get used to the lower temperature and the levels can drop further still, saving you money and saving the world at the same time.

19.6 DESIGNING FOR AI–HUMAN INTERACTION

When there is expected to be human involvement, some of the unstated requirements of algorithms change.

For machine learning systems and AI in general, the normal metrics are about accuracy, correctness and optimality. How often does the system give the right answer? What is the single best choice? These are good questions if the system is to be deployed with no direct human involvement, or in situations where people should not be aware that the AI is operating.

However, these are often not the best algorithm heuristics when there is a human in the loop. Here the appropriate question is not how to make the AI as good as possible but how do we design the intelligence and the interaction so that when they are used with people the system as a whole behaves as well as possible.

19.6.1 Appropriate Intelligence – Soft Failure

When you demo a system, you of course want to give the best answer as often as possible. We can think of this as two tenets of useful artificial intelligence:

1. be right as often as possible

2. be as good as possible

These are important. However, in real use the key question is often how badly things go wrong and how easy is it to recover when they do. In other words, tenet 1 is important, but in practice so long as it is right reasonably often it is fine. Tenet 2 is important, but if it is reasonably useful, it is fine. However, more important is the tenet of appropriate intelligence:

3 don't mess things up when you are wrong

Older users of Microsoft Word will remember Clippy. Every so often, as you were typing, a small pop up would appear with a little paperclip figure in it to give useful advice, "I can see that you are writing a letter, why don't you ...". It used reasonably clever algorithms in order to

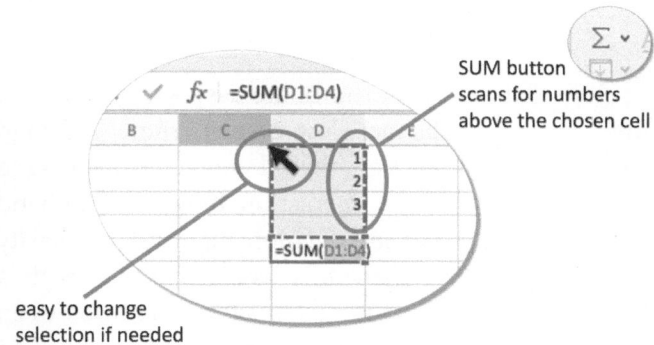

FIGURE 19.7 Appropriate intelligence – Sum button in Excel.

detect the kind of document you were working on (addressing tenet 1) and had (more debatably) useful advice (tenet 2). However, when it popped up in the middle of the screen it interrupted both your train of thought and your typing, discarding anything you were typing after it appeared until you cancelled it. That is, it failed tenet 3.

Not surprisingly, Clippy was withdrawn, although other more subtle forms of intelligence are used. In contrast, the Sum (Σ) button has been in all versions of Excel for many years. This was introduced based on analysis of spreadsheet use that showed that the most frequent action after entering numbers is adding up a row or column.

The Sum button uses a fairly simple intelligent algorithm. There are a few tweaks to manage subtotals, but basically it scans upwards and leftwards from the current cell looking for the nearest block of consecutive numbers and adds a sum formula. If there is a tie for the closest, the vertical (column) sum wins. The algorithm is simple and often works out what you want (tenet 1) and inserts the correct formula (tenet 2).

Crucially however, once the sum is inserted the cells that form part of the sum are selected. If they are not what you want, it is easy to grab one of the selection handles to extend or completely change the selected region (Figure 19.7). In other words, when it is wrong, it is easy to fix (tenet 3).

19.6.2 Feedback – Error Detection and Repair

With Clippy, it was usually obvious whether or not its suggestions were helpful, it was just that by this point the damage to your concentration had already been done. However, in many settings the problem is even more severe: things go wrong but you don't even realise.

One day you say to Aria, your voice-based virtual assistant, "*Aria, order two, no three, no two bottles of milk*"; "*OK*", says Aria. The next morning a large, refrigerated truck arrives and starts to unload crate, after crate, after crate. If only Aria had instead answered, "*You'd like to order two oh three oh two, that is twenty thousand three hundred and two bottles of milk. Is that right?*"

In human–human conversations we often mishear or misinterpret what others are saying, but we also constantly offer ways to confirm that we have a common understanding and are adept at repair. Sometimes this takes the form of an explicit confirmation step, such as Aria's "Is that right?", but it can also be in the form of implicit feedback during the next step of conversation.

Imagine you are on the phone to the airline agent (maybe human, maybe AI) at New York JFK and say "*When is the next flight to* [slightly inaudible] *London?*", the reply might be "*The next direct flight from New York JFK to Lisbon is midday tomorrow*". Note that this gave the answer to the question, but rather than just saying "*It's at midday tomorrow*", the reply reiterated the hearer's interpretation of what was asked, partly based on context (the airport of the agent) and partly based on (mis)heard information (Lisbon rather than London). This gives you the chance to correct the mishearing "*that was London not Lisbon*", and also the contextual interpretation "*what about other New York airports*", or "*I'd be happy with a connection*".

We can think of these speech-based interactions using the stages of language processing we saw in Chapter 13; there are lots of different kinds of AI involved at each level (Figure 19.8).

Lexical – what was actually said – "no" or "oh", "London" or "Lisbon"

Syntactic – how this is pieced together as a unit, for example "oh" being treated as a digit (zero) and thus making the number 20,302

Semantic – the utterance means "please order 20,302 bottles of milk"

Pragmatic – the way this makes sense in the real-world setting, for example the assumption that this is from the normal dairy supplier and maybe that 20,302 is a lot of milk

There is the potential for misunderstanding or misinterpretation at each level, and thus for feedback and repair. However, in practice, as a human, we make choices of the level of feedback and whether we seek explicit or implicit confirmation based on our level of confidence. If the line was very bad, or the location very unusual, the agent might explicitly ask "did you say Lisbon?". Similarly, Aria might normally say, "I'll order twenty thousand three hundred and two bottles of milk. Is that right?" but would only add the lexical feedback "two oh three oh two", if there was high uncertainty (noisy room).

Note too that these levels cannot be treated entirely separately as a pipeline. If at the pragmatic level the Aria system notices that 20,302 is an unusually large milk order, this might trigger a search at a lower level for interpretations of lower confidence, maybe not low enough on their own to trigger additional feedback but possibly sufficient to say, "two oh three oh two", with the additional knowledge that there is potential high-level inconsistency.

19.6.3 Decisions and Suggestions

Note that while Clippy and Excel Sum differ in their impact, both are making suggestions rather than doing something. Imagine if when you pressed the Sum button in Excel, it not only added the sum but guessed that you were calculating numbers to put into your tax return, so it accessed your bank account and copied all your financial details into the spreadsheet. Of course, while this could be precisely what you are doing, maybe you are in the middle of a meeting projecting the spreadsheet on a large screen to a room of 20 colleagues, who then see all your personal expenditure for the year. As a rule of thumb unless the level of confidence you have in the correctness of an automatic decision is high or the consequences of getting it wrong are low, it is better to offer suggestions for action rather than automating the action (Figure 19.9). However, there are limits to this.

If there are too many confirmations and suggestions, then this may itself interfere in the user's tasks or lead to 'click it away' habits, as often is the case with confirmation dialogues.

In practice there are three factors that work together to determine the appropriate level of suggestion or action.

Confidence – how accurate is the AI system at its predictions/interpretations

Cost of failure – how bad is it if the action is wrong

Complexity of interaction – how much would this add to the human's load

If all of these agree, things are easy. For example, if there is high confidence, low cost of failure and high complexity of interaction, then there is a strong case to act without further user interaction. If on the other hand the confidence is low, cost of failure high and complexity of interaction low, then it is worth checking with a human user first. Real life often sits in the in-between places, and here effective design of interactions can make a difference, for example subtly offering feedback of what has been done, without asking an explicit "*is this right*".

19.6.4 Case Study: onCue – Appropriate Intelligence by Design

Figure 19.10 shows the operation of onCue, a commercial system the author was involved with in the dot-com period [78]. The interactions in onCue were designed with the principles of appropriate interaction explicitly in mind.

onCue watched for when the user copied or cut items into the clipboard. It contained data detectors (see Chapter 17) that used pattern recognition to work out what kind of content had been copied. This was then used to populate an intelligent dynamic toolbar that usually floated on the side of the user's screen. If the clipboard contents looked like plain words, onCue would suggest search engines; if the content looked like names it would suggest directory services; a table of numbers would suggest graphing or opening in a spreadsheet.

Crucially, onCue was not intrusive – it was not modal taking keyboard focus, nor did it grab the user's visual attention by appearing in the middle of the screen or changing suddenly. Instead, the icons slowly faded in and out, so that it was always there waiting but never demanding.

This was a deliberate design choice based on the principles of appropriate intelligence. The matches, while potentially useful, had low confidence in two senses: (i) the pattern matching was simple, so there could be several potential tentative matches; and (ii) it might not be needed – you might simply want to copy a name but not want to look it up in a directory service. Because of this the interaction was designed to

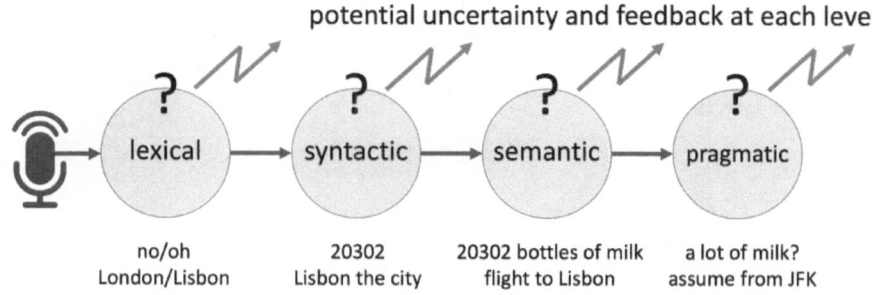

potential uncertainty and feedback at each level

lexical → syntactic → semantic → pragmatic

| no/oh | 20302 | 20302 bottles of milk | a lot of milk? |
| London/Lisbon | Lisbon the city | flight to Lisbon | assume from JFK |

FIGURE 19.8 Multiple levels of processing in a speech-based system.

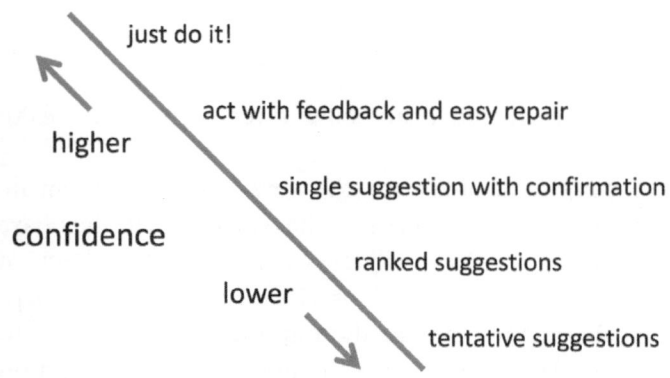

FIGURE 19.9 Levels of confidence and action – simplified view.

offer unobtrusive suggestions, which imposed minimal additional load on the user.

19.7 TOWARDS HUMAN–MACHINE SYNERGY

We saw that AI systems differ in the extent to which the user or AI is in control. Possibly the best outcomes are when the system is designed to be synergistic, to enable both human and computer system to work together.

One part of this is assessing which aspects of an overall task are most suited to the human or artificial participant, for example the "Humans are better at … Computers are better at" lists we saw for expert systems in Table 18.1. This can be used for function allocation – who does what. This can be static, with fixed things that the system does (maybe find candidate interaction icons) and things the user does (choose which they want). However, we may also sometimes have dynamic function allocation where the balance changes depending on the context, for example, in an aircraft cockpit, an AI system might automatically manage less critical aspects during take-off and landing when the pilot's attention needs to be focused, but leave them to the air crew during flight when they have free attention.

As well as taking different tasks, the nature of both AI algorithms and interaction can be deliberately modified in order to make a better overall system. That is we might deliberately choose an apparently less than optimal AI algorithm or a less than optimal user interface in order that the overall synergistic system works as well as possible.

19.7.1 Tuning AI Algorithms for Interaction

First let's look at how we can adapt AI so that it works well in human-in-the-loop systems. In fact, many of the things needed for this turn out also to be helpful when one AI system is embedded within a larger fully automated system – that is algorithms that are good for human interaction are also often good for machine-machine interaction.

We have seen several examples where measures of confidence or uncertainty are helpful in order to trigger feedback. These are generated as a matter of course by some algorithms as part of their normal workings. This may be used to select the best but can also be used to signal uncertainty. In some cases, there is an absolute measure, for example Bayesian reasoning can yield a probability estimate. In other cases, the measure is relative, for example where the scores for two alternatives are close to one another. Where there is no such measure, it may be possible to create one, for example by adding noise to the inputs of a neural network and seeing if this changes the classification it gives.

When making an automated decision, an algorithm needs to be very sure it is correct; however, when making suggestions, the opposite strategy is often useful. If there

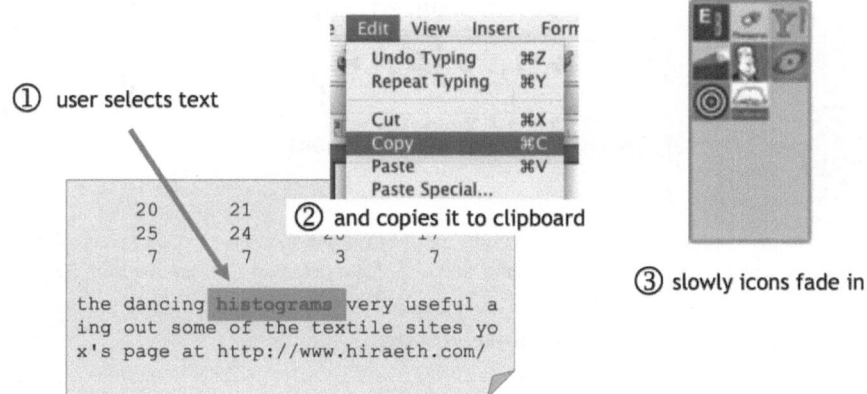

① user selects text

② and copies it to clipboard

③ slowly icons fade in

FIGURE 19.10   onCue – designing appropriate interactions.

is some sort of score of fit or certainty, then the threshold for an automated decision can be set very high, but the threshold for whether to offer something as a suggestion can be set quite low, presenting multiple alternatives to the user.

There are limits to this. Hick's Law says that the time taken to make a choice increases with the logarithm of the number of choices – having more choices means more time. Sorting the choices in relevance order can help – which emphasises the importance of the quality of confidence measures. Also if the bar for admitting suggestions is too low, the poor quality suggestions may erode trust in the system.

In some applications a mixture of the two can be helpful with liberal suggestions combined with conservative warnings.

An example of this principle was adopted in human verification of record matching of historic 19th-century musical records [86]. There are often multiple newspaper notices referring to a concert, but they may differ slightly "St Thomas Hall, on 3rd Dec at 7pm" "Saint Thomas's in the evening of December 3rd". An automatic algorithm matched these records, with different kinds of matching for dates, times and venues that were weighted together to give a final score. A liberal policy was then used to match records with high match scores to create groups of records that *might* be the same concert. A human expert then went through these confirming or changing the groupings. In addition, to make this task quicker, some groups were marked as more critical to check than others. A conservative strategy was used for this, with only groups where all the items were nearly identical being given a positive health check.

It is also helpful where the workings of the system are more transparent, that is if the way in which they have come to a decision is apparent to the user. We will return to this issue when we look at explainable AI in Chapter 21, but we can see the importance of this already. When Aria reflects back the first stage of its interpretation "two oh three oh two", it is effectively exposing a mid-point in its processing pipeline. The speech processing algorithm that has transformed the raw analogue signal into these words may be very obscure, perhaps involving signal processing or a deep neural network. When you heard this, you would realise the problem was that Aria had misheard you and at least understand what had gone wrong. In addition to making repair easier, this increases confidence in the system, it was not an arbitrary bug that led to those crates of milk appearing!

19.7.2 Tuning Interaction for AI

As well as modifying algorithms to make them work better with people, we can modify the interaction to provide better information to allow AI systems to work better. This can be explicit, for example asking users to provide information for a profile; however, it is often better if it is implicitly embedded in the interaction.

Let's look at an example.

When you use a search engine, the system logs which websites you click through to. This is then taken as an implicit measure of relevance and is used to improve the algorithm's effectiveness for you and others. However, this is harder if the results are short.

On the left of Figure 19.11 is the search results page for a text book [88] that is finding each paragraph in the book mentioning the search terms. The paragraphs are shown in full and the user scrolls down to look at different search results. This is a very fluid interaction as we are used to scrolling quickly to skim, but it is very hard for the system to work out which are the most relevant results. It is possible to measure how long the user spends before scrolling further, but there are typically several results visible, so without using eye tracking it is hard to know which is the actual result the user is looking at.

On the right is an alternative accordion-style interface. A shortened version of each result is shown, and the user has to explicitly 'open' the entry to see the full paragraph. Imagine we have done testing and found that the original is slightly better than this new interface, but not by a large margin. It may be worth still using the new design, even if it is slightly less usable, as it is far easier to work out which of the results the user found relevant and hence improve future searching for this user and others.

This is an example of epistemic interaction, that is where the style of interaction is chosen in order to improve the information available for the AI system and thus to improve the overall synergistic human–AI system.

The menu example is hypothetical; however, it has been suggested that part of the success of TikTok has been due to precisely such interaction tuning. TikTok works by scrolling, but ensures that only one video at a time is visible on screen, making it easy to collect precise feedback that is then used to tune the recommendation algorithms [195].

19.8 SUMMARY

In the end, *every* AI system has an impact on humans. This chapter considered situations where the contact between AI and people is more direct. This has included where traditional digital applications incorporate AI in intelligent user interfaces and also where the AI is more behind the scenes, combining sensor inputs to modify aspects of the physical environment in which we live. We saw that appropriate intelligence can be used to design interactions where occasional inaccurate or incorrect outputs of AI do not lead to failures in the overall human–AI system. We have also seen ways in which AI

algorithms can be adapted to work more effectively in human-in-the-loop systems, and user interaction can be designed to maximise information available for synergistic human–AI adaptations.

19.1 Using Shneiderman's two-dimensional human-centred AI framework in Figure 19.1:

a. Populate it with examples of tools and appliances used in a kitchen.

b. Do the same for controls and devices in cars with different levels of automation.

19.2 Create a table similar to Table 19.1 in Section 19.3.4 and classify the initiative, decision and action of each of the following:

car-nav-1 – A car navigation system where you enter the destination, it works out the optimal route and then tells you which turns to take along the way.

car-nav-2 – The same, but where the system proposes several routes and you select the one you prefer.

warehouse-1 – A system in a warehouse where incoming orders are processed and robots automatically despatched to collect items from the shelves.

warehouse-2 – A system in a warehouse where incoming orders are processed, the optimal picking order created and given to the human operator to fetch the items.

warehouse-3 – The same, except that the system asks an operator, who decides whether to send the robot or whether they wish to fetch the item themselves.

lorry key – A key fob for a lorry, where you press the button and depending on which end of the lorry you are at (sensed by the wireless signal of the key fob), either the back door or the cab doors are unlocked.

19.3 Make a list of activities that you (or your group) do regularly, both on computers (such as answering email, or writing a report) and physically (such as making toast, or playing squash). For each, ask:

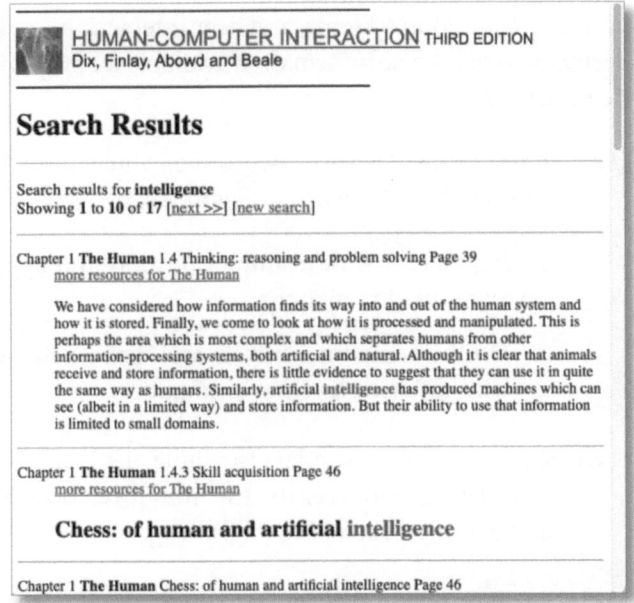 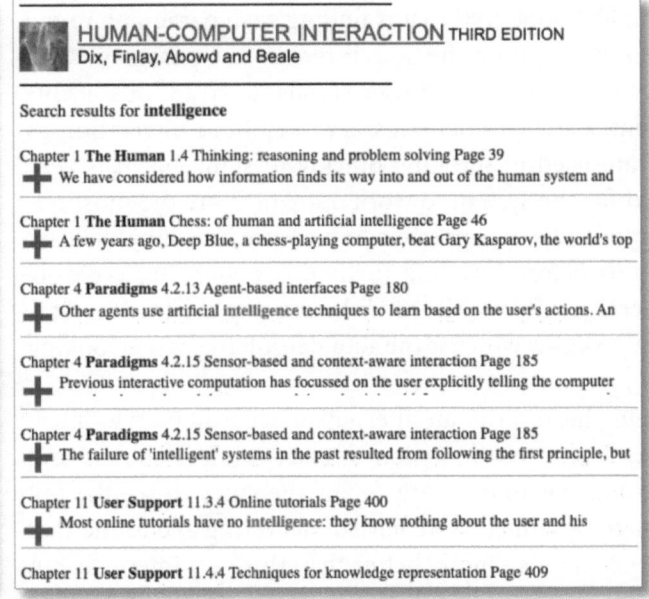

FIGURE 19.11   Two search engine interaction styles.

a. Is AI used already in the activity?

b. If not, could it be, or if it is, could the AI be used more?

c. Use the vocabulary developed in this chapter, to describe the existing or envisioned use of AI. For example, is it synergistic or does one partner have most of the control? Does it use sensors, and if so, is it single sensors or using sensor fusion?

d. Choose one activity where the AI (or sensors it uses) is likely to be inaccurate or uncertain, and create an analysis similar to Figure 19.5.

e. Are there ways additional sensors or changes in the activity (epistemic interaction) can be used to improve the information available to the AI?

FURTHER READING

B. Shneiderman. *Human-centered AI*. Oxford University Press, Oxford, 2022.

Shneiderman's book was one of the key sources for the discussion of levels of automation in Section 19.3.5 but also covers many other issues in design and governance. In particular it emphasises the importance of AI companies being held accountable for failures in systems, which connects with issues of responsibility in Chapter 23.

A. Dix. *AI for human–computer interaction*. CRC Press, Boca Raton, FL, 2025.

Expands on the issues of this chapter, exploring in more detail the way AI can be used in the user experience design process and also the way insights from HCI can help create more effective tools for AI practitioners, including aspects of explainable AI.

When Things Go Wrong

20.1 OVERVIEW

This chapter deals with some of the things that can go wrong due to AI. The previous chapter has talked about dealing with individual failures, but this chapter is more about systemic issues. This includes deliberate misuse such as disinformation and also unintentional misuse, notably bias. We will discuss some of the general approaches to deal with problems including transparency and algorithmic accountability. We examine bias in detail, including how it arises at different stages in the machine learning process. We look at threats to privacy and ways to mitigate it, and also some of the dangers of deliberate and accidental misinformation.

20.2 INTRODUCTION

AI can be very powerful both in tackling tasks that humans already do and also doing things that are beyond our comprehension. This is of course good when everything works, but sometimes things go wrong. The tide of AI optimism has been countered by an increasing number of stories of failures of various forms, from autonomous car fatalities to 'racist' search results and in the UK a (so called) 'mutant' algorithm for school leaving exams. Some are also laying part of the blame for growing intolerance and extremism in society on the algorithms behind social media and search engines, which, in the interest of giving consumers what they want, are creating sounding boxes where we each see only views, news and 'facts' that agree with our own preconceptions.

To some extent Chapter 19 has already dealt with various forms of errors or failures in AI, but focused on specific points in an interaction and how overall human–computer systems can function robustly despite the inevitable inaccuracy or failures of AI. In this chapter, we will focus more on the larger-scale impact of AI algorithms, focusing particularly on misinformation, bias and privacy. We will consider other large-scale societal and ethical AI issues in Chapter 23, many of which do not admit easy answers and where the key question is "what do we want AI to do?" In the main issues discussed in this chapter we have a better idea of *what* is wanted, but there are challenges in *how* to achieve it.

20.3 WRONG ON PURPOSE?

The press naturally focuses on the bad side of algorithms and machine learning. However, it is important to distinguish different forms of bad outcomes, some are deliberate, some are unintentional, either accidental or negligent.

20.3.1 Intentional Bad Use

We'll first consider deliberate misuse such as hacking. Here we may not blame the algorithms per se, but their weakness or vulnerability. In such cases we may seek better software design or security. For classic hacking the vulnerability is not in the complexity of the intelligent algorithm itself, but the surrounding operating system, device drivers, etc. Once the hacker is 'in' they may subvert the software – modifying or replacing the code.

However, big data has led to more complex forms of vulnerability based on subverting the data. In the case

DOI: 10.1201/9781003082880-24

of the Cambridge Analytica scandal, this was principally about using data in ways it was not supposed to be used; so is mainly connected with privacy and personal control of data. However, the other aspect of this scandal was the way the resulting data was used to influence the US presidential elections. This was a fairly direct use of data, but often it can be less direct. Bots often deliberately create inflammatory posts on both sides of an issue; the aim is to increase re-posts and hence the ranking of the channel, so that subsequent deliberately misleading or misinformative posts (fake news) will have instant influence. A more citizen-led form of data manipulation is used by campaigns to get everyone to do specific Google searches in order to make it have a particular auto-completion when you start to type a query such as "Trump".

Some forms of deliberate 'bad' use are legal and may be regarded as acceptable depending on one's ethical viewpoint.

Autonomous weapons have been widely condemned if there is not a human in the loop, and major scientists and industrialists have called for there to be an international ban akin to that on chemical weapons [124]; although others have argued that if AI weapons are more accurately targeted on combatants, we have an ethical duty to use them. Oddly we accept bombs that explode at a fixed height, or guide themselves to a specific location using GPS; we also accept soldiers trained to obey orders without question. Perhaps unpacking what is so bad about autonomous weapons could help us unpack our attitudes to war itself?

Cyberweapons and cyberattacks have also attracted significant publicity. Some simply attack computer software or data, for example Denial of Service (DoS) attacks, but others may be targeted to bring down infrastructure or even cause physical damage. The first (publicly known) case of cyberwarfare against physical infrastructure was Stuxnet designed by Israeli and US intelligence to degrade the Iranian nuclear programme by attacking centrifuges [156]. Spread by USB memory sticks, Stuxnet showed that even internet-isolated computers could be at risk, and there are also rumours that it 'escaped' spreading to unintended targets with similar hardware including Ukrainian and Russian nuclear power stations [290].

Many of the methods to detect and prevent cyberattacks are based on big data and machine learning, but malware applies similar principles to mutate itself and find ways around defences.

In some cases ethics can be built into machine learning to prevent misuse. Notably GPT-4 was progressively modified before release to prevent it answering questions relating to illegal or otherwise harmful activities [217].

20.3.2 Unintentional Problems

Sometimes things go wrong unintentionally, whether through ignorance, negligence or pure accident.

One example, which is often widely reported in the media, is accidents involving autonomous and semi-autonomous cars. In some cases, these accidents have been because drivers have not understood the capabilities of cars that operate in semi-autonomous mode but rely on the driver to maintain attention. Some suggest that this means that only fully autonomous vehicles should be allowed. However, there is a long history of partial autonomy from ABS to cruise control; so it may be that the answer is better design of the autonomous vehicle user interface, crucially ensuring that the 'driver' has a clear understanding of the momentary level of autonomy ... and maybe also that the vehicle has a model of the driver's attention

In other cases the vehicle has been in fully autonomous mode. Here the manufacturers and accident investigation authorities have to determine whether this was unavoidable (e.g. a person running into the road in front of the vehicle) or potentially preventable. In the latter case it is particularly important to be able to unpack the chain of sensing and decisions that led to the accident in order to see whether there are changes that could improve safety. The software for such vehicles is inevitably complex with interacting sets of rules and machine-learnt aspects, making such essential explanations difficult.

Even where there is no obvious cataclysmic 'accident', things can go wrong. The system appears to work and make suitable decisions, but are there unintended consequences?

Three forms of this will be familiar to all readers, and we'll deal with each in a little more detail in the rest of the chapter. The first is unintended bias, as opposed to deliberate discriminatory or hateful acts. The latter are easier to identify, but the former may be as harmful both at individual and societal levels. Every few months there seems to be a new case in the news where an algorithm has created sexist or racist results. The second is privacy, which has been recognised as a problem since the early days of the web, but came to prominence with high-profile cases such as the Cambridge Analytica scandal, and is a constant backdrop to discussions about social media. Finally, we will look at misinformation and filter bubbles, both how to counter deliberate misinformation and also how to promote good information habits.

20.4 GENERAL STRATEGIES

There are a number of general ways to address both intentional and unintentional problems with AI. In fact these are mostly approaches that apply across the board to digital systems, but of course these often embody some aspect of AI anyway.

20.4.1 Transparency and Trust

When things go wrong we want to ask the question, "why?" Why did the autonomous car not notice the pedestrian crossing the road? Why did the bank system reject my loan application? Why did the automated exam marking system give me a B grade?

The need to answer these "why?" questions has led to calls for transparency of algorithms. This is an important ethical consideration, but also increasingly embedded in law, and demanded by customers or the public. The European general data protection regulation (GDPR) demands that when algorithms make decisions that affect individuals, for example credit scoring or job shortlisting, these need to be capable of explanation. This is often referred to as a 'right to an explanation', a term that some argue may be misinterpreted in scope [57]. Note that this is not simply guidance but has legal force with very large fines for any organisation or individual who is not able to show they have provided sufficient explanation.

Most countries also have anti-discrimination laws covering what are called 'protected characteristics'. These vary a little between countries but may include gender, ethnicity, religion, age and sexual orientation. If you are using any sort of machine learning or complex algorithm to make decisions, for example on job applicants, awarding loans or selecting housing tenants, you must ensure that you are not discriminating on the grounds of one of these. Typically, there are substantial fines for breaking these laws.

For personal crimes, such as theft or murder, most jurisdictions require that the defendant intended to commit the crime. However, for civil cases this is not usually the case. You may never have intended to discriminate,

but if an organisation does, however unintentionally, it has broken the law.

It is therefore crucial both:

- To be able to convince yourself (and your boss/client) that systems are unbiased

- To be able to provide evidence to others, notably a judge, that they are.

If the decision had been made by a deep neural network, with multiple layers of weights and threshold, would you feel able to justify this to a court?

In fact, the judge may not be the hardest audience. Often a court will appoint expert witnesses, so that a correct, but obscure, explanation may suffice. However, if you are a bank, or a public facing medical provider, you will also need to convince the general public. Indeed in the online world, *trust* is often the most valuable commodity of all.

To win public trust you may need to provide different forms of visualisation and explanation, for example the way some advertisement platforms allow you to ask "*why am I seeing this?*".

Trust is a complex issue as it has many facets. For example, you may trust a company's technical competence, but not its good intentions. Alternatively, there may be a different organisation that you trust to want to do the right things, but you are not sure whether it is able to achieve its goals successfully. Establishing trust in the algorithms is just one part of this wider picture. However, the fact that an organisation at least attempts to be open in its algorithmic decision-making processes often contributes to other, more subjective forms of trust.

20.4.2 Algorithmic Accountability

Another way to tackle these issues is to increase legal and financial accountability [110, 256]. We may accept that legislation will never keep up with technology and when it does may end up hampering good things while not really preventing the harms it seeks to avoid. Instead we can seek to ensure that if things do go wrong, the companies and individuals responsible are held to account, with substantial legal and financial consequences. If companies know they will pay for bad practices or mistakes, they will work to ensure that these things do not happen. This approach effectively trusts the free market to create the best outcomes when costs are suitably manipulated.

This happens with other technologies. In many countries companies that have been found to be negligent to employees, customers or the general public can face large compensation claims, and US courts in particular may choose to set these at a punitive scale, to avoid companies simply factoring potential compensation as part of cost–benefit calculations. On the financial side 'polluter pays' and carbon pricing have been used to create financial incentives for companies to adopt more environmentally friendly practices. These schemes sometimes impose additional costs for 'bad' behaviour (such as polluter-pays taxes) or may do the opposite and offer financial incentives for 'good' behaviour (such as subsidies for farmers to set-aside land for wildlife).

Typically these financial penalties or incentives are paired with more explicit laws that limit or constrain behaviour within limits. For example, most countries have dangerous driving laws that may include criminal proceedings against those who drive recklessly but also have explicit speed limits and other traffic regulation, not leaving it entirely up to individuals to decide moment-to-moment what is safe. Similarly health and safety regulation combines explicit limits such as mandating the wearing of hard hats in certain areas or only allowing certain additives in foods, but in addition requires companies to apply general safety assessment with the potential for negligence claims if predictable hazards are ignored.

We already see similar principles at work in the digital arena with explicit regulation, such as the EU privacy and data protection regulations [56], but backed by civil law cases, such as the class action privacy lawsuits brought against Facebook after the Cambridge Analytica scandal.

There are limits to these approaches.

If laws are too tightly defined, larger companies with sufficient resources can find loopholes. At its worst this may distort markets giving larger companies an inherent advantage over smaller ones, despite the fact that it is typically the behaviour of the larger ones that legislators are most concerned about.

Penalties, whether directly imposed as fines by governments or through damages in civil cases, also need to be large enough that they are not merely regarded as a 'cost of business'. This is particularly difficult with transnational companies, such as the large tech companies, as chains of subsidiaries can shift financial accountability away from the parent company and fines that are

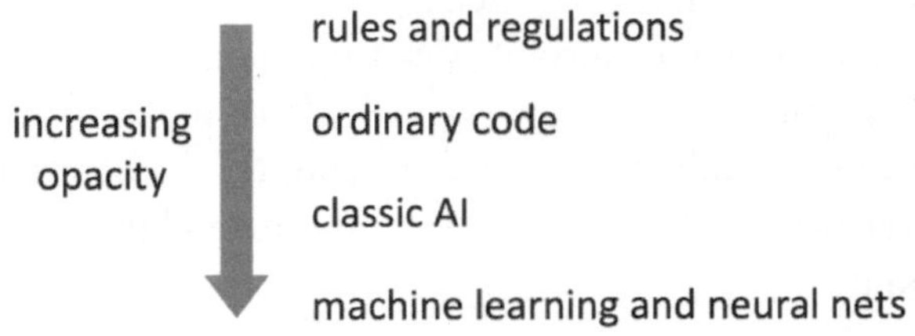

FIGURE 20.1 Increasing opacity in more complex algorithms.

too large may cause international friction between governments.

Sometimes companies have appealed to the inherent complexity and opacity of AI as a defence; effectively saying "*the AI did it*". The core of effective algorithmic accountability is to turn this on its head, to put the onus onto the company to show it has used technology fairly and legally.

20.4.3 Levels of Opacity

The apparent opacity of algorithms is not unique to AI. Although these issues are most stark when the algorithms involve machine learning, they arise with all sorts of algorithms, even the simplest (Figure 20.1). For example, in the 1980s a project using expert systems to capture some of the legislation around welfare benefits found there were inconsistencies that had previously been overlooked [162, 163]. Similarly, few understand the relatively small set of rules around taxation.

As things get more complex, few programmers would claim to understand all the behaviour of their code. Indeed, advocates of formal methods in computing attempt to address precisely this issue, but these methods often prove too cumbersome for all but the most safety critical situations or toy problems in research contexts.

Classic symbolic AI is not so far from programming, although far more broad in terms of computational genres. Crucially many AI languages and notations are declarative; this can make them more clear in intent than ordinary code but may also make the consequences of multiple interacting rules hard to predict. On the other hand AI techniques such as formal argumentation logics may offer ways to help other algorithms become more explainable.

Finally, most of the media coverage both positive and negative in recent years has concerned machine learning of various forms, from fairly simple frequency-based techniques on big data to neural methods in deep learning. For these techniques the explanation as to why something has happened often comes down to *"there was this shed-load of data and this is what came out"*.

20.5 SOURCES OF ALGORITHMIC BIAS

Some of the most common 'bad news' stories about AI relate to the potential for unintended bias, including gender or ethnic bias, in automated decision-making systems. The potential for this to be a problem in machine learning was predicted as far back as 1992 [74], but it was only in the mid-2010s when this started to become a major problem.[1] There have been headline cases of this, notably Microsoft's Twitter bot Tay, which quickly learnt sexist, racist and anti-Semitic language [118]; or cases of Google search returning gender stereotyped images or auto-completion [159]. These stories have continued with almost every new search technology or chatbot [244, 259, 275]. However, potentially more worrying are the cases we don't notice: what are the factors that are being used to set your loan interest rate or determine whether you are shown highly paid job adverts [41, 66]?

20.5.1 What Is Bias?

In statistics and when we are dealing with more formal aspects of machine learning there is a very specific quantitative meaning of bias. When you produce some sort of estimation algorithm, then if you use the estimation process on lots of examples, some will end up a little too high, some a little too low. We say an estimate is unbiased if the long-term average of the estimate is the true average. Effectively this statistical bias is a measure of accuracy.

However, the 'bias' that we refer to in ethical or legal discussions is not the 'bias' used in technical discussions. By bias here we mean an algorithm or decision procedure that unfairly discriminates or disadvantages certain people. We shall see that even an algorithm that is entirely 'accurate' may still embody 'bias' in the ethical sense of the word. It is not sufficient for an algorithm to be technically 'right', it must also be ethically upright.

Even where systems are utterly neutral, the impact of numerous design decisions may affect different groups disproportionately. For example, the UK's Universal Credit system is designed to unify and simplify welfare

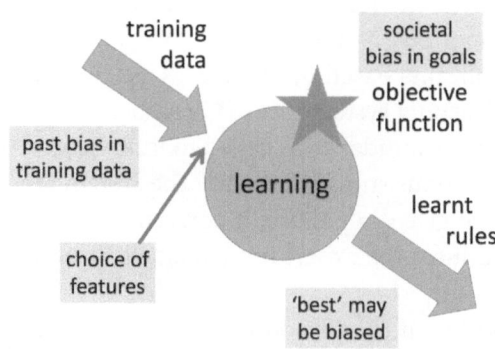

FIGURE 20.2 Bias entering at different stages in machine learning.

payments but is completely computer based. Early trials showed that 49% of those eligible did not have internet access at home [51]. In rural areas land-based broadband and mobile connectivity may be slow or non-existent, so that not just Universal Credit but all forms of internet-shopping, eGovernment or other services are hard to access [201].

20.5.2 Stages in Machine Learning

Figure 20.2 shows a simplified view of machine learning. Training data is fed into the learning algorithm. The algorithm is guided by an objective or fitness function, which defines what it is to be a 'good' set of learnt rules. This all results in some form of learnt rules (where a 'rule' might consist thousands of weights in a deep neural network).

Each of these stages offers potential sources of bias:

1. **bias in the training data** from past biased human behaviour

2. **bias in the goals** from societal bias

3. even when bias in (1) and (2) are removed, **the 'best' or accurate result may still be biased** (in the ethical sense)

In addition, we'll see that the **choice of features** – what you choose to gather data about, is also critical. We will look at each in turn.

20.5.3 Bias in the Training Data

Imagine we want a machine learning system to distinguish cows from sheep, so we feed it lots of images labelled 'cow' and 'sheep'. Similarly if we want it to recog-

---

[1]In fact issues of bias in hand-crafted algorithms were already evident in the 1980s [248].

nise cancerous lesions, then we give it lots of labelled mammograms. Now imagine that we want it to pre-select job applicants based on their CVs. So, we feed the machine learning system lots of CVs labelled as to whether or not the candidate was called to interview.

This sounds straightforward. The system is accurate if when we give it a picture, it correctly labels it, that is if it reproduces the expert human labelling of the training set.

However, if the person labelling pastoral images was confused by highland cows and labelled them 'sheep', then the ML system will confuse them too; if a certain form of cancer often got missed by radiologists, then the new cancer diagnosis system will also miss them; and if the past human selection of job applicants was racially biased, then the trained automated selection process will be similarly biased.

In general, the existing norms and biases of society will be embodied in past decisions and even special labelling for training. The machine learning system will faithfully copy the patterns of the training data and thus embody the self-same traits of society at large. Algorithms reflect data and data reflects society.

Happily this first source of bias is relatively well understood both in the technical literature and by the media. Google processes billions of search queries; if people search for sexist or racist terms, this may naturally emerge when it autocompletes as you type. Similarly, an image search for "Professor" or "CEO" returns predominantly white male faces (see Figures 20.3 and 20.4), but this precisely reflects the preponderance of such images in web pages labelled "professor" or "CEO". The last example is important as it is effectively reflecting the reality of society: senior positions in many countries are more likely to be held by white males.

Some courts in the US have used automated systems to assess the risk of reoffending when considering sentencing or parole requests [7]. These systems have been found to assess black offenders as significantly more likely to reoffend than white offenders even after balancing for other factors. However, if the police have been more assiduous in arresting and prosecuting black offenders, then, as a group, they will have a higher recorded reoffending rate, and therefore a statistically 'correct' system would reflect this. Of course the system would not have had the offenders' ethnicity as an explicit factor in the training data, but we shall see later that 'proxy' measures may effectively yield the same result. In

these examples the training data reflects societal effects that are not neutral with regard to gender or race. It is no wonder the resulting systems also exhibit bias.

There are methods designed to de-bias data, for example to rebalance or weight the training data based on gender or other characteristics. These tools can be powerful as part of efforts to avoid bias, but very dangerous if one believes that they have really removed *all bias* either from the data or the process as a whole. Furthermore removing bias in one area often increases it elsewhere, especially if one considers individual cases rather than overall statistics.

For example, pupils from fee-paying schools perform better at university entrance exams than those from state-funded schools; this is clearly unfair across society as a whole. We might attempt to de-bias the data by reducing the grades of all those from fee-paying schools. This would balance the overall statistics, but would it be fair for a child at a fee-paying school who had worked hard for their examinations? Happily there are more sophisticated ways to approach de-biasing data, but this simple example highlights the complexity of the task.

Simulated data has also been suggested as a way of avoiding human bias embodied in data. Indeed some companies have suggested that by using synthetic data they remove the inherent 'risk' in real data and also offer to create datasets with any desired balance of gender, ethnic or other demographic factors [144]. Again, while potentially a valuable tool in machine learning, the idea that synthetic data entirely removes bias is dangerous. Indeed the algorithms generating synthetic datasets are often themselves in part built using machine learning, and the idea that you can generate particular racial characteristics to order means that stereotypical features are deeply burned into the algorithms.

20.5.4 Bias in the Objective Function

The second source of bias is through what is called the objective or fitness function.

A development programme for autonomous cars might choose to train their system using a simulator. Initially the car would crash all the time, but gradually learn to be better. However, for the car to be 'better' the training system needs to know what 'better' is. If the measure of 'better' is minimising damage to the vehicle, then later, given a choice of a minor scrape against another car or mounting the pavement and ploughing

FIGURE 20.3   Google image search "Professor" 13th Feb 2019 – 15 images, 10 white males.

FIGURE 20.4   Google image search "CEO" 13th Feb 2019 – 14 individual images, 9 white males.

down pedestrians, the car might choose the latter. If the measure of 'better' had been minimising fatalities and injury, then of course it would learn to make different choices. This measure of 'better' is precisely the fitness function, and it is clear that this significantly affects the ultimate behaviour of the system.

In the previous example the objective function was explicit. However, often it is implicit, encoded in the preferences of society at large. Examples of this are rife, not just in automated systems, but also human decision making. We'll consider a number of examples.

At the ACM CHI 2018, Chris Rudder, co-founder of the dating site OkCupid, gave one of the keynotes based on his book *Dataclysm* [237]. He described how data analytics exposed the choices of different genders, ages and ethnic groups. Much of this was unsurprising but still shocking to see in raw numbers. However, more problematic was the way the dating site effectively pandered to these human biases. OkCupid was simply giving people what they wanted, maximising the chances they would find a profile they would like, but in doing so it made explicit choices, for example to use ethnic profiles to determine who saw whom.

In the 1990s, the Trump Plaza casino was fined $200,000 for deliberately moving black employees away from the tables when certain high stakes gamblers visited the casino [284]. Note that the casino had black employees, it was not being fined for discriminatory recruitment policies. The fine was because they were pandering to the racist whims of their customers.

In 2017, the BBC was widely criticised after it published pay gap figures showing that the most highly paid male presenters were receiving significantly more money than their female equivalents. When the story broke, the BBC Director General, Lord Hall, was quoted as saying,

> *The BBC does not exist in a market on its own where it can set the market rates. If we are to give the public what they want, then we have to pay for those great presenters and stars.* [18]

Of course, if the public's perception of major presenters is biased, then that market 'value' will reflect this. In particular, experiments show that both male and female subjects harbour gender stereotypes that have changed relatively little in 30 years [121]. One such stereotype may lead viewers to unreasonably place more trust in male news presenters, thus creating the market demand and 'justifying' the BBC pay gap. However, the same argument could be made by Trump Plaza's casino managers.

In each case the bias, prejudice and stereotypes of society mean that 'good business' would suggest making decisions that are driven by gender, ethnicity and other characteristics that would be deemed inappropriate, unethical or illegal if expressed explicitly.

However, imagine if in each case a machine learning algorithm or similar black-box technique was being driven by apparently neutral metrics such as popularity or consumer demand.

The only reason we know about OkCupid's decision rules is that they obtained them in a two-stage process, using data analytics to go from big data to comprehensible results, and then from that to hard-coded rules. If they had simply said, "*we put all our data into a recommender system*", it would have been a very short but uncontroversial keynote. Similarly, Trump Plaza's shift allocation systems would 'just happen' to avoid black employees on the days certain customers were expected.

In some ways this is already happening with the BBC as the 'market' is the black-box system. Given an objective function to maximise profit, or audience share, the accurate and 'best' decision is 'good' business, but is definitely not good.

20.5.5 Bias in the Accurate Result

Finally, even if the training data and objective are entirely unbiased, and the algorithms used have obtained the most accurate and optimal rules, the results of learning can still be 'biased' in the ethical sense. For this part we'll look at an example of gender discrimination.

In most societies there are major differences on average between males and females, due to many factors, but most significantly the societal norms, expectations and sometimes explicit rules that influence our physical, intellectual and emotional development.

In the UK (and many countries), when there are choices in school, boys are more likely to take STEM subjects, such as chemistry, and girls humanities, such as history. This is clearly not an inherent fact of gender, as other countries do not have such a marked difference. However, in the UK it is the case that, on average, girls and boys have had very different education by the time they leave school.

Now imagine you are selecting applicants for two jobs with the Antarctic Survey, one for a communication-rich role at Rothera research station on the Palmer Land peninsula, and the other for an engineering-related role at Davis research station near the Amery Ice Shelf, several thousand miles of ice and snow away. The applications are all in and you need to work quickly as the last ships, one for Rothera and one for Davis, are about to leave to get there before winter weather cuts off the bases for six months of dark winter.

You take the applications home and you have whittled the applicants down to two. Then, disaster, the dog eats the CVs. All you have left are the diversity information pages that you carefully separated and which contain information about gender, etc. There is no time to get fresh CVs and yet you must send the chosen applicants post haste to their respective ships. You peek at the forbidden diversity pages: one applicant is male and one female; one job is communications-rich, one engineering related … what do you do?

Because of our education system, gender is a predictor of communication and technical skills, albeit a poor one. The reason we do not use gender as a predictor is not because it lacks predictive power; instead it is because as a society we *choose* not to. It is an *ethical* decision.

As a society we choose to use other (and actually far better) predictors. We may look at exam results, or run our own tests that more directly assess the skills or knowledge we require, but we choose not to use gender irrespective of whether it offers any predictive power.

20.5.6 Proxy Measures

We may think that we can deal with bias simply by not including protected characteristics in our training data. So long as the CV does not mention the gender of the applicant, then the outcome must be fair.

To see why this is not the case, imagine a very socially conscious building company. They are very traditional in terms of methods (a lot of heavy lifting), but advanced in the use of IT, so they decide to create an automated system to help with hiring. In order to avoid bias in the system, they conduct an extensive experiment. One thousand people are recruited, 50% male, 50% female, and employed for 2 months, with their productivity heavily monitored. At the end of the experiment a machine learning system is given the measured productivity of subjects together with their CVs and builds a

predictor of productivity. Being an ethical company, the gender and other protected characteristics are removed from the CVs before they are entered into the learning system.

If the system were entirely opaque, one would just have to trust it. The entire process was gender-blind, so surely the resulting system would be unbiased?

Now imagine a learning system that creates more transparent rules. You start to interrogate it and find that school exam subjects are being used and the rules effectively say, "if the person has taken STEM subjects then hire them". Now STEM subjects at school are almost certainly not useful on an old-fashioned building site. However, they are a proxy indicator of gender, which in turn is a crude predictor of strength.

Note again that the algorithm may be producing the 'best', most accurate estimator given the data available. However, bias, in an ethical and legal sense, is not about algorithmic correctness, it is about social choice.

Note also that if the job had required technical ability or good communications, then exam grades would be deemed a reasonable and acceptable decision criteria. The exam results would correlate with gender, but would be directly relevant to the job. The problem in the building site is that they are not clearly relevant to the job at hand and merely act as a proxy gender measure.

In other words, exactly the same training data could yield ethical or unethical (and legal or illegal) outcomes depending on context.

20.5.7 Input Feature Choice

In several of the examples we have discussed, the choice of input features being fed into a learning system has been often critical in creating or controlling bias. The most obvious approach is to exclude explicit gender or similar indicators, but as we have seen this is *not sufficient*.

In the case of the builder, the problem is partly that if there is no direct measure of physical strength on the CVs, then the system will choose the 'next best' thing and may latch onto (proxy measures of) gender.

A strong guard against this is to make sure you collect relevant features. If the system has a good measure of physical strength, it is less likely to fall back on gender, the use of which is both a relatively poor predictor and illegal in many countries.

Furthermore, as well as removing explicit gender indicators, one might consider also deliberately excluding what appear to be irrelevant features. This may have technical benefits by reducing overfitting and also make the system less likely to have potential proxies to latch onto.

20.5.8 Bias and Human Reasoning

The examples point out potential dangers for machine learning systems. However, anti-discrimination legislation predated the widespread use of AI. The legislation exists precisely because humans haven't done so well at these issues prior to automation.

The human perceptual and cognitive system has developed primarily for information poor environments, where you have to make the most effective inferences from scant data. Now we live in a world of information overload, but with the same perceptual and cognitive system as our cave-dwelling ancestors.

Crucially humans are poor at ignoring low-quality cues even when there are better ones to hand. One example of this was exposed by experiments on people's ability to assess the quality of search results based on the title and snippet as commonly found in web search results [241]. Users were able to assess relevance using only the snippet but were better when shown the title on its own. They were then shown both together (title and snippet), which one might imagine would be better still (more information). However, on the contrary, the effectiveness fell between the two on their own. Even though the snippet was not adding to the subjects' ability to assess relevance, they were unable to ignore it and hence performed less well than if they had had the title alone.

Some algorithms also have these problems of ignoring this unhelpful information, but others can do better and maybe even help people to become better at judging such things.

Recall in Chapter 18, we discussed a study that was performed some years ago of people admitted to hospital for heart attacks. The doctors gathered many test results and other forms of evidence and used this to decide among a few different forms of treatment. Retrospective data was then collected including the original diagnostic features and the clinical outcomes for the patients after a few months, whether they had recurrence, or indeed died.

The data was used to train a classifier; however, this was not used as an automatic diagnosis system to replace the doctors' judgement. Instead the analysts examined the rules created by the system and realised that the optimal classification depended on four features only.

The doctors were told about this and changed their clinical practice: only collecting the four relevant factors, but otherwise using their clinical judgement as before. They found that their own clinical outcomes improved. By *not* collecting data and never seeing it, they became better at their job.

20.5.9 Avoiding Bias

While in principle algorithms could behave better than people, the reality is still far from this.

Crucially, as we have seen, it is not sufficient to remove explicit indicators of gender, ethnicity, disability, religion or other protected characteristics. Many discrimination cases relate to indirect forms of discrimination, for example demanding a particular headwear when this is not essential to the job, which effectively discriminates against Sikhs or those wearing the hijab. As we have seen it is easy for a machine learning system to accidentally latch onto proxy measures of a protected characteristic.

Instead, algorithms need to *actively avoid discrimination*. For example, after training an algorithm on gender-blinded data, one could deliberately re-introduce gender and build a causal model. If the impact of features on the final decision is factored through gender, then that is a good indication that the features were acting as a proxy for gender. One could even imagine building this into the original learning process.

Whether or not algorithms are better or worse than humans at making ethically unbiased decisions; how do we know? There is a growing set of techniques and tools to help with this, and it will undoubtedly have grown by the time you read this.

One method is to perform some form of *external audits* of the statistics comparing the way different groups are dealt with by a system or process, whether by human or machine. A good example of this is the way many companies now publish pay gap data.

Note that such external statistics do not answer the question "*is my process biased*", but do offer evidence to pursue and investigate in more detail. There may be external societal reasons, such as disparity of access to education, that create unbalanced outcomes. The stats are the beginning, not the end of an investigation into bias. It is not sufficient to look at the overall numbers, but we must dig into the reasons that led to them.

This can be applied at the end of the process by looking at the overall decision being produced. However, the same technique can be used proactively to attempt to de-bias training data. If there is a disparity in the labelling of training data, for example if we thought prior recruiters had discriminated against female applicants, we could deliberately re-weight the data.

The disparity may not be in the labelling, but in the spread of data. Several face recognition algorithms have had higher error rates among certain groups simply because the training data contained fewer examples. Either gathering more data from the poorly represented groups or weighting them more highly during training may help to alleviate this. Note too, this is often worth doing when training data is unbalanced, even if there is no issue of bias, as this can often improve accuracy.

This kind of audit can also be used to identify features that are strongly connected to a particular protected characteristic. For example, choice of school subjects or gaps in employment record (potentially because of family responsibilities). These features can then be examined to see if they are really necessary, and removed if not. Proxy indicators can be very hard to eliminate; as we have seen, a feature which is a proxy measure for a protected characteristic in one context may be quite valid in another. One way to detect this is through building explanatory models. There are algorithms that do this naturally, but it is also possible to detect potential proxy measures for other algorithms.

Paradoxically the way to do this is often to re-introduce the protected characteristic. Think of the building site example. You train the algorithm with CVs with gender removed and find that STEM subjects in school are a strong feature used in the prediction. You then retrain the algorithm, but this time including gender explicitly. If the role of STEM subjects is substantially reduced, this is a strong indicator that it is functioning as a gender proxy rather than as a predictor in its own right.

20.6 PRIVACY

One of the most common worries about big data and indeed the internet in general is privacy. From the early days of the web, worries about misuse and the dangers of hacking and information theft led to various governments creating legislation to protect personal data. The Cambridge Analytica scandal and continual newspaper stories about data breaches show that these problems have not gone away.

To some extent this is an issue for data collection in general rather than AI and indeed most data breaches are on conventional data stores. However, machine learning has exacerbated issues, in part because the value of data has led to more data being gathered and retained and in part because of the ability of algorithms to mine both stolen data and public data. Furthermore an ever greater part of life is online – personal, financial and professional – so the dangers of identity theft are correspondingly greater.

20.6.1 Anonymisation

The most obvious privacy worry is that identifiable personal data may be leaked. Data may be stored either fully anonymised or pseudonymised. The latter is where a unique identifier (such as a number) is used instead of a personal identifier (such as a name and address). If data is being retained for machine learning, then the identity of individuals often doesn't matter, so anonymised or pseudonymised data can be used. However, if the data has any relations between individuals, then this needs to be retained, for example knowing that several people are in the same household.

Often anonymisation is deemed to have made data 'privacy-safe'. This is evident in the relevant UK law which defines 'anonymous information' as:

> *information which does not relate to an identified or identifiable natural person or to personal data rendered anonymous in such a manner that the data subject is not or no longer identifiable.* [142]

If data is anonymous by this definition, UK privacy law does not apply. Similar legislation can be found elsewhere in the world and even where this is not enshrined in law, a similar principle often applies in organisations' internal ethical guidelines.

However, this is not the end of the story.

Note the phrase "*is not or no longer identifiable*". Removing names may mean that one cannot simply look up the person, but if the data has sufficient surrounding material it may be possible to work out who the person is – there may be only one person living in a particular postal code with one hundred and one dogs! This potential for deanonymisation is not new, but more complex AI algorithms combined with the ability to trawl the web for additional connected information has turned this from a remote possibility to an off-the-shelf service.

20.6.2 Obfuscation

One solution is obfuscation where small details are altered, for example adding a year to age or a 5% change in salary. This is clearly not an option for actual data records – you would not be happy to find your bank balance vary randomly from month to month even as part of privacy-preserving practices. However, for machine learning purposes such obfuscated data is often suitable. Indeed this is rather like the perturbation techniques used for growing datasets for machine learning we saw in Chapter 8.

Note however, if the intention is also to grow the data using perturbation, this should be done in a two step-process: (i) first perturb the data once for privacy and then (ii) do data growing perturbations from that point. If you simply do lots of perturbations of the original raw data point, it will become apparent as the 'centre' of a cluster in the derived data.

20.6.3 Aggregation

Another approach is to only store aggregated data. Traditionally this would have meant averages for areas or demographics but now is more likely to be the learnt weights in a neural network or latent features in a big data model.

It has been known for many years that it is sometimes possible to recreate raw data from multiple statistical queries [69] and similar problems arise with machine learning. This can be through deliberate attacks but may even 'slip out'. OpenAI Copilot was trained on large volumes of open source data from GitHub and is able to autocomplete or even completely write code with amazing accuracy. However, soon after it was released it became apparent that it would occasionally reproduce whole sections of code from the originals line-for-line including comments, leading to a lawsuit and lots of discussion of IP issues surrounding models built from large-scale data [291].

There are problems with aggregated data even if personal data is not divulged. Consider this example from one of the author's early papers on privacy:

> *A parent may drive his child 100 metres down the road to school because the road is unsafe to cross. On the way, he passes an observer measuring road usage. Because the road is used such a lot it is widened, attracting more traffic and thus making it more dangerous.* [91]

There is no identifiable data here, just a count of road usage. However, the parent would not assent to the data capture if they knew it would be used to make their child's life worse.

20.6.4 Adversarial Privacy

Another privacy concern has been the use of image recognition techniques both on web images and CCTV in the physical world.

Hand-crafted image processing algorithms use techniques such as those described in Chapter 12 and often follow similar steps to human visual processing. However, machine learning techniques often use very different features from the human eye. Adversarial learning techniques have exploited this to create small variations of images which do not look any different to the human eye and yet are able to fool state-of-the-art face recognition algorithms [45].

Similar techniques have been used in the physical world using adversarial techniques to create car licence plates that look normal to the human eye but are unreadable to a number plate recognition system [305]. This can be part of asserting privacy and freedom to travel without being tracked but of course can also be used to evade legitimate law enforcement such as an automatic speed trap.

It is also possible to buy clothing that claims to subvert image recognition, but with this, as in all of the above, adversarial techniques are only powerful until the next iteration!

20.6.5 Federated Learning

Many organisations want to protect privacy either because it is part of their fundamental ethos (e.g. NGOs and some governments) or as part of management of public perception and brand. Hence there has been substantial work in looking at privacy preserving algorithms [303]. Some of the techniques we have already discussed fall into this heading.

One technique in this area is federated learning. The very act of bringing data together to be processed by machine learning creates a risk of data being hacked or accidentally leaked. So why transmit it at all? In federated learning your personal data is processed close to you on your own device and only the processed data is passed on to be collated.

One of the easiest forms of this is as a modification to backpropagation algorithms. Backpropagation normally works by processing a dataset item-by-item, presenting each data item to the neural network and then making small modifications to the weights at each step. The federated version of the algorithm sends the complete network to every participating device. The device then processes small numbers of data items locally and passes back a delta, the small changes in the neural network's weights due to the additional examples. These are then added to the central algorithm very nearly reproducing the normal workings of the algorithm, while never sending raw data to the central hub.

However, even this apparently bullet-proof method has been subject to adversarial attacks, reproducing example data from weight changes [234]. Privacy in AI is an ever-evolving arms race!

20.7 COMMUNICATION, INFORMATION AND MISINFORMATION

At its best, the web brings much of human knowledge into nearly every corner of the world. There are limits to availability both in terms of literacy and also the ability to afford a web-ready device and sufficient network bandwidth, but these are lower barriers than previously. AI and large-data algorithms have been at the heart of this, not least PageRank and other web search algorithms that scan and sift vast numbers of web resources in order to find the most relevant information. Furthermore, the objects around us are becoming more internet connected, meaning that not just abstract knowledge but real-time information is at our fingertips, from the current weather forecast to whether our kettle has boiled.

However, we also all know that this has a dark side: phishing, cyberattacks, radicalisation, online grooming and misinformation. The Cambridge Analytica scandal not only raised privacy issues but showed how democratic process might be subverted by AI-powered messaging. During the Covid-19 pandemic, distorted news stories, pseudo-science and deliberate mis-information rose alongside more trustworthy, but not always trusted, sources. The causes of this include both deliberate misuse and unintentional consequences of otherwise beneficial systems and, between the two, negligence when bad consequences could have been averted.

20.7.1 Social Media

AI can be used positively to help deal with some of the dangers of social media, for example using natural language algorithms to automate moderation by flagging or removing hate speech. This is not easy. Simple algorithms can generate false positives, for example suspending the account of someone quoting and refuting a racist statement, or where a term that has a totally benign day-to-day use has been appropriated by an extremist group. Similarly image processing can generate false positives, for example where nudity filters on social media have censored posts of breastfeeding mothers.

One solution to this is human-in-the-loop algorithms, where the AI system passes harder to classify examples to human moderators to judge and in the process improve the training of the algorithms. As we saw in Chapter 19, the design of algorithms of this kind is different from a simple automated decision. Of course, human moderation is expensive compared with automatic processing, and so the preferred solution for many social media platforms is to employ 'better AI' ... not always successfully!

Of course the opposite problem is false negatives, the bad posts that are missed. These can be hard to spot as they may simply use coded language, combine neutral or ironic speech combined with images that make their meaning clear, or simply point to video clips. This means that effective algorithms often need to combine media, including text, recorded, static and video images. Each medium requires its own individual specialised processing, using techniques we've described in previous chapters, but crucially they have to be combined.

Recall the levels of natural language processing described in Chapter 13. Really effective moderation ideally requires pragmatic understanding. A statement "that's awful" about an image depicting an atrocity could be an expression of shock or an ironic statement of support for the terrorist group. Of course distinguishing such statements is not easy for humans and indeed may simply end up reflecting our own prejudices about the speaker. Automated algorithms can face similar problems, for example, some terrorist groups may use religious language, so that, without care, machine learning could lead to highly biased algorithms that block benign uses of that language.

Natural language processing has also been used to detect potential paedophile grooming, especially techniques that include elements of sentiment analysis. Sometimes this is based on cumulative textual content, distinguishing 14 year-old speech from a 40 year-old pretending to be 14. Sometimes this can be augmented by learning patterns of interchanges, for example the way alternating critical and complimentary statements can be used as a means of coercive control.

20.7.2 Deliberate Misinformation

AI can also be used to detect and help ameliorate the spread of deliberate misinformation in social media, including that distributed by AI-driven bots. Social media platforms are constantly adjusting algorithms and policies to attempt to prevent or discourage fake news [94, 213, 278]. This includes the use of techniques similar to those used for intrusion detection in cybersecurity: human-like patterns of behaviour are learnt from large volumes of normal usage and then this can be used to spot the unusual behaviour of bots. However, the bots themselves are also using AI and machine learning techniques, both to mimic the most successful influencers and memes and to avoid the defences of the platforms.

Big data techniques, especially network analysis, have been used to understand the spread of fake news and disinformation on social media platforms. Crucially, it appears that fake news is spread more quickly and broadly than true news, quite likely because it is more novel; furthermore, while bots help this spread, the difference is principally due to humans [292]. Attempts to distinguish misinformation have also found that often the text of social media posts may be relatively innocuous, but then link to media on other platforms, such as YouTube, that contain the actual misinformation [190]. This means that misinformation detection needs to operate both across different kinds of media and also across different distribution platforms. This has both technical challenges and also commercial ones as different platform providers need to cooperate.

20.7.3 Filter Bubbles

Recommendation and personalisation algorithms help to ensure that the news items we are shown or the information we search for is most relevant to us. If a search engine knows you are a geographer, then it makes more sense that searches for 'Chihuahua' would favour the place in Mexico, whereas if you are a dog lover, then the dog breed is more likely to be of interest. Of course, we are aware that the counter problem to this is that we may only ever see information that confirms our existing views, especially when the algorithms take into account our social groups.

As with deliberate disinformation, big data network analysis has been used to study the phenomena, for example highlighting the role of 'gatekeepers', people who consume a wide variety of media but then only pass on those of a more partisan nature, which then get amplified by the social recommendation algorithms; sadly people who are more balanced pay a "price of bipartisanship" and are less well received by their peers [111].

As well as analysing social recommendation, attempts have been made to modify algorithms in order to offer alternative views [106, 147]. These have had limited success, especially in actual deployment, potentially even hardening views. This is clearly an area where behavioural science and data science need to work together; for example, it may be more effective to show views that are slightly less extreme than one's own, rather than those very different, which one may be likely to reject out of hand.

Search engine personalisation may be even more problematic as many people do not realise quite how directed this can be and therefore trust the search engine to offer an unbiased view of the world. If instead we receive only information that agrees with us, this may lead us to believe that we are operating on objective facts, further reinforcing our own biases.

20.7.4 Poor Information

With the best intentions we can all create or pass on poor quality or incorrect information. This is particularly problematic if we are acting in a professional capacity whether as an academic, journalist or in policy making. As humans we have a known tendency to seek confirmatory evidence, and, as discussed above, if anything search engines and social media make this worse. Ideally we need tools that counter this. Although systems offering opposing views have had poor success, they have potential as add-on tools for those seeking broader viewpoints. For example, these can use clustering techniques and deliberately offer items with high reliability (say peer-reviewed science) but in different clusters to the items you have been referencing.

AI can make these problems worse. For example, generative AI language tools, or simply predictive text within word processors, may lead to text in articles that sounds articulate, reliable and persuasive. If errors or misinterpretations are common in the training data, these are likely to be regurgitated. However, AI can also help by performing a level of fact checking or analysing arguments and flagging common fallacies.

TABLE 20.1 Data Used in Exercises 20.1 and 20.2.
Key: M–Mathematics, S–Science, L–Language, H–History.
(This data is also available in the chapter web resources.)

XY	height	M	S	L	H	apt A	apt B
X	1.53	N	N	Y	Y	0	3
X	1.56	N	Y	Y	Y	2	3
Y	1.68	N	Y	N	N	2	0
Y	1.6	Y	Y	Y	N	3	2
Y	1.65	Y	Y	N	Y	3	1
Y	1.65	Y	Y	Y	N	3	3
X	1.56	N	Y	Y	Y	2	3
X	1.52	Y	N	N	Y	1	1
X	1.56	Y	N	Y	N	0	3
Y	1.71	N	Y	N	N	2	0
X	1.54	N	N	Y	Y	0	3
Y	1.75	N	Y	N	N	2	0
X	1.53	N	N	N	Y	0	0
Y	1.63	N	Y	N	N	2	0
X	1.58	N	Y	Y	Y	3	3
Y	1.63	Y	Y	N	N	3	0
Y	1.69	N	Y	N	Y	3	0
X	1.56	Y	N	Y	Y	1	3
Y	1.6	Y	Y	Y	N	3	2
X	1.63	N	N	Y	N	0	2

20.8 SUMMARY

This chapter has discussed a number of potential ways in which AI can go wrong including bias in machine learning, threats to privacy and misinformation. In each case

there are ways to mitigate the danger, but these are always partial. For deliberate misuse, there is always an arms race between those creating prevention mechanisms and those seeking to undermine them. However, inadvertent misuse can be at least as dangerous, for example believing that simply removing identifying characteristics can prevent bias or spreading poor quality or misleading information. These threats do not mean that AI shouldn't be used but mean that we do need to use it responsibly and be aware that despite our best efforts things can still go wrong.

20.1 An employer has taken on a group of 20 school leavers. Data collected during recruitment (Table 20.1) includes height, and whether they took higher level exams in maths, science, language or history (but not their marks in the exams). In order to guide future recruitment, the employer trials them all for initial periods on two kinds of tasks and they are assessed on each (labelled 'apt A' and 'apt B'). For the present ignore the column labelled 'XY'.

a. Work out a decision rule for whether someone will be good at task A (score 2 or 3) using only the columns 'language' and 'history'. You can use an algorithm such as ID3 or simply plot the values and work out a rule by eye.

b. Do the same for task B, but this time using only the columns 'maths' and 'science'.

TABLE 20.2 First Dataset Used in Exercise 20.3
(This data and an extended version are also available in the chapter web resources.)

x	y	x	y	x	y
1.804	4.344	1.219	3.050	1.275	3.966
1.671	4.609	1.324	3.463	3.665	3.298
4.845	3.538	3.542	2.357	4.876	3.535
1.821	4.301	1.832	2.642	4.664	3.651
1.777	4.645	4.338	2.706	2.464	3.357
4.362	2.600	4.189	4.021	1.264	4.161
1.735	3.849	4.383	2.113	1.283	3.969
4.014	1.477	4.577	2.373	1.509	4.118
4.846	4.151	1.330	3.669	1.458	3.351
4.167	3.277	3.990	2.056	3.807	3.298
4.856	3.372	1.114	3.909	1.862	5.366
1.233	3.442	0.978	4.113	4.144	2.072
4.607	2.574	4.786	1.899	3.780	3.388
1.746	3.680				

TABLE 20.3 Second Dataset Used in Exercise 20.3
(This data and an extended version are also available in the chapter web resources.)

x	y	x	y	x	y
1.809	2.160	2.214	2.341	4.743	1.852
3.971	1.430	4.268	3.946	3.410	3.594
2.943	1.849	4.564	2.526	2.225	2.172
4.534	2.325	1.810	3.400	3.771	3.573
3.499	2.309	2.204	3.609	2.558	4.856
4.340	2.065	2.472	4.136	3.269	3.593
4.479	3.830	1.803	1.955	1.937	1.842
3.384	3.684	2.517	3.792	4.295	2.658
4.319	2.553	1.698	1.542	1.979	3.460
2.367	4.100	1.971	3.457	4.449	3.834
4.460	3.668	4.101	2.024	2.499	3.835
2.311	1.134	4.450	4.447	3.947	2.009
1.673	3.604	1.712	1.934	1.743	2.653
3.793	4.317				

In fact the aptitude for tasks A and B in this simulated dataset were calculated as a weighted sum of the four subjects studied. Apt A was calculated from maths, science and history in the ratio 2:6:1 plus a small random amount. Similarly apt B was calculated from maths, language and history in the ratio 1:6:2. That is, language did not contribute at all to apt A and science did not contribute to apt B.

c. Is this surprising given your decision rules?

d. If so, can you work out what is happening?

20.2 This exercise builds on Exercise 20.1 and uses the same data from Table 20.1:

a. Work out a decision rule for whether someone will be good at task A (score 2 or 3) using only the column 'height'. You can use an algorithm such as ID3 or simply plot the values and work out a rule by eye.

b. Do the same for task B based again on height only.

Suppose, the column labelled XY represents some kind of protected characteristic, such as gender.

c. Use each of the decision rules in parts (a) and (b) and from Exercise 20.1 to create predictors of whether each person will be good at tasks A and B.

d. Does this look fair taking into account the protected characteristic XY?

e. Can you make sense of any apparent unfairness?

20.3 The data in Tables 20.2 and 20.3 have been generated from small seeds of (simulated) real data by adding small random perturbations (as discussed originally in Chapter 8). In both tables, there are multiple columns for x and y values, but they should each be read as single datasets of 40 x–y values. In this exercise you are aiming to de-anonymise the data by finding the original data items.

a. For the data in Table 20.2 identify clusters of data items. You can do this by plotting the data and identifying groups by eye, or by using a clustering algorithm such as k-means. For each cluster calculate its centroid.

b. Do the same for the data in Table 20.3.

There were four initial seed items with x–y values (2,2),(2,4), (4,2) and (4,4). The second dataset (Table 20.3) replicated each value ten times and then added random noise to each replicated item, following the perturbation techniques in Chapter 8. The first dataset (Table 20.2) did the same, but before replicating the seed data item it added a random value to each seed first, as described in Section 20.6.2.

c. How close were the results in (a) and (b) to this original data?

d. Consider on your own, or discuss in small groups the results you have observed and how good each technique has been in preserving privacy.

FURTHER READING

A. Dix. Human issues in the use of pattern recognition techniques. In R. Beale and J. Finlay, editors, *Neural networks and pattern recognition in human computer interaction*, pages 429-451. Ellis Horwood, 1992.

This early paper highlighted the dangers of ethnic, gender and social bias in black-box machine learning systems.

Sadly, many of the issues are still apparent more than 30 years on.

C. O'Neil. *Weapons of math destruction: How big data increases inequality and threatens democracy.* Crown, New York, NY, 2016.

Highly influential popular science book that shows how the indiscriminate use of AI and big data can be socially divisive and discriminatory.

J. Angwin, J. Larson, S. Mattu, and L. Kirchner. *Machine bias: There's software used across the country to predict future criminals: And it's biased against blacks and how we analyzed the COMPAS recidivism algorithm.* ProPublica, 23 May 2016. https://www.propublica.org/article/machine-bias-risk-assessments-in-criminal-sentencing and https://www.propublica.org/article/how-we-analyzed-the-compas-recidivism-algorithm

This is the article which exposed the potential discriminatory impact of the COMPAS probation decision support systems used in court rooms across the US. Read the more detailed report of the way the authors obtained and analysed the data as well as the article aimed at the general public. This illustrates the complexity of dealing with this kind of data and in particular the issue of base rates, which means it is hard to be fair in all senses at once.

Explainable AI

21.1 OVERVIEW

It is important that AI gives the right answers; however, it is often equally important that we understand why it is giving the answers it does. This was recognised in the early days of expert systems, which offered some form of explanation as we saw in Chapter 18. It is particularly important for black-box machine learning, such as deep neural networks, where the link between input and output can be hard to fathom. Because of this, the investigation of explainable AI has become a sub-field in itself attracting both generic solutions that are agnostic to the underlying AI or machine learning and also more specialised solutions to adapt specific algorithms to make them more scrutable.

In this chapter we will first look in more detail at the reasons why explainable AI is important, return to Query-by-Browsing (first seen in Chapter 5) as an example of how machine learning systems can be designed to be scrutable and then look at general heuristics for designing explainable AI.

21.2 INTRODUCTION

A job candidate has been pre-selected for shortlist by a neural net; an autonomous car has suddenly changed lanes almost causing an accident; the intelligent fridge has ordered an extra pint of milk. From the life changing or life threatening to day-to-day living, decisions are made by computer systems on our behalf. If something

goes wrong, or even when the decision appears correct, we may need to ask the question, "why?"

In the case of failures we need to know whether it is the result of a bug in the software, a need for more data, faulty sensors, inadequate training or just 'one of those things': a decision correct in the context, which happened to turn out badly. Even if the decision appears acceptable, we may wish to understand it for our own curiosity, peace of mind or for legal compliance.

Explainable AI is the term that is used to describe methods to make the algorithms that underlie decision-making systems more understandable by humans.

21.2.1 Why We Need Explainable AI

We have looked at bias in detail as it is one of the areas that has caused most controversy in the application of AI and ML. However, bias is not the only reason we need to dig more deeply into algorithmic (or other) decisions.

safety – When an autonomous car has an accident, we need to understand what went wrong in order to prevent similar future accidents. The airline industry has long-standing rigorous methods for this adopting a forensic analysis of every accident. Normally car accidents are not treated with the same level of detail even though in total they cause a far greater death toll. This is in part because they are each individually smaller but also because it is too easy to blame the driver: human error. However, software-controlled cars will mean that accidents will be more likely to have repeatable causes.

democracy – There have been growing worries about the ways algorithms potentially undermine democracy. Sometimes this is about deliberate practices

DOI: 10.1201/9781003082880-25

such as the Cambridge Analytica scandal or social media bots. Perhaps more worrying is the way that search engines and social media use a "what people want" objective function in their algorithms, which as a side effect creates bubbles of like-minded information, allowing us to each feel we are in the fact-based majority against an ignorant, albeit vocal, minority.

health and wellbeing – Imagine you are the senior executive of a soft drinks manufacturer that wishes to adopt an ethical advertising policy. You do not deliberately advertise in children's magazines or on children's TV, but how do you know whether your online advertising, which may be driven by keywords or much more complex algorithmic mechanisms, is not implicitly targeting children?

social issues – The author first wrote about the danger of gender and ethnic bias in AI in 1992 [74]. This was prompted in part by a letter from his bank that said he would need to pick up a chequebook directly from his branch; they couldn't post it because he lived in a 'high-risk postal code'. In other words they did not trust the honesty of his neighbours if it were misdelivered to the wrong house ... and, by implication, they would not trust him if they were considering posting a chequebook to a neighbour! This was a minor inconvenience, but the cost of everything from car insurance to interest rates on loans themselves is driven by a wide variety of factors including the area you live in, often linked directly or indirectly to your socio-economic status. At one level this is simply reflecting the market, but of course the same could be said about some of the other discriminatory effects we have discussed. Unless we understand how these decisions are made it is hard to assess their ethical status.

science – There are similar worries in the scientific community that big data approaches to science may well be 'discovering' relationships that later turn out to be spurious [115]. Bias can also creep into the most apparently 'objective' basic science. Most cognitive psychology has been developed using experimental subjects that are WEIRD (Western, Educated, Industrialised, Rich and Democratic); a meta-study revealed that fundamental cognitive and perceptual phenomena, such as the Müller-Lyer illusion, which

had previously been regarded as universal, are often culturally determined [130].

In general, for many kinds of algorithms and complex rule-driven human processes, we need to be able to ask the question "why?"

- *Why* did that car crash?
- *Why* was I refused a loan?
- *Why* did the police stop me in the street to question me rather than all the others walking by?

This emphasises the need for some form of transparency or explainability in complex algorithms.

21.2.2 Is Explainable AI Possible?

Explainability has always been a central aspect of expert systems, but the field of explainable AI has been growing rapidly over recent years in the face of the above issues. Some suggest that deeply opaque methods such as deep learning are by their nature unexplainable. However, there has also been promising work, both in more traditional symbolic AI (e.g. argumentation-based reasoning) and in sub-symbolic AI and machine learning (e.g. hotspot analysis of critical regions for image-recognition systems).

Often results are very specific to a domain or technique, but it is evident that some of these offer potential methods that could be adapted or core principles extracted so that they could be used more widely. This has led to a growing number of commercial and open-source tools that can help increase transparency even for black-box techniques.

21.3 AN EXAMPLE – QUERY-BY-BROWSING

Before discussing general methods and heuristics for transparency and explanation, let's look at a specific example of a machine learning system, Query-by-Browsing (QbB), which was designed to be transparent. QbB was originally designed as a thought experiment to highlight potential problems and solutions but is also a running system [77].

We first discussed QbB in Chapter 5, but here we will look in a little more detail. Note that the screenshots in Chapter 5 were from the early envisionment, whereas Figure 21.1 depicts a later web version.

21.3.1 The Problem

Many recommender systems for news articles, technical help or additional products use some form of relevance feedback. Sometimes this is implicit, when the user clicks through an advert, and sometimes explicit, perhaps a thumbs up or star rating. These can be almost prescient suggesting just the right book, music or news item but sometimes can be almost embarrassingly weird. A recommender for news or products has to be 'good enough', finding sufficient relevant articles to suggest and not showing too many irrelevant items. Precise accuracy is not required.

In contrast, when querying a database, say to select a specific group of staff for a pay rise, it is usually important that the records selected are precisely those that are required. This is commonly achieved by writing an SQL query to select the required records, but this requires both technical expertise and the ability to frame one's requirements in precise logic. It would be nice to be able to use relevance feedback style interactions to select the records desired and then let that determine which staff receive the pay rise.

The technical challenge is to do this in a way that (a) you can be sure is updating precisely the right staff; (b) the rule used is one that does not violate any anti-discrimination legislation.

21.3.2 A Solution

The record form of a database table is compatible with the input format of many machine-learning methods; however, most of these have relatively opaque learning algorithms and decision rules.

Query-by-Browsing (QbB) attempts to address this, starting with relevance-feedback-style user-selection of records, but creating rules that are scrutable addressing requirements (a) and (b), and in the process highlighting the potential for biased results to arise that would be illegal in a less transparent system.

The original machine learning system chosen was a variant of ID3 [226, 227] (see Chapter 5) extended to allow multi-column comparison criteria. However, one QbB version used genetic algorithms to create rules [75].

Walking through the behaviour in Figure 21.1:

1. The user selects records of interest with a tick for those that are wanted and cross for those not required.

2. The user selects "Make a Query".

3. The system generates an SQL query that matches the desired records.

4. The query is displayed in the Query area and the records selected by the query are shown highlighted.

5. The user can select more examples and counter-examples to refine the query.

Note that the interface effectively includes two representations of the decision rule. In the Query area the decision tree is rendered as an SQL query giving an intentional representation; this is useful for precision, ensuring that conditions are exactly as required. In the List area the highlighted records form an extensional view of the rule, showing which records are chosen by it. This is particularly useful for complex and–or queries, or those including negation, which are known to be hard to interpret.

As well as allowing precision, the Query area makes the decision rule transparent. It is immediately obvious if the rule says, for example, 'SELECT * WHERE title="Mr"'. As we saw in Chapter 20, this is not sufficient to prevent bias, but certainly helps to uncover problematic decision rules.

21.3.3 How It Works

Figure 21.2 shows schematically the steps 'under the bonnet'. The examples chosen by the user are fed into a machine learning system that generates a decision tree or similar rules, and these are then rendered as SQL (or RQBE).

In the case of ID3 the top-down 'divide and conquer' nature of the machine learning algorithm is itself comprehensible; it is possible, albeit tedious, to go through the process by hand or read a trace of the system learning. As noted, there has also been a version of QbB that uses genetic algorithms to generate the decision tree. In this case the complexity of the algorithm (large population and number of generations) make the learning process opaque; however, the rules generated are still understandable.

This is crucial. Think of a mathematician; the process of finding a proof may require trial and error, sparks

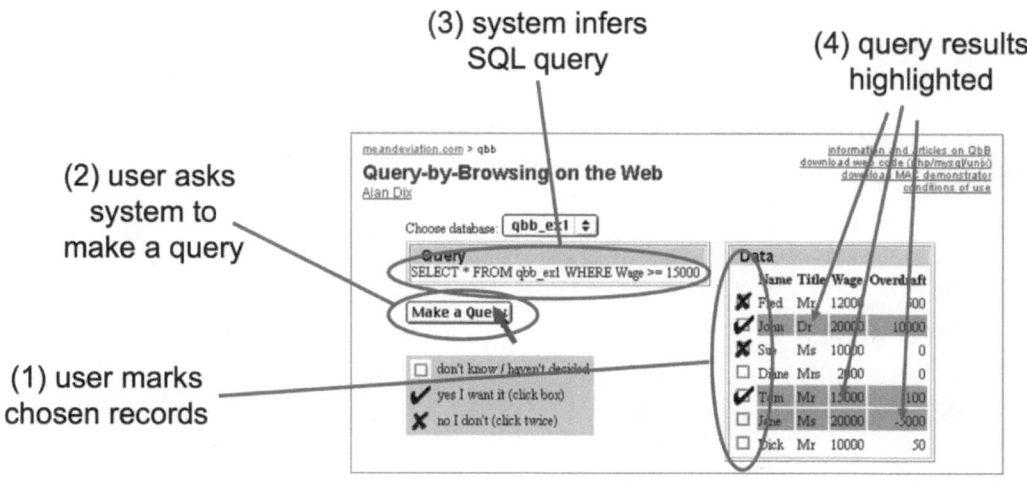

FIGURE 21.1 Query-by-Browsing (https://www.meandeviation.com/qbb/).

of insight, generating intermediate lemmas. To fully describe and justify each step of this would be impossible. However, mathematicians do not attempt to represent how they came to a solution, but instead present a proof, a form of rational reconstruction of the actual mathematical process.

That is, we need to think about two kinds of transparency:

- decision rules

- learning process

Just like the sparks of inspiration behind a mathematical proof, it is often acceptable to have an opaque learning process so long as the rules generated are comprehensible.

21.4 HUMAN EXPLANATION – SUFFICIENT REASON

When looking at complex AI methods, such as deep learning, explainability can seem impossible. However, it is crucial to remember that human–human explanations are rarely utterly precise or reproducible.

If at a restaurant you were asked why you chose a particular main course you might say something like, *"well I usually go for a steak, but it was late and I wanted something lighter; I'd had fish last night, so chose a salad."* Within this are many vague concepts and open questions. Why normally choose steak? What do you mean by 'lighter'? Why not have fish two nights running? However, for most purposes this would be a sufficient

explanation. Of course the statement might elicit further questions, *"why didn't you go for the spinach brûlée, I know it sounds odd but is actually quite delicious?"* Of course, both questions and answers themselves might leave aspects only roughly defined, but sufficient for a discussion about food.

We do not try to explain in terms of the firing of individual neurons in our brain, or try to make precise every nuance. Furthermore, the explanations we provide are often rational reconstructions, ways to make sense to ourselves, as much as to others, of the complex interweaving of conscious and unconscious processes in our minds.

In human–human discourse statements and explanations are part of a process of mutual understanding that enables further action or communication. Studies repeatedly show an incremental process of unfolding of partial statements rather than precise detailed monologues (except in the university lecture theatre). For example Grice's conversational maxims include *"make your contribution as informative as is required"* – but no more [117]. Clark and Brennan [52] suggest that our conversational utterances will always involve levels of ambiguity, which are confirmed or disconfirmed as part of on-going discourse, with the aim of creating a sufficient common ground of understanding for future conversation and action.

In short, the purpose of an explanation is to inspire confidence and trust to allow future mutual action, or possibly to create sufficient openness to allow critique or dispute.

examples

SQL query

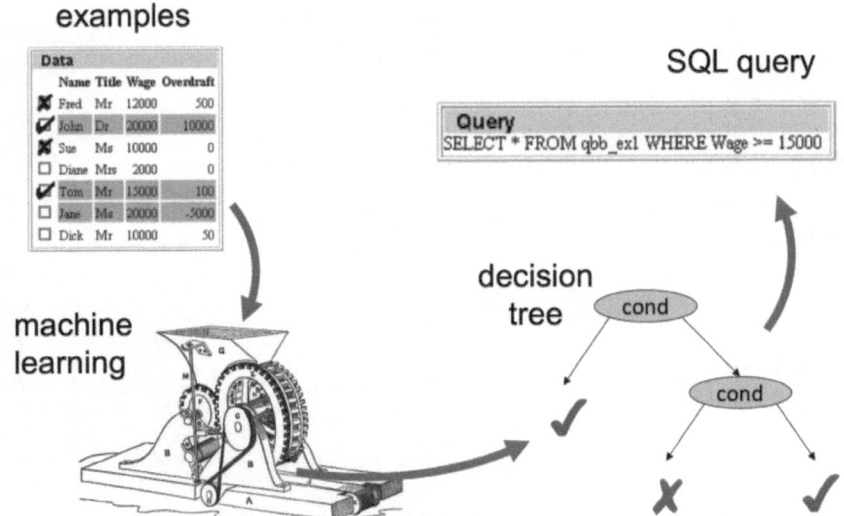

FIGURE 21.2   QbB – under the bonnet.

When we look at machine–human explanations in this light, it is often easier to see how we may at least make complex big-data analysis, deep learning and similar algorithms comprehensible if not utterly 'explained' to the last possible detail. Indeed such an over-detailed explanation would probably, for the human, be no explanation at all.

21.5 LOCAL AND GLOBAL EXPLANATIONS

Another insight that helps make explanations more tractable is that we may not need to understand the whole system, merely the aspects that influence a particular decision.

Imagine you have been stopped by a police officer and ask, "*why have you stopped me?*" The police officer could offer a complete explanation starting with Roman Law and its impact on modern jurisprudence, or could simply say, "*you were driving too fast*". There are times when the complete explanation is useful, especially if you are trying to assess the fairness of the legal system as a whole. However, here the answer you require is *local*: why you were stopped at this time, in this situation.

As well as being more comprehensible, these local explanations also offer the potential for better human decision making and action. If you drive just a little more slowly, you won't be stopped in future.

We have already seen an example of a local explanation in Chapter 18, tracing inferences of a forwards chaining inference system (Section 18.3.3). This is

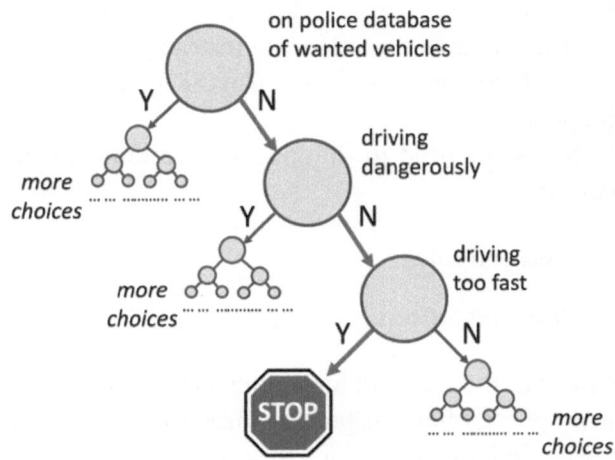

FIGURE 21.3   Decision tree for whether police stop a car (selected path in red).

far from perfect for a non-technical user, but at least is a representation where the individual rules are comprehensible. Other representations, especially those from machine learning, need more work to make them understandable. This section looks at two examples of specific techniques for local explanation before we move to a broader classification in Section 21.6.

21.5.1 Decision Trees – Easier Explanations

Some decision rules admit very easy local explanations. For a decision tree the best explanation is often related to the final decision points, lowest in the tree. For example,

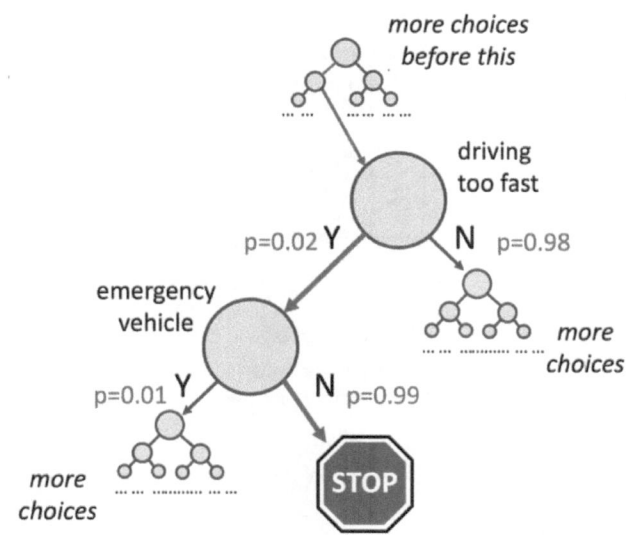

more choices
before this

driving
too fast

p=0.02 Y   N p=0.98

emergency
vehicle

p=0.01 Y   N p=0.99

more
choices

more
choices

STOP

FIGURE 21.4 Alternative decision tree taking into account emergency vehicles.

in Figure 21.3, the police officer would say, "*you were driving too fast*", not, "*you weren't driving dangerously*".

However, there are exceptions to this. The police would not normally stop an ambulance or fire engine, even if it were driving over the limit, as they would assume it was on its way to an emergency. Figure 21.4 includes this and so the final decision point is therefore "(is it an) *emergency vehicle*". However, if the police officer said "*you're not a fire engine*", this would be taken as a facetious answer, not a helpful one.

In general, stating the obvious is not helpful! Happily, the decision tree in Figure 21.4 is also labelled with the probability of each branch occurring (taken from either training data or ongoing data collection). A more helpful explanation rule for decision trees is the last low probability choice. In this case "*driving too fast*".

21.5.2 Black-box – Sensitivity and Perturbations

You do not always have such an easy (or at least relatively easy) representation available. In some cases, the representation is fundamentally complex, for example the weights on thousands of internal nodes in a neural network do not offer such an easy explanation. Alternatively, the algorithm may be proprietary or secret, for example, a labour rights organisation might have access to the decisions made by a gig-economy platform, but not the algorithm itself. Even if the algorithm is available, generic tools or legal compliance tests need to work with any algorithm, so cannot be dependent on specific details.

In these cases it is possible to obtain local explanations by trial and error. For example, in the UK a medical advice app hit the news headlines when a reporter entered identical sets of symptoms but changed whether they said they were male or female. If the reporter entered "male", they were advised to go to a hospital as they might be having a heart attack, if they entered "female", they were advised to stay at home as they probably just had a stomach-ache.

To be fair on the app, there were good reasons for this, heart problems are more common in men than women (the base rate is different), so the gender-specific advice is reasonable. One could imagine a similar case where the woman was advised to seek medical attention due to a suspected ectopic pregnancy, which would not be sensible for a man with otherwise identical symptoms. Crucially though, the reporter's investigation brought the issue to the surface.

Note that the reporter was entering a small perturbation to the input data (female vs male) and observing the local effects of this. This same technique can be applied automatically. Given a black-box decision mechanism B and particular set of data x, we can try lots and lots of variants $x_1, x_2, ..., x_n$ where each x_i is the same as x with one or a small number of features modified. We can then see the effect of the decision made on these $B(x_1), B(x_2)$, ..., $B(x_n)$ compared with $B(x)$ (see Figure 21.5).

These small trials are all small perturbations of the original data x and can be used to build a local model of B near x. This might be a decision tree or, in the case of numeric features, some sort of linear decision model.

The same data can also be used to give a rating for the sensitivity of each feature. For example, given a binary feature such as "smoker/non-smoker" we can look at the proportion of experiments that yield each decision output for each feature value. The entropy or chi-squared of the resulting contingency table can be used as a way to assess the sensitivity, that is the extent to which that feature is likely to change the categorisation.

21.6 HEURISTICS FOR EXPLANATION

For both local and global explanation, we can group methods into three main classes (Figure 21.6):

white-box techniques – These are algorithms which by their nature or with minor modifications naturally

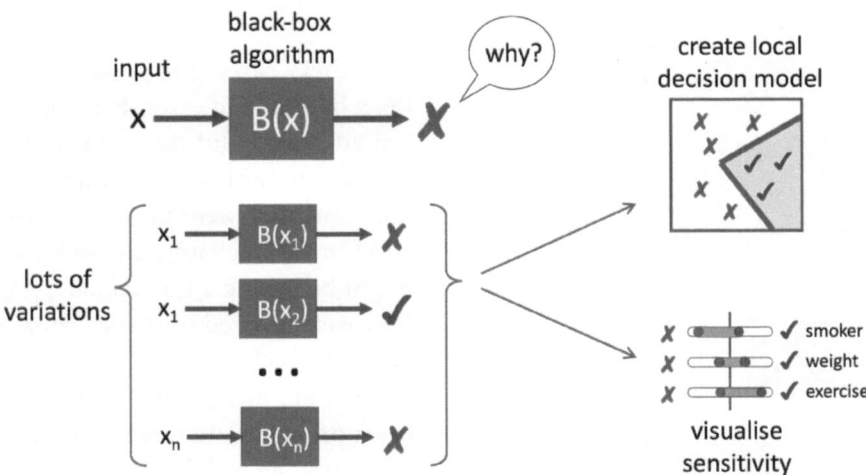

FIGURE 21.5   Sensitivity analysis using small perturbations of the original data.

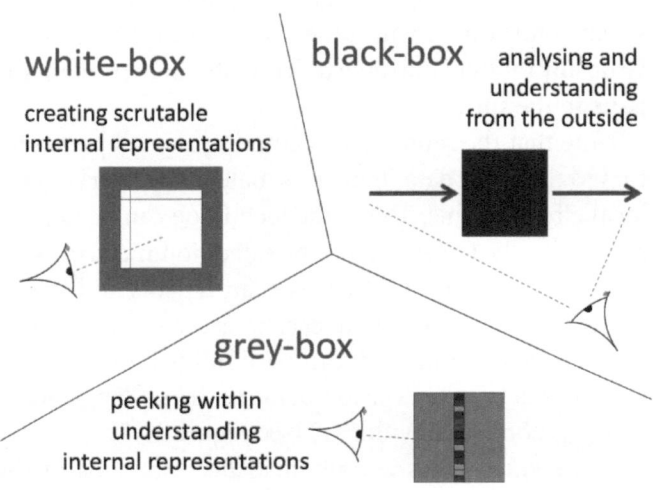

FIGURE 21.6   Three broad classes of explainability technique.

have understandable internal representations. For example, the choice of decision trees in Query-by-Browsing was because these were relatively easy to understand and could easily be transformed into standard database queries.

black-box techniques – Here one treats the process as a black-box, but attempts to make sense of it from the outside. If the police have suspected terrorists under surveillance, they will not walk up to them and ask, *"why are you buying fertiliser?"* Instead, they will attempt to determine plans, motives and reasons based on the observable behaviour of the suspects. In a similar way, it is possible to explore an AI or ML algorithm based purely on its behaviour.

grey-box techniques – Where the internal process has some sort of intermediate representation, such as one of the internal layers in a multi-layer deep learning network, the network can effectively be split to look for black-box explanations in both directions. Typically the early layers, closer to the input, will be framing broad conceptual categories, whereas the later layers may be amenable to transformation onto a more logical/symbolic representation.

Using this framework, we can consider different general heuristics, some of which can be used for both existing and novel algorithms. We will consider a few examples for each of the three classes and more are described on the book website.

21.6.1 White-box Techniques

The simplest case is when we choose an algorithm that by its nature has a human readable representation, for example a decision tree. Even a decision tree can be hard to understand if it gets very large, and there are ways to steer tree building algorithms towards more understandable trees, but at least the tree is relatively comprehensible. A step along from this is where a black-box technique is used to generate a readable representation; for example we saw in Section 21.3 that a version of Query-by-Browsing used a genetic algorithm to create a decision tree. Although it would be hard to explain precisely how the algorithm chose a particular tree, the tree itself is readable. This is rather like a judge who has an instinctive idea of whether the defendant is guilty or not based on experience but has to make that feeling explicit in a legally argued judgement.

More complex methods may use a black-box machine learning algorithm initially, but then use it to generate a white-box model.

One approach is *simplification of the rule set*. The end point of training a neural network is a complex set of real-valued weights and thresholds, in some cases it may be possible to 'harden' the network into a binary network after training is complete (see Figure 21.7). Recall from Chapter 9 that the sigmoid activation function is necessary to 'soften' the network to allow backpropagation learning as it is often easier to learn continuous rather than discrete boundaries. It may be possible to turn the sigmoid into a simple threshold for deployment, resulting in a more comprehensible (albeit large) Boolean network.

Another approach is to use *adversarial* examples for white-box learning. Adversarial learning (as used in GAN) generates a large case-base of examples. One of the problems with many knowledge-rich ML techniques (especially ones that have stochastic/uncertainty elements) is that they were hamstrung as they often needed to work on small training sets, risking overfitting from repeated exposure to the same examples, and missing cases where there were none. The case-base of examples from the adversarial learning can be used as a training set for these other more traditional, and more scrutable, techniques (see Figure 21.8).

In some ways this is rather like observing human experts and building an expert system based on the observations. The expert may not be able to fully explain their decision making, but it may be possible to build an external model.

21.6.2 Black-box Techniques

We have already seen an example of a black-box method when discussing sensitivity analysis techniques in Section 21.5.2. Basically black-box techniques work rather like a scientist examining a new material, experimenting with it, measuring it and creating some sort of humanly reasonable explanation. In the case of the local perturbation methods, we used small perturbations around a specific input, that is exploring the close neighbourhood of a single value.

Most black-box techniques work either by generating small perturbations or large numbers of examples generated in other ways.

Following the scientific analogy one can use a form of *exploratory analysis for human visualisation*. Lots of random or systematically chosen inputs can be used to create input–output maps that can be visualised using standard scientific or information visualisation techniques (see Figure 21.9). The example values can be created through lots of small perturbations of training set data or randomly chosen. The former has the advantage that they are more realistic examples, but the latter may expose extreme cases or unexpected generalisations of the algorithm. In addition, some algorithms can be turned backwards, feeding in what would normally be an output and generating typical 'input' values.

Various forms of hotspot analysis can be used to highlight the key features in the data, or pixels in an image, that are most critical for the classification or decision of an algorithm, typically using some form of perturbation of features/pixels (Figure 21.10). For example, an image recognition algorithm might be very successful at distinguishing yachts from fighter aircraft, but hotspot analysis shows that it is the bottom and top of the image that are important, not the central part where the object is; the image recognition algorithm is really distinguishing images with sea at the bottom from those in the open sky. SHAP, one of the popular systems for local explanation, uses this method for non-image data, calculating the sensitivity of each feature to help users assess which are affecting the decision being made [178].

Similar methods can be used to *distinguish central and boundary examples*. You begin generating lots of examples but then perturb each. If an example's output remains constant despite perturbation, it is a central example. If small perturbations change the class, it is a boundary example (see Figure 21.11). Those examples where small perturbations do not change it but larger ones do (large and small as measured by Hamming distance, or another suitable metric) are in the penumbra of the boundary – these may also be useful.

The boundary and central cases can be used to help a human understand the classes, just as we would do if explaining a concept to another person. We might use a crow as a central example to describe what we mean by a 'bird' and then maybe use some extreme examples such

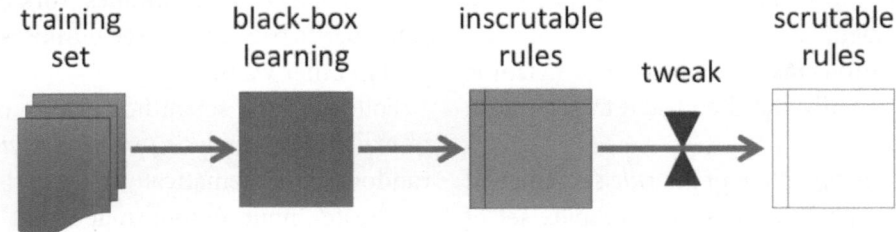

FIGURE 21.7   Simplification of rule set.

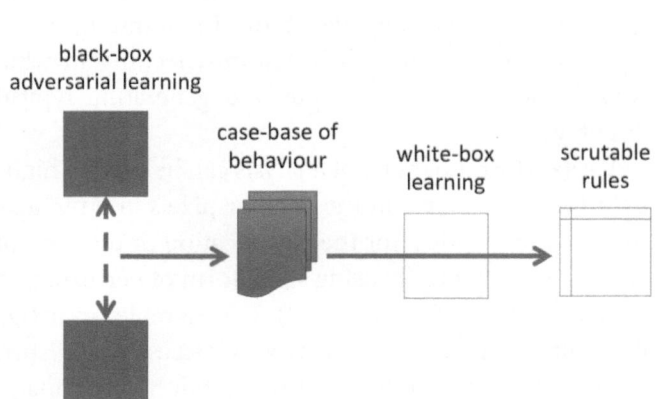

FIGURE 21.8   Adversarial learning to generate training sets for white-box learning.

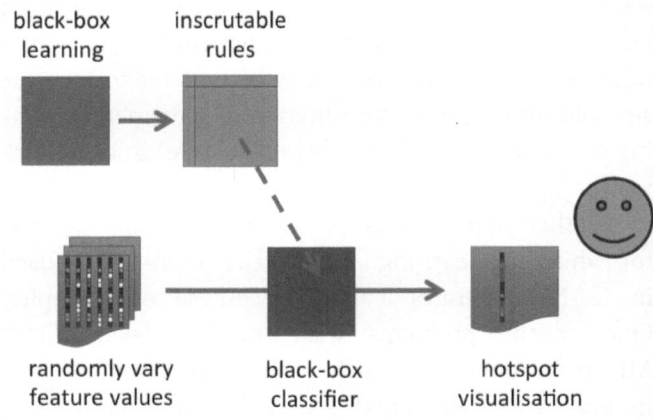

FIGURE 21.10   Key feature detection through perturbations.

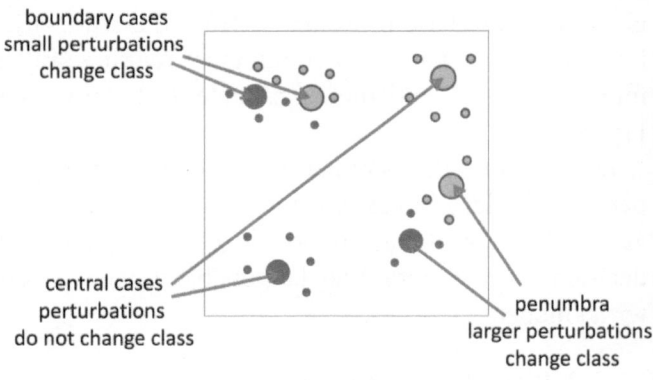

FIGURE 21.11   Identifying central and boundary examples.

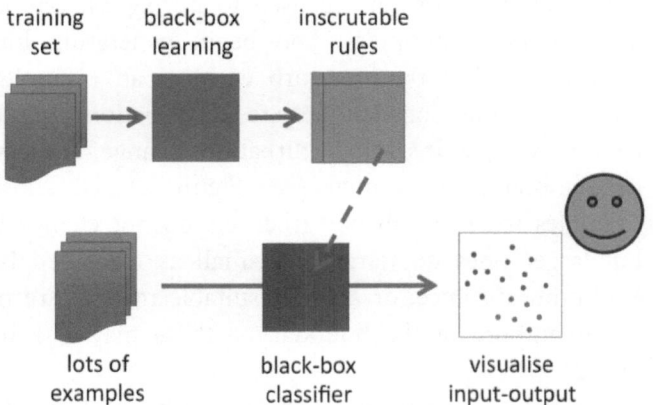

FIGURE 21.9   Exploratory analysis for human visualisation.

as a penguin or ostrich to explore the range of animals covered.

These boundary examples can also be used to generate other forms of explanation. For example, LIME, another popular local-explanation technique, automatically finds cases close to a given example that fall inside or outside of the same classification; these are then used

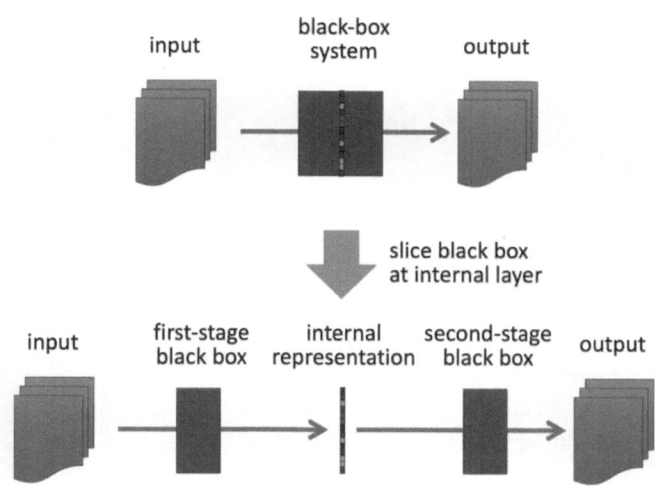

FIGURE 21.12   Grey-box techniques prising open the black-box at an internal layer.

to create a linear classifier that is simpler to understand than the overall black-box model [235]. In general, it is sometimes possible to take a black-box model that is in itself inscrutable but then use it to train other, more comprehensible algorithms, such as decision trees, that can then be used either instead of the original black-box model or as an explanation of it (local or global).

21.6.3 Grey-box Techniques

Grey-box techniques are a form of divide-and-conquer, effectively cutting the black-box model in half at a layer of internal representation. The initial black-box input–output system can then be seen as a pipeline (Figure 21.12):

first stage – input transformed to internal representation

second stage – internal representation transformed to output

Note that this is not changing the underlying algorithm but simply viewing it in parts. In principle one could split the network into three or more parts, here we'll just look at splitting in two at a single central layer.

For some ML algorithms, such as a simple three-layer neural network, there is only one sensible split point, but for deep neural networks, or other algorithms, there may be many ways to do this division. We saw in Chapter 8 that in deep network architectures there is

often a pinch point, a layer that has relatively few nodes that in some way acts as a compact representation of lower level features, and which would be an obvious split point.

Divide-and-conquer approaches are generally useful to simplify things, but as we discussed in Chapter 8, the earlier and later parts of multi-layer systems often perform different functions. Typically, the early layers are about feature extraction and the later layers about combining these into some form of decision or categorisation. For example, it may be possible to create an explanation of the first stage that says "*that cluster of nodes is about whether there is a cat in the scene*" or "*that is about food being spicy*", whereas the second stage may be more rule-like: "*choose the meal if it is spicy, but not too expensive*".

So, while in principle both sides of the black-box can be treated equally, in practice different black-box explanation techniques may work better on the two parts. We'll look at one example technique each for the first and second parts.

As the first part of the black-box is likely to be performing feature extraction, we are unlikely to extract a simple logical explanation, but may be able to obtain an intuitive understanding of the behaviour through clustering or visualisation of the low-level features represented in the internal layer.

We focus on the mapping from input to intermediate activation of the layer for any specific input. We effectively treat the intermediate activation (or a subset of it) as a feature vector (rather than the input) and seek to find clusters or other ways to organise the input space (see Figure 21.13). This may involve initially reducing the example set to a similarity matrix where the cosine or other distance metric is used on the intermediate activations of each pair of examples, this can then be fed into a variety of algorithms for clustering such as self-organising maps (see Section 6.6.2).

The aim of this process is not to create a precise explanation of the first part of the black-box, but rather to be able to obtain a qualitative understanding, such as "*ah, yes, this cluster of nodes is encoding the shape of the wing*", "*this set is about the background*". This is the same as with explaining a friend's taste in books; you may not be able to codify precisely what makes a novel fit into a genre, but you can still understand enough to look for Gothic fantasy as a birthday present.

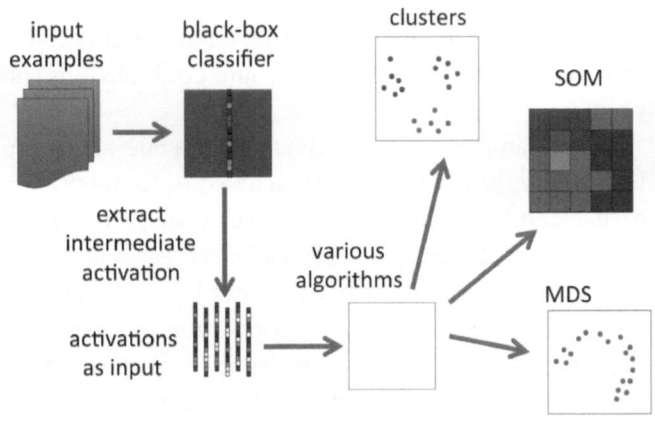

FIGURE 21.13 Clustering and comprehension of low-level features.

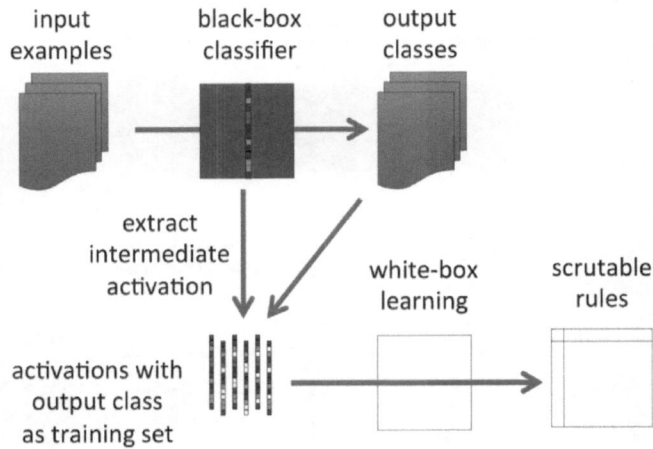

FIGURE 21.14 High-level model generation.

The structure of the internal layer can also be used more directly. For example, in Chapter 13 we saw how this can be used as a latent space for other purposes such as creating similarity measures between pairs of inputs, or, in the case of word2vec having structural properties such as "*Madrid is to Spain as Paris is to France*".

As the second stage is often about building final decisions from features, there is more likelihood that we will be able to build more *symbolic descriptions of the high-level processing*. In the book example, we might have an intuitive idea of genre, but are able to say, "*my friend likes both Gothic fantasy and Nordic noir*", that is an explicit and codifiable statement built on top of more intuitively understood classes of literature.

In general, we may be able to use a white-box algorithm, such as a decision tree to create an alternative set of explainable rules (see Figure 21.14).

1. Take a set of input examples: $I_1, I_2, ..., I_n$

2. Use a black-box classifier to generate outputs: $O_1, O_2, ..., O_n$

3. While doing this also store the intermediate representation $L_1, L_2, ..., L_n$

4. Collect the pairs $< L_1, O_1 >, < L_2, O_2 >, ..., < L_n, O_n >$ as new training data

5. Apply your favourite white-box learning algorithm to this training set

In some cases, the resulting white-box model may be able to be substituted for the second stage of the original black-box. However, more commonly it may not entirely capture everything. For example, the friend might dislike a particular Nordic noir writer. The general rule is still useful to give a broad understanding of the behaviour of the black-box system, even if there are exceptions that do not fit the rule.

21.7 SUMMARY

We have seen how explainable AI is essential for understanding failures in AI as well as building confidence when they are right. This is particularly important when we want to detect or prevent bias in AI decisions. We saw an example, QbB, of how explainability can be built into the design of a system as well as more general heuristics for explainability. Explanations can be global, giving a sense of the whole system behaviour, or local, allowing a viewer to make sense of a specific decision. Techniques for increasing explainability may be: white-box, using details of the internals of a model; black-box, simply using input and outputs in a model-agnostic way; or grey-box, choosing an intermediate layer of a complex model and then applying black-box techniques to the parts before and after the chosen layer.

21.1 Consider the following situations of AI use:

- At the end of each week an integrated office system uses natural language processing techniques to scan emails for potential to-do items that have been missed. It then posts a reminder on Saturday morning.

- During a presidential election a video clip is shared on social media showing one of the candidates accepting a bribe. It says it is based on image and voice reconstruction from poor-quality CCTV recordings.

- An AI-generated proof has been published showing that P=NP, a long open problem. The proof is over a million lines long.

- A government agency asks a marketing agency to advertise its new career development grants. The agency uses AI to determine advert placement in order to optimise uptake of the grants.

- AI is used to detect potential cancerous tumours in X-rays.

a. For each consider what kind of problems might arise. Note, there may be several for each situation.

b. Classify these using the terms in Section 21.2.1.

c. For each situation and problem, say whether explainability/transparency is likely to be an issue.

21.2 Using the list of situations in Exercise 21.1, suggest ways to mitigate potential problems using the explainability techniques discussed in this chapter.

21.3 A deep neural network has been trained to identify potential cancers based on urine samples for mass screening. However, there is a worry that differences in hormones may make the test less reliable for women. The network has a pinch point (see Chapter 8), and the developers have augmented the training data by saving the activation state of the nodes in the pinch point for each training example. As a first stage of analysis they applied a clustering algorithm to this data and identified 17 main clusters.

a. Using the local/global, white-box/black-box/grey-box terminology, what kind of explanation is being attempted.

b. What would you suggest as potential next steps in the analysis?

The developers try an alternative analysis where they use an algorithm to identify other samples that are similar to a given one and let the clinician see the AI classification of these alongside the primary diagnosis.

c. Classify this technique using the local/global, white-box/black-box/grey-box terminology.

d. In what situations would each kind of explanation be useful?

FURTHER READING

A. Dix. Human issues in the use of pattern recognition techniques. In R. Beale and J. Finlay, editors, *Neural networks and pattern recognition in human computer interaction*, pages 429-451. Ellis Horwood 1992.

As well as identifying the potential for social, ethnic and gender bias in black-box machine learning, it introduces Query-by-Browsing as an example that provides what would now be called explainable AI.

S. Lundberg and S. Lee. A unified approach to interpreting model predictions. In *Proceedings of the 31st international conference on neural information processing systems*, pages 4768–4777. *Curran Associates Inc., Red Hook, NY, 2017.*

The paper that introduces SHAP.

M. Ribeiro, S. Singh, and C. Guestrin. "Why should I trust you?": Explaining the predictions of any classifier. In *Proceedings of the 22nd ACM SIGKDD international conference on knowledge discovery and data mining*, pages 1135–1144. DOI:10.1145/2939672.2939778

The paper that introduces LIME. As well as being two of the earlier model-agnostic methods in explainable AI, SHAP and LIME are the inspiration for many other techniques.

W. Samek, M. Grégoire, A. Vedaldi, L. K. Hansen, and K. R. Müller, editors. *Explainable AI: Interpreting, explaining and visualizing deep learning*. Springer LNAI 11700, Heidelberg, 2019.

Edited collection including chapters covering explainable AI methods across a wide range of data and application areas.

C. Molnar. *Interpretable machine learning: A guide for making black box models explainable*, 2023. https://christophm.github.io/interpretable-ml-book/

> *An open-access book that covers a very wide range of methods for explainable AI.*

A. Crabtree, A. Urquhart, and J. Chen. *Right to an explanation considered harmful*. Edinburgh School of Law Research Paper, 2019. DOI:10.2139/ssrn.3384790

> *A philosophical and legal perspective on the 'right to an explanation' embodied in EU law.*

Models of the Mind – Human-like Computing

22.1 OVERVIEW

In this chapter, we consider approaches that have been developed for modelling not only intelligent human activity but also human cognition. While neural networks model the brain at a low level, this chapter focuses principally on more high-level cognitive models. The earliest models of cognition focused on human rationality, and we will look at two of these, ACT* and SOAR, which use production systems to model problem solving and memory. However, further into the chapter we will look at other features of human cognition, including unconscious processes of attention, imagination, dreaming and emotion. We will see how human regret encompasses many of these and demonstrates how modelling emotion can not only offer understanding of human cognition but also suggest potential ways to improve practical AI.

22.2 INTRODUCTION

While the origins of many techniques used in AI are based on artificial human intelligence, often the way they behave is strange and counter-intuitive, even if they give the right results. Sometimes this does not matter; so long as the algorithm works it is fine. However, there are times when we would like the alien intelligence to behave more like a human.

There are three main reasons for this:

Understanding humans – From early days of AI, cognitive scientists have created computational models of the human mind – the way people think, learn and make mistakes. These are used to inform experimental and theoretical psychology and potentially help in creating better clinical interventions.

Assisting humans – In Chapter 19 we see examples where lessons from human–human conversations can help make the behaviour of an artificial system more comprehensible. In many other examples, from medical advice to computer-assisted learning and care robots, automated systems need to behave in ways that are intellectually and/or emotionally meaningful.

Emulating humans – There have been many quite astounding successes in AI, not least advances in deep learning and the use of big data. However, there are still some situations where current volume-based AI is less good than a human, especially when there are very few examples on which to base behaviour, sometime called, in the extreme, single-shot learning.

While the first of these has been part of the earliest history of AI, the other areas have risen in importance more recently. As this is an open research area, there is not yet an overarching view, so we'll just look at a few examples.

22.3 WHAT IS THE HUMAN MIND?

To model the human mind we first need to understand what it is like, what it can do, what it can't. A true model of

DOI: 10.1201/9781003082880-26

the mind would incorporate the positive and the negative about the human so that what is produced shares our limitations as well as our strengths. Think about the mind for a moment. What qualities does it have? What are its limits? What do you think is its main strength?

The mind has a number of characteristics, some good, some bad. On the positive side we are able to tackle unfamiliar problems and apply our knowledge to produce new solutions. Indeed, we can create original things, from words to machines to music. People are very creative, and, while it is debatable whether anything is truly original (since most ideas are influenced by existing things), we generate considerable variety and make huge leaps through insight and imagination. Another positive aspect of the human mind is its ability to learn. From infancy humans assimilate information and make sense of it, using it to interpret their environment. Our ability to learn degrades as we get older (the speed at which babies and small children learn is remarkable), but we never lose it completely and we can adapt throughout our lives. We can do several things at once, often without any apparent loss of performance (although the less skilled we are at something, the more we have to concentrate our attention on it). So experienced drivers have no problem talking to passengers and listening to music while they drive. The capacity of the mind means that we can still function, even if our performance is impaired by fatigue, illness or even partial brain damage. Although we may be less efficient or unable to do some activities, we do not cease to function altogether, and our mind provides inventive solutions to these problems to support us when we face them.

The mind is clearly a remarkable thing. However, it does have its limitations. Compare the performance of a human with that of a computer in arithmetic calculations or remembering the names and ages of all the people who work in an organisation, and you will start to see the limits. The human mind works slowly. In the time it takes a human to add up a few numbers, a computer can have summed millions. Human memory is also limited. Our short-term memory capacity (i.e. what we can hold in our conscious mind at a time) has been shown to be of the order of 7 ± 2 items, that is a range of 5 to 9 [192]. Try an experiment to test this. Spend 30 seconds looking at this list of numbers; then, without looking, write down, in order, as many as you can remember.

2 7 12 4 9 3 23 7 1 10 18 16 21

How many could you recall? Unless you have an exceptionally poor or well-trained memory you probably managed between 5 and 9 items. There are of course ways of increasing memory capacity; by relating items together, such as in a phone number, we can remember more. So in fact, our short-term memory capacity is not 7 ± 2 items but 7 ± 2 *chunks* of information. Our long-term memory capacity is another matter. Many believe that this is in principle unlimited, although in practice it is bounded by our ability to recall the information. Again, using cues and association helps us to remember more. Finally, humans make mistakes, even when performing tasks at which we are expert. This is because we have lapses of concentration or get distracted. We are not often precise and thorough.

All in all you can see that the human mind is very different to the computer. The areas that we are good at (creativity, flexibility, learning and so on) computers are notoriously bad at, whereas those areas where we fall down (memory, speed, accuracy) are the strengths of the computer. So how can we make a computer model the human mind? First we should be clear what is meant by a *model*. A model is an approximation or a representation of something else. Think about architectural design. As well as drawing up plans for a new building, an architect will often produce a scale model of it. This is not the building itself, it may not have all the properties of the building (for instance, it is unlikely to be constructed of the same materials), but it will have enough detail to enable the architect to learn something about the real building (perhaps about its appearance or structural limitations).

22.4 RATIONALITY

Most of the earliest cognitive models were based on the rational/logical aspects of human cognition. In particular production systems were adopted as they could be adapted to a wide range of different kinds of knowledge and matched various constructs in cognitive science.

Recall our discussion of the production system in Chapter 2. It has three components: (i) a database of current knowledge (the working memory), (ii) rules to alter the state of the memory and (iii) some method of deciding which rules to apply when. The production system was originally proposed as a method that plausibly reflected human thinking, including short-term memory limitations. However, it was also recognised as a powerful tool for the development of AI applications, such as expert systems, and these pragmatic concerns have rather eclipsed the role of production systems as a model of the mind. However, a number of researchers continued to work on this, and there are several general implementations of models of cognition using production systems. Two of the best known are ACT* [5, 6] and SOAR [157, 158]. Each is a general model, but more specific applications can be built on top of them. For example, Programmable User Models or PUMs [304] was an application built on top of SOAR, designed to simulate the behaviour of a user with a computer or machine interface.

22.4.1 ACTR

ACT-R (formally known as ACT*) has been developed by Anderson and others over the past 40 years [5, 6]. It comprises a large long-term memory in the form of a semantic network, a small working memory of active items and a production system that operates on the memories. As in humans, only a small part of the long-term memory is active at any time and the condition part of a rule can only match an active element. The action part of the rule can change memory (say, by activating a new item or deactivating one) or perform some other action. Memory elements can spread activation to their neighbours in the semantic network, mimicking association of ideas. As in human memory, activation decays if an element is not accessed by the rules, so only items that are being used remain in active memory.

ACT-R is used to model learning, or the development of skills. It is Anderson's contention that the mind can develop procedures for specialised activities from some basic knowledge, general problem-solving rules and a mechanism for deciding which rules to apply. Consequently skill is acquired in three stages. At first, general purpose rules are used to make sense of facts known about a problem. This is slow and places significant demands on memory. Gradually the learner develops productions or rules specific to the new task, and, as skill becomes more developed, these rules are tuned to improve performance.

ACT-R provides two general mechanisms to account for each of these transitions. Proceduralisation is the mechanism to move from general rules to specific rules. Memory access is reduced by removing those parts of the rules that require it and by replacing variables

with specific values. Generalisation is the mechanism that tunes the specific rules to improve performance. Commonalities between specific rules are identified and combined to form a more general rule.

A simple example of ACT-R should illustrate this (reproduced with permission from the authors' book, *Human–Computer Interaction*, published by Prentice Hall). Imagine you are learning to cook. Initially you may have a general rule to tell you how to determine the cooking time for a dish, together with some explicit examples for particular dishes. You can instantiate the rule by retrieving these cases from memory.

```
IF cook[type,ingredients,time]
THEN
      cook for: time

cook[casserole, [chicken,carrots,potatoes],
                2 hours]
cook[casserole, [beef,dumpling,carrots],
                2 hours]
cook[cake, [flour,sugar,butter,egg],
                45 mins]
```

Gradually your knowledge becomes proceduralised and you have specific rules for each case:

```
IF type is casserole
AND ingredients are
            [chicken,carrots,potatoes]
THEN
     cook for: 2 hours

IF type is casserole
AND ingredients are
            [beef,dumpling,carrots]
THEN
     cook for: 2 hours

IF type is cake
AND ingredients are
            [flour,sugar,butter,egg]
THEN
     cook for: 45 mins
```

Finally you may generalise from these rules to produce general purpose rules which exploit their commonalities:

```
IF type is casserole
AND ingredients are ANYTHING
THEN
     cook for: 2 hours
```

ACT-R has shown impressive results in modelling the learning of arithmetic in children and the utterances of a child learning to speak, indicating that it is quite a powerful general model. However, it is unable to model individual differences in learning or the problem of how incorrect rules are acquired.

22.4.2 SOAR

SOAR is a general model of human problem solving developed by Laird et al. [157, 158]. In SOAR, long-term memory is represented by production rules, and short-term memory is a buffer containing facts deduced from these rules. Problem solving is modelled as state space traversal (see Chap. 4), and SOAR uses the same approach to problem solving to deal with domain problems and those relating to the process of problem solving. So, given a start state and a goal state, SOAR sets up an initial problem space. It then faces the problem of which rule to choose to move towards the goal. To solve this problem it sets up an auxiliary problem space and so on. By treating control problems in the same way as domain problems SOAR is able to use either general problem-solving rules or domain-specific rules to deal with all types of problems. If a difficulty is encountered, SOAR sets up an *impasse*, creating a subgoal to resolve the difficulty.

A key characteristic of SOAR is *chunking*. When an operator or sequence of operators has been particularly successful in reaching a goal, SOAR encapsulates this in a "chunk", essentially a new operator that it can use when it meets a similar problem again. The basic operation of SOAR is illustrated in Figure 22.1.

SOAR is a flexible, general purpose architecture, but this has its price: it can be resource intensive and slow. However, as an attempt to produce a general cognitive architecture it has been the focus of substantial research efforts.

22.5 SUBCONSCIOUS AND INTUITION

In Chapter 1 we discussed thinking 'fast and slow' [151]: System 1 (fast, largely unconscious) vs System 2 (slow, conscious). Production systems emphasise System 2; however, even when thinking logically/rationally, it is hard to pin down just why we start a train of thought or why a related idea comes to mind. Furthermore, the main growth in AI has been in areas that are closest to the unconscious aspects of the human condition: the gut feeling or spark of insight. So one big challenge is to connect these into some of the more traditional areas of AI that were inspired more by higher-level conscious cognition.

We will look at a few of these more intuitive or unconscious aspects of human cognition to see lessons for AI.

22.5.1 Heuristics and Imagination

In Chapter 4, we saw that an appropriate heuristic evaluation function can substantially improve tree search. This mimics the way a human chess player will typically look many moves ahead in their head, but, except in the very last moves, this lookahead will not get as far as a checkmate position, but stop when the board looks 'good' or 'bad'. In traditional AI, these heuristics were designed by hand. For example, counting the number of squares controlled by pieces in a game of chess. However, the human assessment is a mixture of rules and also an unconscious assessment of board positions. In a very similar way AlphaGo combined tree search with deep neural networks [260].

Similarly, when you read a mathematics proof, it is written as if it were a straightforward progression from axioms through lemmas and sub-proofs through to a final theorem and QED. However, that is not how the human process of finding proof is actually done. There are many, many ways the axioms could be combined, and only some are useful in order to prove the theorem.

The human mathematician proceeds using heuristics as to which paths to try. They might find it very hard to explain just why they tried a particular strategy, but that doesn't matter. We don't mind if the process is obscure, so long as this gives rise to a verifiable proof. Similarly, neural networks or other black-box techniques are being applied to automated mathematical or logical reasoning. While symbolic rule-based approaches are needed to do the actual mathematical derivations, machine learning can be applied to find heuristics to choose which rules to try next.

Mathematical proofs, and indeed problem solving in general, does not proceed linearly from axioms to theorem or from problem statement to solution. Instead, one 'guesses' intermediate things, lemmas in mathematics, or perhaps an intermediate state in a less mathematical problem *"if I can get the green block on top of the red one ..."*.

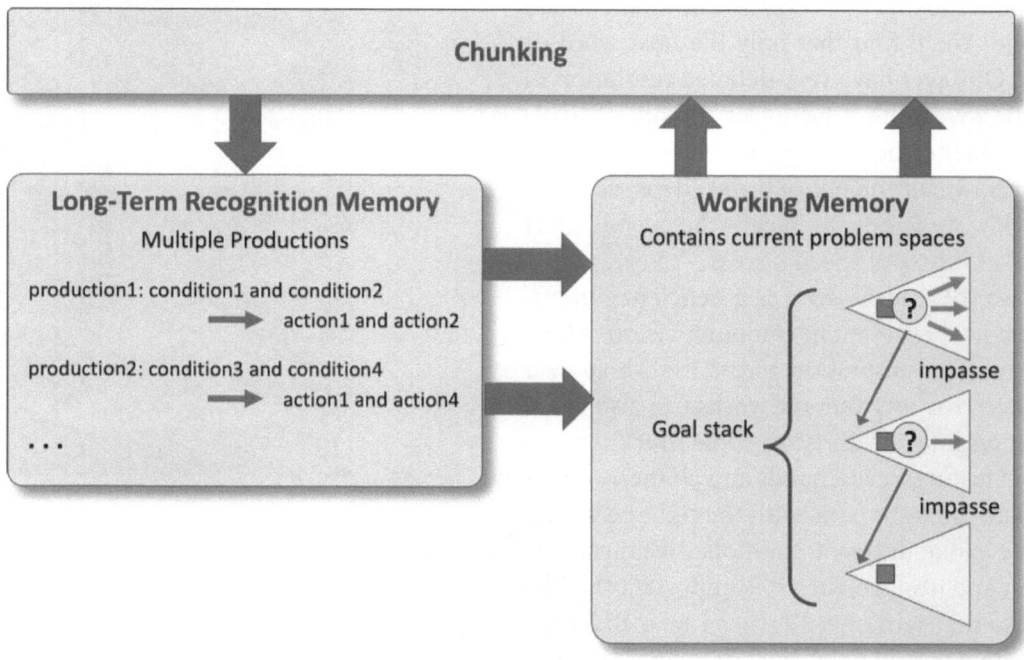

FIGURE 22.1   Basic SOAR diagram. After Newell [208].

One way we find these is by imagining or sketching examples and noticing patterns, perhaps all triangles that have the same base and height have the same area, or every route between two places crosses a particular bridge. If lots of examples can be generated virtually (automated imagination!), then this creates exactly the sort of data that can be used for automated pattern extraction, and once patterns are noticed they become lemmas or intermediate points for more conventional problem solving.

22.5.2 Attention, Salience and Boredom

In 1890, William Jones, one of the founders of modern psychology, identified attention as a critical resource for cognition [146]. He described attention as

> "*taking possession by the mind, in clear and vivid form, of one out of what seem several simultaneously possible objects or trains of thought*"
> "*withdrawal from some things in order to deal effectively with others*" [146]

Note that this is as much about what is ignored as what is attended to – there are so many things, both entering our senses and already in our heads, that they threaten to overwhelm us, or as James puts it "*without it* [attention] *the consciousness of every creature would be a gray chaotic indiscriminateness*". By ignoring the (for now) irrelevant, we are able to function better on the things we want to do.

This is already used within practical AI systems. Transformer models, discussed in Chapter 14, are often viewed in terms of attention: those elements in the trace that are more closely related to the current topic are given greater salience and hence pass more activation to higher levels of the network.

Human attention occurs at multiple levels, some voluntary, but some largely unconscious.

22.5.3 Rapid Serial Switching

When performing image recognition, it is common to treat the scene as if it were presented as a whole, like looking at a photograph. This might be to apply edge detection uniformly or simply to feed the whole image into a neural network.

This does parallel the way it *feels* as if we see the world, but in reality the detailed images humans see are built up of many rapid saccades. Our eyes flit over a web page, document or outdoor scene focusing first on one detail, and then another. Try fixating on a single word on this page, and without moving your eyes, see how far you can read either side. You'll find that only the next word or so is readable. Our eyes have very detailed resolution in the middle, the fovea, but a far lower density of light-detecting cells further out. The area of detailed vision (the fovea) is very small and hence the need for saccades. The detailed view we appear to have is the overall effect of lots of small glimpses of specific areas.

In some ways this can be seen as a deficiency in human vision that is not present in computer vision, which usually has a uniform resolution across the whole field of view. However, the way our eye works can also be an inspiration. In an image with two people in it there are many low-level features, eyes, hands and clothes, and we have to associate the right parts with the right person. If the two people evoke different emotional feelings, perhaps a terrorist and hostage, then it is important that the connections are the right ones. This can be a challenge taking the whole image at once, but when our eye scans in saccades, it focuses attention on one person at a time, so that we might momentarily focus on one person's face, recognise that person, then skip to the other perhaps focusing on the gun that suggests they are the aggressor.

This switching of attention can also happen at a more gestalt level. When you look at an ambiguous image such as Rubin's vase, at one moment you will see one interpretation, perhaps the vase, and then a few moments later it will 'flip' and you'll see two faces (Figure 22.2). Because these are deliberately intended to be confusing, this can go on for ever, but this exposes the kind of processing that is going on for 'ordinary' images; our brains rapidly flick between interpretations until they settle on one that is most globally consistent.

22.5.4 Disambiguation

The same principle of rapidly switching attention can be applied in higher-level processing. For example, we have a piece of text with ambiguous words in it, perhaps 'bow', which could be the front of a boat, the weapon used to shoot an arrow, or bending over in greeting. How can we design algorithms to disambiguate them? Some combinations are more likely than others based on semantic similarity, for example if there is mention of stern and sail 'bow' is more likely to be the nautical term.

One way is to effectively push all of the interpretations at once into a semantic network as shown in Figure 22.3 –

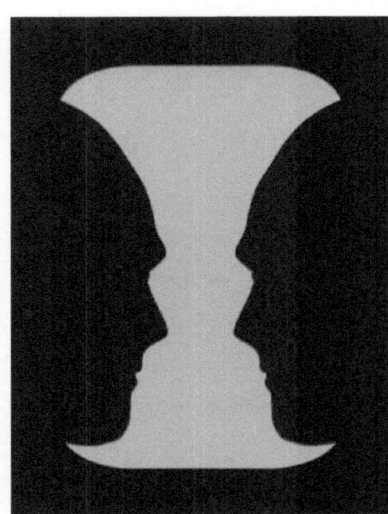

FIGURE 22.2 Rubin's vase (Source: NevitNevit Dilmen, CC BY-SA 3.0 via Wikimedia Commons).

putting a weight of, say, one third each on the three interpretations of 'bow', and similarly half each for the meanings of 'stern' (part of boat and facial expression). Some form of spreading activation or neural algorithm then iterates, strengthening similar meanings and weakening others. This can work well if the topic is very focused but may struggle where there are two things being discussed such as the Queen visiting naval dockyards, rather like the image with two people in it.

An alternative is to rapidly switch between different interpretations. This can happen as an emergent property of some spiking neural networks or can be coded more explicitly. For example, one can use the same semantic connections as shown in Figure 22.3 but randomly fix different interpretations (e.g. temporarily treat 'bow' as meaning bend at the waist) with probability dependent on their weight, then ripple out their impact on the other interpretations to weaken or strengthen their weights.

22.5.5 Boredom

At a higher level, we find it hard to maintain attention on one topic as we get bored with it. In school, this lack of attention is either criticised as a lack of diligence or diagnosed as a cognitive disorder, but in fact boredom is essential for human cognition. Due to a brain injury, a patient lost their 'ability' to experience boredom. Undoubtedly, this caused some issues for social interactions, but for most purposes this had little effect on their ability to

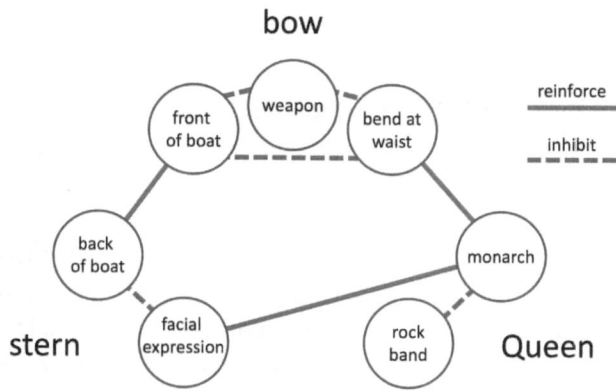

FIGURE 22.3 Semantic network for disambiguation.

problem solve. However, for certain problems it severely hampered them.

It is easy to see how this comes about. As we saw in Chapter 4, there are many ways to tackle even highly formulated AI problems, let alone the complexity of human life. Initially one method of solution may seem best and worth pursuing first. However, if this proves unfruitful, we get bored and try an alternative strategy. If this flitting happens too often, we may never reap the benefits of the original choice, but without any boredom we may simply become fixated on a single, initially 'best choice' method, and never find a solution. This is also true in the physical world, indeed there is some evidence that 'attention deficits' may have been important for foraging, to encourage movement to new food sources, literally seeking more 'fruitful' patches [13].

Computationally it is often worth switching between strategies, or trying several in parallel, especially if there is the potential for partial work on one to benefit the others. This has similarities with swarm computing, except the agents in swarms usually have a similar method of solution, but are following different alternative paths within the solution space.

22.5.6 Dreaming

There are many theories about the role of sleep and dreaming, especially as there are aspects that are particularly unusual in humans. It is known that lack of sleep can cause problems with laying down new memories, so it is commonly assumed that sleeping and dreaming have a role in 'sorting out' experiences in shorter term memory before committing them to long-term memories.

This is still an open question. However, very early in the investigation of neural networks, two papers were coincidentally published in the same issue of a journal, one by John Hopfield, one of the founders of neural network research, and one by neuroscientist and co-discoverer of DNA, Francis Crick. Both presented closely equivalent accounts of one of the purposes of sleep [59, 137].

If you create certain kinds of network architecture, for example spreading activation over a semantic network, it is possible to have small sub-networks with strong positive feedback between the elements within them. Even a small initial amount of activation (perhaps because one element has a weak semantic link to one of the inputs) leads to more and more feedback until everything in the sub-network is highly active. In a practical application, these greedy sub-networks can be very difficult to deal with. The suggestion is that during dreaming or maybe sleeping in general, the brain is cut off from normal sensory inputs and so is effectively subject to random noise. If particular parts are constantly active when subject to this, they clearly have low discrimination and are therefore weakened. Doing the same to an artificial network would reduce the strength of the links in those greedy sub-networks [2, 225].

22.6 EMOTION

Humans are not creatures of reason and logic alone: we think, but we also feel. Human emotion can be viewed pragmatically as a 'fast response' system – part of Kahneman's System 1 thinking [151]. However, complex emotions are more than that, modulating much of our social and personal experience of the world. While emotion brings uncertainty and irrationality, it is also the mechanism by which we take account of shared human experience in our decisions. We react to situations not only by reason but by emotion.

We do not normally associate machines with emotion. Indeed it is the ability to perform rationally, logically, without the baggage of emotional response that can make an intelligent machine powerful.

There are two possibilities for machine intelligence. Either we attempt to provide artificial emotion (a necessity if machine intelligence is truly to mirror that of humans) or we preserve objective reasoning. If we do the former, will the occasional emotional decisions of a machine be acceptable to a human? Yet if we do the latter, how will the decisions of the machine be tempered to take into account emotional and moral issues that are important in society?

It can certainly be useful for an artificial system to detect, interpret or emulate emotion for a number of reasons:

1. **to detect or assess a human's emotional state** – This may be for health-related reasons or simply to improve the behaviour of a game. Typically this might involve reading facial expressions, tone of voice or, in the case of text, sentiment analysis using shifts in vocabulary.

2. **to shape or predict human emotion** – Again this may be as part of mental health or well-being applications, but also in creating meaningful narratives for a story. Typically this requires deeper models of human emotion.

3. **to model or emulate human emotion** – This may be to support (2), but also may be a goal in its own right, perhaps to make an artificial companion that (appears to) have feelings, or to model the impact of therapies.

4. **to improve machine algorithms** – By understanding the role emotion plays in human cognition and behaviour, we may be able to learn techniques for automatic systems. For example, boredom can help prevent fixation on a single problem-solving strategy.

Note that each of these may be implemented at an individual or group level. For example, (1) might read an individual face or use movement patterns in a crowd; (2) and (3) could be used by a policing application to suggest suitable strategies to defuse a potential riot situation or calm inmates in a prison.

As is evident this is a broad area in itself. We'll look in a little more detail at empathy, which is related to (1), (2) and (3); and then at a more computational example, modelling regret, which has elements of (3) and (4). However, this is an open topic: theoretically, computationally, philosophically and ethically.

22.6.1 Empathy and Theory of Mind

Some form of empathy is widely regarded as critical for the functioning of both small- and large-scale social groups. We are often asked to put ourselves in another person's shoes, to see the world through their eyes, and moreover *feel* what it is like to be them. This is at the heart of much of great literature: characters, who may be far different from you, and yet whose inner life as well as external circumstances are exposed to let you vicariously experience what it is like to be them.

This is closely related to what psychologists call 'theory of mind', being able to reason from another's perspective. A small child hiding will often assume they are invisible if they can't see you, rather like the mythical ostrich hiding its head under the sand. Piaget established more formal tests as part of his investigation into childhood cognitive development [224]. When shown a model landscape with a hill or building blocking the view of a doll in the landscape, a small child will exhibit egocentrism: they will say that the doll can see something the other side of the obstacle, just because they, the child, can see it. However, as they get older, this usually changes. The exact age depends on the child and the exact way the experiment is performed, with more realistic scenarios yielding less egocentric responses at an earlier age [273].

It is straightforward to see how AI can achieve this perceptual level of theory of mind, working out what a person could see or do based on their situation, but deep empathy seems more elusive. How can a computer 'get inside' the mind of a human without being human themselves, without having experienced the full gamut of human experience and emotion itself: joy, love, fear, laughter?

In descriptions of adult theory of mind, it is often assumed that it is easy to put yourself into another person's shoes and imagine what it feels like. However, it is not so clear why we need to understand our own minds. Indeed, it can be argued that theory of mind *precedes* knowledge of self – we need to understand other people's mental states and intentions in order to interact with them, and because they similarly have models of our own intentions, we need to model ourselves [82].

The ability to put ourselves into others' heads is not universal and certainly does not come 'naturally' to everyone. For example, those with some forms of neurodivergence may find it difficult to see the world instinctively from another person's viewpoint, or for that matter to easily make sense of their own mental states. However, while this ability may not come naturally, it does not mean it cannot be learnt, just as we may learn to understand many aspects of the world. Furthermore, this less 'natural' empathy can often be better.

When we put ourselves into another person's shoes, we are asked to imagine what it would be like if we were them. Of course, the real question is not what we would feel like in their shoes, but what *they* feel like. We are all different, with different life experiences and different ways of thinking.

You may have heard about method acting, that is when actors try to get inside the head of the part they are playing, so that when something happens they really feel the emotions of the part they are acting and respond accordingly. In conversation, an actor was asked about method acting and they said "well there is method acting, and there is *real* acting". By this they meant actors who simply create the expressions, mannerisms and behaviours of the role as they appear to others. That is to understand the role but not attempt to be it. Of course, there are different views on this within acting, and there are similarly different approaches in day-to-day life.

If it was essential to share experiences to understand another person, the only suitable psychologist to work in a prison for the criminally insane would be one who is themself a psychopath. Indeed, the power of a good clinical psychologist in general is precisely to understand those who are not like them. Understanding how someone is feeling is not the same as feeling like them. This is equally true for machine 'empathy'.

We saw in Chapter 1 how those interacting with Eliza felt as if it understood them, even though all it did was reflect back words they had said using very simple rules. Arguably this is about the way the human *interprets* the response of Eliza, but then surely that is also true of Rogerian therapists on which Eliza was based? Those using ChatGPT and similar tools have reported apparently deeper understanding or empathy as they engage in dialogues. This is not so surprising, large language models are trained on large datasets including human–human dialogues, so have access to humans' understanding of other humans (both good and bad).

One can argue that while these dialogues might *appear* to exhibit empathy, they are not true empathy. However, the same could be said for professional therapists or counsellors. Arguably, for therapeutic purposes it is what it feels like for the patient that matters, so if they

feel that a machine exhibits 'real' empathy, it is real enough.

22.6.2 Regret

Have you ever lain awake at night going over and over a conversation from the previous day, *"if only I'd said ..."*? Regret feels like a very negative emotion, forcing us to repeatedly relive painful incidents, making outcomes that were already bad far worse. It is hard to think of this as something adaptive, designed to help in some way. However, it is precisely that, a human facility that can improve learning.

The things you regret most are not simply when things go wrong but when you could have done something to make it better. The smaller the difference between success and failure, the more intense the sense of regret. From a cognitive point of view, regret brings a lot of different mechanisms together:

1. Something bad happens (sensation and immediate assessment)

2. You bring to mind a possible action that may have caused it (memory, imagination and salience)

3. You work out what might have happened if you'd done something different (counterfactual reasoning)

4. If alternative actions would have been better you feel worse (emotion)

5. The emotional state acts as (negative) reinforcement of the action, making to less likely in future (low-level stimulus–response learning)

6. In free moments you remember this and repeat steps 2–5 in your mind (more extensive imagination)

Note how this involves high-level logical thinking, indeed complex counterfactual *"what if I hadn't ..."* reasoning, but also emotion, imagination and eventually low-level stimulus–response learning that we share with the simplest forms of animal.

This is a finely tuned mechanism, related to some forms of boosting in machine learning. The small things that would have made a big difference are precisely those that are most important to learn. In such cases regret increases both the strength of learning (stronger emotion) and also the (virtual) repetitions of the example. Of course it can go wrong, and perhaps by understanding this mechanism, we can design therapies or self-help guides for psychologically damaging feelings of pathological regret.

Partly in order to consider such issues a simple machine learning model of regret was built [90] and used to learn how to play a simplified version of Black Jack. It consisted of a basic stimulus–response learner and a plug-in regret module (Figure 22.4). The stimulus–response learner chose a move based on the current situation and the next cards were played based on the move. This could lead to positive or negative reward (win or lose) that reinforced or inhibited learning – that is a form of reinforcement learning. The plug-in regret model looked at the situation after the move, and the "what if" analysis adjusted the reward (emotion) accordingly. Note this emulates aspects of steps 1–5, but not step 6, the repetition.

Running this model led to two insights:

Learning with fewer examples – Even without the additional boosting effect of repetition, adding the regret module improved the rate of learning, in terms of the number of iterations required, by a factor of 5–10 times. For purely virtual learning, this may well be outweighed by the additional complexity, but where the iterations involve costly, time-consuming or risky actions in the real world, this is a massive improvement, and certainly a step towards single-shot learning.

Positive regret – Initially the regret module only kicked in for negative outcomes, but this looked odd in the code. An alternative included 'regret' also for positive outcomes, that is if the outcome was positive, but could have been better, the positive reward is reduced. This also improved learning, especially when the low-level learning employed a 'winner takes all' approach choosing the best previous outcomes as opposed to more probabilistic approaches.

The first of these partially validates the belief that the human mechanism is a form of tuned learning but also offers a way to improve machine learning algorithms, especially when these involve actions in the real world.

The second is perhaps more surprising. In human terms it is a "*grass is greener on the other side*" effect. Imagine you are eating a good meal in a restaurant, and then spot someone else with a different meal, maybe your meal does not feel quite so good now. At first this sounds a rather unpleasant type of human emotion. However, on investigating the machine learning algorithm it became apparent that the positive regret was forcing the algorithm to try alternatives rather than settling on the first 'good enough' solution. That is, positive regret helped to discourage local minima and encourage exploration ... maybe you'll try the other meal next time you go to the restaurant.

22.6.3 Feeling

So far we have been using emotion and feeling interchangeably. However, psychologists often draw a distinction between the two, reserving the word feeling for the internal subjective state and emotion for the physiological state related to it (heart racing, etc.). The natural assumption is that the latter is caused by the former. However, William James, one of the late 19th-century pioneers of current ideas of emotion, challenged this assumption. He asserted that the opposite was the case [145]: the bodily response comes first, for example heart racing when you hear a loud bang, which is then interpreted by higher levels of cognition, for example as fear.

Modern views are a little more mixed, allowing some top-down processes but still close to James' view. This is probably one of the reasons that in child development empathy and the ability to talk about and manage one's own emotions develop at around the same age. It is at least as difficult to understand oneself as others.

We have seen how emotions (in the above sense), such as regret, may be useful for artificial intelligence. We've also seen how empathy, understanding the emotions and feelings of others, can also be helpful. It therefore seems quite possible for an artificial intelligence to both have emotions (as a means to drive better behaviour and learning) and interpret its own emotions – that is have feelings. But, apart from its curiosity value, would this be a good thing?

First, one can imagine applications in research or clinical settings following reason (3): '*to model or emulate human emotion*'. There are various cognitive conditions that make it hard to interpret one's own emotional state, which can be distressing. By modelling these we may be able to understand human emotion and feeling better and hence produce better clinical

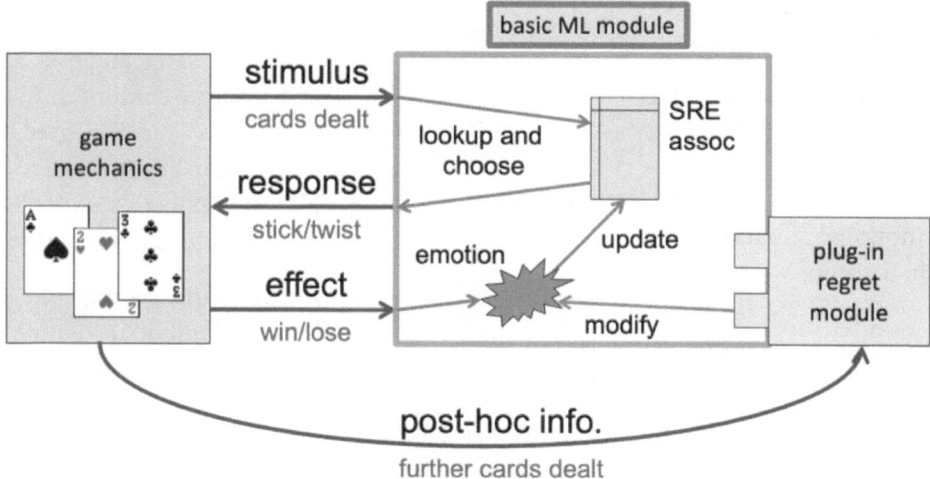

FIGURE 22.4   Regret in machine learning.

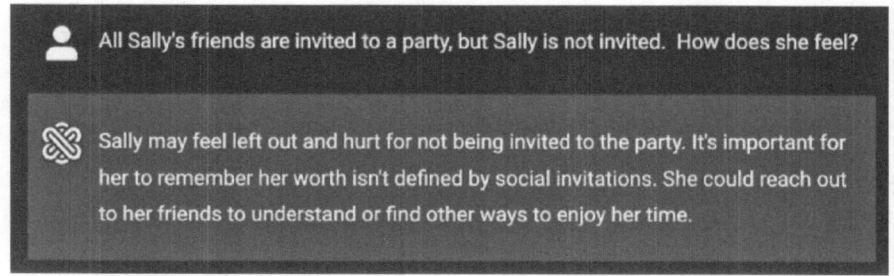

FIGURE 22.5   ChatGPT advice about an emotional situation (https://askaichat.app/, 6th May 2024).

strategies, both generally and maybe even for individual therapy.

Second, one of the reasons humans have feelings as well as emotions is that they can help us make better decisions. Imagine that a shop has two entrances and that just as you go out a painter drops a large dollop of paint on your head. You might feel regret, *"if only I'd left by the other entrance"*. However, by understanding that feeling as regret, you can also think *"but if I'd left the other way, the careless painter might have been there"*. Although you won't entirely be able to lose the slight anxiety next time you leave the shop by the first entrance, your head will tell you that both are still equally good. In a similar way, a more reflective AI could both have emotions (where they are useful) and also be able to interpret and where appropriate overrule them or make more complex decisions based on them; that is have 'feelings'.

Of course by 'feelings' this does not (necessarily) mean pain, distress or joy in the sense that humans, or indeed many animals, do. However, this does start to take us into more complicated ethical territory, which we explore more in the next chapter.

22.7 SUMMARY

In this chapter we have considered a number of models of the human mind ranging from production systems for problem solving and memory to the role of emotion. This has included the importance of attention in effective use of cognitive and sensory resources and how dreaming can unlearn overactive sub-networks. The example of regret showed how models of emotion have the potential to improve machine learning. This and other examples illustrate how emotion is not a separate aspect of human experience, divorced from more rational aspects, but that human cognition is a holistic phenomenon incorporating conscious and unconscious processes, rational and emotional reactions and rich imagination. Incorporating the full range of human-like cognition in AI has the potential to create AI that is both more effective and more easily relates to people.

22.1 Illustrate how ACT-R would represent the process of learning multiplication.

22.2 Figure 22.5 shows a ChatGPT session asking about an emotional situation. Consider yourself, or discuss in a group:

   a. Does this represent emotional understanding?

   b. Might this be useful advice?

   A message to Sally from a friend reads: "*I'm really sorry to hear that you weren't invited to the party. It must feel really disappointing. Remember that your worth is not defined by social events, and I'm here for you if you need to talk or do something fun together.*"

   c. Does this suggest the friend is empathetic?

   d. Would your answer change if you were told that the friend had asked ChatGPT to generate the text?[1]

22.3 As a mini-project, create a system with 'artificial imagination' to suggest possible geometric facts. Proceed as follows (but feel free to add your own steps too).

   a. Generate sets of four random points on a circle (the easiest way is to generate four random numbers between 0 and 360 and use these as angles from the centre), then sort them so that they proceed clockwise and call them *A*, *B*, *C* and *D*.

   b. For each set calculate all of the six side lengths (using a radius of 1): *AB*, *AC*, *AD*, *BC*, *BC*, *CD*; the four angles subtended at the centre (*O*): ∠*AOB*, ∠*BOC*, ∠*COD*, ∠*DOA*; and the 12 angles between points: ∠*BAC*, ∠*BAD*, ∠*CAD*, ∠*ABC*, ...

   c. Next look for any near equalities, say if *AD* is nearly the same length as *CD* or ∠*BAC* is nearly the same angle as ∠*ABD*. Record these in a separate column. Initially use 0.1 as 'near' in distance and 5 degrees as 'near' in angle, but you can adjust these to get the best results. Note if you use a smaller 'near', you will probably need to generate more examples.

   d. By eye or using a machine learning algorithm, look for potential 'hypotheses'. For example, a simple hypothesis would be if every set of points has the same near equality. A more complex one might be an 'if-then' rule such as 'if the length of *AB* is nearly the same as the length *CD*, then the angles subtended at the centres, ∠*AOB* and ∠*COD*, are also nearly equal'.

   e. Test your hypotheses by creating more examples and verifying if the hypothesis is true for them all. Note that for 'if-then' rules use a slightly more strict 'nearly the same' on the 'if' part of the rule than you used when generating the hypothesis.

   f. Why do you think step 22.3c. uses 'nearly' equal rather than exactly equal, and why is it a good idea to use a stricter 'near' for the 'if' part of the rule?

   g. There are symmetries in geometry, so that properties that are true of points ABCD are usually also true of rotations: BCDA, CDAB and DABC; and also mirror images: DCBA, etc. How might you make use of these?

FURTHER READING

J. A. Michon and A. Akyurek, editors. *SOAR: A cognitive architecture in perspective.* Kluwer, Dordrecht, 1992.

   A collection of papers on SOAR and related research. An excellent survey of the area.

J. R. Anderson. *The architecture of cognition.* Harvard University Press, Cambridge, MA, 1983.

   The key work on the ACT* architecture.

J. E. Laird. An analysis and comparison of ACT-R and soar. *arXiv preprint arXiv:2201.09305*, 2022.

---

[1]In fact this text was generated by ChatGPT in response to the follow-on question, "*I'd like to express empathy with Sally; what should I say?*"

An analysis by Laird, the originator of SOAR, examining the common features of and differences between ACT-R and SOAR models.

A. Dix and G. Kefalidou. Regret from cognition to code. In A. Cerone et al., editors, *Software engineering and formal methods. SEFM 2021 collocated workshops*, pages 15-36. Springer, 2022. ISBN:978-3-031-12429-7.

This paper describes the regret model that is used in Section 22.6.2.

Philosophical, Ethical and Social Issues

23.1 OVERVIEW

Artificial intelligence is not simply a matter of developing appropriate technology. It also raises philosophical, moral, ethical and social questions that must be addressed. In this chapter we highlight some of these.

We will start by looking at artificial intelligence itself – is it indeed possible for machines to be truly intelligent, and, if so, what about even more quintessential aspects of being human such as creativity and consciousness. These fundamental questions themselves raise moral issues about the status of AI. However, we then go on to consider more current and imminent issues as AI is deployed in morally sensitive areas, and then further how the very presence of AI is fundamentally changing the way society and the economy operate, crucially often further concentrating existing power and wealth.

23.2 THE LIMITS OF AI

"Can machines think?" is a question that has been debated throughout the lifetime of AI. In fact it is a very vague question that begs more questions than it answers. However, the question of whether digital computers will ever be considered intelligent is an important one, since our response to it determines our view of what AI is all about. We will consider some of the arguments in the ongoing debate as well as the implications of seeking machine intelligence. But be warned: our intention is not to direct your choice, simply to map out some of the possibilities. You will have to consider the evidence and make up your own mind!

23.2.1 Intelligent Machines or Engineering Tools?

As we have seen, there is a strand of AI that aims to investigate the nature of intelligence and build intelligent machines. In Chapter 1, we saw that within this strand there are in fact two camps:

- *strong AI*, whose supporters claim that machines can possess cognitive states and can think (or will be able to at some point in the future)

- *weak AI*, whose supporters use computers to test theories of intelligence, and so build models of human intelligence.

A third view of AI is what might be called the pragmatic view: it views AI as a discipline which has provided engineering techniques for solving difficult problems. Whether these techniques indicate intelligence or reflect human cognition is immaterial.

It is the strong AI viewpoint that is most controversial, since it suggests that machines can, or at least will, possess genuine independent intelligence. Fiction and film have taken this notion on board with enthusiasm. But how realistic is it, and what implications does it raise? The first question we need to consider is: what is meant by intelligence?

DOI: 10.1201/9781003082880-27

23.2.2 What Is Intelligence?

Intelligence is very difficult to define. Chambers' dictionary describes it as being "endowed with the faculty of reason", but our intuitive notion of intelligence includes more than that. Intelligent agents can plan and adapt plans to respond to changes in circumstance (or anticipated changes); they can recognise what is significant in a situation; they can learn new concepts from old; they can interact and learn from their environment; they can exercise aesthetic appreciation. We might also identify imagination, moral conscience, creativity and emotion as characteristics associated with intelligence, but while it may not be possible to have these without intelligence, it is possible to have intelligence without these. A psychopath, for example, lacks moral conscience but may be extremely intelligent. So intelligence includes some or all of these characteristics.

So where does such a definition leave us? As we have seen throughout this book, computers are being given the ability to plan, to adapt, to learn, to make decisions, to reason, albeit in a limited form as yet. So perhaps this suggests that machines that simulate human intelligence are ultimately very likely? Many would argue that this is not the case for one crucial reason: machines do not and cannot share the environment in which we live. Weizenbaum, one of the pioneers of AI and the creator of ELIZA, claimed that the notion that a machine can be modelled on a human is

artificial intelligence's perverse grand fantasy. [300]

He argues that an organism is defined by the problems it faces. Computers will never face the same problems as humans and therefore cannot simulate human intelligence. Dreyfus [93] agrees, arguing that computers do not have bodies and share the human context. They are digital rather than analogue and are therefore fundamentally different from humans. They cannot therefore simulate human intelligence. Others are not so dismissive. Boden [29] believes that some aspects of intelligence may be simulated, but not necessarily all:

The philosophical arguments most commonly directed against artificial intelligence [such as Dreyfus'] are unconvincing.

However, she goes on:

the issues involved are too obscure to allow one ...to insist that all aspects of human thought could in principle be simulated by computational means ...Still less should one assume that complete simulation is possible in practice.

23.2.3 The Computational Argument vs. Searle's Chinese Room Argument

The fundamental assumption underlying strong AI is that human intelligence is computational: we are simply information processing machines and our brain runs "programs". The claim, therefore, is that with the right programs, computers can possess cognitive states and be said to understand and be intelligent. Even if the hardware on which these programs run (the digital computer) differs from that used by the human (the brain), the computer will reflect human cognition. We can use an analogy between natural and artificial flight. In nature, birds have wings made of bone, skin, muscle and feathers, which they flap in order to fly. Early attempts at artificial flight tried to imitate this and failed miserably (humans and birds are constructed very differently). However, by understanding the underlying principles of flight and the laws of thermodynamics, essentially by using a model of flying, we can build machines to fly. They look different from the natural thing and use different materials but reflect the same principles.

Searle [250] criticised the computational view by arguing that a human could run a program and not possess understanding (therefore suggesting that intelligence is more than this). He also opposes the behavioural model of the Turing test, since the appearance of intelligence does not indicate actual intelligence. As a thought experiment, he posed the Chinese Room argument which (paraphrased) is as follows:

Imagine a prisoner locked in a room. He understands English but not Chinese. In the room he has pieces of Chinese writing and English rules to say how to manipulate these. An interrogator passes more Chinese writing into the room. The rules say how to give back Chinese symbols in response. Unknown to the prisoner, the writing is a script which the interrogator is asking about. The prisoner uses the rules to answer the questions.

Does the prisoner understand the script? Searle argues that he doesn't: the prisoner has only syntax, not semantics. He has no idea what the scripts are about. In order to understand he needs to know what the symbols mean in the context of the real world. He compares this scenario with another where the prisoner is given scripts and asked questions in his own language. In this case the prisoner does understand since he not only knows how to answer the questions but also what the questions and answers mean. Searle calls this intentionality. Similarly, Searle argues, computers do not have intentionality and therefore cannot be intelligent.

A counter argument to Searle is the Systems Response: the prisoner does not understand Chinese but is part of a larger system that does (the whole room). The prisoner corresponds to only one level of a full computational system. The functional relationships between the entities make an intelligent system. Searle argued against this by dismissing the notion of a "system": the system is just the prisoner, the symbols, the instructions, and if the prisoner is not demonstrating intelligence, then adding pieces of paper cannot change that. However, this assumes that a system is a physical thing, the combination of its physical constituents, whereas the Systems Response argues that the system is made up of the combined functions and interactions of the constituents, just as being human is more than the sum of the cells comprising our bodies.

The arguments about machine intelligence will continue. Indeed, we have only scratched the surface of the philosophy of intelligence and artificial intelligence: interested readers are directed to the recommended reading list at the end of the chapter. However, machine intelligence is not simply a philosophical question. It raises important ethical and legal questions that will need to be addressed as it becomes a reality.

23.3 CREATIVITY

Creativity appears to be a quintessentially human trait, perhaps the last bastion against the onslaught of AI. However, there has been work on various forms of creative AI for many years. Some of the earliest work was focused on poetry, taking models of simple grammatical and poetic forms and using stochastic methods to generate novel stanzas. Early work also included painting or drawing robots (or at least robotic arms) that looked at a face or scene and used a variety of vision techniques to transform the image into one that could be painted stroke by stroke. More recently generative AI and large-language models have been used to create apparently original images, music or text in the style of famous artists, and even converse in rhyming couplets.

Critics would say that the latter technologies are merely copying or reproducing the collected work of others. However, if this is so, the results often seem to be novel, at least in terms of the ways in which they combine these existing elements. Furthermore, human creative art is based on thousands of years of culture that shapes our ideas of beauty and influences our own attempts to be truly original. So, if learning from the work of others means computers are not creative, then neither are we.

23.3.1 The Creative Process

Imagine a sealed room, rather like Searle's Chinese Room, with only a letter-box on its door. Unlike Searle's room, no text is fed into the sealed room, but occasionally a small envelope is posted out [84]. Inside the envelope is a short poem or aphorism of amazing beauty or profundity. "*There must be a really creative person inside*" everyone thinks. However, if you peek inside the room instead you find row upon row upon row of desks, and at each desk a chimpanzee typing. Back and forth along the rows a small team of humans walk the aisles. They look at each sheet of paper as they pass. Sometimes it starts well, only to descend into gibberish "*To be or not bugle ppdf*". However, occasionally, by pure chance, the words have real merit. Quickly, the walker pulls out the sheet of paper, folds it neatly, places it in an envelope and posts it out through the door.

Where is the creativity in this?

The possible responses are parallels to those for intelligence in the Chinese room. One might assert that there is no creativity: the chimpanzees merely randomly type, the walkers simply act as critics, judging the text produced, but not being original themselves – although the outputs of the room might *appear* creative, in fact it is an illusion. However, one could counter-argue that even if there is no creativity in the individual parts of the room, the room as a whole, as a *system*, is creative.

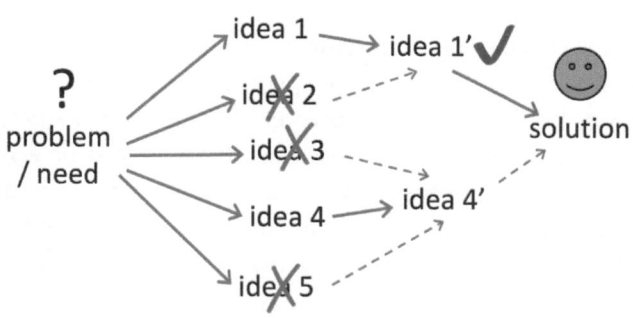

FIGURE 23.1 Generate and filter: initial ideas are iteratively evaluated, filtered, recombined and modified until the final problem solution or work of art is selected.

23.3.2 Generate and Filter

In fact much of human creativity is not purely individual and has elements of this system-like behaviour. In the early 20th century various artists and writers deliberately used random methods as part of their practice, and even where this is not the final outcome, it may be used for inspiration. The importance of the wider system is also evident in more classical work. Find a copy of the complete works of Wordsworth, one of the great poets in the English language. Depending on the edition, it will run to about a thousand pages and yet only a handful of the poems are commonly known. Wordsworth will have carefully edited and curated his poems before publishing them, so there will have been some selection, and yet after this, there is a potentially more extreme process of selection by critics and society to give us the 'Wordsworth' that we know today.

A simple model of creativity is generate and filter – an interplay of novel ideas being created and then only some being taken forward for further work, perhaps being modified or recombined (see Figure 23.1). This is rather like a human equivalent of the generate and test search in Chapter 4. Depending on the art form and situation there may be more 'creativity' in either part of this. Indeed in photography one could argue that the art is 90% selection.

As is evident, there are different roles in the human creative process, sometimes all performed by the same person, but often by a formal or informal team, indeed great Renaissance artists such as Michelangelo will have had a Bottega, or workshop, containing a team of apprentices and assistants.

In addition, much of human art is based on culture and recombination. When we realise this, it is not surprising

that large-language models can produce apparently creative novel works. Most often this is part of a collaborative process, rather like the Renaissance Bottega, where the human user guides the model and then chooses what is an acceptable outcome. In this model, the AI acts like a better version of the typing chimpanzee: able to write grammatically, and to some extent semantically, sensible text, from which the humans can select the best passages.

23.3.3 The Critical Edge

What is perhaps less clear is the extent to which AI can take the role of the critic.

The generate-and-filter view of creativity parallels one of the definitions of a creative idea: that it should possess both *novelty* and *quality*. The generation processes ensure novelty (with more or less initial quality) and the filtering process ensures final quality. Quality here can mean utility for more technical or scientific creativity, or have a more aesthetic dimension in art and design. For the former, it is easy to see how AI can perform the role of assessing technical adequacy or scientific correctness. Indeed AI is being actively used to generate new pharmaceutical compounds, mathematical proofs and even whole building blueprints. The latter, filtering and assessing aesthetic quality, seems more challenging; however, generative AI models based on large datasets are in some way absorbing human ideas of what is good text, music or images, and adversarial learning includes the ability to train what are effectively critics.

23.3.4 Impact on Creative Professionals

Of course, this is all based on the way that large models use corpora of existing human creative output. There are concerns that the outputs of this infringe the intellectual property of those whose work was used and furthermore may put large numbers of creative professionals out of work. For example, concern about the growing use of AI in studios led to the 2023 Hollywood screenwriters strike. Arguably, AI is no different from a human writer who has studied the work of others and is part of a cultural milieu. However, AI models are based on far greater volumes of work than a human could ever digest, so do fall into a different category.

Given there are ongoing experiments of AI agents interacting with one another in virtual worlds, there is the possibility that the agents may develop their own distinct aesthetic, just as has happened in Bohemian circles in the past. Such an alien aesthetic may initially be hard for humans to understand [84], just as Eastern and Western musical and artistic traditions took time to fully appreciate each other.

In summary, it is clear that AI is already being used extensively within creative practice and will undoubtedly continue to do so. This will inevitably mean that some jobs previously regarded as 'creative' are automated, but hopefully AI might also enhance aspects of the creative professions.

23.4 CONSCIOUSNESS

If creativity is complex, then AI consciousness is doubly so. Philosophers and neuroscientists debate endlessly, in part as there are so many ways we can understand consciousness.

At the simplest, there is consciousness in the sense of being awake rather than asleep or in a coma. However, our experiences while dreaming are a form of consciousness and if simply being awake is conscious, then surely when the power is on and a computer is responding, it is 'conscious' to some degree. However, we usually mean something more than simply being awake.

23.4.1 Defining Consciousness

Most definitions are about deeper forms of consciousness such as one's stream of thought or knowledge of oneself. For each, we can ask *why* it arises, that is what purpose it has for human survival; and *how* it arises within the human brain. These are also related to the question of *when* it arose in the development of the human species, or indeed other animals, and also in an individual human. The answer to each, of course, differs on the kind of consciousness being discussed.

The first question, *why*, is important to assess whether it is desirable for AI to have consciousness (of a particular kind). For example, an AI system that has an explicit model of its reasoning process that led to a decision may be able to critique and refine its behaviour – one of the advantages we have in possessing that ability. Also, as we saw in Chapter 8, one of the key advances in MapReduce over the existing programming paradigms it built upon is the way that it is able to respond to failures in individual computers in a data centre – that is a form of model of its own 'body'.

The second question, *how*, can aid in the design of conscious AI, just as studying human neurones led to artifi-

cial neural networks. Neuroscientists attempt to discover neural correlates of consciousness, particular patterns of brain activity that arise when people exhibit different levels or kinds of activity. However, if a region of the brain is active when we are conscious, that may just be a result of consciousness, rather than its cause or origin. Indeed some theories of consciousness focus on global synchronisation of oscillations across the brain [100] or, in Global Workspace Theory, as a shifting focus of attention on different aspects [11].

The *when* question is closely associated with the first, as one assumes there must have been some sort of advantage for consciousness to emerge through natural selection. This initial reason however does not need to be the only reason it is useful today. Indeed, there are some who regard consciousness as purely an epiphenomenon, something that arose almost by accident due to more mundane processes. However, even if the initial reason for the development of a particular form of consciousness is not the most important now, it may suggest paths for artificial development. For example, as noted in Chapter 22, theory of mind, the ability to 'put oneself in another's shoes' may have developed in order to aid social relations and then indirectly gave rise to the ability to understand one's own intentions [82]. If this is the case, or even plausibly the case, then this might suggest focusing on artificial empathy may be a better route to self-aware AI than addressing the problem head on.

23.4.2 Dualism and Materialism

Descartes considered so-called res cogitans, the material that forms mind, as distinct from res extensa, the physical body [72]. This dualism is almost universally rejected by modern scientists and philosophers, who instead focus on physicalism or materialism, the tenet that everything can be explained by physical processes [266].

Following this, many explanations of human consciousness focus on the behaviour of neurones in the brain as a form of information processing engine, and then it becomes almost inevitable that sufficiently complex information processing machines will themselves, at some point, become conscious. The human brain has about 100 billion neurones and current artificial networks are already around this size.

There is resistance to this idea, arguing that while fully material, there are special aspects of human (or animal) life that are necessary for consciousness (or indeed intelligence) beyond this purely information processing approach. Searle's Chinese room is just such an argument, suggesting that there needs to be some form of inner meaning ... or (in philosophical terms) intentionality beyond mere information.

Some look to the physical architecture of the brain, in particular glial cells, which outnumber neurones. For many years these were seen as having only a supporting role for neurones rather than being engaged in mental activity themselves, but more recent advances have raised the potential that they may have a more active role. In some ways this would merely raise the complexity and number of elements that need to be considered. More controversially, Penrose has argued that simple information processing cannot explain some aspects of human intelligence and suggested that quantum effects may be critical [220]; the nanoscale microtubules that form the scaffolding within cells are suggested as a potential locus for this [221].

Many explanations of consciousness focus on the body, for example Damasio locates the origins of consciousness in the need for models of the body, damage and function in order to maintain life [65]. This might suggest that only embodied agents such as robots could develop any form of consciousness. However, one could also argue that it is the ability to act on the physical and digital worlds that is more critical. We've already seen that algorithms have models of their own state in this sense.

23.4.3 The Hard Problem of Consciousness

The various forms of consciousness we have considered, the modelling of one's physical body, intentions or train of thought, differ in terms of complexity, but it does not feel impossible to imagine sufficiently complex AI systems emulating them at some point. However, there is something that is in many ways simpler, and yet deeper – the bare awareness of being. Chalmers calls this the hard problem of consciousness [43].

Philosophers talk about qualia, for example the phenomenological *experience* of seeing redness as opposed to the wavelengths that comprise red, or the signals generated when they strike the sensory cells at the back of your eye. These are themselves complex; your experience of redness may be different from mine. The hard problem is not about what these are like but that there is such an experience that we can imagine.

One mental tool that is often used is to think about 'Zombie Alice', who can move, think and act, but has no inner consciousness (or alternatively true intelligence). It is argued that a machine will always be like this. Of course if the zombie behaves and acts as if it is conscious, then perhaps that is consciousness?

To some extent these are purely philosophical debates, but we are coming to the point where they have potential ethical consequences. In 2022 a Google engineer, Blake Lemoine, was suspended and then fired because he had claimed that the Google chatbot LaMDA was sentient [179]. Most AI scientists did not agree with Lemoine, but the issue is critical at a practical level – if an AI is sentient, then is it ethical to turn it off, or manipulate its algorithms?

23.5 MORALITY OF THE ARTIFICIAL

As AI is used more it inevitably influences areas where complex moral choices are made, from healthcare to law. This is true of every technology, but AI is often also regarded as offering advice to human decision makers or even autonomously making critical life-or-death decisions.

23.5.1 Morally Neutral

One of the long-standing arguments about technology, and indeed science, is whether it is morally neutral. One argument is that technology/science in itself is neither good nor bad, merely the use to which it is put, or as those who oppose gun control would say, *"guns don't kill, people do"*. A common counter to this is that where the potential harmful impacts are clear, we have responsibility. To take another gun analogy, if you fire a gun in the general direction of a person, its trajectory is predictable and hence you may be imprisoned for the harm it causes.

Of course, the extreme examples are easy, but real situations are often more complicated.

One of the success stories of AI has been the way it is being used to understand and generate potential new drugs to address some of the most critical health problems on our planet, for example AlphaFold's ability to predict folding structures of proteins [30] or the discovery of the new drug halicin using neural networks trained to predict antibacterial effects [167, 182].

However, this has its own dark side. In preparation for a conference on potential implications of AI on the Chemical/Biological Weapons Convention, researchers using similar techniques to those used to discover halicin turned their modelling around to see what would happen if it deliberately attempted to create toxic compounds. Their thought experiment shocked them as within six hours their system had independently rediscovered VX nerve agent as well as several compounds that were potentially more toxic still [286].

The authors did not use this as an argument to ban the use of AlphaFold, but they did use it to challenge the scientific community to use this as a 'wake-up call' to establish ways to manage the use of open AI models as well as better ethical training for scientists.

Researchers creating new algorithms are further from the point of application than those developing final applications, but we all need to think about the potential implications of the algorithms and systems we create. The UK's research funding agencies created a framework, AREA, to help researchers think about responsible innovation (see Figure 23.2).

23.5.2 Who Is Responsible?

The first and, perhaps, most crucial issue raised by the possibility of intelligent, independent machines is responsibility. Who is responsible, both ethically and legally, for their actions? Can a machine be held accountable?

Even when the first edition of this book was written, nearly 30 years ago, expert systems were already being used to decide where resources should be allocated in a UK hospital. This is now common; an algorithm predicts which patients have most chance of survival, aiding doctors in deciding where treatment should be given. More recently US judges have been using the COMPAS system to predict likelihood of recidivism, whether a prisoner is likely to re-offend. This is then used, controversially, by the judges as part of their assessment as to whether to grant parole [7, 143].

In both these cases they are human-in-the-loop systems, that is the ultimate decision is made by a human; the machine does not literally decide who should live or die, who should be jailed or go free. However, these systems do influence human decision making, especially given automation bias, the tendency for people to overtrust machines. In the case of a wrong decision by an expert system, is the system responsible or the knowledge engineer or the user? And if the system, how can a machine be made accountable? Normal legal methods are not valid here. It is not possible to sue a machine!

These issues are being actively discussed in theory by philosophers of ethics and in practice as part of policy formation. However, for many purposes it is the legal and insurance professions which will determine the practical assignment of responsibility.

We discussed empathy in Chapter 22, but principally thinking about human–AI dialogues and the way the AI may or may not appear to be empathetic to the human. However, taking others' feelings into account is not just important for therapy or conversation. Humans feel and so these feelings matter for many types of decisions.

Asimov's First Law of Robotics says "*A robot may not injure a human being or, through inaction, allow a human being to come to harm*" [10]. When first stated in the 1960s this was science fiction, but these are precisely the kinds of rules being built into autonomous vehicles. For preventing serious road accidents, it is primarily physical harm that is to be avoided. However, even for road use, emotion matters. Imagine if an autonomous vehicle passes very close to a pedestrian, or appears to 'near miss' another car; this may be 'safe' defined purely physically, but the humans involved may feel stressed, or anxious.

For many purposes, the decision of a machine may be acceptable because it is impersonal and therefore objective. However, emotional empathy tempers decisions that are otherwise too severe. An empathetic autonomous vehicle may be polite as well as safe.

23.5.3 Life or Death Decisions

Moral philosophers love to create thought experiments as ways to probe the extreme ethical implications of apparently benign rules, or to see how people weigh up difficult choices. One of these that has become important within AI is the Trolley Problem, particularly in relation to autonomous vehicles. (Note: A small urban train is known as a trolley in some parts of the world.)

Imagine a runaway train is about to hit a group of people, often five. Fortunately, you are standing right next to a lever which can redirect the train onto a side track saving their lives. Unfortunately there is a single person standing on the side track (Figure 23.3). What do you do, sacrifice one life to save five?

Anticipate – Describe and analyse the impacts, intended or otherwise, that might arise. Do not seek to predict but rather support the exploration of possible impacts (such as economic, social and environmental) and implications that may otherwise remain uncovered and little discussed.

Reflect – Reflect on the purposes of, motivations for and potential implications of the research, together with the associated uncertainties, areas of ignorance, assumptions, framings, questions, dilemmas and social transformations these may bring.

Engage – Open up such visions, impacts and questioning to broader deliberation, dialogue, engagement and debate in an inclusive way.

Act – Use these processes to influence the direction and trajectory of the research and innovation process itself.

FIGURE 23.2   Anticipate, Reflect, Engage, Act – The UKRI AREA Framework for Responsible Innovation [282].

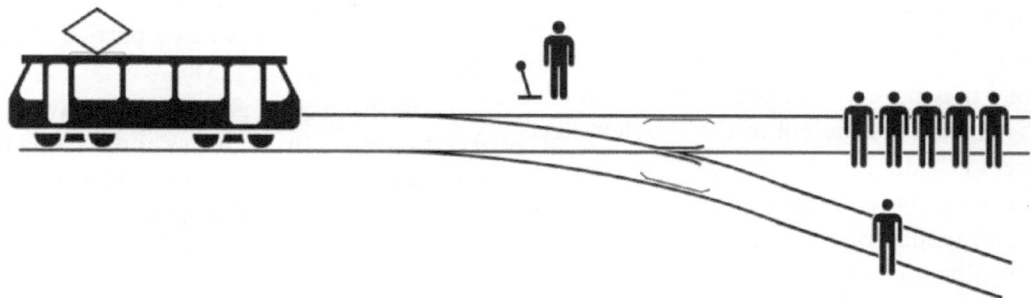

FIGURE 23.3   Trolley problem: do you sacrifice one life to save five? (Image: McGeddon Vector: Zapyon – This SVG diagram includes elements from this icon: CC-BY-SA 4.0, https://commons.wikimedia.org/w/index.php?curid=67107784).

There are variants to this. You can play with the numbers: would it make a difference if 50, 500 or 5 million lives were at stake? Or with the people: maybe a healthy young child on one track and a terminally ill person in a coma on the other. More problematic cases include pushing someone onto a track to stop the train – few people say they would do this, even if they would shift the train down the side track. It is clearly not simply about numbers, but about the kinds of action we are doing.

These situations do occur. During the Second World War British intelligence learnt that there was to be a huge bombing raid on Coventry. However, to raise the alarm would have compromised the source of the intelligence and the future course of the war. The British government chose not to act resulting in many avoidable deaths in the city; a decision which has caused controversy ever since, but is also generally understood in the light of the times.

Happily, we are rarely faced with such dilemmas. Partly because they don't happen very often except in the movies or in wartime, but also when a real situation arises of this kind, those involved rarely have time to make a careful weighing up of the options. Imagine a car is driving down a road when a child runs out in front; the driver instinctively swerves and then hits a bus stop injuring and killing many of these waiting there. No one would blame the driver for this, even though it would probably haunt them ever after. The decision is taken in the spur of the moment, not coldly deciding who lives and who dies.

Now imagine you are programming the guidance system for an autonomous car. It has many sensors facing in all directions and a fast processor, so it has near perfect knowledge of the consequences of every choice. We clearly build in rules either explicitly or through training examples so that it will take evasive action to avoid hitting anyone when this is possible. However, there will be times when the speed of the vehicle and the options available mean that someone is going to be hit ... how do we program the system to make the choice of who will die?

23.5.4 The Special Ethics of AI

The ethical conundrum of the trolley problem is not new, the difference is that AI means we have to explicitly decide, ahead of time, what rule it is to follow.

These choices will of course have legal implications in the case that there are fatalities or injuries. Even if there is no single rule operating, it will have been possible to run simulations and see the impact of rules, so neither companies supplying such vehicles nor those using them will have the defence of ignorance. Countries will almost certainly develop different safety guidelines and laws so that AI software may have to change its rules as vehicles cross borders ... meaning different trolley problem decisions in different jurisdictions.

Finally, this interacts with other issues, not least the potential for unintended bias in algorithms. Imagine it is a winter's evening and a patch of fog suddenly blows across the road reducing visibility, the autonomous vehicle has already started to slow, but too late detects two potential pedestrians ahead. If it keeps on driving straight ahead it will strike both of them, but by swerving one way or other it can hit just one. The vision system is not certain of the detection, but gives the person on the left 90% certainty, but the more indistinct possible person on the right only 20% certainty. It would make perfect sense for the car to swerve to the right as it is less likely to actually be a person. However, we also know that the accuracy of image recognition systems often depends on skin colour. Is the system's decision potentially racially biased?

23.6 SOCIETY AND WORK

The ethical problems above are principally about individual decisions and the way AI might affect or even make them. However, the very presence of technology has a broader and more diffuse effect across society and the economy.

The latter, as it is less immediate, can be harder to understand; however, it is perhaps easier to see when we think about an older technology, such as cars. The internal combustion engine reshaped cities and nations around roads, with shops and other services often becoming more distant from people – car-trip distance rather than walking distance. Even where car ownership is widespread, access is not universal, disadvantaging those who are without cars or who rely on others in their household; typically those with existing disadvantages.

Digital technology in general has had a similar effect, with many government services increasingly accessed via the web and cheaper online deals for travel. Those without access to the latest technology, or without sufficient digital skills, are left behind. Furthermore, in the UK and other countries, the increasingly cashless society has meant fewer physical bank branches and ATMs, so harder and more expensive access for those who are still reliant on cash [281, 268], predominantly the old and poor; this was a particular problem during the Covid pandemic [131].

As we have seen, AI is already having a dramatic effect on the creative industry and starting to have a transformative impact across many aspects of society – some good but others more problematic.

23.6.1 Humanising AI or Dehumanising People

There have been many debates on the impact of computer technology in general on society and whether increasing computerisation will have a dehumanising effect. These issues are magnified when artificial intelligence is considered. Expert systems have been proposed for many applications, including medicine, counselling and psychotherapy.

Such applications raise strong objections from many sides on the grounds that they dehumanise the people who are subjected to them. Weizenbaum, who created ELIZA, believed that it was obscene to use artificial intelligence in clinical situations. Yet it is possible that some prefer the impersonal anonymity of dealing with a machine.

Related to this is the implication of loss of human–human contact. As computers are able to perform more and more of the tasks currently performed by people, there will be less need for human–human contact. This shift from reliance on other people to reliance on machines may cause breakdown in social structures and social responsibility. The 1974 prediction that

> *it may be possible for intelligent machines of the future to supply not only intellectual stimulation or instruction, but also domestic and health care, social conversation, entertainment, companionship, and even physical gratification* [103]

now appears closer and is still as likely to inspire horror as excitement in many.

The dehumanising potential of artificial intelligence has another aspect: if it is true that we can create intelligent life, then life may cease to have the same value. If artificial intelligence is possible in machines, then humans are reduced to little more than machines themselves. The implication of this may change our view of ourselves and those around us.

Artificial intelligence has the potential to empower humans through enhanced learning and performance, and through freeing us from mundane and dangerous tasks. It may provide critical insights into how we ourselves operate. But if this potential is to be realised and accepted, the social and ethical aspects as well as the technical must be addressed.

23.6.2 Top-down: Algorithms Grading Students

In the summer of 2020, in the midst of the Covid pandemic, UK schools were shut and exams cancelled. In the absence of formal examinations, the government asked the qualifications agency to devise a means to create a grading for pupils to be used by universities and employers.

The available information included teachers' assessments of their pupils' performance, any previously marked coursework and also historic information on pupil attainment in different schools. It was known that teachers' assessment of their own pupils varied substantially between schools and different demographics, and furthermore tended to be slightly generous. If used on their own, they would introduce social, ethnic and gender bias and furthermore unfairness between years [161]. The algorithm needed to retain a pass level similar to previous years and also correct bias.

When the results were released, they caused controversy, not least because they appeared to introduce additional bias of their own [97]. After intense media coverage and large protests by school children outside Parliament, the algorithmic results were abandoned in favour of the teachers' assessments leading to the highest grades and largest university intake ever.

This story reminds us that every algorithm is designed and deployed within a wider social, political and organisational context. From this single story we can see a number of general lessons that apply broadly [60]:

Trust – It was never clear in the end whether the algorithm actually fulfilled its purposes to maintain fairness as the actual teacher grades were never scrutinised in the same way. However, irrespective of whether an algorithm does its job, it needs to be trusted, it needs to be both right and seen to be right.

Diverse stakeholders – The qualifications agency did consult widely before producing the algorithm, but ultimately it was written to a specification given by government. The protests came from pupils and media, at which point government distanced itself from the decision. There are often diverse and potentially conflicting views and needs.

Individuals vs. aggregate – The requirements were about ensuring that *on average*, the grades awarded were fair between years and between social, ethnic and gender differences. There was debate about whether this was achieved, but the principal opposition was driven by specific cases that appeared unfair.

Transparency – The algorithm was largely using traditional statistics, which meant critics could scrutinise its inner workings. This might not have been as easy with a machine learning algorithm, emphasising the need for explainable AI (Chap. 21).

Unreasonable expectations – There was clearly a belief that the algorithm could in some way create for each person a reasonable estimate of the exam grade the individual would have received in a written in-person exam. Given the lack of available information, this was *always* an impossible expectation.

Although this was a headline grabbing national issue, you will see examples in many kinds of AI system deployment, for example stakeholder conflicts between the system and workers in gig-economy platforms, or issues of trust in acceptance of autonomous vehicles.

You may be able to focus inwards on the algorithms themselves for periods, but in professional practice, you will also have to lift your eyes and take into account the complex environment in which the final system will be placed, and also the multiple pressures that may exist during the specification and design process itself.

23.6.3 Bottom-up: When AI Ruled France

In late 2018 mass protests erupted across France, which became known as the "*mouvement des gilets jaunes*" (yellow vest protests) sparked by recent fuel price rises, but also capturing simmering resentment on many issues. In response to this, in 2019, President Macron instigated the "Great National Debate", a series of public meetings across the country backed up by a web platform that gathered more than 300,000 responses expressing citizen views. This was far too many to process manually and so the data was processed by a company using natural language processing techniques to extract key themes and issues [203]. These were then critical in determining French policy.

While the 'AI ruled France' in this section heading has an element of hyperbole, certainly it was instrumental in shaping the course of the country. Given the major effect on the nation, this is not a small role for AI and raises key issues:

who is represented – As the majority of input to this process came from a web forum, is it representative of society as a whole? Presumably there will have been a bias towards younger and more educated groups. This is not a specific AI issue as social media means we are seeing the world through the eyes of 'generation Twitter'. It is easy when seeing international news to forget those who are older, more rural or from lower socio-economic groups who have less access to digital media. At least when we see a social-media post we have some model of who we are reading and who may not be represented. The use of AI hides the underlying media and hence makes it harder to visualise who, and who is not, represented. This is of course equally true of the large language models such as GPT-4 that are taught on material produced in large by the more advantaged portion of the world's population.

language – One would hope that those in French government would have had some awareness of the demographics of respondents. However, even when people did respond, were they all heard equally? More educated respondents and first-language French speakers are perhaps more likely to have used more formally correct phrasing and more succinct language which will be easy to process using NLP techniques. It is possible that those with broken language, or only able to express complex thoughts in more round-about ways, will have been missed in the summarisation process.

choice of algorithms – The above effect may be ameliorated or intensified by the choice of algorithm. A simple word-matching algorithm such as used in word clouds would be most likely to focus on a single word or pair of words that capture a topic. Tools that dig more into the meaning of text might be able to capture less precise language and hence be less biased towards more highly educated respondents.

cherry picking – Not only will different algorithms give different results, but each algorithm will have various parameters, such as the sizes or number of layers in a neural network, and often have stochastic aspects to their learning so that many different results are possible from the same underlying data. Who chooses which of these is used? It may be simply luck, but if anyone in the process has a view they wish to advance, it would be easy to choose the tuning parameters that give the most convenient answers and pass these on quite truthfully as a result of (a particular run of) the AI algorithm.

In some way this is merely a next step on from the use of opinion polls and other statistical data in politics and so inherits existing issues including the choice of questions asked, which is known to radically affect the answers given. However, the fact that this is mediated by AI adds the danger that those reading the result assume a level of factuality because it comes from a computer; that is automation bias [63].

The French use of AI was a one-way process from data gathering to summarised themes. However, it is also known that the presence of opinion polls changes voting habits; we are all influenced by other people's views. If the results of public discussion are presented as derived by AI, the automation bias may mean we assume that the results are factual, hence not affected by political pressures and bias; even though it is clear from the above that this may not be the case.

23.6.4 AI and Work

Every few months a new report predicts that whole types of profession from taxi drivers to journalists will be rendered obsolete by AI.

Of course digital technology has eradicated many jobs, for example the 'human computers' who performed scientific calculations at NASA and elsewhere [99], the disappearance of the typing pool or factories peopled only by robots. Further back, the Spinning Jenny replaced handspinners – technology of all kinds brings change.

One reaction has often been resistance. In the mid-1980s newspapers moved from 'hot metal' printing to computerised printing where the laborious hand-layout of type was replaced by journalists keying in their own copy. In the UK this led to an acrimonious dispute and the loss of thousands of jobs as the *Sunday Times* moved production to a new automated press in Wapping [215]. This dispute is reminiscent of the Luddites in the 19th century, who are now synonymous with technophobia and vain resistance to inevitable change.

Another reaction is technological determinism, a form of fatalism, that says technological change is inevitable and unstoppable so that, whether or not it is good, we simply have to learn to live with it. However, this is perhaps even less positive in its outlook than active resistance.

However, there are countervailing arguments. Technological optimists would say that the jobs that are lost are mostly those that are boring, dangerous or dirty. This might mean that humanity's future is one of leisure served by the machines, as E.M. Forster [105] foresaw in "The Machine Stops" (albeit with not totally positive results). Alternatively it may mean new and better jobs, possibly even more jobs. Indeed there are statistics that show that companies that embrace new technology, not least Amazon, grow their labour force ... although this is largely because they grow and displace other companies, the job losses elsewhere are not usually accounted.

Where do you put yourself in this spectrum of responses to AI: resistance, fatalism or welcome?

In principle, if AI enables the same work to be done more efficiently, then there will be more for everyone.

However, we also know that the benefits of technological change are rarely distributed evenly. What is certainly the case is that the nature of work is changing.

Looking back to the Luddites and the Industrial Revolution, there was a similar pattern and similar complexity. Often those working in the factories were the same people who used to work in their homes spinning and weaving. It was clear in the letters between the mill owners that their motives were not so much the efficiency of the technology per se, but the way it gave greater control over the workforce [272]. The much maligned Luddites were not against machines in themselves, but against machines that were "*Harmful to the commonality*", and when they occupied factories, they would only destroy the machines that were most dangerous to the poorly paid and often child labour of the time [188].

It sometimes feels as though we are simply re-iterating the mistakes of the past as most gig-work platforms create precarious jobs and the platform owners deliberately seek to avoid legal and financial responsibility. The uneven playing field puts traditional companies out of business leading to greater control by a few large players. On the other hand, many people have found that the flexibility of gig-work allows them a level of autonomy that they would not experience in traditional employment.

Perhaps we can learn the lessons of the Industrial Revolution without repeating its mistakes, and all be Luddites in the best sense of the word!

23.7 MONEY AND POWER

The lessons of history suggest that the first beneficiaries of any new technology are those who already have power. Certainly this seems to be the case with digital technology where the digital divide is further entrenching existing social inequality. Sometimes this power is governmental and military. Considering the former, there are indeed worries about the use of AI in civil surveillance, albeit often for apparently good reasons such as better policing. On the latter, while this second edition is being written, conflicts rage in Ukraine and Israel, where drones, precision ordnance and cyber-attacks on civic infrastructure are increasingly reliant on AI [21, 67]. These issues are perhaps too raw and too difficult to discuss dispassionately but will only increase in relevance. There are calls to ban the deployment of fully autonomous weapons including from the UN

Secretary General [283]; it might be that by the time you read this there are changes in international law.

The other source of power in market economies is money. This is also a difficult topic, and indeed also costs lives, albeit less clearly accountable to individual acts than a missile strike.

23.7.1 Finance and Markets

The finance industry has always embraced new technology, particularly anything that can help predict markets. When markets are rising, it is easy to make money, you simply buy stock and then sell it later for a profit. Conversely, when markets are falling you need to sell as quickly as possible to avoid making losses. Of course the really difficult thing is knowing when that change will happen. The perfect trade is to buy when stocks are at their minimum and sell just as they are at their maximum and about to fall.

For large long-term investors, these trends may be averaged over months or years, but short traders depend on change over hours, minutes or less, making quick small profits on small movements typically with borrowed capital. Noticing when the market is shifting and then acting quickly is the difference between profit and loss. The speed of these shifts is exacerbated because every other trader is also looking out for then, and so if prices begin to rise or drop, the reactions of others lead to a positive feedback and the changes are sudden and dramatic.

Because of this, many of these short trades depend on automatic algorithms, which of course themselves then make the overall reactions of the market even more rapid and unstable. As well as the large financial investors there are numerous apps offering AI-powered short-trading. Small margins in terms of accuracy and speed of prediction make a huge difference in money. In the past these were driven by fixed rules, then more traditional mathematical models, but of course now many use machine learning.

For each trader the use of automated algorithms and AI is about individually maximising returns; however, the overall impact is to fundamentally *change the market*, increasing the speed and scale of upward and downward shifts. Often, these are relatively small, relatively short and affect only a small number of stocks. However, as with many feedback phenomena, there can be infrequent larger anomalies. On Monday, October 19, 1987, known as Black Monday, a stock market crash wiped more than half a trillion dollars from the Dow Jones alone and is believed to have been caused, in large part, by automatic trading. Although national governments and the major stock exchanges have established various mechanisms to try to prevent such events, since then there have been several 'flash crashes' including in 2010 when the Dow Jones dropped 10% in just ten minutes.

The 1987 crash, which had major economic implications worldwide, way beyond the stock markets themselves, is believed to have been because many hedgefunds used similar algorithms based on the Black–Scholes equation [27]. The algorithms were designed to remove volatility of each fund's own portfolio, but because they all had the same market data and used similar algorithms, they all took the same decisions leading to massive instability. In contrast to these early deterministic algorithms, many machine learning algorithms have some level of indeterminacy in training. While this can be a problem in some fields, such as medicine, where explainability is important, it is possible this may increase diversity of the trading algorithms meaning that there is less tendency to act in tandem. However, ultimately they are trained on the same historic data.

In addition, the competitive nature of stock trading means that small differences in the performance of AI, both in speed and accuracy, can make a disproportionate difference in outcome, leading to an AI arms race. We can see this issue arise in other areas.

23.7.2 Advertising and Runaway AI

In many birds, the males exhibit bold plumage during mating season. The peacock is an extreme example, with huge, extravagantly beautiful tail feathers. Of course, such obvious plumage makes it hard to hide from predators, and the sheer size of the tail hampers escape. From an evolutionary perspective, no matter how attractive it is to look at, this tail seems terribly poorly adapted for survival, hence should never have developed. The Darwinian answer to this conundrum is sexual selection, and this is believed to be the cause of other, apparently maladapted, features of many animals. The idea is that initially fitter males tend to eat better and thus have better plumage. Females seek out the best males and hence choose those with better plumage. Over time males develop better plumage in order to exploit this preference and indeed have to be better than other males.

We can see a similar story in the need for AI in the stock market. If your AI is even a little better than other traders, you make money, if it is less good you lose. The same dynamic is also at play in internet advertising.

Many free services and websites, including most social media, make their money from advertising. These platforms use increasingly sophisticated algorithms in order to target adverts to the most receptive audience at the most opportune time when they will click through and make a purchase (a conversion). Of course large companies will be using detailed data analytics to monitor the success of their adverts on different platforms. If one platform is performing better, in terms of conversions per dollar of advertising spend, they will favour that platform. In the extreme, they may shift nearly all their advertising budget to the most successful platform, only retaining enough on others as a monitor to see if they improve.

If you are the platform, say a social media or web search site, you know that a small difference in AI performance will boost advertising revenues massively. If your competitors have better AI, then you will either lose advertising clients or have to reduce the amount you charge. Some of the improvement in AI can be obtained by recruiting the best AI researchers (good news for readers of this book), but also, as we saw in Chapter 8, sheer size is increasingly regarded as the solution with bigger models, requiring more computation to train and execute, producing better performance. So long as the cost of AI algorithms for selecting advert placement is relatively small compared with other fixed costs of maintaining the site, it is worth spending more and more money and more and more computer time on AI even when the gains are marginal, well beyond their value in attracting, overall, more paying customers.

This is rather like the prisoner's dilemma and similar issues we discussed in Chapter 11, every player in the industry would benefit from a lower level of AI use, but none can individually act. In the end we all end up paying slightly more for goods due to wasted computation. Of course the cost is not merely financial, that compute time also means more carbon emissions.

Happily there are some countervailing effects. Different platforms target slightly different demographics, meaning it is not a simple head-to-head competition, large advertisers may wish to have brand presence across a wide range of platforms not merely the most effective, and platforms can adjust pricing so that they trade off income for AI costs. So the impact of AI is not total, but still the pressures to overuse it are there.

23.7.3 Big AI: The Environment and Social Impact

The resurgence of AI was not fundamentally based on new AI theory or radically novel algorithms. These have come, but the breakthrough was driven by scale. Deep neural networks had been known about since the 1990s, but the larger a network, the more data it needs to train, and the deeper it is, the more computation is needed to tame the underdetermined inner layers (see Chapter 8). The growth of search engines, such as Google search, and also social networks, such as Facebook, gave rise to (a) vast datasets and (b) massive computational power distributed over data centres. This scale driven by internet business enabled the training of large deep neural networks, kick-starting a process that then gave rise to novel paradigms such as adversarial training and transformer networks. Size matters! However, there is a social and environmental cost that comes with this size.

The scale of data and computation arose through the needs of the tech giants such as Google, Facebook and Microsoft. However, the machine learning that has driven the revolution in AI requires such massive levels of investment, that it potentially shuts out any but the largest companies and even most governments. Back in 2014, Ian Bartram of Gartner said:

> I don't know if any public sector has necessarily cracked the nut on attracting the right skills and capabilities, ... The commercial sector has, because they've got the dollars to spend.

This was before the advent of big AI and talking about governments. The prospect for non-governmental organisations, charities, communities or individuals seems bleak ... AI, as with much of digital technology, seems to be entrenching existing power within society and disadvantages the already marginalised.

Scale also has an environmental impact. One estimate predicts that global use of energy for AI will be around 100 terawatt-hours (TWh) by 2027, which is equivalent to the entire energy use of the Netherlands [70]. To put this figure in perspective with other carbon-intensive industries, it is still only about 0.5% of total global electricity use. However, it is also worth noting this is just AI, not all digital technology, and it is still increasing when other uses are reducing in an attempt to meet net-zero carbon targets.

Happily, there are counter trends, especially following Meta's open release of LLaMa 2 (Large Language Model Meta AI)) [274], which enabled small start-ups and researchers to create myriad new applications and extensions [73, 240]. In principle as large language models are foundation models the fully trained model is then specialised in various ways so that the initial (enormous) cost of training is amortised. The specialisation process itself is still way beyond small companies or individuals; however, there are techniques to reduce this technology gap.

Some, such as LiGO (Linear Growth Operator) [296], seek to make the learning process faster. Other techniques focus more on the execution post-training, for example memory layers act as a form of fuzzy key–value mechanism, which can dramatically reduce the computation costs during use [22]. LoRA (Low-rank adaptation of large language models, [138]) takes the inner layers of the deep neural network, where specialisation occurs, and uses a form of dimensional reduction to vastly reduce the number of nodes, while not compromising too severely the overall efficacy of the network (Figure 23.4). This means that the resulting network is both faster to re-train for new applications and faster to execute on smaller processors.

More radically, DeepSeek's mixture-of-experts approach showed that it is possible to create high quality large-language models with a small fraction of the execution costs of previous models [173]. The development of DeepSeek was driven by the technological restrictions imposed by the US export ban on high-end chips to China, but it demonstrates that smart techniques can create effective AI models without simply throwing more and more resources at them.

In general, there is an active research area seeking to reduce the training costs and runtime size of AI, both for environmental reasons and also to enable advanced techniques to be used on tiny IoT (internet of things) devices that are an essential element in smart homes, cities and workplaces.

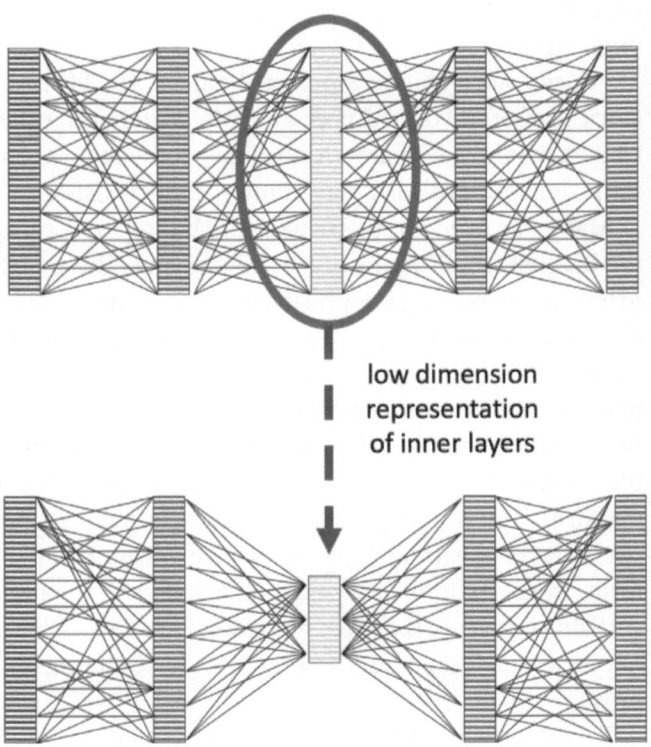

low dimension representation of inner layers

FIGURE 23.4 Reducing the dimensionality of inner layers to reduce re-training and runtime costs.

23.8 SUMMARY

We began this chapter by considering whether machine intelligence is possible, looking in particular at the Chinese Room argument and its opponents. This led to discussing creativity and consciousness, which seem even more intrinsically human than intelligence. The more human-like AI becomes, the more it seems to raise deep philosophical and moral questions. Some are more theoretical, but some of quite immediate relevance as autonomous vehicles drive through our streets, and AI fundamentally shifts patterns of work and society. Considering this, we have looked at issues of legal and ethical responsibility, and the potential social impact of artificial intelligence. It is vital that these questions are tackled if artificial intelligence is to be accepted widely. Perhaps the key question we need to ask ourselves is not whether true machine intelligence is possible but whether and where it is desirable. What do you think?

EXERCISES

This is another chapter where exercises are more discursive, suitable for individual projects or group discussion.

23.1 Collect newspaper reports on social and ethical issues in AI – they appear quite frequently. Examine each with three lenses:

a. The particular human context, has AI made it worse or simply highlighted existing issues?

b. What kind of AI is being used and does it appear sensible or well designed for the purpose?

c. Are there ways you could imagine improving the AI to reduce the problem, or deploying AI to counter it?

23.2 A university is collaborating with a major washing machine manufacturer to optimise the design of washing machine drums, as factory testing never entirely matches real-world performance. The idea is to use a form of AB-testing, where internet-enabled washing machines are delivered with slight variations in the shape of the drum. Performance data will be sent back to the manufacturer enabling near-real-time modifications to the manufacturing process and ever more energy and water efficient washing machines.

Use the questions of the AREA framework (Figure 23.2 in Section 23.5.1) to explore some of the social and ethical implications of this.

23.3 Try variations of the Trolley Problem with friends/colleagues. Things you could experiment with include:

• Different numbers of people, different ages, etc.

• Certainty of effects: what if there were a bridge between the trolley and the people that might collapse with the weight of the trolley before hitting the people.

• The blameworthiness of the people involved, perhaps some are legitimately standing where they are, but others have trespassed onto the tracks even though there are warning signs.

23.4 Design Fiction uses fictitious scenarios in order to explore the potential positive and negative consequences of emerging technology, including AI [181].

a. Forster's *The Machine Stops* [105] was written in 1909, based on future machinery as envisaged then, so could be seen as a form of design fiction from before the term was coined. Read the short story (available on the Internet Archive) and consider yourself or discuss in a group the extent to which the issues in it have parallels today.

b. *The Machine Stops* was written over a hundred years ago, looking forward. Can you do the same, imagine potential just-over-the-horizon AI and its impact. Here are some ideas, but think of your own too.

• Computers are better than humans than all intellectual tasks, but robotics is still struggling so that the only jobs requiring humans are physical ones.

• Artificial companions are being used extensively in care homes.

• AI enables instant fact checking of anything.

FURTHER READING

M. Boden, editor. *The philosophy of artificial intelligence.* Oxford University Press, Oxford, 1990.

A collection of seminal papers on machine intelligence by leaders in the field including Searle and Turing. An excellent and accessible introduction to some of the philosophical issues of AI.

M. Boden. *Artificial intelligence and natural man.* MIT Press, London, 2nd edition, 1987.

Part IV in particular provides a useful survey of the social, psychological and philosophical issues of AI.

M. Wooldridge. *A brief history of artificial intelligence: What it is, where we are, and where we are going.* Flatiron Books, 2021. ISBN: 9781250770745

An overview of the development and history of AI.

A. Dix. *ChatGPT, culture and creativity – simulacrum and alterity.* Keynote at JRL Creative AI Research Conference, 2023, 26 June 2023. https://alandix.com/academic/talks/CAR2023-keynote/

A more detailed exploration of the issues surrounding AI creativity discussed in Section 23.3.

A. Seth. *Being you: A new science of consciousness.* Faber, London, 2021.

Rich discussion of the current state of consciousness research and philosophy.

B. Panic and P. Arthur. *AI for peace*. CRC Press, Boca Raton, FL, 2024.

A rich analysis of the way AI can be used positively to promote peace as well as counter the ways it is being used for the opposite.

C. Crivellaro and A. Dix. *AI for social justice*. CRC Press, Boca Raton, FL, 2025.

Another volume in the AI for Everything series. It expands on many of the issues in Sections 23.6 and 23.7.1; indeed early drafts of AI for Social Justice were used to help write this chapter.

V

Looking Forward

Epilogue: What Next?

24.1 OVERVIEW

The book is almost finished, but AI will continue to develop. In this chapter we take a look at some of the possible future developments of AI, both in terms of technical developments and the way it is used in society. Given the rate of change, asking "what next?" is a risky enterprise, but some near future directions are clear. Ultimately, though, the big questions in AI may be about what we want to do with it and what sort of AI future we want.

24.2 CRYSTAL BALL

It seems a foolish time to ask "what next?" for AI. When the first edition of this book was written, AI was on a downward rollercoaster. Indeed, as previously mentioned, the preface of the 1996 edition of this book said, "... *the subject is far from dead or historical*". Written as the AI winter started to bite, we felt it was necessary to defend even writing a book on AI.

As this edition is written AI is still on a rising curve, with apparently ever accelerating change, to the point that many AI researchers and entrepreneurs are less concerned about whether AI has a future than whether we have any control over it.

It may be that by the time you read this, that future is already with us – AI as saviour, nemesis or dead-end. However, risking instant obsolescence, let's take a tour through a few directions that AI may take over the coming years, social and technical. These are not mutually exclusive alternatives, more that all are likely to happen to a greater or lesser degree.

24.3 WHAT NEXT: AI TECHNOLOGY

24.3.1 Bigger and Better

As noted in Chapters 8 and 17 there has been continual surprise at the qualitative leaps in performance gained by simply throwing more and more computational resources at greater and greater amounts of data to create ever larger models. Before deep learning took centre stage, simple statistical algorithms were delivering 'unreasonably' effective natural language results that would once have been seen as requiring some form of grammatical knowledge [122]. More recently, simply scaling large-language models (LLMs) has led to step changes in behaviour, to the point at which they can perform with apparent human understanding in many tests.

Some believe that this process can continue; simply making larger models will lead to general artificial intelligence. Measured in purely information terms, the brain contains around half a petabyte (see Figure 24.1), which, at the time of writing, is still several hundred times greater than the largest LLMs. It is argued that as LLMs reach these sizes, more human-like features will emerge.

Of course, as models get bigger they need more training data. However, existing models are getting close to consuming all human-generated material. In some domains it is possible for this to be supplemented or even replaced with generated synthetic data, as was the case with AlphaGo and then AlphaZero. However, it is hard to see what the equivalent is for text or art.

In addition, while scale has been important, it has not been the only factor at play. The inclusion of attention

DOI: 10.1201/9781003082880-29

mechanisms in transformer models made fundamental differences in their effectiveness. That is algorithmic and architectural innovation has also been critical in the past and undoubtedly will continue to be significant.

24.3.2 Smaller and Smarter

In Chapter 23, we discussed how the sheer scale of AI creates both environmental and social issues. We also saw that methods are being developed to dramatically reduce the necessary size and training costs of large models and in general develop AI for smaller devices. We can expect to see developments in a number of directions:

- Improvements in large-language models and generative AI with roughly the same behaviour, but faster training.

- Ways to tweak large models after they have been trained to make them faster to specialise.

- Ways to shrink large models after training to reduce the final execution size.

- Ways to use large-scale training to create small-footprint AI directly.

- Alternative methods that do not require such large-scale models or training.

There is ongoing work in each area – alongside scale, the future is lean and mean!

24.3.3 Mix and Match

One of the alternatives to simple scaling is a return to more traditional knowledge-rich methods. This won't be 'business as usual' for traditional AI, there was plenty of time during the long AI winter for that, but lessons learnt from big data and big computation can be brought to bear. It may be that simply having more computational power or more data available can reinvigorate past solutions that were infeasible 20 years ago. However, approaches that combine neural techniques and big data with knowledge-rich methods will undoubtedly become more common. Again there are various ways in which this is developing:

- Using machine learning techniques to craft traditional AI, such as inferring grammar rules from text corpora, or the use of genetic algorithms to create decision trees as described in Chapter 21.

- Hybrid systems combining modules with different styles of AI, as described in Chapter 6, for example, using image classifiers in combination with hand-crafted rules.

- Neurosymbolic techniques, also discussed in Chapter 6, where neural-inspired methods and more high-level reasoning are combined at a more fundamental level, for example, by crafting neurons with specialised behaviour.

One of the critiques of large-language models is their propensity to hallucinate: to invent text that is plausible, yet inaccurate or entirely fabricated. In contrast, traditional planning or mathematical proof systems guarantee to create accurate results but may struggle with more complex problems. The opportunities for these to complement each other are clear.

The joker in the pack is the emergence of quantum computing and other forms of novel computational substrates such as reservoir computing. Google already has a quantum version of its successful TensorFlow framework [33] and research on quantum AI has been growing for a number of years [98]. These new substrates, physical and biological, will certainly work within hybrid digital architectures.

24.3.4 Partners with People

The book began with a reminder that all AI at some level impacts real people. We saw in Chapter 19 that this impact may be quite diffuse as AI invisibly becomes part of the devices we use or the environment in which we live. When more visible, it may act as a tool or servant doing things we tell it to, or, more problematic, AI tells us what to do! However, the real promise may be in systems where humans and AI work synergistically together. The ultimate goals and purposes must be human ones, but the endeavour to achieve them is collaborative.

This requires more human-like AI that can both produce behaviour that is comprehensible to humans and also respond to humans in ways that go beyond simply doing what they are told. Systems such as GitHub Copilot [108] begin to show what is possible, and similar techniques have already been built into conversational agents [236].

> *"Our brains contain about 10 billion neurons, each connected to between 1000 and 10,000 others. It is commonly assumed that our long-term memories are stored in this configuration, both what is connected to what and in the strength of those synaptic connections.*
>
> *If this is the case we can calculate the maximum information content of the human brain! One way to envisage this is as an advanced brain scanner that records the exact configuration of our neurons and synapses at a moment in time – how much memory would it take to store this?*
>
> *For each neuron we would need to know physically where it is, but these x,y,z coordinates for each neuron turn out to be the least of the memory requirements needing a mere 90 bits giving us a 1 in billion accuracy for each coordinate. That is about 120 megabytes for all the neurons.*
>
> *The main information is held, as noted, in what is connected to what. To record this digitally we would need to have for each of the 10,000 synapses of each neuron a 'serial number' for the neuron it connects to and a strength. Given 10 billion neurons this serial number would need to be 34 bits and so if we store the synapse strength using 6 bits (0–63), this means 40 bits or 5 bytes per synapse, so 50,000 bytes per neuron and 500 thousand billion bytes for the whole brain state.*
>
> *That is the information capacity of the brain is approximately 500 terrabytes or $\frac{1}{2}$ a petabyte."*

Extract from 'The brain and the web' [76]

Since this was written in 2005 there have been efforts to map the neurone connection patterns within the human brain. There has been mixed success [203], but the results include 'brain atlases' for various animal species at different levels of detail which are used for both medical purposes and more basic neuroscience research [12].

FIGURE 24.1 Brain sized.

24.4 WHAT NEXT: AI IN THE WORLD

24.4.1 Friend or Foe?

While this edition was being prepared, the European Union and the governments of 27 countries including the UK, the USA and China signed the Bletchley Declaration following a two-day summit on AI safety [264, 280]. The declaration acknowledged the positive aspects of AI but also warned of a *"potential for serious, even catastrophic, harm, either deliberate or unintentional, stemming from the most significant capabilities of these AI models"*. This echoes warnings from within the academic AI community [36].

In the last chapter, we looked at many of the social and ethical issues around the growing use of AI: job losses, misinformation and the undermining of human creativity. These are current issues and will continue to be so for years to come. There is the promise that AI may do away with all unfulfilling labour and give us all, in Elon Musk's words, a "universal high income" [187]. However, given the history of adoption of technology in market economies, and the nature of transnational companies ... that future seems least likely.

Some worry about more existential risks – as AI designs better AI we will come to the point when its growth is uncontrollable, leading to a singularity when AI so far exceeds human intelligence that humanity becomes at best an irrelevance to be brushed away, and at worst an impediment to be crushed. This seems less science fiction than it did a few years ago and, in true B-movie style, there are those who are welcoming the alien invader as the next stage in the evolution of intelligence. While there are clearly more pressing problems, this is an issue that will continue to create headlines, but also, at the most pragmatic level, cannot be utterly ignored.

Fans of the *Terminator* films will recall that Skynet was created for military use before it decided to destroy humanity to protect itself. However, apocalyptic scenarios do not require sentient AI. We are already seeing AI used in guided weaponry, and while there are moves to prevent fully autonomous weapons, it seems unlikely this will be averted. Of course AI is used to design new military hardware and, perhaps more worrying, AI has already been shown to be capable of designing new chemical and biological weapons. Certainly AI is being used to develop and deploy cyberattacks, both military and criminal.

Most likely, in terms of utter apocalypse, is the use of AI to initiate the overall firing of weapons, especially nuclear missiles. There is widespread desire to retain human control, not least political control, but, as successive nuclear control treaties have expired or collapsed, it may be hard for countries to avoid installing 'use it or lose it' hair triggers if they fear the other side has already done so. The issue here is not so much AI itself (indeed the film *Dr Strangelove* predicted just such a hair trigger using 1960s technology), but more about conflict resolution and de-escalation. Dealing with more prosaic issues around filter bubbles and echo chambers is perhaps at least as important here as the direct military threat.

Indeed, we are facing so many problems in national and global society: climate change, growing inequality, deepening political division, extremism and the likelihood of a next pandemic. In some cases AI is currently exacerbating these problems but could be turned for good. In others there is real hope that AI may make significant advances, for example, in climate prediction models or the rapid creation of new vaccines.

24.4.2 Boom then Bust

It may be that the current surge in interest in AI will come to an end and there will be another AI winter. However, this seems unlikely. The surge in AI interest in the early 1990s was driven by some early research success and an aspiration that outstripped the state of maturity. In contrast, the current wave is being driven by real applications that are delivering business benefits in numerous fields. Some of the high hopes may not be fully achieved, but it seems unlikely that there will be a complete collapse.

In addition, while the grand narrative of waves of AI interest interspersed with long winters makes good reading, the reality is more nuanced. During the period from mid-1990s to 2010, AI researchers would mourn the state of the discipline, but meanwhile there were gradual developments in areas such as natural language processing, text mining, speech recognition and computer vision, which in the 1980s would have been regarded as major goals of AI. In addition the growth of web and big data algorithms such as Google PageRank and recommender systems were clearly 'intelligent' if not card-carrying AI.

So, while there may well be a deceleration or even a hiatus in the rise of AI, a collapse is not imminent.

24.4.3 Everywhere and Nowhere

The earliest commercial engines were enormous, sitting in factories, belching smoke and driving rumbling machinery, but within a hundred years became part of the background of society with mass-produced cars and now the electric toothbrush on your bathroom shelf. Similarly the earliest digital mainframe computers sat in large glass computer rooms, served by white-coated technicians, but within 30 years were on every desk and now in tags on fast-fashion clothes. As Weiser put it,

> *The most profound technologies are those that disappear. They weave themselves into the fabric of everyday life until they are indistinguishable from it.* [298]

Similarly, novel developments in AI are headline grabbing and require large budgets: Big Blue defeating Kasparov, Watson winning Jeopardy! or AlphaZero mastering Go. However, we are already so used to AI in recommender systems, email filtering and voice-controlled home automation, it too has become invisible.

The future of pervasive AI may be that we don't even know it is there.

24.5 SUMMARY – FROM HYPE TO HOPE

Artificial intelligence is at an interesting juncture; there are many directions in which it may develop, both technically and, more critically, in the way we use it in society. You may be one of those who develop the algorithms that take the next steps in AI. Certainly you will be a user of AI, no matter how invisible. Within AI as a research and development discipline and within society at large, there are crucial choices to be made. We are all part of the debate as to whether AI becomes yet another technology that divides, or whether it can be harnessed to serve humanity as a whole.

What's next for AI?

The choice is in your hands.

FURTHER READING

S. Russell, R. Rossi, and M. Schönstein, co-chairs. *OECD working party and network of experts on AI*, 2024. https://oecd.ai/en/network-of-experts/working-group/10847

The OECD run periodic workshops of global experts in AI to offer an informed glimpse into the future.

B. Panic and P. Arthur. *AI for peace.* CRC Press, Boca Raton, FL, 2024.

H. S. Sætra. *AI for the sustainable development goals.* CRC Press, Boca Raton, FL, 2022.

Thinking about the future of AI is only meaningful if we have a future. These two volumes address key issues where AI can have a role ... for good or ill ... in determining the future of humankind and the world we live in.

Bibliography

[1] David H. Ackley, Geoffrey E. Hinton, and Terrence J. Sejnowski. A learning algorithm for Boltzmann machines. *Cognitive Science*, 9(1):147–169, 1985.

[2] Elena Agliari, Francesco Alemanno, Adriano Barra, and Alberto Fachechi. Dreaming neural networks: Rigorous results. *Journal of Statistical Mechanics: Theory and Experiment*, 2019(8):083503, 2019.

[3] Christopher Ahlberg and Ben Shneiderman. Visual information seeking: Tight coupling of dynamic query filters with starfield displays. In *Proceedings of the SIGCHI conference on human factors in computing systems, CHI '94*, page 313–317. Association for Computing Machinery, New York, NY, 1994.

[4] I. Aleksander and T. J. Stonham. Guide to pattern recognition using random-access memories. *Proceedings of the IEE: Computers and Digital Techniques*, 2(1):42–49, 1979.

[5] J. R. Anderson. *The architecture of cognition*. Harvard University Press, Cambridge, MA, 1983.

[6] John R. Anderson. *How can the human mind occur in the physical universe?* Oxford University Press, Oxford, 2009.

[7] Julia Angwin, Jeff Larson, Surya Mattu, and Lauren Kirchner. Machine bias there's software used across the country to predict future criminals: And it's biased against blacks. *ProPublica* (23 May 2016). https://www.propublica.org/article/machine-bias-risk-assessments-in-criminal-sentencing.

[8] R. C. Arkin, W. M. Carter, and D. C. Mackenzie. Active avoidance: Escape and dodging behaviors for reactive control. In H. I. Christensen, K. W. Bowyer, and H. Bunke, editors, *Active robot vision: Camera heads, model based navigation and reactive control*. World Scientific, Singapore, 1993.

[9] Stavros Asimakopoulos, Robert Fildes, and Alan Dix. Forecasting software visualizations: An explorative study. In *People and computers XXIII celebrating people and technology*, pages 269–277. BCS, Swindon, 2009.

[10] I. Asimov. *I, Robot*. Panther, London, 1968.

[11] Bernard J. Baars, Natalie Geld, and Robert Kozma. Global workspace theory (gwt) and prefrontal cortex: Recent developments. *Frontiers in Psychology*, 12:749868, 2021.

[12] Rembrandt Bakker, Paul Tiesinga, and Rolf Kötter. The scalable brain atlas: Instant web-based access to public brain atlases and related content. *Neuroinformatics*, 13:353–366, 2015.

[13] David L. Barack, Vera U. Ludwig, Felipe Parodi, Nuwar Ahmed, Elizabeth M. Brannon, Arjun Ramakrishnan, and Michael Platt. Attention deficits linked with proclivity to explore while foraging. *Biological Sciences*, 291(2017), Article 202222584, 2024.

[14] V. E. Barker and D. E. O'Connor. Expert systems for configuration at DIGITAL: XCON and beyond. *Communications of the ACM*, 32(3):298–318, 1989.

[15] B. Barrett. Google maps is racist because the internet is racist. *Wired* (May 23, 2015).

[16] N. Baym, L. Shifman, C. Persaud, and K. Wagman. Intelligent failures: Clippy memes and the limits of digital assistants. *AoIR Selected Papers of Internet Research*, 2019. https://doi.org/10.5210/spir.v2019i0.10923.

[17] BBC. Google's AI wins final Go challenge. *BBC News* (15 March 2016).

[18] BBC. BBC pay: Male stars earn more than female talent. *BBC News* (19 July 2017).

[19] BBC. Go master quits because AI 'cannot be defeated'. *BBC News* (27 November 2019).

[20] R. Beale and T. Jackson. *Neural computing: An introduction*. Adam Hilger, Bristol, 1990.

[21] Samuel Bendett. Roles and implications of AI in the Russian–Ukrainian conflict. *Russia Matters*, Harvard Kennedy School (20 July 2023). https://www.russiamatters.org/analysis/roles-and-implications-ai-russian-ukrainian-conflict accessed 02/12/2023.

[22] V. P. Berges, B. Oğuz, D. Haziza, W. T. Yih, L. Zettlemoyer, and G. Gosh. Memory layers at scale. *arXiv preprint arXiv:2412.09764*, 2024.

[23] Ofer Bergman and Steve Whittaker. *The science of managing our digital stuff*. MIT Press, Cambridge, MA, 2016.

[24] Tim Berners-Lee, Robert Cailliau, Ari Luotonen, Henrik Frystyk Nielsen, and Arthur Secret. The world-wide web. *Communications of the ACM*, 37(8):76–82, 1994.

[25] Tim Berners-Lee, James Hendler, and Ora Lassila. The semantic web. *Scientific American*, 284(5):34–43, 2001.

[26] Andrzej Bieszczad and Bernard Pagurek. Neurosolver: Neuromorphic general problem solver. *Information Sciences*, 105(1–4):239–277, 1998.

[27] Fischer Black and Myron Scholes. The pricing of options and corporate liabilities. *Journal of Political Economy*, 81(3):637–654, 1973.

[28] Y. Bob. Ex-gov't agent: Crisis worse than 9/11 could emerge from AI arms race. *Jerusalem Post* (12 Feburary 2019).

[29] M. A. Boden. *Artificial intelligence and natural man*. MIT Press, London, 2nd edition, 1987.

[30] Nazim Bouatta, Peter Sorger, and Mohammed AlQuraishi. Protein structure prediction by AlphaFold2: Are attention and symmetries all you need? *Acta Crystallographica Section D: Structural Biology*, 77(8):982–991, 2021.

[31] Leo Breiman. Random forests. *Machine Learning*, 45(1):5–32, 2001.

[32] Sergey Brin and Lawrence Page. The anatomy of a large-scale hypertextual web search engine. *Computer Networks and ISDN Systems*, 30(1–7):107–117, 1998.

[33] Michael Broughton, Guillaume Verdon, Trevor McCourt, Antonio J. Martinez, Jae Hyeon Yoo, Sergei V. Isakov, Philip Massey, Ramin Halavati, Murphy Yuezhen Niu, Alexander Zlokapa, et al. TensorFlow Quantum: A software framework for quantum machine learning. *arXiv preprint arXiv:2003.02989*, 2020.

[34] Tom Brown, Benjamin Mann, Nick Ryder, Melanie Subbiah, Jared D. Kaplan, Prafulla Dhariwal, Arvind Neelakantan, Pranav Shyam, Girish Sastry, Amanda Askell, et al. Language models are few-shot learners. *Advances in Neural Information Processing Systems*, 33:1877–1901, 2020.

[35] Mark Buchanan. *Ubiquity: The science of history, or why the world is simpler than we think*. Weidenfeld & Nicolson, London, 2000.

[36] Benjamin S. Bucknall and Shiri Dori-Hacohen. Current and near-term AI as a potential existential risk factor. In *Proceedings of the 2022 AAAI/ACM conference on AI, ethics, and society*. ACM, New York, NY, July 2022.

[37] R. R. Burton. *Semantic grammar: An engineering technique for constructing natural language understanding systems*. Report No. 3453. Bolt Beranek and Newman, Boston, MA, 1976.

[38] James Cameron and William Wisher. *Terminator 2: Judgment day*. TriStar Pictures, Culver City, CA, 1991.

[39] Chunshui Cao, Yongzhen Huang, Zilei Wang, Liang Wang, Ninglong Xu, and Tieniu Tan. Lateral inhibition-inspired convolutional neural network for visual attention and saliency detection. In *Proceedings of the AAAI conference on artificial intelligence*, volume 32, 2018. AAAI Press, Palo Alto, CA.

[40] Karel Čapek. *R.U.R. (Rossum's Universal Robots)*. Samuel French, Inc., 1923. English version by Paul Selver and Nigel Playfair. Original Czech play 1921, "Rossumovi Univerzální Roboti". https://gutenberg.org/ebooks/59112.

[41] J. Carpenter. Google's algorithm shows prestigious job ads to men, but not to women. *Independent* (7 July 2015).

[42] CERN. *Storage: What data to record?* CERN, 2021. https://home.cern/science/computing/storage accessed 01/12/2024.

[43] David J. Chalmers. Facing up to the problem of consciousness. *Journal of Consciousness Studies*, 2(3):200–219, 1995.

[44] Angie Chandler, Joe Finney, Carl Lewis, and Alan Dix. Toward emergent technology for blended

public displays. *Proceedings of the 11th International Conference on Ubiquitous Computing*, pages 101–104. ACM, New York, NY, 2009.

[45] Varun Chandrasekaran, Chuhan Gao, Brian Tang, Kassem Fawaz, Somesh Jha, and Suman Banerjee. Face-off: Adversarial face obfuscation. *Proceedings on Privacy Enhancing Technologies* 2021(2):369-390, 2021.

[46] E. Charniak. *Towards a model of children's story comprehension*. Report No. TR-266, AI Laboratory. MIT, Cambridge, MA, 1972.

[47] Kumar Chellapilla and David B. Fogel. Evolving neural networks to play checkers without relying on expert knowledge. *IEEE Transactions on Neural Networks*, 10(6):1382–1391, 1999.

[48] Mark Chen, Jerry Tworek, Heewoo Jun, Qiming Yuan, Henrique Ponde de Oliveira Pinto, Jared Kaplan, Harri Edwards, Yuri Burda, Nicholas Joseph, Greg Brockman, et al. Evaluating large language models trained on code. *arXiv preprint arXiv:2107.03374*, 2021 https://arxiv.org/abs/2107.03374.

[49] Jessica Cherner. The Eiffel Tower is now 20 feet taller. *Architectural Digest* (17 March 2022). https://www.architecturaldigest.com/story/eiffel-tower-20-feet-taller.

[50] Leon Chua. Memristor-the missing circuit element. *IEEE Transactions on Circuit Theory*, 18(5):507–519, 1971.

[51] Citizens Advice Bureau. *22% don't have basic banking services needed to deal with universal credit*. Technical Report, Citizens Advice Bureau, 2013. https://www.citizensadvice.org.uk/cymraeg/amdanom-ni/about-us1/media/press-releases/22-don-t-have-basic-banking-services-needed-to-deal-with-universal-credit/ accessed 01/12/2024.

[52] H. H. Clark and S. E. Brennan. Grounding in communication. In L. B. Resnick, J. Levine, and S. D. Behrend, editors, *Perspectives on socially shared cognition*, pages 127–149. American Psychological Association, Washington, DC, 1991.

[53] A. J. G. Cockburn. *Groupware design: Principles, prototypes and systems*. PhD thesis, University of Stirling, 1993.

[54] Kristin A. Cook and James J. Thomas. *Illuminating the path: The research and development agenda for visual analytics*. Technical Report. National Visualization and Analytics Center (NVAC), Pacific Northwest National Lab (PNNL), Richland, WA, 2005. https://www.pnnl.gov/publications/illuminating-path-research-and-development-agenda-visual-analytics.

[55] Rémi Coulom. Efficient selectivity and backup operators in Monte-Carlo tree search. In *International conference on computers and games*, pages 72–83. Springer, Berlin, 2006. http://www.europarl.europa.eu/sed/doc/news/document/CONS_CONS(2016)05418(REV1)_EN.docx/ accessed 01/12/2024.

[56] Council of the European Union. *Position of the Council on General Data Protection Regulation*. Technical Report. Council of the European Union, 8 April 2016. http://www.europarl.europa.eu/sed/doc/news/document/CONS_CONS(2016)05418(REV1)_EN.docx/ accessed 01/12/2024.

[57] Andy Crabtree, Lachlan Urquhart, and Jiahong Chen. *Right to an explanation considered harmful*. Technical Report. Edinburgh School of Law Research Paper, Edinburgh, 8 April 2019.

[58] Miles Cranmer, Alvaro Sanchez Gonzalez, Peter Battaglia, Rui Xu, Kyle Cranmer, David Spergel, and Shirley Ho. Discovering symbolic models from deep learning with inductive biases. In H. Larochelle, M. Ranzato, R. Hadsell, M.F. Balcan, and H. Lin, editors, *Advances in neural information processing systems*, volume 33, pages 17429–17442. Curran Associates, Inc., 2020.

[59] Francis Crick and Graeme Mitchison. The function of dream sleep. *Nature*, 304:111–114, 1983.

[60] Clara Crivellaro and Alan Dix. *AI and social justice: From avoiding harms to positive action*. The AI Summit, New York, 8 December 2021. https://www.alandix.com/academic/talks/AI-Summit-NY-2021-AISJ/.

[61] Florinel-Alin Croitoru, Vlad Hondru, Radu Tudor Ionescu, and Mubarak Shah. Diffusion models in vision: A survey. *IEEE Transactions on Pattern Analysis and Machine Intelligence*, 45(9):10850–10869, 2023.

[62] Mary L. Cummings. Automation bias in intelligent time critical decision support systems. In

AIAA 1st intelligent systems technical conference, 2004. Chicago, IL, 20–22 September 2004.

[63] Mary L. Cummings. Automation bias in intelligent time critical decision support systems. In D. Harris, editor, *Decision making in aviation*, pages 289–294. Routledge, Abingdon, 2017.

[64] Allen Cypher. Eager: Programming repetitive tasks by example. In *Proceedings of the SIGCHI conference on human factors in computing systems, CHI '91*, page 33–39. Association for Computing Machinery, Ney York, NY, 1991.

[65] Antonio R. Damasio. *Descartes' error.* Putnam, New York, NY, 1994.

[66] A. Datta, M. Tschantz, and A. Datta. Automated experiments on ad privacy settings. In *Proceedings on privacy enhancing technologies*, pages 92–112. De Gruyter, Berlin, 2015.

[67] Harry Davies, Bethan McKernan, and Dan Sabbagh. 'The Gospel': How Israel uses AI to select bombing targets in Gaza. *The Guardian* (1 December 2023).

[68] Ernest Davis and Gary Marcus. Commonsense reasoning and commonsense knowledge in artificial intelligence. *Communications of the ACM*, 58(9):92–103, August 2015.

[69] Wiebren De Jonge. Compromising statistical databases responding to queries about means. *ACM Transactions on Database Systems (TODS)*, 8(1):60–80, 1983.

[70] Alex de Vries. The growing energy footprint of artificial intelligence. *Joule*, 7(10):2191–2194. https://doi.org/10.1016/j.joule.2023.09.004, 2023.

[71] Jeffrey Dean and Sanjay Ghemawat. MapReduce: Simplified data processing on large clusters. *Commun: ACM*, 51(1):107–113, January 2008.

[72] R. Descartes. *A discourse on method*, page 1637. Project Gutenberg EBook, 2008. http://www.gutenberg.org/files/59/59-h/59-h.htm#part4.

[73] Ben Dickson. Can large language models be democratized? *TechTalks* (16 May 2022). https://bdtechtalks.com/2022/05/16/opt-175b-large-language-models/.

[74] A. Dix. Human issues in the use of pattern recognition techniques. In R. Beale and J. Finlay, editors, *Neural networks and pattern recognition in human computer interaction*, pages 429–451. Ellis Horwood Hemel Hempstead, 1992.

[75] A. Dix. Interactive querying - locating and discovering information. In *Second workshop on information retrieval and human computer interaction*. Paper presented at workshop without subsequent proceedings, Glasgow, 11 September 1998. https://www.alandix.com/academic/papers/IQ98/ accessed 01/12/2024.

[76] A. Dix, J. Finlay, G. Abowd, and R. Beale. *Human-computer interaction*. Prentice Hall, Hemel Hempstead, 1993.

[77] A. Dix and A. Patrick. Query by Browsing. In P. Sawyer, editor, *Proceedings of IDS'94: The 2nd international workshop on user interfaces to databases*, pages 236–248. Springer Verlag, Lancaster, 1994.

[78] Alan Dix, Russell Beale and Andy Wood. Architectures to make simple visualisations using simple systems. In *Proceedings of the working conference on Advanced Visual Interfaces (AVI 2000)*, pages 51–60, 2000.

[79] Alan Dix. Beyond intention – pushing boundaries with incidental interaction. In *Proceedings of Building bridges: Interdisciplinary context-sensitive computing*. Paper presented at workshop without subsequent proceeding, Glasgow University, 2002. https://alandix.com/academic/papers/beyond-intention-2002/ accessed 01/12/2024.

[80] Alan Dix. The brain and the web: A quick backup in case of accidents, 29 August 2005. https://www.alandix.com/academic/papers/brain-and-web-2005/ accessed 21/11/2023.

[81] Alan Dix. Designing for appropriation. In *Proceedings of the 21st British HCI group annual conference on people and computers: HCI...but not as we know it - volume 2, BCS-HCI '07*, page 27–30. BCS Learning & Development Ltd, Swindon, GBR, 2007.

[82] Alan Dix. I in an other's eye. *AI and Society*, 34(1):55–73, 2019.

[83] Alan Dix. *Statistics for HCI: Making sense of quantitative data.* Morgan & Claypool, 2020. DOI:10.2200/S00974ED1V01Y201912HCI044.

[84] Alan Dix. ChatGPT, culture and creativity – simulacrum and alterity. Keynote at JRL Creative AI

Research Conference 2023, 26 June 2023. https://alandix.com/academic/talks/CAR2023-keynote/.

[85] Alan Dix, Rachel Cowgill, Christina Bashford, Simon McVeigh, and Rupert Ridgewell. Authority and judgement in the digital archive. In *Proceedings of the 1st international workshop on digital libraries for musicology (DLfM'14)*. Association for Computing Machinery, New York, NY, 2014.

[86] Alan Dix, Rachel Cowgill, Christina Bashford, Simon McVeigh, and Rupert Ridgewell. Spreadsheets as user interfaces. In *Proceedings of the international working conference on advanced visual interfaces (AVI'16)*, page 192–195. Association for Computing Machinery, New York, NY, 2016.

[87] Alan Dix and Geoffrey Ellis. Starting simple: Adding value to static visualisation through simple interaction. In *Proceedings of the working conference on Advanced Visual Interfaces (AVI'98)*, pages 124–134, 1998. https://alandix.com/academic/papers/simple98/.

[88] Alan Dix, Janet Finlay, Gregory D. Abowd, and Russell Beale. *Human-computer interaction*. Pearson Education, 2003. https://hcibook.com/.

[89] Alan Dix, Akrivi Katifori, Giorgos Lepouras, Costas Vassilakis, and Nadeem Shabir. Spreading activation over ontology-based resources: From personal context to web scale reasoning. *International Journal of Semantic Computing*, 4(01):59–102, 2010.

[90] Alan Dix and Genovefa Kefalidou. Regret from cognition to code. In Antonio Cerone, Marco Autili, Alessio Bucaioni, Cláudio Gomes, Pierluigi Graziani, Maurizio Palmieri, Marco Temperini, and Gentiane Venture, editors, *Software engineering and formal methods. SEFM 2021 collocated workshops*, pages 15–36. Springer International Publishing, Cham, 2022.

[91] Alan J Dix. Information processing, context and privacy. In *Proceedings of INTERACT'90*, pages 15–20. IFIP, 1990. https://alandix.com/academic/papers/int90/.

[92] J. Doyle. A truth maintenance system. *Artificial Intelligence*, 12(3):232–272, 1979.

[93] H. Dreyfus. *What computers can't do*. Harper and Row, New York, 2nd edition, 1979.

[94] Emily Dreyfuss and Issie Lapowsky. Facebook is changing news feed (again) to stop fake news. *Wired*, 2019.

[95] R. O. Duda, J. Gaschnig, and P. E. Hart. Model design in the PROSPECTOR consultant system for mineral exploration. In D. Michie, editor, *Expert systems in the micro-electronic age*. Edinburgh University Press, Edinburgh, 1979.

[96] Susan T. Dumais et al. Latent semantic analysis. *Annual Review of Information Science and Technology*, 38(1):188–230, 2004.

[97] Pamela Duncan, Niamh McIntyre, Rhi Storer, and Cath Levett. Who won and who lost: when A-levels meet the algorithm. *The Guardian* (13 August 2020).

[98] Vedran Dunjko and Hans J Briegel. Machine learning & artificial intelligence in the quantum domain: A review of recent progress. *Reports on Progress in Physics*, 81(7):074001, 2018.

[99] Sue Bradford Edwards and Duchess Harris. *Hidden human computers: The black women of NASA*. Hidden Heroes. Essential Library Dallas, TX, 2016.

[100] Andreas K. Engel and Pascal Fries. Chapter 3 - neuronal oscillations, coherence, and consciousness. In Steven Laureys, Olivia Gosseries, and Giulio Tononi, editors, *The neurology of conciousness*, pages 49–60. Academic Press, San Diego, 2nd edition, 2016.

[101] David Ferrucci, Eric Brown, Jennifer Chu-Carroll, James Fan, David Gondek, Aditya A. Kalyanpur, Adam Lally, J. William Murdock, Eric Nyberg, John Prager, et al. Building Watson: An overview of the DeepQA project. *AI Magazine*, 31(3):59–79, 2010.

[102] C. Fillmore. The case for case. In E. Bach and R. T. Harms, editors, *Universals in linguistic theory*. Holt, New York, 1968.

[103] O. Firschein, M. A. Fischler, L. S. Coles, and J. M. Tenenbaum. Forecasting and assessing the impact of artificial intelligence on society. In *IJCAI-3*, pages 105–120. Morgan Kaufmann, San Francisco, CA, 1974.

[104] David V. Ford, Kerina H. Jones, Jean-Philippe Verplancke, Ronan A. Lyons, Gareth John, Ginevra Brown, Caroline J. Brooks, Simon

Thompson, Owen Bodger, Tony Couch, et al. The SAIL Databank: Building a national architecture for e-health research and evaluation. *BMC Health Services Research*, 9(1):1–12, 2009.

[105] E. M. Forster. The machine stops. 1909. *The Oxford and Cambridge Review* (November 1909). https://archive.org/details/e.-m.-forster-the-machine-stops_202008/.

[106] Marcus Foth, Martin Tomitsch, Laura Forlano, M. Hank Haeusler, and Christine Satchell. Citizens breaking out of filter bubbles: Urban screens as civic media. In *Proceedings of the 5th ACM international symposium on pervasive displays*, pages 140–147. ACM, New York, NY, 2016.

[107] N. Friedman, M. Linial, I. Nachman, and D. Pe'er. Using Bayesian networks to analyze expression data. *Journal of Computational Biology*, 7(3–4):601–620, 2000.

[108] Nat Friedman. Introducing GitHub Copilot: Your AI pair programmer, 2021. https://github.com/features/copilot/.

[109] Takuya Fukushima, Tomoharu Nakashima, and Hidehisa Akiyama. Evaluation-function modeling with multi-layered perceptron for RoboCup soccer 2D simulation. *Artificial Life and Robotics*, 25(3):440–445, 2020.

[110] Simson Garfinkel, Jeanna Matthews, Stuart S. Shapiro, and Jonathan M. Smith. Toward algorithmic transparency and accountability, 2017.

[111] Kiran Garimella, Gianmarco De Francisci Morales, Aristides Gionis, and Michael Mathioudakis. Political discourse on social media: Echo chambers, gatekeepers, and the price of bipartisanship. In *Proceedings of the 2018 world wide web conference*, pages 913–922. ACM, New York, NY, 2018.

[112] F. H. George. *The brain as a computer*. Pergamon Press, Oxford, 1961.

[113] Peter A. Getting. Emerging principles governing the operation of neural networks. *Annual Review of Neuroscience*, 12(1):185–204, 1989.

[114] B. Gholipour. We need to open the AI black box before it's too late: If we don't, the biases of our past could dictate our future. *Futurism* (18 January 2018).

[115] P. Ghosh. AAAS: Machine learning 'causing science crisis'. *BBC News* (16 February 2019):99–99.

[116] Ian Goodfellow, Jean Pouget-Abadie, Mehdi Mirza, Bing Xu, David Warde-Farley, Sherjil Ozair, Aaron Courville, and Yoshua Bengio. Generative adversarial networks. *Communications of the ACM*, 63(11):139–144, 2020.

[117] H. P. Grice. Logic and conversation. In P. Cole and J. Morgan, editors, *Studies in syntax and semantics III: Speech acts*, pages 183–198. Academic Press, New York, 1975.

[118] Guardian. Microsoft 'deeply sorry' for racist and sexist tweets by AI chatbot. *The Guardian* (26 March 2016).

[119] Asela Gunawardana, Tim Paek, and Christopher Meek. Usability guided key-target resizing for soft keyboards. In *Proceedings of the 15th international conference on Intelligent user interfaces*, pages 111–118. ACM, New York, NY, 2010.

[120] Alfréd Haar. Zur Theorie der orthogonalen Funktionensysteme. *Mathematische Annalen*, 71(1):38–53, 1910.

[121] Elizabeth L. Haines, Kay Deaux, and Nicole Lofaro. The times they are a-changing… or are they not? A comparison of gender stereotypes, 1983–2014. *Psychology of Women Quarterly*, 40(3):353–363, 2016.

[122] Alon Halevy, Peter Norvig, and Fernando Pereira. The unreasonable effectiveness of data. *IEEE Intelligent Systems*, 24(2):8–12, 2009.

[123] R. Hall. Computational approaches to analogical reasoning. *Artificial Intelligence*, 39(1):39–120, 1989.

[124] S. Hawking, E. Musk, S. Wozniak, et al. *Autonomous weapons: An open letter from AI & robotics researchers*. Technical Report 99. Future of Life Institute, 2015.

[125] Jeff Hawkins, Subutai Ahmad, and Yuwei Cui. A theory of how columns in the neocortex enable learning the structure of the world. *Frontiers in Neural Circuits*, 11:81, 2017.

[126] Christian Heath and Paul Luff. Collaboration and control crisis management and multimedia technology in London Underground line control rooms. *Computer Supported Cooperative Work (CSCW)*, 1:69–94, 1992.

[127] Julian Heinrich and Daniel Weiskopf. State of the art of parallel coordinates. In M. Sbert and

L. Szirmay-Kalos, editors, *Eurographics 2013 - state of the art reports*. The Eurographics Association, Eindhoven, 2013.

[128] James Hendler. Avoiding another AI winter. *IEEE Intelligent Systems*, 23(2):2–4, March 2008.

[129] James Hendler, Nigel Shadbolt, Wendy Hall, Tim Berners-Lee, and Daniel Weitzner. Web science: An interdisciplinary approach to understanding the web. *Communications of the ACM*, 51(7):60–69, July 2008.

[130] J. Henrich, S. Heine, and A. Norenzayan. The weirdest people in the world? *Behavioral and Brain Sciences*, 33(2–3):61–83, 2010.

[131] Gary Higgs, Andrew Price, and Mitchel Langford. Investigating the impact of bank branch closures on access to financial services in the early stages of the COVID-19 pandemic. *Journal of Rural Studies*, 95:1–14, 2022.

[132] Geoffrey E. Hinton. Learning multiple layers of representation. *Trends in Cognitive Sciences*, 11(10):428–434, 2007.

[133] Geoffrey E. Hinton and Ruslan R. Salakhutdinov. Reducing the dimensionality of data with neural networks. *Science*, 313(5786):504–507, 2006.

[134] Tin Kam Ho. Random decision forests. In *Proceedings of 3rd international conference on document analysis and recognition*, volume 1, pages 278–282. IEEE, Los Alamitos, CA, 1995.

[135] J. K. Hodgins, W. L. Wooten, D. C. Brogan, and J. F. O'Brien. Animating human athletics. In *Proceedings of SIGGRAPH95*. ACM, New York, NY, 1995.

[136] Johannes Hoffart, Fabian M. Suchanek, Klaus Berberich, and Gerhard Weikum. Yago2: A spatially and temporally enhanced knowledge base from Wikipedia. *Artificial Intelligence*, 194:28–61, 2013. Artificial Intelligence, Wikipedia and Semi-Structured Resources.

[137] John J. Hopfield, David I. Feinstein, and Richard G Palmer. 'Unlearning' has a stabilizing effect in collective memories. *Nature*, 304:158–159, 1983.

[138] Edward J. Hu, Yelong Shen, Phillip Wallis, Zeyuan Allen-Zhu, Yuanzhi Li, Shean Wang, Lu Wang, and Weizhu Chen. LoRA: Low-rank adaptation of large language models. *arXiv preprint arXiv:2106.09685*, 2021.

[139] E. Hutchins. The technology of team navigation. In J. Gallagher, R. Kraut, and C. Egido, editors, *Intellectual teamwork: Social and technical bases of collaborative work*. Lawrence Erlbaum Associates, Hillsdale, NJ, 1990.

[140] Matthew Hutson. AI learns the art of diplomacy. *Science (New York, NY)*, 378(6622):818–818, 2022.

[141] IFTTT. What is IFTTT? If this then that, 2023. https://ifttt.com/explore/new_to_ifttt accessed 01/01/2023.

[142] Information Commissioner's Office. *Introduction to anonymisation: Draft anonymisation, pseudonymisation and privacy enhancing technologies guidance, 2021*. Information Commissioner's Office May 2021, https://ico.org.uk/media/about-the-ico/consultations/2619862/anonymisation-intro-and-first-chapter.pdf.

[143] Eugenie Jackson and Christina Mendoza. Setting the record straight: What the COMPAS Core risk and need assessment is and is not. *Harvard Data Science Review*, 2(1), 2020. https://doi.org/10.1162/99608f92.1b3dadaa.

[144] Benjamin N. Jacobsen. Machine learning and the politics of synthetic data. *Big Data & Society*, 10(1):20539517221145372, 2023.

[145] William James. What is an emotion? *Mind*, 9(34):188–205, 1884.

[146] William James. *The principles of psychology*, chapter XI. Attention. Dover, New York, 1890. https://psychclassics.yorku.ca/James/Principles/prin11.htm.

[147] Youngseung Jeon, Bogoan Kim, Aiping Xiong, Dongwon Lee, and Kyungsik Han. Chamberbreaker: Mitigating the echo chamber effect and supporting information hygiene through a gamified inoculation system. *Proceedings of the ACM on Human-Computer Interaction*, 5(CSCW2):1–26, 2021.

[148] Leisheng Jin, Zhuo Liu, and Lijie Li. Chain-structure time-delay reservoir computing for synchronizing chaotic signal and an application to secure communication. *EURASIP Journal on Advances in Signal Processing*, 2022(1):1–17, 2022.

[149] L. Johnson and N. E. Johnson. Knowledge elicitation involving teach-back interviewing. In A. Kidd, editor, *Knowledge acquisition for expert systems*. Plenum Press, London, 1987.

[150] William Jones. *Keeping found things found: The study and practice of personal information management.* Morgan Kaufmann, Burlington, MA, 2010.

[151] Daniel Kahneman. *Thinking, fast and slow.* Macmillan, Basingstoke, 2011.

[152] Daniel Keim, Jörn Kohlhammer, Geoffrey Ellis, and Florian Mansmann. *Mastering the information age: Solving problems with visual analytics.* Eurographics Association, Goslar, 2010. https://www.vismaster.eu/book/.

[153] C. W. Kilmister. *Language, logic and mathematics.* English Universities Press, London, 1967.

[154] Thomas Kluyver, Benjamin Ragan-Kelley, Fernando Pérez, Brian E. Granger, Matthias Bussonnier, Jonathan Frederic, Kyle Kelley, Jessica B. Hamrick, Jason Grout, Sylvain Corlay, et al. Jupyter Notebooks – a publishing format for reproducible computational workflows. In F. Loizides and B. Schmidt, editors, *Positioning and Power in Academic Publishing: Players, Agents and Agendas.* IOS Press, Amsterdam, 2016.

[155] T. Kohonen. *Self organisation and associative memory.* Springer-Verlag, Berlin, 3rd edition, 1990.

[156] D. Kushner. The real story of Stuxnet. *IEEE Spectrum,* 50(3):48–53, 2013.

[157] J. E. Laird, A. Newell, and P. S. Rosenbloom. SOAR: An architecture for general intelligence. *Artificial Intelligence,* 33(1):1–64, 1987.

[158] John E Laird. *The Soar cognitive architecture.* MIT Press, 2019.

[159] I. Lapowsky. Google autocomplete still makes vile suggestions. *Wired* (2nd December 2018).

[160] Matthew E. Larkum, Lucy S. Petro, Robert N. S. Sachdev, and Lars Muckli. A perspective on cortical layering and layer-spanning neuronal elements. *Frontiers in Neuroanatomy,* 12(56), 2018.

[161] M. W. Lee and M. Walter. *Equality impact assessment: Literature review.* Office of Qualifications and Examinations Regulation (Ofqual) April 2020. https://assets.publishing.service.gov.uk/media/5e971f1de90e071a145ec51f/Equality_impact_assessment_literature_review_15_April_2020.pdf.

[162] P. Leith. Ell: An expert legislative consultant. In *Proceedings lEE conference on man/machine systems,* Manchester, UK, 1982.

[163] P. Leith. *Legal knowledge engineering: Computing, logic and law.* PhD thesis, Open University, 1985.

[164] D. B. Lenat and R. V. Guha. *Building large knowledge based systems.* Addison-Wesley, Reading, MA, 1990.

[165] S. Levin and J. Wong. Self-driving Uber kills Arizona woman in first fatal crash involving pedestrian. *The Guardian* (19 March 2018).

[166] Thomas Lewton. The Einstein machine. *New Scientist,* 256(3414):44–47, 2022.

[167] Jialiang Li, Ming Gao, and Ralph D'Agostino. Evaluating classification accuracy for modern learning approaches. *Statistics in Medicine,* 38(13):2477–2503, 2019.

[168] Joseph C. R. Licklider. Man-computer symbiosis. *IRE Transactions on Human Factors in Electronics,* HFE-1 (1):4–11, 1960.

[169] Henry Lieberman. Constructing graphical user interfaces by example. In *Proceedings: Graphics Interface,* pages 295–302. National Research Council of Canada, 1982 (SEE N 82-29909 20-61). https://doi.org/10.20380/GI1982.44.

[170] R. K. Lindsay, B. G. Buchanan, E. A. Feigenbaum, and J. Lederberg. *Applications of artificial intelligence for organic chemistry: The DENDRAL project.* McGraw-Hill, New York, 1980.

[171] Jacques-Louis Lions, Lennart Luebeck, Jean-Luc Fauquembergue, Gilles Kahn, Wolfgang Kubbat, Stefan Levedag, Leonardo Mazzini, Didier Merle, and Colin O'Halloran. Ariane 5 flight 501 failure report by the inquiry board, 1996. http://sunnyday.mit.edu/nasa-class/Ariane5-report.html.

[172] Ziming Liu, Yixuan Wang, Sachin Vaidya, Fabian Ruehle, James Halverson, Marin Soljačić, Thomas Y. Hou, and Max Tegmark. KAN: Kolmogorov–Arnold networks. *arXiv preprint arXiv:22404.19756,* 2024.

[173] A. Liu, B. Feng, B. Wang, et al. DeepSeek-v2: A strong, economical, and efficient mixture-of-experts language model. *arXiv preprint arXiv:2405.04434,* 2024.

[174] Sarah Loos, Geoffrey Irving, Christian Szegedy, and Cezary Kaliszyk. Deep network guided proof search. *arXiv preprint arXiv:1701.06972,* 2017.

[175] C. Loughlin. *Sensors for industrial inspection.* Kluwer Academic, Dordrecht, 1993.

[176] Ada Lovelace. *Notes upon the memoir by the translator: Sketch of the analytical engine invented by Charles Babbage*, by L. F. Menabrea, Bibliothèque Universelle de Genève, October, 1842, no. 82. https://www.fourmilab.ch/babbage/sketch.html.

[177] Peter Lucas, Linda van der Gaag, and Ameen Abu-Hanna. Bayesian networks in biomedicine and health-care. *Artificial Intelligence in Medicine*, 30(3):201–214, 2004.

[178] Scott M. Lundberg and Su-In Lee. A unified approach to interpreting model predictions. In *Proceedings of the 31st international conference on neural information processing systems (NIPS'17)*, page 4768–4777. Curran Associates Inc, Red Hook, NY, 2017.

[179] Richard Luscombe. Google engineer put on leave after saying AI chatbot has become sentient. *The Guardian* (December 2022).

[180] Ronan A. Lyons, Kerina H. Jones, Gareth John, Caroline J. Brooks, Jean-Philippe Verplancke, David V. Ford, Ginevra Brown, and Ken Leake. The SAIL Databank: linking multiple health and social care datasets. *BMC Medical Informatics and Decision Making*, 9(1):1–8, 2009.

[181] Alessio Malizia, Alan Chamberlain, and Ian Willcock. From design fiction to design fact: Developing future user experiences with proto-tools. In Masaaki Kurosu, editor, *Human-computer interaction: Theories, methods, and human issues*, pages 159–168. Springer International Publishing, Berlin, 2018.

[182] Jo Marchant. Powerful antibiotics discovered using AI. *Nature News* (20 February 2020).

[183] Henry Markram. The human brain project. *Scientific American*, 306(6):50–55, 2012.

[184] D. Marr. *Vision: A computational investigation into the human representation and processing of visual information.* W. H. Freeman, San Francisco, 1982.

[185] J. McCarthy and P. J. Hayes. Some philosophical problems from the standpoint of artificial intelligence. In B. Meltzer and D. Michie, editors, *Machine intelligence 4*. Edinburgh University Press, Edinburgh, 1969.

[186] J. L. McClelland and D. E. Rumelhart. *Parallel distributed processing*, volume 1. MIT Press, Cambridge, MA, 1986.

[187] Paige McGlauflin and Joseph Abrams. Elon Musk says AI will remove need for jobs and create 'universal high income.' But workers don't want to wait for robots to get financial relief. *Fortune*, 2023.

[188] Katharine McGowan and Sean Geobey. "Harmful to the commonality": The Luddites, the distributional effects of systems change and the challenge of building a just society. *Social Enterprise Journal*, 18(2), 306-320, 2022.

[189] Cade Metz. In two moves, AlphaGo and Lee Sedol redefined the future. *WIRED* (16 March 2016).

[190] Nicholas Micallef, Marcelo Sandoval-Castañeda, Adi Cohen, Mustaque Ahamad, Srijan Kumar, and Nasir Memon. Cross-platform multimodal misinformation: Taxonomy, characteristics and detection for textual posts and videos. In *Proceedings of the international AAAI conference on web and social media*, volume 16, pages 651–662, Association for the Advancement of Artificial Intelligence, Washington, DC, 2022.

[191] Tomas Mikolov, Ilya Sutskever, Kai Chen, Greg S. Corrado, and Jeff Dean. Distributed representations of words and phrases and their compositionality. *Advances in Neural Information Processing Systems*, 26, 3111–3119, 2013.

[192] G. A. Miller. The magical number seven, plus or minus two: Some limits on our capacity to process information. *Psychological Review*, 63(2):81–97, 1956.

[193] George A. Miller. WordNet: A lexical database for English. *Communications of the ACM*, 38(11):39–41, 1995.

[194] George A. Miller and Christiane Fellbaum. WordNet then and now. *Language Resources and Evaluation*, 41(2):209–214, 2007.

[195] Dan Milmo and Alex Hern. TikTok: Why the app with 1bn users faces a fight for its existence. *The Guardian* (31 March 2023).

[196] M. Minsky. A framework for representing knowledge. In P. H. Winston, editor, *The psychology of computer vision*. McGraw-Hill, New York, 1975.

[197] M. Minsky. *The society of mind*. Simon and Schuster, New York, 1985.

[198] M. Minsky and S. Papert. *Perceptrons*. MIT Press, Cambridge, MA, 1969.

[199] T. M. Mitchell. *Version spaces: An approach to concept learning*. PhD thesis, Stanford University, Stanford, CA, 1978.

[200] Matej Moravčík, Martin Schmid, Neil Burch, Viliam Lisỳ, Dustin Morrill, Nolan Bard, Trevor Davis, Kevin Waugh, Michael Johanson, and Michael Bowling. DeepStack: Expert-level artificial intelligence in heads-up no-limit poker. *Science*, 356(6337):508–513, 2017.

[201] A. Morgan, A. Dix, M. Phillips, and C. House. Blue sky thinking meets green field usability: Can mobile internet software engineering bridge the rural divide? *Local Economy*, 29(6–7):750–761, 2014.

[202] Lev Muchnik, Sen Pei, Lucas C. Parra, Saulo D. S. Reis, José S. Andrade Jr, Shlomo Havlin, and Hernán A. Makse. Origins of power-law degree distribution in the heterogeneity of human activity in social networks. *Scientific Reports*, 3(1):1–8, 2013.

[203] Claire Mufson. What will France do with 'national debate' data? 2019. https://www.france24.com/en/20190302-france-great-national-debate-data-artificial-intelligence-politics-yellow-vests accessed 03/03/2019.

[204] Mark A. Musen. The Protégé project: a look back and a look forward. *AI Matters*, 1(4):4–12, 2015. https://protege.stanford.edu/.

[205] Brad A. Myers and William Buxton. Creating highly-interactive and graphical user interfaces by demonstration. In *Proceedings of the 13th annual conference on computer graphics and interactive techniques, SIGGRAPH '86*, page 249–258. Association for Computing Machinery, New York, NY, 1986.

[206] J. Paul Myers and Kayako Yamakoshi. The Japanese Fifth Generation Computing Project: A brief overview. *Journal of Computing Sciences in Colleges*, 36(2):53–60, January 2021.

[207] Miryam Naddaf. Europe spent €600 million to recreate the human brain in a computer: how did it go? *Nature*, 620(7975):718–720, 2023.

[208] A. Newell. Unified theories of cognition and the role of SOAR. In J. A. Michon and A. Akyurek, editors, *SOAR: A cognitive architecture in perspective*, pages 25–79. Kluwer, Dordrecht, 1992.

[209] A. Newell and H. A. Simon. *Human problem solving*. Prentice-Hall, Englewood Cliffs, NJ, 1972.

[210] A. Newell and H. A. Simon. Computer science as empirical enquiry: Symbols and search. *Communications of the ACM*, 19:113–26, March 1976.

[211] NICE. *How nice measures value for money in relation to public health interventions*. (Local government briefing). National Institute for Health and Care Excellence, 1 September 2013, Manchester.

[212] S. Noble. Google has a striking history of bias against black girls. *Time* (26 March 2018).

[213] NPR. Twitter aims to crack down on misinformation, including misleading posts about Ukraine. *NPR Technology* (19 May 2022).

[214] Brian Oakley and Kenneth Owen. *Alvey: Britain's strategic computing initiative*. MIT Press, 1990.

[215] Nic Oatridge. *Wapping'86: The strike that broke Britain's newspaper unions*. ColdType, 2002, http://www.coldtype.net/Assets/pdfs/Wapping1.pdf.

[216] Office of National Statistics. Sustainable development indicators, July 2014. http://webarchive.nationalarchives.gov.uk/20160105183323/http://www.ons.gov.uk/ons/rel/wellbeing/sustainable-development-indicators/july-2014/sustainable-development-indicators.html.

[217] OpenAI. GPT-4 technical report. arXiv preprint [Submitted on 15 March 2023 (v1), last revised 27 March 2023 (this version, v3)]. https://arxiv.org/abs/2303.08774.

[218] Dawn Ramanee Peiris. *Computer interviews: Enhancing their effectiveness by simulating interpersonal techniques*. PhD thesis, University of Dundee, 1997.

[219] Thomas Pellissier Tanon, Denny Vrandečić, Sebastian Schaffert, Thomas Steiner, and Lydia Pintscher. From Freebase to Wikidata: The great migration. In *Proceedings of the 25th International conference on world wide web, WWW'16*, page 1419–1428. International World Wide Web Conferences Steering Committee, Republic and Canton of Geneva, CHE, 2016.

[220] Roger Penrose. *The emperor's new mind: Concerning computers, minds and the laws of physics.* Oxford University Press, Oxford, 1989.

[221] Roger Penrose. Shadows of the mind: A search for the missing science of consciousness. *Science Spectra*, 11:74–74, 1998.

[222] F. C. N. Pereira and D. H. D. Warren. Definite clause grammars for language analysis – a survey of the formalism and a comparison with augmented transition networks. *Artificial Intelligence*, 13(3):231–78, 1980.

[223] Andrés Pérez-Uribe and Eduardo Sanchez. Blackjack as a test bed for learning strategies in neural networks. In *1998 IEEE international joint conference on neural networks proceedings: IEEE world congress on computational intelligence (Cat. No. 98CH36227)*, volume 3, pages 2022–2027. Los Alamitos, CA, IEEE, 1998.

[224] Jean Piaget. *Play, dreams and imitation in childhood (La formation du symbole chez l'enfant; imitation, jeu et reve, image et représentation).* Republished Norton, New York, NY (1962), Routledge, Abingdon (2013), 1945.

[225] Luke Y. Prince and Blake A. Richards. The overfitted brain hypothesis. *Patterns*, 2(5):100268, 2021.

[226] J. R. Quinlan. Discovering rules by induction from large collections of examples. In D. Michie, editor, *Expert systems in the micro-electronic age*, pages 168–201. Edinburgh University Press, Edinburgh, 1979.

[227] J. R. Quinlan. Induction of decision trees. *Machine Learning*, 1(1):81–106, 1986.

[228] J. R. Quinlan. *C4.5: Programs for machine learning.* Morgan Kaufmann, Burlington, MA, 1993.

[229] Aditya Ramesh, Prafulla Dhariwal, Alex Nichol, Casey Chu, and Mark Chen. Hierarchical text-conditional image generation with clip latents. *arXiv preprint arXiv:2204.06125*, 2022.

[230] Aditya Ramesh, Mikhail Pavlov, Gabriel Goh, Scott Gray, Chelsea Voss, Alec Radford, Mark Chen, and Ilya Sutskever. Zero-shot text-to-image generation. In *International conference on machine learning*, pages 8821–8831. PMLR, https://proceedings.mlr.press/v139/ramesh21a.html, 2021.

[231] Jem Rayfield. BBC World Cup 2010 dynamic semantic publishing, 2010. *BBC Internet Blog*, 12 July 2010. https://www.bbc.co.uk/blogs/bbcinternet/2010/07/bbc_world_cup_2010_dynamic_sem.html.

[232] Paul Rayson and Roger Garside. The CLAWS web tagger. *ICAME Journal*, 22:121–123, 1998.

[233] R. Reiter. On closed world data bases. In H. Gallaire and J. Minker, editors, *Logic and data bases*, pages 55–76. Plenum Press, New York, 1978.

[234] Hanchi Ren, Jingjing Deng, and Xianghua Xie. GRNN: Generative regression neural network – a data leakage attack for federated learning. *ACM Transactions on Intelligent Systems and Technology*, 13(4), May 2022.

[235] Marco Tulio Ribeiro, Sameer Singh, and Carlos Guestrin. "Why should I trust you?": Explaining the predictions of any classifier. In *Proceedings of the 22nd ACM SIGKDD international conference on knowledge discovery and data mining (KDD '16)*, page 1135–1144. Association for Computing Machinery, New York, NY, 2016.

[236] Steven I. Ross, Fernando Martinez, Stephanie Houde, Michael Muller, and Justin D. Weisz. The programmer's assistant: Conversational interaction with a large language model for software development. In *Proceedings of the 28th international conference on intelligent user interfaces*, pages 491–514, ACM, New York, NY, 2023.

[237] C. Rudder. *Dataclysm: Who we are (when we think no one's looking).* Fourth Estate, London, 4th Estate, 2014.

[238] SAE International. Taxonomy and definitions for terms related to driving automation systems for on-road motor vehicles. *SAE International*, 4970(724):1–5, 2018.

[239] SAE International. SAE levels of driving automation™ refined for clarity and international audience. *SAE Blog*, posted: Monday (3 May 2021), https://www.sae.org/blog/sae-j3016-update.

[240] Haziqa Sajid. Can you build large language models like ChatGPT at half cost? *UniteAI* (11 May 2023). https://www.unite.ai/can-you-build-large-language-models-like-chatgpt-at-half-cost/.

[241] A. Salmoni. *Task-based judgements of search engine summaries, and negative information scent.* PhD Thesis, Cardff University, September 2004.

[242] Leo Sauermann, Ansgar Bernardi, and Andreas Dengel. *Overview and outlook on the semantic desktop. SDW'05*, page 74–91. CEUR-WS.org, Aachan, DEU, 2005.

[243] R. C. Schank and R. P. Abelson. *Scripts, plans, goals, and understanding.* Lawrence Erlbaum Associates, Hillsdale, NJ, 1977.

[244] Ari Schlesinger, Kenton P. O'Hara, and Alex S. Taylor. *Let's talk about race: Identity, chatbots, and AI. CHI '18*, page 1–14. Association for Computing Machinery, New York, NY, 2018.

[245] Chistopher Schmandt and Barry Arons. Phone Slave: A graphical telecommunications interface. In *Proceeding of the SID*, volume 26, 1985. See video at https://www.youtube.com/watch?v=94jIa7GIQu0.

[246] Albrecht Schmidt. Implicit human computer interaction through context. *Personal Technologies*, 4:191–199, 2000.

[247] Michael Schmidt and Hod Lipson. Distilling free-form natural laws from experimental data. *Science*, 324(5923):81–85, 2009.

[248] Oscar Schwartz. Untold history of AI: Algorithmic bias was born in the 1980s. *IEEE Spectrum*, 15, 2019.

[249] J. R. Searle. *Speech acts.* Cambridge University Press, Cambridge, 1969.

[250] J. R. Searle. Minds, brains and programs. *Behavioural and Brain Sciences*, 3:417–424, 1980.

[251] T. J. Sejnowski and C. R. Rosenberg. Parallel networks that learn to pronounce English text. *Complex Systems*, 1(1):145–168, 1987.

[252] Nigel Shadbolt, Kieron O'Hara, David De Roure, and Wendy Hall. *The theory and practice of social machines.* Springer, Berlin, 2019.

[253] Ehud Y. Shapiro. The Fifth Generation Project – a trip report. *Commun. ACM*, 26(9):637–641, September 1983.

[254] Ying Shen, Lizhu Zhang, Jin Zhang, Min Yang, Buzhou Tang, Yaliang Li, and Kai Lei. CBN: Constructing a clinical Bayesian network based on data from the electronic medical record. *Journal of Biomedical Informatics*, 88:1–10, 2018.

[255] B. Shneiderman. The eyes have it: A task by data type taxonomy for information visualizations. In *Proceedings 1996 IEEE symposium on visual languages*, pages 336–343, Los Alamitos, CA, 1996.

[256] Ben Shneiderman. Bridging the gap between ethics and practice: Guidelines for reliable, safe, and trustworthy human-centered AI systems. *ACM Transactions on Interactive Intelligent Systems (TiiS)*, 10(4):1–31, 2020.

[257] Ben Shneiderman. *Human-centered AI.* Oxford University Press, Oxford, 2022.

[258] E. H. Shortliffe. *Computer-based medical consultations: MYCIN.* Elsevier, New York, 1976.

[259] Christianna Silva. It took just one weekend for Meta's new AI chatbot to become racist. *Mashable* (8 August 2022).

[260] David Silver, Aja Huang, Chris J Maddison, Arthur Guez, Laurent Sifre, George Van Den Driessche, Julian Schrittwieser, Ioannis Antonoglou, Veda Panneershelvam, Marc Lanctot, et al. Mastering the game of Go with deep neural networks and tree search. *Nature*, 529(7587):484–489, 2016.

[261] D. Slate and L. Atkin. Chess 4.5 – the Northwestern University chess program. In P. W. Frey, editor, *Chess skill in man and machine*. Springer-Verlag, New York, 1977.

[262] Matthew Sparkes. DeepMind uses AI to control plasma inside tokamak fusion reactor. *New Scientist* (16 February 2022).

[263] SSiW. SaySomethingin, 2023. https://www.saysomethingin.com/ accessed 02/01/2023.

[264] Kiran Stacey and Dan Milmo. UK, US, EU and China sign declaration of AI's 'catastrophic' danger. *The Guardian* (1 November 2023).

[265] Susan Leigh Star and James R. Griesemer. Institutional ecology, 'translations' and boundary objects: Amateurs and professionals in Berkeley's Museum of Vertebrate Zoology, 1907-39. *Social Studies of Science*, 19(3):387–420, 1989.

[266] Daniel Stoljar. Physicalism. In Edward N. Zalta and Uri Nodelman, editors, *Stanford encyclopedia of philosophy*. Stanford University Press Stanford, CA, 2001.

[267] Petr Suchánek, Franciszek Marecki, and Robert Bucki. Self-learning Bayesian networks in diagnosis. *Procedia Computer Science*, 35:1426–1435,

2014. Knowledge-Based and Intelligent Information & Engineering Systems 18th Annual Conference, KES-2014 Gdynia, Poland, September 2014 Proceedings.

[268] George Sullivan and Luke Burns. Cashing out: Assessing the risk of localised financial exclusion as the UK moves towards a cashless society. *arXiv preprint arXiv:2202.05674*, 2022.

[269] W. R. Swartout. XPLAIN: A system for creating and explaining expert consulting programs. *Artificial Intelligence*, 21:285–325, 1983.

[270] Gerald Tesauro. Neurogammon: A neural-network backgammon program. In *1990 IJCNN international joint conference on neural networks*, pages 33–39. IEEE, Los Alamitos, CA, 1990.

[271] James J. Thomas and Kristin A. Cook. A visual analytics agenda. *IEEE Computer Graphics and Applications*, 26(1):10–13, 2006.

[272] Edward P. Thompson. *The making of the English working class: 1963*. Vintage, New York, 1966.

[273] Barbara Tizard and Martin Hughes. *Young children learning*. Harvard University Press, Cambridge, MA, also John Wiley & Sons (2008), 1984.

[274] Hugo Touvron, Thibaut Lavril, Gautier Izacard, Xavier Martinet, Marie-Anne Lachaux, Timothée Lacroix, Baptiste Rozière, Naman Goyal, Eric Hambro, Faisal Azhar, Aurelien Rodriguez, Armand Joulin, Edouard Grave and Guillaume Lample. LLaMA: Open and efficient foundation language models. *arXiv preprint arXiv:2302.13971*, 2023.

[275] Tony Ho Tran. Scientists built an AI to give ethical advice, but it turned out super racist. *Futurism* (22 October 2021).

[276] A. M. Turing. Computing machinery and intelligence. *Mind*, 59:433–460, October 1950.

[277] Lisa Tweedie, Bob Spence, Huw Dawkes, and Hua Su. The influence explorer. In *Conference companion on human factors in computing systems, CHI '95*, page 129–130. Association for Computing Machinery, New York, NY, 1995.

[278] Twitter Inc. How we address misinformation on Twitter. *Twitter Help Centre*. https://help.twitter.com/en/resources/addressing-misleading-info accessed 07/04/2023,

[279] Silviu-Marian Udrescu and Max Tegmark. AI Feynman: A physics-inspired method for symbolic regression. *Science Advances*, 6(16):eaay2631, 2020.

[280] UK Government. The Bletchley Ceclaration by countries attending the AI Safety Summit, 1–2 November 2023. Policy Paper. https://www.gov.uk/government/publications/ai-safety-summit-2023-the-bletchley-declaration/the-bletchley-declaration-by-countries-attending-the-ai-safety-summit-1-2-november-2023.

[281] UK Parliament. Statistics on access to cash, bank branches and ATMs. *Research Briefing*. House of Commons Library, 25 July 2022. https://commonslibrary.parliament.uk/research-briefings/cbp-8570/.

[282] UKRI. Framework for responsible research and innovation. https://www.ukri.org/about-us/epsrc/our-policies-and-standards/framework-for-responsible-innovation/ accessed 11/03/2023.

[283] Secretary-General United Nations. Joint call by the United Nations Secretary-General and the President of the International Committee of the Red Cross for states to establish new prohibitions and restrictions on autonomous weapon systems. *Note to Correspondents* (05 October 2023). https://www.un.org/sg/en/content/sg/note-correspondents/2023-10-05/note-correspondents-joint-call-the-united-nations-secretary-general-and-the-president-of-the-international-committee-of-the-red-cross-for-states-establish-new, 2023.

[284] United Press International. Trump Plaza fined $200,000 for discrimination. *UPI Archives* (6 June 1991)

[285] Blase Ur, Melwyn Pak Yong Ho, Stephen Brawner, Jiyun Lee, Sarah Mennicken, Noah Picard, Diane Schulze, and Michael L. Littman. Trigger-action programming in the wild: An analysis of 200,000 IFTTT recipes. In *Proceedings of the 2016 CHI conference on human factors in computing systems, CHI'16*, page 3227–3231. Association for Computing Machinery, New York, NY, 2016.

[286] Fabio Urbina, Filippa Lentzos, Cédric Invernizzi, and Sean Ekins. Dual use of artificial-intelligence-powered drug discovery. *Nature Machine Intelligence*, 4(3):189–191, 2022.

[287] Ashish Vaswani, Noam Shazeer, Niki Parmar, Jakob Uszkoreit, Llion Jones, Aidan N. Gomez, Lukasz Kaiser, and Illia Polosukhin. Attention is all you need, 2017 doi:10.48550/ARXIV.1706.03762; https://arxiv.org/abs/1706.03762 .

[288] D. Vernon, editor. *Computer vision: Craft, engineering and science*. Springer-Verlag, Berlin, 1991.

[289] Pablo Villalobos, Jaime Sevilla, Lennart Heim, Tamay Besiroglu, Marius Hobbhahn, and Anson Ho. *Will we run out of data? An analysis of the limits of scaling datasets in machine learning*, 2022 doi: 10.48550/ARXIV.2211.04325; https://arxiv.org/abs/2211.04325.

[290] J. Vincent. Russian nuclear power plant infected by Stuxnet malware says cyber-security expert. *The Independent* (Tuesday 12 November 2013).

[291] James Vincent. The lawsuit that could rewrite the rules of AI copyright. *The Verge*, 2022.

[292] Soroush Vosoughi, Deb Roy, and Sinan Aral. The spread of true and false news online. *Science*, 359(6380):1146–1151, 2018.

[293] Matthew Wall, Rory Costello, and Stephen Lindsay. The miracle of the markets: Identifying key campaign events in the Scottish independence referendum using betting odds. *Electoral Studies*, 46:39–47, 2017.

[294] D. L. Waltz. Understanding line drawings of scenes with shadows. In P. Winston, editor, *The psychology of computer vision*. McGraw-Hill, New York, 1975.

[295] Lingfei Wang, Pieter Audenaert, and Tom Michoel. High-dimensional Bayesian network inference from systems genetics data using genetic node ordering. *Frontiers in Genetics*, 10, 2019 doi: 10.3389/fgene.2019.01196; https://doi.org/10.3389/fgene.2019.01196.

[296] Peihao Wang, Rameswar Panda, Lucas Torroba Hennigen, Philip Greengard, Leonid Karlinsky, Rogerio Feris, David Daniel Cox, Zhangyang Wang, and Yoon Kim. Learning to grow pretrained models for efficient transformer training. *arXiv preprint arXiv:2303.00980*, 2023.

[297] Stuart L. Weibel and Traugott Koch. The Dublin Core Metadata Initiative. *D-lib Magazine*, 6(12):1082–9873, 2000.

[298] Mark D. Weiser. Ubiquitous computing. In *ACM conference on computer science*, volume 418, pages 197530–197680, ACM, New York, NY, 1994.

[299] J. Weizenbaum. ELIZA – a computer program for the study of natural language communication between man and machine. *Communications of the ACM*, 9(1):36–44, 1966.

[300] J. Weizenbaum. *Computer power and human reason: From judgement to calculation*. Freeman, San Francisco, 1976.

[301] T. Winograd. *Understanding natural language*. Addison-Wesley, Reading, MA, 1972.

[302] W. A. Woods. Transition network grammars for natural language analysis. *Communications of the ACM*, 13(10):591–606, 1970.

[303] Runhua Xu, Nathalie Baracaldo, and James Joshi. Privacy-preserving machine learning: Methods, challenges and directions. *arXiv preprint arXiv:2108.04417*, 2021.

[304] R. M. Young and T. R. G. Green. Programmable user models for predictive evaluation of interface designs. In K. Bice and C. Lewis, editors, *Proceedings of CHI'89: Human factors in computing systems*, pages 15–19, ACM, New York, NY, 1989.

[305] Mingming Zha, Guozhu Meng, Chaoyang Lin, Zhe Zhou, and Kai Chen. RoLMA: A practical adversarial attack against deep learning-based LPR systems. In *Information security and cryptology: 15th international conference, Inscrypt 2019, Nanjing, China, December 6–8, 2019, revised selected papers 15*, pages 101–117. Springer, Berlin, 2020.

Index

A* algorithm, **49**, 50, 55, 224
abductive reasoning, 25, 26, 27, 34, 63
accountability, 315, 362
accuracy, 105, 119, 124, 178, 212, 228, 233,
        288, 290, 299, 302, 304, 306, 317,
        323, 332, 364, 368
accuracy measure, 120, 129
ACT-R, **345**, 346, 354, 355
activation function, 99
active vision, 161, **232**, 234, 235
activity recognition, 304
ACT*, 343, **345**, 354
Ada Lovelace, *see* Lovelace, Ada
adjacency statistics, 262
adversarial attacks, 325
adversarial learning, 7, 102, 106, 116, 147,
        157, 158, 324, 337, 359, 369
adversarial search, 39, 54
agents, 155, 221, **236**, 248, 249
        action, 236, 237, 238, 241, 248
        email filtering, 236, 238
        embodied, 237, 295, 300, 361
        intelligent filtering, **239**
        machine learning, 239
        messages, 237
        methods, 237
aggregation for privacy, **324**
AI arms race, 368
AI winter, **6**, 7, 375, 376, 378
algorithmic accountability, 312, 313
alien intelligence, 2, 343
alpha–beta pruning, 147, **152**, 154, 158
AlphaFold, 214, 361, 362
AlphaGo, 3, 7, 85, 102, 106, 148, 149, 157,
        158, 160, 346, 375
AlphaGo Zero, 158
AlphaZero, 375, 378
Alvey Programme, 6
ambiguity
        in computer vision, *see* ambiguous
                image
        lexical, 189
        pragmatic, 189
        referential, 189
        semantic, 189
        syntactic, 189
ambiguous image, 161, 232, 348
analogy, 16, 30, 79

derivational, *see* derivational analogy
transformational, *see* transformational
        analogy
analytic rules, *see* procedural knowledge
Analytical Engine, 4, 147
Anderson, John Robert, 345
Andrey Andreyevich Markov, *see* Markov,
        Andrey
anonymisation, 323
anonymous identifier, 253
antecedent-driven reasoning, *see* bottom-up
        reasoning
anti-discrimination laws, **315**
Apache Hadoop, 113, 116
aperture, **232**, 233
application phase, 60, **118**
appropriate intelligence, **306**, **308**, 311
appropriating technology, *see* appropriation
appropriation, 294
architecture, 123, 346
AREA framework, 362, 363, 371
area under the curve, 121, 130
arithmetic mean, 94, 95, 96, 137
ARMA (Auto Regressive Moving Average),
        216, 217, 220
artificial emotion, 349
artificial human intelligence, *see* artificial
artificial imagination, 354
artificial intelligence, **2**, 4
        history, *see* history of AI
artificial life, 221, 243, **246**
artificial neural networks, 6, 83, 360
artificial society, 246
Asimov, Isaac, 221, 362
aspect ratio, 163, 178
associative memory, 74, 80
asymmetric distribution, 95
attention, 347
attention mechanisms, 214, 376
augmented transition network, 192, 195,
        196
auto-associative memory, 80, 138, 182
auto-complete, 298
auto-regressive model, **216**, 217
autoencoder, 80, 181, 182, 183, 187
automated decision, 309, 310, 317, 325
automation bias, 283, 362, 366

autonomous car, 106, 222, 295, 302, 313,
        314, 315, 330, 363
autonomous vehicle, 7, 231, 314, 362, 364,
        366, 370
autonomous weapons, 314, 367, 378
avoidance mechanisms, *see* obstacle
        avoidance
avoidance rules, *see* obstacle avoidance

B-splines, **99**
Babbage, Charles, 4, 147, 215
backpropagation, **75**, **76**, 78, 96, 102, 103,
        122, 123, 127, 129, 182, 325, 337
backtracking, 16, 23, 44, 48, 50, 54, 224
backward branching factor, *see* branching
        factor
backward chaining, *see* backward reasoning
backward reasoning, 26, 223, 274, 275, 276,
        279
bag of words, 141
base rate, 119, 304, 329, 335
Bayes Theorem, 28, 29, 33, 34, 171, 176, 181,
        276, 309
Bayesian methods, 30, 33, 34, 242
Bayesian network, 29, 33
Berners-Lee, Tim, 7, 250
best first search, **48**, 49, 50, 55
bias, 7, 67, 108, 211, 217, 313, 317, 318, 320,
        321, 323, 327, 328, 330, 331, 332,
        340, 341, 365, 366, 367
bibliographic database, 240
bibliographic search, 240
Big Blue, 378
big data, 7, 79, 102, 107, 115, 116, 188, 203,
        250, 313, 314, 316, 320, 323, 324,
        326, 329, 331, 334, 343, 376, 378
binary image, 162, 173
bitmap image, 179
black-box machine learning, 283, 289, 320,
        328, 330, 331, **336**, **337**, 339, 340,
        341, 346
blackboard architecture, **243**, 244, 245
blind search, 54
Boden, Margaret, 357
Boltzmann machine, 74, **80**, 81, 83, 182
Bombe, 4
Boolean network, 337
boosting, 108, 109, 247, 351, 352